Fall,

Christoph Schiller

MOTION MOUNTAIN

The Adventure of Physics
Volume I

Fall, Flow and Heat

Edition 31, available as free pdf
with films at www.motionmountain.net

Editio trigesima prima.

Proprietas scriptoris © Chrestophori Schiller
tertio anno Olympiadis trigesimae secundae.

Omnia proprietatis iura reservantur et vindicantur.
Imitatio prohibita sine auctoris permissione.
Non licet pecuniam expetere pro aliqua, quae
partem horum verborum continet; liber
pro omnibus semper gratuitus erat et manet.

Thirty-first edition.

Copyright © 1990–2019 by Christoph Schiller,
from the third year of the 24th Olympiad
to the third year of the 32nd Olympiad.

All rights reserved. Reproduction prohibited without
permission from the author. In particular, you are
not allowed to make money with anything containing
any part of this text. However, the pdf file of this text
was, is and remains free for everybody, as explained
on www.motionmountain.net.

To Britta, Esther and Justus Aaron

τῷ ἐμοὶ δαίμονι

Die Menschen stärken, die Sachen klären.

PREFACE

> Primum movere, deinde docere.*
> Antiquity

This book series is for anybody who is curious about motion in nature. How do things, people, animals, images and empty space move? The answer leads to many adventures; this volume presents the best ones about *everyday* motion. Carefully observing everyday motion allows us to deduce six essential statements: everyday motion is continuous, conserved, relative, reversible, mirror-invariant – and lazy. Yes, nature is indeed lazy: in every motion, it minimizes change. This text explores how these six results are deduced and how they fit with all those observations that seem to contradict them.

In the structure of modern physics, shown in Figure 1, the results on everyday motion form the major part of the starting point at the bottom. The present volume is the first of a six-volume overview of physics. It resulted from a threefold aim I have pursued since 1990: to present motion in a way that is simple, up to date and captivating.

In order to be *simple*, the text focuses on concepts, while keeping mathematics to the necessary minimum. Understanding the concepts of physics is given precedence over using formulae in calculations. The whole text is within the reach of an undergraduate.

In order to be *up to date*, the text is enriched by the many gems – both theoretical and empirical – that are scattered throughout the scientific literature.

In order to be *captivating*, the text tries to startle the reader as much as possible. Reading a book on general physics should be like going to a magic show. We watch, we are astonished, we do not believe our eyes, we think, and finally we understand the trick. When we look at nature, we often have the same experience. Indeed, every page presents at least one surprise or provocation for the reader to think about. Numerous interesting challenges are proposed.

The motto of the text, *die Menschen stärken, die Sachen klären*, a famous statement on pedagogy, translates as: 'To fortify people, to clarify things.' Clarifying things – and adhering only to the truth – requires courage, as changing the habits of thought produces fear, often hidden by anger. But by overcoming our fears we grow in strength. And we experience intense and beautiful emotions. All great adventures in life allow this, and exploring motion is one of them. Enjoy it.

Christoph Schiller

---

* 'First move, then teach.' In modern languages, the mentioned type of *moving* (the heart) is called *motivating*; both terms go back to the same Latin root.

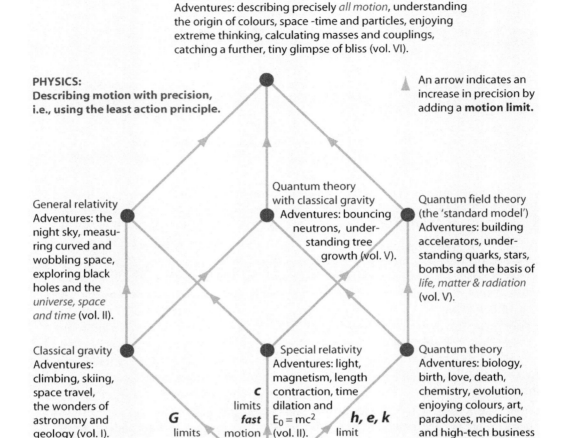

FIGURE 1 A complete map of physics, the science of motion, as first proposed by Matvei Bronshtein (b. 1907 Vinnytsia, d. 1938 Leningrad). The Bronshtein cube starts at the bottom with everyday motion, and shows the connections to the fields of modern physics. Each connection increases the precision of the description and is due to a limit to motion that is taken into account. The limits are given for uniform motion by the gravitational constant G, for fast motion by the speed of light c, and for tiny motion by the Planck constant h, the elementary charge e and the Boltzmann constant k.

Using this book

Marginal notes refer to bibliographic references, to other pages or to challenge solutions. In the colour edition, marginal notes, pointers to footnotes and links to websites are typeset in green. Over time, links on the internet tend to disappear. Most links can be recovered via www.archive.org, which keeps a copy of old internet pages. In the free pdf edition of this book, available at www.motionmountain.net, all green pointers and links are clickable. The pdf edition also contains all films; they can be watched directly in Adobe Reader.

Solutions and hints for *challenges* are given in the appendix. Challenges are classified as easy (e), standard student level (s), difficult (d) and research level (r). Challenges for which no solution has yet been included in the book are marked (ny).

Advice for learners

Learning allows us to discover what kind of person we can be. Learning widens knowledge, improves intelligence and provides a sense of achievement. Therefore, learning from a book, especially one about nature, should be efficient and enjoyable. Avoid bad learning methods like the plague! Do not use a marker, a pen or a pencil to highlight or underline text on paper. It is a waste of time, provides false comfort and makes the text unreadable. And do not learn from a screen. In particular, never, ever, learn from the internet, from videos, from games or from a smartphone. Most of the internet, almost all videos and all games are poisons and drugs for the brain. Smartphones are dispensers of drugs that make people addicted and prevent learning. Nobody putting marks on paper or looking at a screen is learning efficiently or is enjoying doing so.

In my experience as a pupil and teacher, one learning method never failed to transform unsuccessful pupils into successful ones: if you read a text for study, summarize every section you read, *in your own words and images, aloud*. If you are unable to do so, read the section again. Repeat this until you can clearly summarize what you read in your own words and images, aloud. And *enjoy* the telling aloud! You can do this alone or with friends, in a room or while walking. If you do this with everything you read, you will reduce your learning and reading time significantly; you will enjoy learning from good texts much more and hate bad texts much less. Masters of the method can use it even while listening to a lecture, in a low voice, thus avoiding to ever take notes.

Advice for teachers

A teacher likes pupils and likes to lead them into exploring the field he or she chose. His or her enthusiasm is the key to job satisfaction. If you are a teacher, before the start of a lesson, picture, feel and tell yourself how you enjoy the topic of the lesson; then picture, feel and tell yourself how you will lead each of your pupils into enjoying that topic as much as you do. Do this exercise consciously, every day. You will minimize trouble in your class and maximize your teaching success.

This book is not written with exams in mind; it is written to make teachers and students *understand* and *enjoy* physics, the science of motion.

Feedback

The latest pdf edition of this text is and will remain free to download from the internet. I would be delighted to receive an email from you at fb@motionmountain.net, especially on the following issues:

Challenge 1 s
— What was unclear and should be improved?
— What story, topic, riddle, picture or film did you miss?

Also help on the specific points listed on the www.motionmountain.net/help.html web page is welcome. All feedback will be used to improve the next edition. You are welcome to send feedback by mail or by sending in a pdf with added yellow notes, to provide illustrations or photographs, or to contribute to the errata wiki on the website. If you would like to translate a chapter of the book in your language, please let me know.

On behalf of all readers, thank you in advance for your input. For a particularly useful contribution you will be mentioned – if you want – in the acknowledgements, receive a reward, or both.

Support

Your donation to the charitable, tax-exempt non-profit organisation that produces, translates and publishes this book series is welcome. For details, see the web page www.motionmountain.net/donation.html. The German tax office checks the proper use of your donation. If you want, your name will be included in the sponsor list. Thank you in advance for your help, on behalf of all readers across the world.

The paper edition of this book is available, either in colour or in black and white, from www.amazon.com, in English and in certain other languages. And now, enjoy the reading.

Contents

15 1 Why should we care about motion?

Does motion exist? 16 • How should we talk about motion? 18 • What are the types of motion? 20 • Perception, permanence and change 25 • Does the world need states? 27 • Galilean physics in six interesting statements 29 • Curiosities and fun challenges about motion 30 • First summary on motion 33

34 2 From motion measurement to continuity

What is velocity? 35 • What is time? 40 • Clocks 44 • Why do clocks go clockwise? 48 • Does time flow? 48 • What is space? 49 • Are space and time absolute or relative? 52 • Size – why length and area exist, but volume does not 52 • What is straight? 59 • A hollow Earth? 60 • Curiosities and fun challenges about everyday space and time 61 • Summary about everyday space and time 74

75 3 How to describe motion – kinematics

Throwing, jumping and shooting 78 • Enjoying vectors 80 • What is rest? What is velocity? 82 • Acceleration 85 • From objects to point particles 85 • Legs and wheels 89 • Curiosities and fun challenges about kinematics 92 • Summary of kinematics 97

98 4 From objects and images to conservation

Motion and contact 99 • What is mass? 100 • Momentum and mass 102 • Is motion eternal? – Conservation of momentum 108 • More conservation – energy 110 • The cross product, or vector product 115 • Rotation and angular momentum 116 • Rolling wheels 121 • How do we walk and run? 122 • Curiosities and fun challenges about mass, conservation and rotation 123 • Summary on conservation in motion 133

135 5 From the rotation of the earth to the relativity of motion

How does the Earth rotate? 145 • Does the Earth move? 150 • Is velocity absolute? – The theory of everyday relativity 156 • Is rotation relative? 158 • Curiosities and fun challenges about rotation and relativity 158 • Legs or wheels? – Again 168 • Summary on Galilean relativity 172

173 6 Motion due to gravitation

Gravitation as a limit to uniform motion 173 • Gravitation in the sky 174 • Gravitation on Earth 178 • Properties of gravitation: G and g 182 • The gravitational potential 186 • The shape of the Earth 188 • Dynamics – how do things move in various dimensions? 189 • The Moon 190 • Orbits – conic sections and more 192 • Tides 197 • Can light fall? 201 • Mass: inertial and gravitational 202 • Curiosities and fun challenges about gravitation 204 • Summary on gravitation 224

226 7 Classical mechanics, force and the predictability of motion

Should one use force? Power? 227 • Forces, surfaces and conservation 231 • Friction and motion 232 • Friction, sport, machines and predictability 234 • Complete states – initial conditions 237 • Do surprises exist? Is the future determined? 238 • Free will 241 • Summary on predictability 242 • From predictability to global descriptions of motion 242

248   8   MEASURING CHANGE WITH ACTION
       The principle of least action 253 • Lagrangians and motion 256 • Why is motion so often bounded? 257 • Curiosities and fun challenges about Lagrangians 261 • Summary on action 264

266   9   MOTION AND SYMMETRY
       Why can we think and talk about the world? 270 • Viewpoints 271 • Symmetries and groups 272 • Multiplets 273 • Representations 275 • The symmetries and vocabulary of motion 276 • Reproducibility, conservation and Noether's theorem 280 • Parity inversion and motion reversal 284 • Interaction symmetries 285 • Curiosities and fun challenges about symmetry 285 • Summary on symmetry 286

288   10   SIMPLE MOTIONS OF EXTENDED BODIES – OSCILLATIONS AND WAVES
       Oscillations 288 • Damping 289 • Resonance 291 • Waves: general and harmonic 293 • Water waves 295 • Waves and their motion 300 • Why can we talk to each other? – Huygens' principle 304 • Wave equations 305 • Why are music and singing voices so beautiful? 307 • Measuring sound 310 • Is ultrasound imaging safe for babies? 313 • Signals 314 • Solitary waves and solitons 316 • Curiosities and fun challenges about waves and oscillation 318 • Summary on waves and oscillations 331

333   11   DO EXTENDED BODIES EXIST? – LIMITS OF CONTINUITY
       Mountains and fractals 333 • Can a chocolate bar last forever? 333 • The case of Galileo Galilei 335 • How high can animals jump? 337 • Felling trees 338 • Little hard balls 339 • The sound of silence 340 • How to count what cannot be seen 340 • Experiencing atoms 342 • Seeing atoms 344 • Curiosities and fun challenges about solids and atoms 345 • Summary on atoms 351

354   12   FLUIDS AND THEIR MOTION
       What can move in nature? – Flows of all kinds 354 • The state of a fluid 357 • Laminar and turbulent flow 361 • The atmosphere 364 • The physics of blood and breath 367 • Curiosities and fun challenges about fluids 369 • Summary on fluids 382

383   13   ON HEAT AND MOTION REVERSAL INVARIANCE
       Temperature 383 • Thermal energy 387 • Why do balloons take up space? – The end of continuity 389 • Brownian motion 391 • Why stones can be neither smooth nor fractal, nor made of little hard balls 394 • Entropy 395 • Entropy from particles 398 • The minimum entropy of nature – the quantum of information 399 • Is everything made of particles? 400 • The second principle of thermodynamics 402 • Why can't we remember the future? 404 • Flow of entropy 404 • Do isolated systems exist? 405 • Curiosities and fun challenges about reversibility and heat 405 • Summary on heat and time-invariance 413

415   14   SELF-ORGANIZATION AND CHAOS – THE SIMPLICITY OF COMPLEXITY
       Appearance of order 418 • Self-organization in sand 420 • Self-organization of spheres 422 • Conditions for the appearance of order 422 • The mathematics of order appearance 423 • Chaos 424 • Emergence 426 • Curiosities and fun challenges about self-organization 427 • Summary on self-organization and chaos 433

434   15   FROM THE LIMITATIONS OF PHYSICS TO THE LIMITS OF MOTION

Research topics in classical dynamics 434 • What is contact? 435 • What determines precision and accuracy? 436 • Can all of nature be described in a book? 436 • Something is wrong about our description of motion 437 • Why is measurement possible? 438 • Is motion unlimited? 438

440 A **Notation and conventions**

The Latin alphabet 440 • The Greek alphabet 442 • The Hebrew alphabet and other scripts 444 • Numbers and the Indian digits 445 • The symbols used in the text 446 • Calendars 448 • People Names 450 • Abbreviations and eponyms or concepts? 450

452 B **Units, measurements and constants**

SI units 452 • The meaning of measurement 455 • Curiosities and fun challenges about units 455 • Precision and accuracy of measurements 458 • Limits to precision 459 • Physical constants 459 • Useful numbers 467

468 C **Sources of information on motion**

474 **Challenge hints and solutions**

522 **Bibliography**

555 **Credits**

Acknowledgements 555 • Film credits 556 • Image credits 556

563 **Subject index**

Fall, Flow and Heat

In our quest to learn how things move,
the experience of hiking and other motion
leads us to introduce the concepts of
velocity, time, length, mass and temperature.
We learn to use them to *measure change*
and find that nature minimizes it.
We discover how to float in free space,
why we have legs instead of wheels,
why disorder can never be eliminated,
and why one of the most difficult open issues
in science is the flow of water through a tube.

Chapter 1
WHY SHOULD WE CARE ABOUT MOTION?

> "All motion is an illusion.
> Zeno of Elea**

Wham! The lightning striking the tree nearby violently disrupts our quiet forest walk and causes our hearts to suddenly beat faster. In the top of the tree we see the fire start and fade again. The gentle wind moving the leaves around us helps to restore the calmness of the place. Nearby, the water in a small river follows its complicated way down the valley, reflecting on its surface the ever-changing shapes of the clouds.

Motion is everywhere: friendly and threatening, terrible and beautiful. It is fundamental to our human existence. We need motion for growing, for learning, for thinking, for remaining healthy and for enjoying life. We use motion for walking through a forest, for listening to its noises and for talking about all this. Like all animals, we rely on motion to get food and to survive dangers. Like all living beings, we need motion to reproduce, to breathe and to digest. Like all objects, motion keeps us warm.

Motion is the most fundamental observation about nature at large. It turns out that *everything* that happens in the world is some type of motion. There are no exceptions. Motion is such a basic part of our observations that even the origin of the word is lost in the darkness of Indo-European linguistic history. The fascination of motion has always made it a favourite object of curiosity. By the fifth century BCE in ancient Greece, its study had been given a name: *physics*. Ref. 1

Motion is also important to the human condition. What can we know? Where does the world come from? Who are we? Where do we come from? What will we do? What should we do? What will the future bring? What is death? Where does life lead? All these questions are about motion. And the study of motion provides answers that are both deep and surprising.

Motion is mysterious. Though found everywhere – in the stars, in the tides, in our eyelids – neither the ancient thinkers nor myriads of others in the 25 centuries since then have been able to shed light on the central mystery: *what is motion?* We shall discover that the standard reply, 'motion is the change of place in time', is correct, but inadequate. Just recently a full answer has finally been found. This is the story of the way to find it. Ref. 2

Motion is a part of human experience. If we imagine human experience as an island, then destiny, symbolized by the waves of the sea, carried us to its shore. Near the centre of

---
** Zeno of Elea (*c.* 450 BCE), one of the main exponents of the Eleatic school of philosophy.

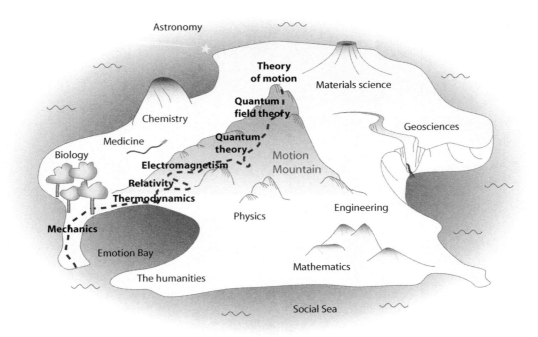

FIGURE 2 Experience Island, with Motion Mountain and the trail to be followed.

the island an especially high mountain stands out. From its top we can see over the whole landscape and get an impression of the relationships between all human experiences, and in particular between the various examples of motion. This is a guide to the top of what I have called Motion Mountain (see Figure 2; a less symbolic and more exact version is given in Figure 1). The hike is one of the most beautiful adventures of the human mind. The first question to ask is:

Does motion exist?

> *Das Rätsel* gibt es nicht. Wenn sich eine Frage überhaupt stellen läßt, so *kann* sie beantwortet werden.*
> Ludwig Wittgenstein, *Tractatus*, 6.5

To sharpen the mind for the issue of motion's existence, have a look at Figure 3 or Figure 4 and follow the instructions. In all cases the figures seem to rotate. You can experience similar effects if you walk over cobblestone pavement that is arranged in arched patterns or if you look at the numerous motion illusions collected by Kitaoka Akiyoshi at www.ritsumei.ac.jp/~akitaoka. How can we make sure that real motion is different from these or other similar illusions?

Many scholars simply argued that motion does not exist at all. Their arguments deeply influenced the investigation of motion over many centuries. For example, the Greek

---

* '*The riddle* does not exist. If a question can be put at all, it *can* also be answered.'

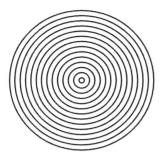 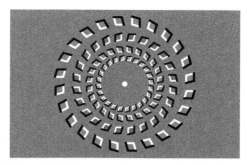

FIGURE 3 Illusions of motion: look at the figure on the left and slightly move the page, or look at the white dot at the centre of the figure on the right and move your head back and forward.

philosopher Parmenides (born *c.* 515 BCE in Elea, a small town near Naples) argued that since nothing comes from nothing, change cannot exist. He underscored the *permanence* of nature and thus consistently maintained that all change and thus all motion is an illusion.

Heraclitus (*c.* 540 to *c.* 480 BCE) held the opposite view. He expressed it in his famous statement πάντα ῥεῖ 'panta rhei' or 'everything flows'.* He saw change as the essence of nature, in contrast to Parmenides. These two equally famous opinions induced many scholars to investigate in more detail whether in nature there are *conserved* quantities or whether *creation* is possible. We will uncover the answer later on; until then, you might ponder which option you prefer.

Parmenides' collaborator Zeno of Elea (born *c.* 500 BCE) argued so intensely against motion that some people still worry about it today. In one of his arguments he claims – in simple language – that it is impossible to slap somebody, since the hand first has to travel halfway to the face, then travel through half the distance that remains, then again so, and so on; the hand therefore should never reach the face. Zeno's argument focuses on the relation between *infinity* and its opposite, finitude, in the description of motion. In modern quantum theory, a related issue is a subject of research up to this day.

Zeno also stated that by looking at a moving object at a *single* instant of time, one cannot maintain that it moves. He argued that at a single instant of time, there is no difference between a moving and a resting body. He then deduced that if there is no difference at a single time, there cannot be a difference for longer times. Zeno therefore questioned whether motion can clearly be distinguished from its opposite, *rest*. Indeed, in the history of physics, thinkers switched back and forward between a positive and a negative answer. It was this very question that led Albert Einstein to the development of general relativity, one of the high points of our journey. In our adventure, we will explore all known differences between motion and rest. Eventually, we will dare to ask whether single instants of time do exist at all. Answering this question is essential for reaching the top of Motion Mountain.

When we explore quantum theory, we will discover that motion is indeed – to a certain extent – an illusion, as Parmenides claimed. More precisely, we will show that motion is observed only due to the limitations of the human condition. We will find that we experience motion only because

---

* Appendix A explains how to read Greek text.

FIGURE 4 Zoom this image to large size or approach it closely in order to enjoy its apparent motion (© Michael Bach after the discovery of Kitaoka Akiyoshi).

— we have a finite size,
— we are made of a large but finite number of atoms,
— we have a finite but moderate temperature,
— we move much more slowly than the speed of light,
— we live in three dimensions,
— we are large compared with a black hole of our own mass,
— we are large compared with our quantum mechanical wavelength,
— we are small compared with the universe,
— we have a working but limited memory,
— we are forced by our brain to approximate space and time as continuous entities, and
— we are forced by our brain to approximate nature as made of different parts.

If any one of these conditions were not fulfilled, we would not observe motion; motion, then, would not exist! If that were not enough, note that none of the conditions requires human beings; they are equally valid for many animals and machines. Each of these conditions can be uncovered most efficiently if we start with the following question:

How should we talk about motion?

> Je hais le mouvement, qui déplace les lignes,
> Et jamais je ne pleure et jamais je ne ris.
> Charles Baudelaire, *La Beauté*.*

Like any science, the approach of physics is twofold: we advance with *precision* and with *curiosity*. Precision is the extent to which our description matches observations. Curios-

---
* Charles Baudelaire (b. 1821 Paris, d. 1867 Paris) *Beauty*: 'I hate movement, which changes shapes, and never do I weep and never do I laugh.'

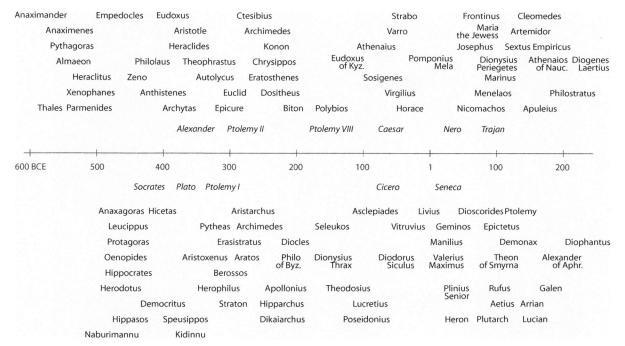

FIGURE 5 A time line of scientific and *political* personalities in antiquity. The last letter of the name is aligned with the year of death. For example, Maria the Jewess is the inventor of the bain-marie process and Thales is the first mathematician and scientist known by name.

ity is the passion that drives all scientists. Precision makes meaningful communication possible, and curiosity makes it worthwhile. Take an eclipse, a beautiful piece of music or a feat at the Olympic games: the world is full of fascinating examples of motion.

If you ever find yourself talking about motion, whether to understand it more precisely or more deeply, you are taking steps up Motion Mountain. The examples of Figure 6 make the point. An empty bucket hangs vertically. When you fill the bucket with a certain amount of water, it does not hang vertically any more. (Why?) If you continue adding water, it starts to hang vertically again. How much water is necessary for this last transition? The second illustration in Figure 6 is for the following puzzle. When you pull a thread from a reel in the way shown, the reel will move either forwards or backwards, depending on the angle at which you pull. What is the limiting angle between the two possibilities?

High precision means going into fine details. Being attuned to details actually *increases* the pleasure of the adventure.* Figure 7 shows an example. The higher we get on Motion Mountain, the further we can see and the more our curiosity is rewarded. The views offered are breathtaking, especially from the very top. The path we will follow – one of the many possible routes – starts from the side of biology and directly enters the forest that lies at the foot of the mountain.

---

* Distrust anybody who wants to talk you *out* of investigating details. He is trying to deceive you. Details are important. Be vigilant also during *this* journey.

FIGURE 6 How much water is required to make a bucket hang vertically? At what angle does the reel (drawn incorrectly, with too small rims) change direction of motion when pulled along with the thread? (© Luca Gastaldi).

Challenge 5 s

FIGURE 7 An example of how precision of observation can lead to the discovery of new effects: the deformation of a tennis ball during the c. 6 ms of a fast bounce (© International Tennis Federation).

Intense curiosity drives us to go straight to the limits: understanding motion requires exploration of the largest distances, the highest velocities, the smallest particles, the strongest forces, the highest precision and the strangest concepts. Let us begin.

What are the types of motion?

> Every movement is born of a desire for change.
> Antiquity

A good place to obtain a general overview on the types of motion is a large library. Table 1 shows the results. The domains in which motion, movements and moves play a role are indeed varied. Already the earliest researchers in ancient Greece – listed in Figure 5 – had the suspicion that all types of motion, as well as many other types of change, are related. Three categories of change are commonly recognized:

1. *Transport.* The only type of change we call motion in everyday life is material transport, such as a person walking, a leaf falling from a tree, or a musical instrument playing. Transport is the change of position or orientation of objects, fluids included. To a large extent, the behaviour of people also falls into this category.

2. *Transformation.* Another category of change groups observations such as the dissolution of salt in water, the formation of ice by freezing, the rotting of wood, the cooking of food, the coagulation of blood, and the melting and alloying of metals.

TABLE 1 Content of books about motion found in a public library.

Motion topics	Motion topics
motion pictures and digital effects	motion as therapy for cancer, diabetes, acne and depression
motion perception Ref. 11	motion sickness
motion for fitness and wellness	motion for meditation
motion control and training in sport and singing	motion ability as health check
perpetual motion	motion in dance, music and other performing arts
motion as proof of various gods Ref. 12	motion of planets, stars and angels Ref. 13
economic efficiency of motion	the connection between motional and emotional habits
motion as help to overcome trauma	motion in psychotherapy Ref. 14
locomotion of insects, horses, animals and robots	motion of cells and plants
collisions of atoms, cars, stars and galaxies	growth of multicellular beings, mountains, sunspots and galaxies
motion of springs, joints, mechanisms, liquids and gases	motion of continents, bird flocks, shadows and empty space
commotion and violence	motion in martial arts
motions in parliament	movements in art, sciences and politics
movements in watches	movements in the stock market
movement teaching and learning	movement development in children Ref. 15
musical movements	troop movements Ref. 16
religious movements	bowel movements
moves in chess	cheating moves in casinos Ref. 17
connection between gross national product and citizen mobility	

These changes of colour, brightness, hardness, temperature and other material properties are all transformations. Transformations are changes not visibly connected with transport. To this category, a few ancient thinkers added the emission and absorption of light. In the twentieth century, these two effects were proven to be special cases of transformations, as were the newly discovered appearance and disappearance of matter, as observed in the Sun and in radioactivity. *Mind change*, such as change of mood, of health, of education and of character, is also (mostly) a type of transformation. Ref. 18

3. *Growth.* This last and especially important category of change, is observed for animals, plants, bacteria, crystals, mountains, planets, stars and even galaxies. In the nineteenth century, changes in the population of systems, *biological evolution*, and in the twentieth century, changes in the size of the universe, *cosmic evolution*, were added to this category. Traditionally, these phenomena were studied by separate sciences. Independently they all arrived at the conclusion that growth is a combination of transport and transformation. The difference is one of complexity and of time scale. Ref. 19

FIGURE 8 An example of transport, at Mount Etna (© Marco Fulle).

At the beginnings of modern science during the Renaissance, only the study of transport was seen as the topic of physics. Motion was equated to transport. The other two domains were neglected by physicists. Despite this restriction, the field of enquiry remains large, covering a large part of Experience Island. Early scholars differentiated types of transport by their origin. Movements such as those of the legs when walking were classified as *volitional*, because they are controlled by one's will, whereas movements of external objects, such as the fall of a snowflake, which cannot be influenced by will-power, were classified as *passive*. Young humans, especially young male humans, spend considerable time in learning elaborate volitional movements. An example is shown in Figure 10.

The complete distinction between passive and volitional motion is made by children by the age of six, and this marks a central step in the development of every human towards a precise description of the environment.* From this distinction stems the historical but now outdated definition of physics as the science of motion of non-living things.

The advent of machines forced scholars to rethink the distinction between volitional and passive motion. Like living beings, machines are self-moving and thus mimic volitional motion. However, careful observation shows that every part in a machine is moved by another, so their motion is in fact passive. Are living beings also machines? Are human actions examples of passive motion as well? The accumulation of observations in the last 100 years made it clear that volitional movement** indeed has the same physical properties as passive movement in non-living systems. A distinction between the two types

---

* Failure to pass this stage completely can result in a person having various strange beliefs, such as believing in the ability to influence roulette balls, as found in compulsive players, or in the ability to move other bodies by thought, as found in numerous otherwise healthy-looking people. An entertaining and informative account of all the deception and self-deception involved in creating and maintaining these beliefs is given by JAMES RANDI, *The Faith Healers*, Prometheus Books, 1989. A professional magician, he presents many similar topics in several of his other books. See also his www.randi.org website for more details.

** The word 'movement' is rather modern; it was imported into English from the old French and became popular only at the end of the eighteenth century. It is never used by Shakespeare.

FIGURE 9 Transport, growth and transformation (© Philip Plisson).

FIGURE 10 One of the most difficult volitional movements known, performed by Alexander Tsukanov, the first man able to do this: jumping from one ultimate wheel to another (© Moscow State Circus).

of motion is thus unnecessary. Of course, from the emotional viewpoint, the differences are important; for example, *grace* can only be ascribed to volitional movements.

Since passive and volitional motion have the same properties, through the study of motion of non-living objects we can learn something about the human condition. This is most evident when touching the topics of determinism, causality, probability, infinity, time, love and death, to name but a few of the themes we will encounter during our adventure.

In the nineteenth and twentieth centuries other classically held beliefs about motion fell by the wayside. Extensive observations showed that all transformations and all growth phenomena, including behaviour change and evolution, are also examples of transport. In other words, over 2 000 years of studies have shown that the ancient classification of observations was useless:

▷ All change is transport.

And

▷ Transport and motion are the same.

In the middle of the twentieth century the study of motion culminated in the experimental confirmation of an even more specific idea, previously articulated in ancient Greece:

▷ Every type of change is due to the motion of particles.

It takes time and work to reach this conclusion, which appears only when we relentlessly pursue higher and higher precision in the description of nature. The first five parts of this adventure retrace the path to this result. (Do you agree with it?)

The last decade of the twentieth century again completely changed the description of motion: the particle idea turns out to be limited and wrong. This recent result, reached through a combination of careful observation and deduction, will be explored in the last part of our adventure. But we still have some way to go before we reach that result, just below the summit of our journey.

In summary, history has shown that classifying the various types of motion is not productive. Only by trying to achieve maximum precision can we hope to arrive at the fundamental properties of motion. *Precision, not classification, is the path to follow.* As Ernest Rutherford said jokingly: 'All science is either physics or stamp collecting.'

In order to achieve precision in our description of motion, we need to select specific examples of motion and study them fully in detail. It is intuitively obvious that the most precise description is achievable for the *simplest* possible examples. In everyday life, this is the case for the motion of any non-living, solid and rigid body in our environment, such as a stone thrown through the air. Indeed, like all humans, we learned to throw objects long before we learned to walk. Throwing is one of the first physical experiments we performed by ourselves. The importance of throwing is also seen from the terms derived from it: in Latin, words like *subject* or 'thrown below', *object* or 'thrown in front', and *interjection* or 'thrown in between'; in Greek, the act of throwing led to terms like *symbol* or 'thrown together', *problem* or 'thrown forward', *emblem* or 'thrown into', and – last but not least – *devil* or 'thrown through'. And indeed, during our early childhood, by throwing stones, toys and other objects until our parents feared for every piece of the household, we explored the perception and the properties of motion. We do the same here.

> Die Welt ist unabhängig von meinem Willen.*
> Ludwig Wittgenstein, *Tractatus*, 6.373

---

* 'The world is independent of my will.'

FIGURE 11 How do we distinguish a deer from its environment? (© Tony Rodgers).

Perception, permanence and change

> Only wimps study only the general case; real scientists pursue examples.
> Beresford Parlett

Human beings enjoy perceiving. Perception starts before birth, and we continue enjoying it for as long as we can. That is why television, even when devoid of content, is so successful. During our walk through the forest at the foot of Motion Mountain we cannot avoid perceiving. Perception is first of all the ability to *distinguish*. We use the basic mental act of distinguishing in almost every instant of life; for example, during childhood we first learned to distinguish familiar from unfamiliar observations. This is possible in combination with another basic ability, namely the capacity to *memorize* experiences. Memory gives us the ability to experience, to talk and thus to explore nature. Perceiving, classifying and memorizing together form *learning*. Without any one of these three abilities, we could not study motion.

Children rapidly learn to distinguish *permanence* from *variability*. They learn to *recognize* human faces, even though a face never looks exactly the same each time it is seen. From recognition of faces, children extend recognition to all other observations. Recognition works pretty well in everyday life; it is nice to recognize friends, even at night, and even after many beers (not a challenge). The act of recognition thus always uses a form of *generalization*. When we observe, we always have some general idea in our mind. Let us specify the main ones.

Sitting on the grass in a clearing of the forest at the foot of Motion Mountain, surrounded by the trees and the silence typical of such places, a feeling of calmness and tranquillity envelops us. We are thinking about the essence of perception. Suddenly, something moves in the bushes; immediately our eyes turn and our attention focuses. The nerve cells that detect motion are part of the most ancient part of our brain, shared with birds and reptiles: the brain stem. Then the cortex, or modern brain, takes over to analyse the type of motion and to identify its origin. Watching the motion across our field of vision, we observe two invariant entities: the fixed landscape and the moving animal. After we recognize the animal as a deer, we relax again.

How did we distinguish, in case of Figure 11, between landscape and deer? Perception involves several processes in the eye and in the brain. An essential part for these processes is motion, as is best deduced from the flip film shown in the lower left corners of these pages. Each image shows only a rectangle filled with a mathematically random pattern. But when the pages are scanned in rapid succession, you discern a shape – a square – moving against a fixed background. At any given instant, the square cannot be distinguished from the background; there is no visible object at any given instant of time. Nevertheless it is easy to perceive its motion.* Perception experiments such as this one have been performed in many variations. Such experiments showed that detecting a moving square against a random background is nothing special to humans; flies have the same ability, as do, in fact, all animals that have eyes.

The flip film in the lower left corner, like many similar experiments, illustrates two central attributes of motion. First, motion is perceived only if an *object* can be distinguished from a *background* or *environment*. Many motion illusions focus on this point.** Second, motion is required to *define* both the object and the environment, and to distinguish them from each other. In fact, the concept of space is – among others – an abstraction of the idea of background. The background is extended; the moving entity is localized. Does this seem boring? It is not; just wait a second.

We call a localized entity of investigation that can change or move a *physical system* – or simply a system. A system is a recognizable, thus permanent part of nature. Systems can be objects – also called 'physical bodies' – or radiation. Therefore, images, which are made of radiation, are aspects of physical systems, but not themselves physical systems. These connections are summarized in Table 2. Now, are holes physical systems?

In other words, we call the set of localized aspects that remain invariant or permanent during motion, such as size, shape, colour etc., taken together, a (physical) *object* or a (physical) *body*. We will tighten the definition shortly, to distinguish objects from images.

We note that to specify permanent moving objects, we need to distinguish them from the environment. In other words, right from the start we experience motion as a *relative* process; it is perceived in relation and in opposition to the environment.

The conceptual distinction between localized, isolable objects and the extended environment is important. True, it has the appearance of a circular definition. (Do you agree?) Indeed, this issue will keep us busy later on. On the other hand, we are so used to our ability of isolating local systems from the environment that we take it for granted. However, as we will discover later on in our walk, this distinction turns out to be logically and experimentally impossible!*** The reason for this impossibility will turn out to be fascinating. To discover the impossibility, we note, as a first step, that apart from moving entities and the permanent background, we also need to describe their relations. The

---

* The human eye is rather good at detecting motion. For example, the eye can detect motion of a point of light even if the change of angle is smaller than that which can be distinguished in a fixed image. Details of this and similar topics for the other senses are the domain of perception research.
** The topic of motion perception is full of interesting aspects. An excellent introduction is chapter 6 of the beautiful text by Donald D. Hoffman, *Visual Intelligence – How We Create What We See*, W.W. Norton & Co., 1998. His collection of basic motion illusions can be experienced and explored on the associated www.cogsci.uci.edu/~ddhoff website.
*** Contrary to what is often read in popular literature, the distinction *is* possible in quantum theory. It becomes impossible only when quantum theory is unified with general relativity.

TABLE 2 Family tree of the basic physical concepts.

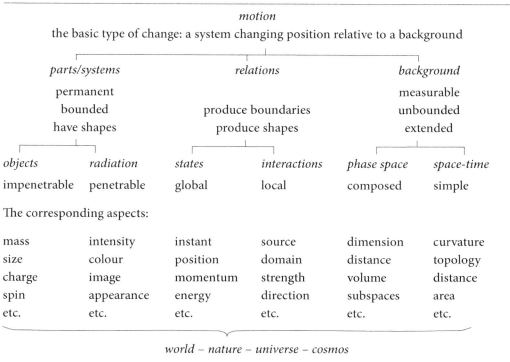

necessary concepts are summarized in Table 2.

> " Wisdom is one thing: to understand the thought which steers all things through all things. "
> Heraclitus of Ephesus

Ref. 24

Does the world need states?

> " Das Feste, das Bestehende und der Gegenstand sind Eins. Der Gegenstand ist das Feste, Bestehende; die Konfiguration ist das Wechselnde, Unbeständige.* "
> Ludwig Wittgenstein, *Tractatus*, 2.027 – 2.0271

What distinguishes the various patterns in the lower left corners of this text? In everyday life we would say: the situation or configuration of the involved entities. The situation somehow describes all those aspects that can differ from case to case. It is customary to call the list of all *variable* aspects of a set of objects their *(physical) state of motion*, or simply their *state*. How is the state characterized?

---

* 'The fixed, the existent and the object are one. The object is the fixed, the existent; the configuration is the changing, the variable.'

The configurations in the lower left corners differ first of all in *time*. Time is what makes opposites possible: a child is in a house and the same child is outside the house. Time describes and resolves this type of contradiction. But the state not only distinguishes situations in time: the state contains *all* those aspects of a *system* – i.e., of a group of objects – that set it apart from all *similar* systems. Two similar objects can differ, at each instant of time, in their
— position,
— velocity,
— orientation, or
— angular velocity.

These properties determine the state and pinpoint the *individuality* of a physical system among *exact copies* of itself. Equivalently, the state describes the relation of an object or a system with respect to its environment. Or, again, equivalently:

▷ The *state* describes all aspects of a system that depend on the observer.

The definition of state is not boring at all – just ponder this: Does the *universe* have a state? And: is the list of state properties just given *complete*?

In addition, physical systems are described by their permanent, *intrinsic properties*. Some examples are
— mass,
— shape,
— colour,
— composition.

Intrinsic properties do not depend on the observer and are independent of the state of the system. They are permanent – at least for a certain time interval. Intrinsic properties also allow to distinguish physical systems from each other. And again, we can ask: What is the *complete* list of intrinsic properties in nature? And does the universe have intrinsic properties?

The various aspects of objects and of their states are called (physical) *observables*. We will refine this rough, preliminary definition in the following.

Describing nature as a collection of permanent entities and changing states is the starting point of the study of motion. Every observation of motion requires the distinction of permanent, intrinsic properties – describing the objects that move – and changing states – describing the way the objects move. Without this distinction, there is no motion. Without this distinction, there is not even a way to *talk* about motion.

Using the terms just introduced, we can say

▷ Motion is the change of state of permanent objects.

The exact separation between those aspects belonging to the object, the permanent *intrinsic properties*, and those belonging to the state, the varying *state properties*, depends on the precision of observation. For example, the length of a piece of wood is not permanent; wood shrinks and bends with time, due to processes at the molecular level. To be precise, the length of a piece of wood is not an aspect of the object, but an aspect of its state. Precise observations thus *shift* the distinction between the object and its state;

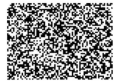

the distinction itself does not disappear – at least not in the first five volumes of our adventure.

In the end of the twentieth century, neuroscience discovered that the distinction between changing states and permanent objects is not only made by scientists and engineers. Also nature makes the distinction. In fact, nature has hard-wired the distinction into the brain! Using the output signals from the visual cortex that processes what the eyes observe, the adjacent parts on the *upper* side of the human brain – the *dorsal stream* – process the *state* of the objects that are seen, such their distance and motion, whereas the adjacent parts on the *lower* side of the human brain – the *ventral stream* – process *intrinsic properties*, such as shape, colours and patterns.

In summary, states are indeed required for the description of motion. So are permanent, intrinsic properties. In order to proceed and to achieve a *complete* description of motion, we thus need a complete description of the possible states and a complete description of the intrinsic properties of objects. The first approach that attempt this is called Galilean physics; it starts by specifying our *everyday* environment and the motion in it as precisely as possible.

Galilean physics in six interesting statements

The study of everyday motion, Galilean physics, is already worthwhile in itself: we will uncover many results that are in contrast with our usual experience. For example, if we recall our own past, we all have experienced how important, delightful or unwelcome *surprises* can be. Nevertheless, the study of everyday motion shows that there are *no* surprises in nature. Motion, and thus the world, is *predictable* or *deterministic*.

The main surprise of our exploration of motion is that there are no surprises in nature. Nature is predictable. In fact, we will uncover six aspects of the predictability of everyday motion:

1. *Continuity.* We know that eyes, cameras and measurement apparatus have a finite resolution. All have a smallest distance they can observe. We know that clocks have a smallest time they can measure. Despite these limitations, in everyday life all movements, their states, as well as space and time themselves, are *continuous*.

2. *Conservation.* We all observe that people, music and many other things in motion stop moving after a while. The study of motion yields the opposite result: motion never stops. In fact, three aspects of motion do not change, but are *conserved*: momentum, angular momentum and energy are conserved, separately, in all examples of motion. No exception to these three types of conservation has ever been observed. (In contrast, mass is often, but not always conserved.) In addition, we will discover that conservation implies that motion and its properties are the same at all places and all times: motion is *universal*.

3. *Relativity.* We all know that motion differs from rest. Despite this experience, careful study shows that there is no intrinsic difference between the two. Motion and rest depend on the observer. Motion is *relative*. And so is rest. This is the first step towards understanding the theory of relativity.

4. *Reversibility.* We all observe that many processes happen only in one direction. For example, spilled milk never returns into the container by itself. Despite such obser-

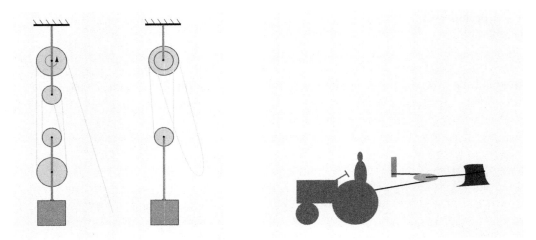

FIGURE 12 A block and tackle and a differential pulley (left) and a farmer (right).

vations, the study of motion will show us that all everyday motion is *reversible*. Physicists call this the invariance of everyday motion under *motion reversal*. Sloppily, but incorrectly, one sometimes speaks of 'time reversal'.

5. *Mirror invariance.* Most of us find scissors difficult to handle with the left hand, have difficulties to write with the other hand, and have grown with a heart on the left side. Despite such observations, our exploration will show that everyday motion is *mirror-invariant* (or *parity-invariant*). Mirror processes are always possible in everyday life.

6. *Change minimization.* We all are astonished by the many observations that the world offers: colours, shapes, sounds, growth, disasters, happiness, friendship, love. The variation, beauty and complexity of nature is amazing. We will confirm that all observations are due to motion. And despite the appearance of complexity, all motion is simple. Our study will show that all observations can be summarized in a simple way: Nature is lazy. All motion happens in a way that *minimizes change*. Change can be measured, using a quantity called 'action', and nature keeps it to a minimum. Situations – or states, as physicists like to say – evolve by minimizing change. Nature is lazy.

These six aspects are essential in understanding motion in sport, in music, in animals, in machines or among the stars. This first volume of our adventure will be an exploration of such movements. In particular, we will confirm, against all appearances of the contrary, the mentioned six key properties in all cases of everyday motion.

Curiosities and fun challenges about motion*

In contrast to most animals, sedentary creatures, like plants or sea anemones, have no legs and cannot move much; for their self-defence, they developed *poisons*. Examples of such plants are the stinging nettle, the tobacco plant, digitalis, belladonna and poppy;

---

* Sections entitled 'curiosities' are collections of topics and problems that allow one to check and to expand the usage of concepts already introduced.

poisons include caffeine, nicotine, and curare. Poisons such as these are at the basis of most medicines. Therefore, most medicines exist essentially because plants have no legs.

∗ ∗

A man climbs a mountain from 9 a.m. to 1 p.m. He sleeps on the top and comes down the next day, taking again from 9 a.m. to 1 p.m. for the descent. Is there a place on the path that he passes at the same time on the two days?

Challenge 12 s

∗ ∗

Every time a soap bubble bursts, the motion of the surface during the burst is the same, even though it is too fast to be seen by the naked eye. Can you imagine the details?

Challenge 13 s

∗ ∗

Is the motion of a ghost an example of motion?

Challenge 14 s

∗ ∗

Can something stop moving? How would you show it?

Challenge 15 s

∗ ∗

Does a body moving forever in a straight line show that nature or space is infinite?

Challenge 16 s

∗ ∗

What is the length of rope one has to pull in order to lift a mass by a height h with a block and tackle with four wheels, as shown on the left of Figure 12? Does the farmer on the right of the figure do something sensible?

Challenge 17 s

In the past, block and tackles were important in many machines. Two particularly useful versions are the *differential* block and tackle, also called a differential pulley, which is easy to make, and the *Spanish burton*, which has the biggest effect with the smallest number of wheels. There is even a so-called *fool's tackle*. Enjoy their exploration.

Ref. 25

Challenge 18 e

All these devices are examples of the *golden rule of mechanics*: what you gain in force, you loose in displacement. Or, equivalently: force times displacement – also called (physical) *work* – remains unchanged, whatever mechanical device you may use. This is one example of conservation that is observed in everyday motion.

∗ ∗

Can the universe move?

Challenge 19 s

∗ ∗

To talk about precision with precision, we need to measure precision itself. How would you do that?

Challenge 20 s

∗ ∗

Would we observe motion if we had no memory?

Challenge 21 s

∗ ∗

What is the lowest speed you have observed? Is there a lowest speed in nature?

Challenge 22 s

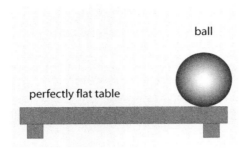

FIGURE 13 What happens?

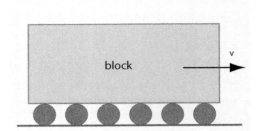

FIGURE 14 What is the speed of the rollers? Are other roller shapes possible?

∗ ∗

According to legend, Sissa ben Dahir, the Indian inventor of the game of *chaturanga* or chess, demanded from King Shirham the following reward for his invention: he wanted one grain of wheat for the first square, two for the second, four for the third, eight for the fourth, and so on. How much time would all the wheat fields of the world take to produce the necessary grains?

∗ ∗

When a burning candle is moved, the flame lags behind the candle. How does the flame behave if the candle is inside a glass, still burning, and the glass is accelerated?

∗ ∗

A good way to make money is to build motion detectors. A motion detector is a small box with a few wires. The box produces an electrical signal whenever the box moves. What types of motion detectors can you imagine? How cheap can you make such a box? How precise?

∗ ∗

A perfectly frictionless and spherical ball lies near the edge of a perfectly flat and horizontal table, as shown in Figure 13. What happens? In what time scale?

∗ ∗

You step into a closed box without windows. The box is moved by outside forces unknown to you. Can you determine *how* you are moving from inside the box?

∗ ∗

When a block is rolled over the floor over a set of cylinders, as shown in Figure 14, how are the speed of the block and that of the cylinders related?

∗ ∗

Do you dislike formulae? If you do, use the following three-minute method to change the situation. It is worth trying it, as it will make you enjoy this book much more. Life is short; as much of it as possible, like reading this text, should be a pleasure.

1. Close your eyes and recall an experience that was *absolutely marvellous*, a situation when you felt excited, curious and positive.
2. Open your eyes for a second or two and look at page 280 – or any other page that contains many formulae.
3. Then close your eyes again and return to your marvellous experience.
4. Repeat the observation of the formulae and the visualization of your memory – steps 2 and 3 – three more times.

Then leave the memory, look around yourself to get back into the here and now, and test yourself. Look again at page 280. How do you feel about formulae now?

∗ ∗

In the sixteenth century, Niccolò Tartaglia* proposed the following problem. Three young couples want to cross a river. Only a small boat that can carry two people is available. The men are extremely jealous, and would never leave their brides with another man. How many journeys across the river are necessary?

∗ ∗

Cylinders can be used to roll a flat object over the floor, as shown in Figure 14. The cylinders keep the object plane always at the same distance from the floor. What cross-sections *other* than circular, so-called *curves of constant width*, can a cylinder have to realize the same feat? How many examples can you find? Are objects different than cylinders possible?

∗ ∗

Hanging pictures on a wall is not easy. First puzzle: what is the best way to hang a picture on one nail? The method must allow you to move the picture in horizontal position after the nail is in the wall, in the case that the weight is not equally distributed. Second puzzle: Can you hang a picture on a wall – this time with a long rope – over two nails in such a way that pulling either nail makes the picture fall? And with three nails? And *n* nails?

First summary on motion

Motion, the change of position of physical systems, is the most fundamental observation in nature. Everyday motion is predictable and deterministic. Predictability is reflected in six aspects of motion: continuity, conservation, reversibility, mirror-invariance, relativity and minimization. Some of these aspects are valid for *all* motion, and some are valid only for *everyday* motion. Which ones, and why? We explore this now.

---

* Niccolò Fontana Tartaglia (1499–1557), important Renaissance mathematician.

Chapter 2
FROM MOTION MEASUREMENT TO CONTINUITY

> "Physic ist wahrlich das eigentliche Studium des Menschen.**
> Georg Christoph Lichtenberg

The simplest description of motion is the one we all, like cats or monkeys, use throughout our everyday life: *only one thing can be at a given spot at a given time.* This general description can be separated into three assumptions: matter is *impenetrable* and *moves*, time is made of *instants*, and space is made of *points*. Without these three assumptions (do you agree with them?) it is not even possible to define velocity. We thus need points embedded in continuous space and time to talk about motion. This description of nature is called *Galilean physics*, or sometimes *Newtonian physics*.

Galileo Galilei (1564–1642), Tuscan professor of mathematics, was the central founder of modern physics. He became famous for advocating the importance of observations as checks of statements about nature. By requiring and performing these checks throughout his life, he was led to continuously increase the accuracy in the description of motion. For example, Galileo studied motion by measuring change of position with a self-constructed stopwatch. Galileo's experimental aim was to measure all what is measurable about motion. His approach changed the speculative description of ancient Greece into the experimental physics of Renaissance Italy.***

After Galileo, the English alchemist, occultist, theologian, physicist and politician Isaac Newton (1643–1727) continued to explore with vigour the idea that different types of motion have the same properties, and he made important steps in constructing the concepts necessary to demonstrate this idea.****

Above all, the explorations and books by Galileo popularized the fundamental experimental statements on the properties of speed, space and time.

** 'Physics truly is the proper study of man.' Georg Christoph Lichtenberg (b. 1742 Ober-Ramstadt, d. 1799 Göttingen) was an important physicist and essayist.
*** The best and most informative book on the life of Galileo and his times is by Pietro Redondi (see the section on page 335). Galileo was born in the year the pencil was invented. Before his time, it was impossible to do paper and pencil calculations. For the curious, the www.mpiwg-berlin.mpg.de website allows you to read an original manuscript by Galileo.
**** Newton was born a year after Galileo died. For most of his life Newton searched for the philosopher's stone. Newton's hobby, as head of the English mint, was to supervise personally the hanging of counterfeiters. About Newton's lifelong infatuation with alchemy, see the books by Dobbs. A misogynist throughout his life, Newton believed himself to be chosen by god; he took his Latin name, *Isaacus Neuutonus*, and formed the anagram *Jeova sanctus unus*. About Newton and his importance for classical mechanics, see the text by Clifford Truesdell.

FIGURE 15 Galileo Galilei (1564–1642).

FIGURE 16 Some speed measurement devices: an anemometer, a tachymeter for inline skates, a sport radar gun and a Pitot–Prandtl tube in an aeroplane (© Fachhochschule Koblenz, Silva, Tracer, Wikimedia).

What is velocity?

> There is nothing else like it.
> Jochen Rindt*

Velocity fascinates. To physicists, not only car races are interesting, but any moving entity is. Therefore, physicists first measure as many examples as possible. A selection of measured speed values is given in Table 3. The units and prefixes used are explained in detail in Appendix B. Some speed measurement devices are shown in Figure 16.

Everyday life teaches us a lot about motion: objects can overtake each other, and they can move in different directions. We also observe that velocities can be added or changed smoothly. The precise list of these properties, as given in Table 4, is summarized by mathematicians in a special term; they say that velocities form a *Euclidean vector space*.** More details about this strange term will be given shortly. For now we just note that in describing nature, mathematical concepts offer the most accurate vehicle.

When velocity is assumed to be an Euclidean vector, it is called *Galilean* velocity. Velocity is a profound concept. For example, velocity does not need space and time measurements to be defined. Are you able to find a means of measuring velocities without

---

* Jochen Rindt (1942–1970), famous Austrian Formula One racing car driver, speaking about speed.
** It is named after Euclid, or Eukleides, the great Greek mathematician who lived in Alexandria around 300 BCE. Euclid wrote a monumental treatise of geometry, the Στοιχεῖα or *Elements*, which is one of the milestones of human thought. The text presents the whole knowledge on geometry of that time. For the first time, Euclid introduces two approaches that are now in common use: all statements are deduced from a small number of basic *axioms* and for every statement a *proof* is given. The book, still in print today, has been the reference geometry text for over 2000 years. On the web, it can be found at aleph0.clarku.edu/~djoyce/java/elements/elements.html.

TABLE 3 Some measured velocity values.

Observation	Velocity
Growth of deep sea manganese crust	80 am/s
Can you find something slower?	Challenge 36 s
Stalagmite growth	0.3 pm/s
Lichen growth	down to 7 pm/s
Typical motion of continents	10 mm/a = 0.3 nm/s
Human growth during childhood, hair growth	4 nm/s
Tree growth	up to 30 nm/s
Electron drift in metal wire	1 µm/s
Sperm motion	60 to 160 µm/s
Speed of light at Sun's centre Ref. 29	1 mm/s
Ketchup motion	1 mm/s
Slowest speed of light measured in matter on Earth Ref. 30	0.3 m/s
Speed of snowflakes	0.5 m/s to 1.5 m/s
Signal speed in human nerve cells Ref. 31	0.5 m/s to 120 m/s
Wind speed at 1 and 12 Beaufort (light air and hurricane)	< 1.5 m/s, > 33 m/s
Speed of rain drops, depending on radius	2 m/s to 8 m/s
Fastest swimming fish, sailfish (*Istiophorus platypterus*)	22 m/s
2009 Speed sailing record over 500 m (by trimaran Hydroptère)	26.4 m/s
Fastest running animal, cheetah (*Acinonyx jubatus*)	30 m/s
Speed of air in throat when sneezing	42 m/s
Fastest throw: a cricket ball thrown with baseball technique while running	50 m/s
Freely falling human, depending on clothing	50 to 90 m/s
Fastest bird, diving *Falco peregrinus*	60 m/s
Fastest badminton smash	70 m/s
Average speed of oxygen molecule in air at room temperature	280 m/s
Speed of sound in dry air at sea level and standard temperature	330 m/s
Speed of the equator	434 m/s
Cracking whip's end	750 m/s
Speed of a rifle bullet	1 km/s
Speed of crack propagation in breaking silicon	5 km/s
Highest macroscopic speed achieved by man – the *Helios II* satellite	70.2 km/s
Speed of Earth through universe	370 km/s
Average speed (and peak speed) of lightning tip	600 km/s (50 Mm/s)
Highest macroscopic speed measured in our galaxy Ref. 32	$0.97 \cdot 10^8$ m/s
Speed of electrons inside a colour TV tube	$1 \cdot 10^8$ m/s
Speed of radio messages in space	299 792 458 m/s
Highest ever measured group velocity of light	$10 \cdot 10^8$ m/s
Speed of light spot from a lighthouse when passing over the Moon	$2 \cdot 10^9$ m/s
Highest *proper* velocity ever achieved for electrons by man	$7 \cdot 10^{13}$ m/s
Highest possible velocity for a light spot or a shadow	no limit

TABLE 4 Properties of everyday – or Galilean – velocity.

VELOCITIES CAN	PHYSICAL PROPERTY	MATHEMATICAL NAME	DEFINITION
Be distinguished	distinguishability	element of set	Vol. III, page 284
Change gradually	continuum	real vector space	Page 80, Vol. V, page 364
Point somewhere	direction	vector space, dimensionality	Page 80
Be compared	measurability	metricity	Vol. IV, page 236
Be added	additivity	vector space	Page 80
Have defined angles	direction	Euclidean vector space	Page 81
Exceed any limit	infinity	unboundedness	Vol. III, page 285

measuring space and time? If so, you probably want to skip to the next volume, jumping 2000 years of enquiries. If you cannot do so, consider this: whenever we measure a quantity we assume that everybody is able to do so, and that everybody will get the same result. In other words, we define *measurement* as a comparison with a standard. We thus implicitly assume that such a standard exists, i.e., that an example of a 'perfect' velocity can be found. Historically, the study of motion did not investigate this question first, because for many centuries nobody could find such a standard velocity. You are thus in good company.

How is velocity measured in everyday life? Animals and people estimate their velocity in two ways: by estimating the frequency of their own movements, such as their steps, or by using their eyes, ears, sense of touch or sense of vibration to deduce how their own position changes with respect to the environment. But several animals have additional capabilities: certain snakes can determine speeds with their infrared-sensing organs, others with their magnetic field sensing organs. Still other animals emit sounds that create echoes in order to measure speeds to high precision. Other animals use the stars to navigate. A similar range of solutions is used by technical devices. Table 5 gives an overview.

Velocity is not always an easy subject. Physicists like to say, provokingly, that what cannot be measured does not exist. Can you measure your own velocity in empty interstellar space?

Velocity is of interest to both engineers and evolution scientist. In general, self-propelled systems are faster the larger they are. As an example, Figure 17 shows how this applies to the cruise speed of flying things. In general, cruise speed scales with the sixth root of the weight, as shown by the trend line drawn in the graph. (Can you find out why?) By the way, similar *allometric scaling* relations hold for many other properties of moving systems, as we will see later on.

Some researchers have specialized in the study of the lowest velocities found in nature: they are called geologists. Do not miss the opportunity to walk across a landscape while listening to one of them.

Velocity is a profound subject for an additional reason: we will discover that all its seven properties of Table 4 are only approximate; *none* is actually correct. Improved experiments will uncover exceptions for every property of Galilean velocity. The failure of

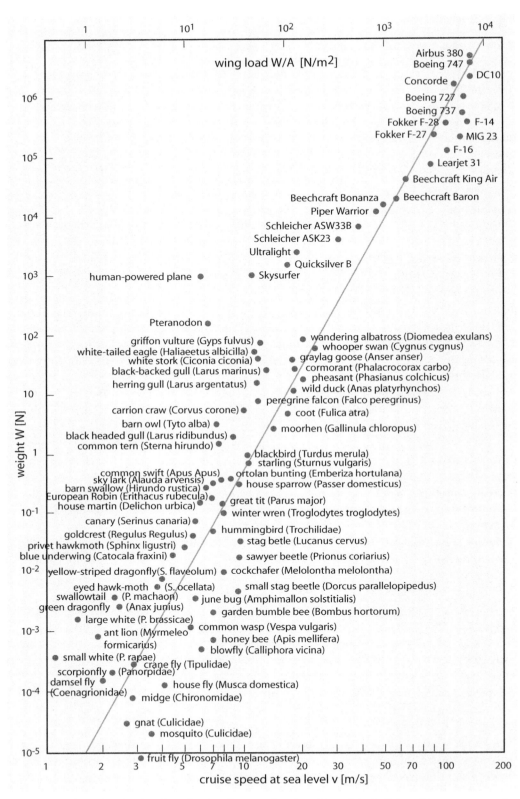

FIGURE 17 How wing load and sea-level cruise speed scales with weight in flying objects, compared with the general trend line (after a graph © Henk Tennekes).

TABLE 5 Speed measurement devices in biological and engineered systems.

MEASUREMENT	DEVICE	RANGE
Own running speed in insects, mammals and humans	leg beat frequency measured with internal clock	0 to 33 m/s
Own car speed	tachymeter attached to wheels	0 to 150 m/s
Predators and hunters measuring prey speed	vision system	0 to 30 m/s
Police measuring car speed	radar or laser gun	0 to 90 m/s
Bat measuring own and prey speed at night	doppler sonar	0 to 20 m/s
Sliding door measuring speed of approaching people	doppler radar	0 to 3 m/s
Own swimming speed in fish and humans	friction and deformation of skin	0 to 30 m/s
Own swimming speed in dolphins and ships	sonar to sea floor	0 to 20 m/s
Diving speed in fish, animals, divers and submarines	pressure change	0 to 5 m/s
Water predators and fishing boats measuring prey speed	sonar	0 to 20 m/s
Own speed relative to Earth in insects	often none (grasshoppers)	n.a.
Own speed relative to Earth in birds	visual system	0 to 60 m/s
Own speed relative to Earth in aeroplanes or rockets	radio goniometry, radar	0 to 8000 m/s
Own speed relative to air in insects and birds	filiform hair deflection, feather deflection	0 to 60 m/s
Own speed relative to air in aeroplanes	Pitot–Prandtl tube	0 to 340 m/s
Wind speed measurement in meteorological stations	thermal, rotating or ultrasound anemometers	0 to 80 m/s
Swallows measuring prey speed	visual system	0 to 20 m/s
Bats measuring prey speed	sonar	0 to 20 m/s
Macroscopic motion on Earth	Global Positioning System, Galileo, Glonass	0 to 100 m/s
Pilots measuring target speed	radar	0 to 1000 m/s
Motion of stars	optical Doppler effect	0 to 1000 km/s
Motion of star jets	optical Doppler effect	0 to 200 Mm/s

the last three properties of Table 4 will lead us to special and general relativity, the failure of the middle two to quantum theory and the failure of the first two properties to the unified description of nature. But for now, we'll stick with Galilean velocity, and continue with another Galilean concept derived from it: time.

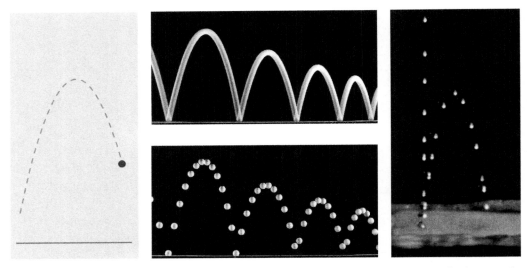

FIGURE 18 A typical path followed by a stone thrown through the air – a parabola – with photographs (blurred and stroboscopic) of a table tennis ball rebounding on a table (centre) and a stroboscopic photograph of a water droplet rebounding on a strongly hydrophobic surface (right, © Andrew Davidhazy, Max Groenendijk).

> Without the concepts *place*, *void* and *time*, change cannot be. [...] It is therefore clear [...] that their investigation has to be carried out, by studying each of them separately.
> Aristotle* *Physics*, Book III, part 1.

What is time?

> Time is an accident of motion.
> Theophrastus**

> Time does not exist in itself, but only through the perceived objects, from which the concepts of past, of present and of future ensue.
> Lucretius,*** *De rerum natura*, lib. 1, v. 460 ss.

In their first years of life, children spend a lot of time throwing objects around. The term 'object' is a Latin word meaning 'that which has been thrown in front.' Developmental psychology has shown experimentally that from this very experience children extract the concepts of time and space. Adult physicists do the same when studying motion at university.

When we throw a stone through the air, we can define a *sequence* of observations. Figure 18 illustrates how. Our memory and our senses give us this ability. The sense of

---

* Aristotle (b. 384/3 Stageira, d. 322 BCE Euboea), important Greek philosopher and scientist, founder of the *Peripatetic school* located at the Lyceum, a gymnasium dedicated to Apollo Lyceus.
** Theophrastus of Eresos (c. 371 – c. 287) was a revered Lesbian philosopher, successor of Aristoteles at the Lyceum.
*** Titus Lucretius Carus (c. 95 to c. 55 BCE), Roman scholar and poet.

TABLE 6 Selected time measurements.

Observation	Time
Shortest measurable time	10^{-44} s
Shortest time ever measured	10 ys
Time for light to cross a typical atom	0.1 to 10 as
Shortest laser light pulse produced so far	200 as
Period of caesium ground state hyperfine transition	108.782 775 707 78 ps
Beat of wings of fruit fly	1 ms
Period of pulsar (rotating neutron star) PSR 1913+16	0.059 029 995 271(2) s
Human 'instant'	20 ms
Shortest lifetime of living being	0.3 d
Average length of day 400 million years ago	79 200 s
Average length of day today	86 400.002(1) s
From birth to your 1000 million seconds anniversary	31.7 a
Age of oldest living tree	4600 a
Use of human language	0.2 Ma
Age of Himalayas	35 to 55 Ma
Age of oldest rocks, found in Isua Belt, Greenland and in Porpoise Cove, Hudson Bay	3.8 Ga
Age of Earth	4.6 Ga
Age of oldest stars	13.8 Ga
Age of most protons in your body	13.8 Ga
Lifetime of tantalum nucleus ^{180m}Ta	10^{15} a
Lifetime of bismuth ^{209}Bi nucleus	$1.9(2) \cdot 10^{19}$ a

hearing registers the various sounds during the rise, the fall and the landing of the stone. Our eyes track the location of the stone from one point to the next. All observations have their place in a sequence, with some observations preceding them, some observations simultaneous to them, and still others succeeding them. We say that observations are perceived to happen at various *instants* – also called 'points in time' – and we call the sequence of all instants *time*.

An observation that is considered the smallest part of a sequence, i.e., not itself a sequence, is called an *event*. Events are central to the definition of time; in particular, starting or stopping a stopwatch are events. (But do events really exist? Keep this question in the back of your head as we move on.)

Sequential phenomena have an additional property known as stretch, extension or duration. Some measured values are given in Table 6.* *Duration* expresses the idea that sequences *take* time. We say that a sequence takes time to express that other sequences can take place in parallel with it.

How exactly is the concept of time, including sequence and duration, deduced from observations? Many people have looked into this question: astronomers, physicists,

---

* A year is abbreviated a (Latin 'annus').

watchmakers, psychologists and philosophers. All find:

▷ Time is deduced by comparing motions.

This is even the case for children and animals. Beginning at a very young age, they develop the concept of 'time' from the comparison of motions in their surroundings. Grown-ups take as a standard the motion of the Sun and call the resulting type of time *local time*. From the Moon they deduce a *lunar calendar*. If they take a particular village clock on a European island they call it the *universal time coordinate* (UTC), once known as 'Greenwich mean time.'* Astronomers use the movements of the stars and call the result *ephemeris time* (or one of its successors). An observer who uses his personal watch calls the reading his *proper time*; it is often used in the theory of relativity.

Not every movement is a good standard for time. In the year 2000, an Earth rotation did not take 86 400 seconds any more, as it did in the year 1900, but 86 400.002 seconds. Can you deduce in which year your birthday will have shifted by a whole day from the time predicted with 86 400 seconds?

All methods for the definition of time are thus based on comparisons of motions. In order to make the concept as precise and as useful as possible, a *standard* reference motion is chosen, and with it a standard sequence and a standard duration is defined. The device that performs this task is called a *clock*. We can thus answer the question of the section title:

▷ Time is what we read from a clock.

Note that all definitions of time used in the various branches of physics are equivalent to this one; no 'deeper' or more fundamental definition is possible.** Note that the word 'moment' is indeed derived from the word 'movement'. Language follows physics in this case. Astonishingly, the definition of time just given is final; it will never be changed, not even at the top of Motion Mountain. This is surprising at first sight, because many books have been written on the nature of time. Instead, they should investigate the nature of motion!

▷ Every clock reminds us that in order to understand time, we need to understand motion.

But this is the aim of our walk anyhow. We are thus set to discover all the secrets of time as a side result of our adventure.

Time is not only an aspect of observations, it is also a facet of personal experience. Even in our innermost private life, in our thoughts, feelings and dreams, we experience sequences and durations. Children learn to relate this internal experience of time with

---

* Official UTC is used to determine the phase of the power grid, phone and internet companies' bit streams and the signal to the GPS system. The latter is used by many navigation systems around the world, especially in ships, aeroplanes and mobile phones. For more information, see the www.gpsworld.com website. The time-keeping infrastructure is also important for other parts of the modern economy. Can you spot the most important ones?

** The oldest clocks are sundials. The science of making them is called *gnomonics*.

TABLE 7 Properties of Galilean time.

INSTANTS OF TIME	PHYSICAL PROPERTY	MATHEMATICAL NAME	DEFINITION
Can be distinguished	distinguishability	element of set	Vol. III, page 284
Can be put in order	sequence	order	Vol. V, page 364
Define duration	measurability	metricity	Vol. IV, page 236
Can have vanishing duration	continuity	denseness, completeness	Vol. V, page 364
Allow durations to be added	additivity	metricity	Vol. IV, page 236
Don't harbour surprises	translation invariance	homogeneity	Page 238
Don't end	infinity	unboundedness	Vol. III, page 285
Are equal for all observers	absoluteness	uniqueness	

external observations, and to make use of the sequential property of events in their actions. Studies of the origin of psychological time show that it coincides – apart from its lack of accuracy – with clock time.* Every living human necessarily uses in his daily life the concept of time as a combination of sequence and duration; this fact has been checked in numerous investigations. For example, the term 'when' exists in all human languages.

Time is a concept *necessary* to distinguish between observations. In any sequence of observations, we observe that events succeed each other smoothly, apparently without end. In this context, 'smoothly' means that observations that are not too distant tend to be not too different. Yet between two instants, as close as we can observe them, there is always room for other events. Durations, or *time intervals*, measured by different people with different clocks agree in everyday life; moreover, all observers agree on the order of a sequence of events. Time is thus *unique* in everyday life. One also says that time is *absolute* in everyday life.

Time is necessary to distinguish between observations. For this reason, all observing devices that distinguish between observations, from brains to dictaphones and video cameras, have internal clocks. In particular, all animal brains have internal clocks. These brain clocks allow their users to distinguish between present, recent and past data and observations.

When Galileo studied motion in the seventeenth century, there were as yet no stopwatches. He thus had to build one himself, in order to measure times in the range between a fraction and a few seconds. Can you imagine how he did it?

If we formulate with precision all the properties of time that we experience in our daily life, we are lead to Table 7. This concept of time is called *Galilean time*. All its properties can be expressed simultaneously by describing time with the help of *real numbers*. In fact, real numbers have been constructed by mathematicians to have exactly the same properties as Galilean time, as explained in the chapter on the brain. In the case of Ga-

---

* The brain contains numerous clocks. The most precise clock for short time intervals, the internal interval timer of the brain, is more accurate than often imagined, especially when trained. For time periods between a few tenths of a second, as necessary for music, and a few minutes, humans can achieve timing accuracies of a few per cent.

lilean time, every instant of time can be described by a real number, often abbreviated t. The duration of a sequence of events is then given by the difference between the time values of the final and the starting event.

We will have quite some fun with Galilean time in this part of our adventure. However, hundreds of years of close scrutiny have shown that *every single* property of Galilean time listed in Table 7 is approximate, and none is strictly correct. This story is told in the rest of our adventure.

Clocks

> "The most valuable thing a man can spend is time."
> Theophrastus

A *clock* is a moving system whose position can be read.

There are many types of clocks: stopwatches, twelve-hour clocks, sundials, lunar clocks, seasonal clocks, etc. A few are shown in Figure 19. Most of these clock types are also found in plants and animals, as shown in Table 8.

Interestingly, there is a strict rule in the animal kingdom: large clocks go slow. How this happens is shown in Figure 20, another example of an *allometric scaling 'law'*.

A clock is a moving system whose position can be read. Of course, a *precise* clock is a system moving as regularly as possible, with as little outside disturbance as possible. Clock makers are experts in producing motion that is as regular as possible. We will discover some of their tricks below. We will also explore, later on, the limits for the precision of clocks.

Is there a perfect clock in nature? Do clocks exist at all? We will continue to study these questions throughout this work and eventually reach a surprising conclusion. At this point, however, we state a simple intermediate result: since clocks do exist, somehow there is in nature an intrinsic, natural and *ideal* way to measure time. Can you see it?

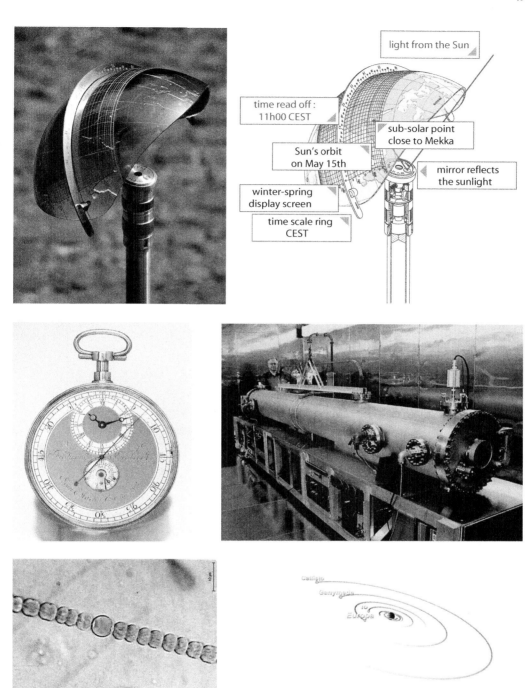

FIGURE 19 Different types of clocks: a high-tech sundial (size c. 30 cm), a naval pocket chronometer (size c. 6 cm), a caesium atomic clock (size c. 4 m), a group of cyanobacteria and the Galilean satellites of Jupiter (© Carlo Heller at www.heliosuhren.de, Anonymous, INMS, Wikimedia, NASA).

TABLE 8 Examples of biological rhythms and clocks.

LIVING BEING	OSCILLATING SYSTEM	PERIOD
Sand hopper (*Talitrus saltator*)	knows in which direction to flee from the position of the Sun or Moon	circadian
Human (*Homo sapiens*)	gamma waves in the brain	0.023 to 0.03 s
	alpha waves in the brain	0.08 to 0.13 s
	heart beat	0.3 to 1.5 s
	delta waves in the brain	0.3 to 10 s
	blood circulation	30 s
	cellular circhoral rhythms	1 to 2 ks
	rapid-eye-movement sleep period	5.4 ks
	nasal cycle	4 to 9 ks
	growth hormone cycle	11 ks
	suprachiasmatic nucleus (SCN), circadian hormone concentration, temperature, etc.; leads to jet lag	90 ks
	skin clock	circadian
	monthly period	2.4(4) Ms
	built-in aging	3.2(3) Gs
Common fly (*Musca domestica*)	wing beat	30 ms
Fruit fly (*Drosophila melanogaster*)	wing beat for courting	34 ms
Most insects (e.g. wasps, fruit flies)	winter approach detection (diapause) by length of day measurement; triggers metabolism changes	yearly
Algae (*Acetabularia*)	Adenosinetriphosphate (ATP) concentration	
Moulds (e.g. *Neurospora crassa*)	conidia formation	circadian
Many flowering plants	flower opening and closing	circadian
Tobacco plant	flower opening clock; triggered by length of days, discovered in 1920 by Garner and Allard	annual
Arabidopsis	circumnutation	circadian
	growth	a few hours
Telegraph plant (*Desmodium gyrans*)	side leaf rotation	200 s
Forsythia europaea, F. suspensa, F. viridissima, F. spectabilis	Flower petal oscillation, discovered by Van Gooch in 2002	5.1 ks

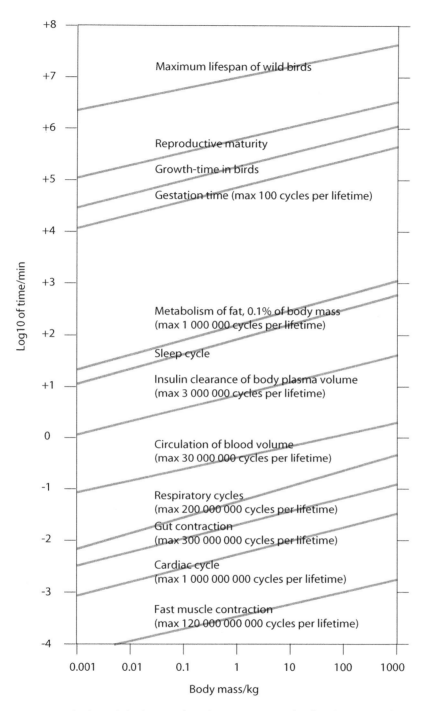

FIGURE 20 How biological rhythms scale with size in mammals: all scale more or less with a quarter power of the mass (after data from the EMBO and Enrique Morgado).

Why do clocks go clockwise?

> What time is it at the North Pole now?

Most rotational motions in our society, such as athletic races, horse, bicycle or ice skating races, turn anticlockwise.* Mathematicians call this the positive rotation sense. Every supermarket leads its guests anticlockwise through the hall. Why? Most people are right-handed, and the right hand has more freedom at the outside of a circle. Therefore thousands of years ago chariot races in stadia went anticlockwise. As a result, all stadium races still do so to this day, and that is why runners move anticlockwise. For the same reason, helical stairs in castles are built in such a way that defending right-handers, usually from above, have that hand on the outside.

On the other hand, the clock imitates the shadow of sundials; obviously, this is true on the northern hemisphere only, and only for sundials on the ground, which were the most common ones. (The old trick to determine south by pointing the hour hand of a horizontal watch to the Sun and halving the angle between it and the direction of 12 o'clock does not work on the southern hemisphere – but there you can determine north in this way.) So every clock implicitly continues to state on which hemisphere it was invented. In addition, it also tells us that sundials on walls came in use much later than those on the floor.

Does time flow?

> Wir können keinen Vorgang mit dem 'Ablauf der Zeit' vergleichen – diesen gibt es nicht –, sondern nur mit einem anderen Vorgang (etwa dem Gang des Chronometers).**
> Ludwig Wittgenstein, *Tractatus*, 6.3611

> Si le temps est un fleuve, quel est son lit?***

The expression 'the flow of time' is often used to convey that in nature change follows after change, in a steady and continuous manner. But though the hands of a clock 'flow', time itself does not. Time is a concept introduced specially to describe the flow of events around us; it does not itself flow, it *describes* flow. Time does not advance. Time is neither linear nor cyclic. The idea that time flows is as hindering to understanding nature as is the idea that mirrors exchange right and left.

The misleading use of the incorrect expression 'flow of time' was propagated first by some flawed Greek thinkers and then again by Newton. And it still continues. Aristotle, careful to think logically, pointed out its misconception, and many did so after him. Nevertheless, expressions such as 'time reversal', the 'irreversibility of time', and the much-abused 'time's arrow' are still common. Just read a popular science magazine

---

* Notable exceptions are most, but not all, Formula 1 races.
** 'We cannot compare any process with 'the passage of time' – there is no such thing – but only with another process (say, with the working of a chronometer).'
*** 'If time is a river, what is his bed?'

chosen at random. The fact is: time cannot be reversed, only motion can, or more precisely, only velocities of objects; time has no arrow, only motion has; it is not the flow of time that humans are unable to stop, but the motion of all the objects in nature. Incredibly, there are even books written by respected physicists that study different types of 'time's arrows' and compare them with each other. Predictably, no tangible or new result is extracted.

> ▷ Time does *not* flow. Only bodies flow.

Time has no direction. Motion has. For the same reason, colloquial expressions such as 'the start (or end) of time' should be avoided. A motion expert translates them straight away into 'the start (or end) of motion'.

What is space?

> " The introduction of numbers as coordinates [...] is an act of violence [...].
> Hermann Weyl, *Philosophie der Mathematik und Naturwissenschaft.**

Whenever we distinguish two objects from each other, such as two stars, we first of all distinguish their positions. We distinguish positions with our senses of sight, touch, proprioception and hearing. Position is therefore an important aspect of the physical state of an object. A position is taken by only one object at a time. Positions are limited. The set of all available positions, called *(physical) space*, acts as both a container and a background.

Closely related to space and position is *size*, the set of positions an object occupies. Small objects occupy only subsets of the positions occupied by large ones. We will discuss size in more detail shortly.

How do we deduce space from observations? During childhood, humans (and most higher animals) learn to bring together the various *perceptions* of space, namely the visual, the tactile, the auditory, the kinaesthetic, the vestibular etc., into one self-consistent set of experiences and description. The result of this learning process is a certain concept of space in the brain. Indeed, the question 'where?' can be asked and answered in all languages of the world. Being more precise, adults derive space from distance measurements. The concepts of length, area, volume, angle and solid angle are all deduced with their help. Geometers, surveyors, architects, astronomers, carpet salesmen and producers of metre sticks base their trade on distance measurements.

> ▷ Space is formed from all the position and distance relations between objects using metre sticks.

Humans developed metre sticks to specify distances, positions and sizes as accurately as possible.

---

* Hermann Weyl (1885–1955) was one of the most important mathematicians of his time, as well as an important theoretical physicist. He was one of the last universalists in both fields, a contributor to quantum theory and relativity, father of the term 'gauge' theory, and author of many popular texts.

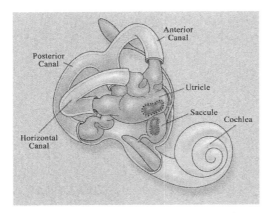 

FIGURE 21 Two proofs that space has three dimensions: the vestibular labyrinth in the inner ear of mammals (here a human) with three canals and a knot (© Northwestern University).

Metre sticks work well only if they are straight. But when humans lived in the jungle, there were no straight objects around them. No straight rulers, no straight tools, nothing. Today, a cityscape is essentially a collection of straight lines. Can you describe how humans achieved this?

Once humans came out of the jungle with their newly built metre sticks, they collected a wealth of results. The main ones are listed in Table 9; they are easily confirmed by personal experience. In particular, objects can take positions in an apparently *continuous* manner: there indeed are more positions than can be counted.* Size is captured by defining the distance between various positions, called *length*, or by using the field of view an object takes when touched, called its *surface area*. Length and area can be measured with the help of a metre stick. (Selected measurement results are given in Table 10; some length measurement devices are shown in Figure 23.) The length of objects is independent of the person measuring it, of the position of the objects and of their orientation. In daily life the sum of angles in any triangle is equal to two right angles. There are no limits to distances, lengths and thus to space.

Experience shows us that space has three dimensions; we can define sequences of positions in precisely three independent ways. Indeed, the inner ear of (practically) all vertebrates has three semicircular canals that sense the body's acceleration in the three dimensions of space, as shown in Figure 21.** Similarly, each human eye is moved by three pairs of muscles. (Why three?) Another proof that space has three dimensions is provided by shoelaces: if space had more than three dimensions, shoelaces would not be useful, because knots exist only in three-dimensional space. But why does space have three dimensions? This is one of the most difficult question of physics. We leave it open for the time being.

It is often said that thinking in four dimensions is impossible. That is wrong. Just try. For example, can you confirm that in four dimensions knots are impossible?

Like time intervals, length intervals can be described most precisely with the help

---

* For a definition of uncountability, see page 287 in Volume III.
** Note that saying that space has three dimensions *implies* that space is continuous; the mathematician and philosopher Luitzen Brouwer (b. 1881 Overschie, d. 1966 Blaricum) showed that dimensionality is only a useful concept for continuous sets.

TABLE 9 Properties of Galilean space.

Points, or positions in space	Physical property	Mathematical name	Definition
Can be distinguished	distinguishability	element of set	Vol. III, page 284
Can be lined up if on one line	sequence	order	Vol. V, page 364
Can form shapes	shape	topology	Vol. V, page 363
Lie along three independent directions	possibility of knots	3-dimensionality	Page 81, Vol. IV, page 235
Can have vanishing distance	continuity	denseness, completeness	Vol. V, page 364
Define distances	measurability	metricity	Vol. IV, page 236
Allow adding translations	additivity	metricity	Vol. IV, page 236
Define angles	scalar product	Euclidean space	Page 81
Don't harbour surprises	translation invariance	homogeneity	
Can beat any limit	infinity	unboundedness	Vol. III, page 285
Defined for all observers	absoluteness	uniqueness	Page 52

of *real numbers*. In order to simplify communication, standard *units* are used, so that everybody uses the same numbers for the same length. Units allow us to explore the general properties of *Galilean space* experimentally: space, the container of objects, is continuous, three-dimensional, isotropic, homogeneous, infinite, Euclidean and unique – or 'absolute'. In mathematics, a structure or mathematical concept with all the properties just mentioned is called a three-dimensional *Euclidean space*. Its elements, *(mathematical) points*, are described by three real parameters. They are usually written as

$$(x, y, z) \tag{1}$$

and are called *coordinates*. They specify and order the location of a point in space. (For the precise definition of Euclidean spaces, see below.)

What is described here in just half a page actually took 2000 years to be worked out, mainly because the concepts of 'real number' and 'coordinate' had to be discovered first. The first person to describe points of space in this way was the famous mathematician and philosopher René Descartes*, after whom the coordinates of expression (1) are named *Cartesian*.

Like time, space is a *necessary* concept to describe the world. Indeed, space is automatically introduced when we describe situations with many objects. For example, when many spheres lie on a billiard table, we cannot avoid using space to describe the relations between them. There is no way to avoid using spatial concepts when talking about nature.

Even though we need space to talk about nature, it is still interesting to ask why this is possible. For example, since many length measurement methods do exist – some are

---

* René Descartes or Cartesius (b. 1596 La Haye, d. 1650 Stockholm), mathematician and philosopher, author of the famous statement 'je pense, donc je suis', which he translated into 'cogito ergo sum' – I think therefore I am. In his view this is the only statement one can be sure of.

FIGURE 22 René Descartes (1596–1650).

listed in Table 11 – and since they all yield consistent results, there must be a *natural* or *ideal* way to measure distances, sizes and straightness. Can you find it?

As in the case of time, each of the properties of space just listed has to be checked. And again, careful observations will show that each property is an approximation. In simpler and more drastic words, *all* of them are wrong. This confirms Weyl's statement at the beginning of this section. In fact, his statement about the violence connected with the introduction of numbers is told by every forest in the world. The rest of our adventure will show this.

> " Μέτρον ἄριστον.* "
>
> Cleobulus

Are space and time absolute or relative?

In everyday life, the concepts of Galilean space and time include two opposing aspects; the contrast has coloured every discussion for several centuries. On the one hand, space and time express something invariant and permanent; they both act like big *containers* for all the objects and events found in nature. Seen this way, space and time have an existence of their own. In this sense one can say that they are fundamental or *absolute*. On the other hand, space and time are tools of description that allow us to talk about relations between objects. In this view, they do not have any meaning when separated from objects, and only result from the relations between objects; they are derived, relational or *relative*. Which of these viewpoints do you prefer? The results of physics have alternately favoured one viewpoint or the other. We will repeat this alternation throughout our adventure, until we find the solution. And obviously, it will turn out to be a third option.

Size – why length and area exist, but volume does not

A central aspect of objects is their size. As a small child, under school age, every human learns how to use the properties of size and space in their actions. As adults seeking precision, with the definition of *distance* as the difference between coordinates allows us to define *length* in a reliable way. It took hundreds of years to discover that this is *not* the case. Several investigations in physics and mathematics led to complications.

The physical issues started with an astonishingly simple question asked by Lewis

---

* 'Measure is the best (thing).' Cleobulus (Κλεοβουλος) of Lindos, (*c.* 620–550 BCE) was another of the proverbial seven sages.

TABLE 10 Some measured distance values.

Observation	Distance
Galaxy Compton wavelength	10^{-85} m (calculated only)
Planck length, the shortest measurable length	10^{-35} m
Proton diameter	1 fm
Electron Compton wavelength	2.426 310 215(18) pm
Smallest air oscillation detectable by human ear	11 pm
Hydrogen atom size	30 pm
Size of small bacterium	0.2 μm
Wavelength of visible light	0.4 to 0.8 μm
Radius of sharp razor blade	5 μm
Point: diameter of smallest object visible with naked eye	20 μm
Diameter of human hair (thin to thick)	30 to 80 μm
Total length of DNA in each human cell	2 m
Largest living thing, the fungus *Armillaria ostoyae*	3 km
Longest human throw with any object, using a boomerang	427 m
Highest human-built structure, Burj Khalifa	828 m
Largest spider webs in Mexico	c. 5 km
Length of Earth's Equator	40 075 014.8(6) m
Total length of human blood vessels (rough estimate)	$4 to 16 \cdot 10^4$ km
Total length of human nerve cells (rough estimate)	$1.5 to 8 \cdot 10^5$ km
Average distance to Sun	149 597 870 691(30) m
Light year	9.5 Pm
Distance to typical star at night	10 Em
Size of galaxy	1 Zm
Distance to Andromeda galaxy	28 Zm
Most distant visible object	125 Ym

Richardson:* How long is the western coastline of Britain?

Following the coastline on a map using an odometer, a device shown in Figure 24, Richardson found that the length l of the coastline depends on the scale s (say 1 : 10 000 or 1 : 500 000) of the map used:

$$l = l_0 \, s^{0.25} \tag{2}$$

(Richardson found other exponentials for other coasts.) The number l_0 is the length at scale 1 : 1. The main result is that the larger the map, the longer the coastline. What would happen if the scale of the map were increased even beyond the size of the original? The length would increase beyond all bounds. Can a coastline really have *infinite* length? Yes, it can. In fact, mathematicians have described many such curves; nowadays, they are called *fractals*. An infinite number of them exist, and Figure 25 shows one example.**

---

* Lewis Fray Richardson (1881–1953), English physicist and psychologist.
** Most of these curves are *self-similar*, i.e., they follow scaling 'laws' similar to the above-mentioned. The

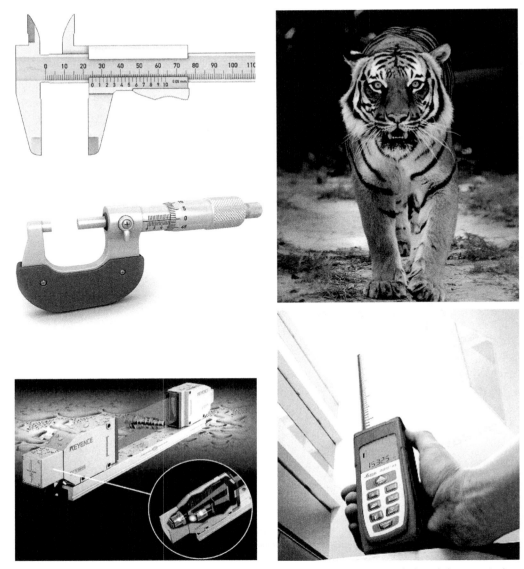

FIGURE 23 Three mechanical (a vernier caliper, a micrometer screw, a moustache) and three optical (the eyes, a laser meter, a light curtain) length and distance measurement devices (© www.medien-werkstatt.de, Naples Zoo, Keyence, and Leica Geosystems).

Can you construct another?

Length has other strange properties. The mathematician Giuseppe Vitali was the first

term 'fractal' is due to the mathematician Benoît Mandelbrot and refers to a strange property: in a certain sense, they have a non-integral number D of dimensions, despite being one-dimensional by construction. Mandelbrot saw that the non-integer dimension was related to the exponent e of Richardson by $D = 1 + e$, thus giving $D = 1.25$ in the example above. The number D varies from case to case. Measurements yield a value $D = 1.14$ for the land frontier of Portugal, $D = 1.13$ for the Australian coast and $D = 1.02$ for the South African coast.

TABLE 11 Length measurement devices in biological and engineered systems.

MEASUREMENT	DEVICE	RANGE
Humans		
Measurement of body shape, e.g. finger distance, eye position, teeth distance	muscle sensors	0.3 mm to 2 m
Measurement of object distance	stereoscopic vision	1 to 100 m
Measurement of object distance	sound echo effect	0.1 to 1000 m
Animals		
Measurement of hole size	moustache	up to 0.5 m
Measurement of walking distance by desert ants	step counter	up to 100 m
Measurement of flight distance by honey bees	eye	up to 3 km
Measurement of swimming distance by sharks	magnetic field map	up to 1000 km
Measurement of prey distance by snakes	infrared sensor	up to 2 m
Measurement of prey distance by bats, dolphins, and hump whales	sonar	up to 100 m
Measurement of prey distance by raptors	vision	0.1 to 1000 m
Machines		
Measurement of object distance by laser	light reflection	0.1 m to 400 Mm
Measurement of object distance by radar	radio echo	0.1 to 50 km
Measurement of object length	interferometer	0.5 µm to 50 km
Measurement of star, galaxy or quasar distance	intensity decay	up to 125 Ym
Measurement of particle size	accelerator	down to 10^{-18} m

FIGURE 24 A curvimeter or odometer (photograph © Frank Müller).

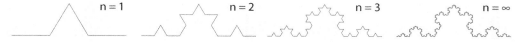

FIGURE 25 An example of a fractal: a self-similar curve of *infinite* length (far right), and its construction.

to discover that it is possible to cut a line segment of length 1 into pieces that can be reassembled – merely by shifting them in the direction of the segment – into a line segment of length 2. Are you able to find such a division using the hint that it is only possible using infinitely many pieces?

To sum up

> ▷ Length exists. But length is well defined only for lines that are straight or nicely curved, but not for intricate lines, or for lines that can be cut into infinitely many pieces.

We therefore avoid fractals and other strangely shaped curves in the following, and we take special care when we talk about infinitely small segments. These are the central assumptions in the first five volumes of this adventure, and we should never forget them! We will come back to these assumptions in the last part of our adventure.

In fact, all these problems pale when compared with the following problem. Commonly, area and volume are defined using length. You think that it is easy? You're wrong, as well as being a victim of prejudices spread by schools around the world. To define area and volume with precision, their definitions must have two properties: the values must be *additive*, i.e., for finite and infinite sets of objects, the total area and volume must be the sum of the areas and volumes of each element of the set; and the values must be *rigid*, i.e., if we cut an area or a volume into pieces and then rearrange the pieces, the value must remain the same. Are such definitions possible? In other words, do such concepts of volume and area exist?

For areas in a plane, we proceed in the following standard way: we define the area A of a rectangle of sides a and b as $A = ab$; since any polygon can be rearranged into a rectangle with a finite number of straight cuts, we can then define an area value for all polygons. Subsequently, we can define area for nicely curved shapes as the limit of the sum of infinitely many polygons. This method is called *integration*; it is introduced in detail in the section on physical action.

However, integration does not allow us to define area for arbitrarily bounded regions. (Can you imagine such a region?) For a complete definition, more sophisticated tools are needed. They were discovered in 1923 by the famous mathematician Stefan Banach.* He proved that one can indeed define an area for any set of points whatsoever, even if the border is not nicely curved but extremely complicated, such as the fractal curve previously mentioned. Today this generalized concept of area, technically a 'finitely additive isometrically invariant measure,' is called a *Banach measure* in his honour. Mathematicians sum up this discussion by saying that since in two dimensions there is a Banach measure, there is a way to define the concept of area – an additive and rigid measure – for any set of points whatsoever.** In short,

> ▷ Area exists. Area is well defined for plane and other nicely behaved surfaces, but not for intricate shapes.

---

* Stefan Banach (b. 1892 Krakow, d. 1945 Lvov), important mathematician.
** Actually, this is true only for sets on the plane. For curved surfaces, such as the surface of a sphere, there are complications that will not be discussed here. In addition, the problems mentioned in the definition of length of fractals also reappear for area if the surface to be measured is not flat. A typical example is the area of the human lung: depending on the level of details examined, the area values vary from a few up to over a hundred square metres.

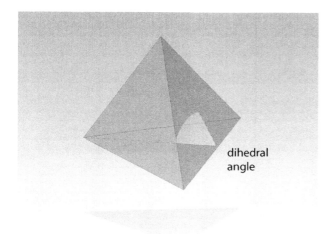

FIGURE 26 A polyhedron with one of its dihedral angles (© Luca Gastaldi).

What is the situation in *three* dimensions, i.e., for volume? We can start in the same way as for area, by defining the volume V of a rectangular polyhedron with sides a, b, c as $V = abc$. But then we encounter a first problem: a general polyhedron cannot be cut into a cube by straight cuts! The limitation was discovered in 1900 and 1902 by Max Dehn.* He found that the possibility depends on the values of the edge angles, or dihedral angles, as the mathematicians call them. (They are defined in Figure 26.) If one ascribes to every edge of a general polyhedron a number given by its length l times a special function $g(\alpha)$ of its dihedral angle α, then Dehn found that the sum of all the numbers for all the edges of a solid does not change under dissection, provided that the function fulfils $g(\alpha + \beta) = g(\alpha) + g(\beta)$ and $g(\pi) = 0$. An example of such a strange function g is the one assigning the value 0 to any rational multiple of π and the value 1 to a basis set of irrational multiples of π. The values for all other dihedral angles of the polyhedron can then be constructed by combination of rational multiples of these basis angles. Using this function, you may then deduce for yourself that a cube cannot be dissected into a regular tetrahedron because their respective Dehn invariants are different.**

Despite the problems with Dehn invariants, a rigid and additive concept of volume for polyhedra does exist, since for all polyhedra and, in general, for all 'nicely curved' shapes, the volume can be defined with the help of integration.

Now let us consider general shapes and general cuts in three dimensions, not just the 'nice' ones mentioned so far. We then stumble on the famous *Banach–Tarski theorem* (or paradox). In 1924, Stefan Banach and Alfred Tarski*** proved that it is possible to cut one sphere into five pieces that can be recombined to give *two* spheres, each the size of the original. This counter-intuitive result is the Banach–Tarski theorem. Even worse,

---

* Max Dehn (b. 1878 Hamburg, d. 1952 Black Mountain), mathematician, student of David Hilbert.
** This is also told in the beautiful book by M. AIGLER & G. M. ZIEGLER, *Proofs from the Book*, Springer Verlag, 1999. The title is due to the famous habit of the great mathematician Paul Erdős to imagine that all beautiful mathematical proofs can be assembled in the 'book of proofs'.
*** Alfred Tarski (b. 1902 Warsaw, d. 1983 Berkeley), influential mathematician.

FIGURE 27 Straight lines found in nature: cerussite (picture width approx. 3 mm, © Stephan Wolfsried) and selenite (picture width approx. 15 m, © Arch. Speleoresearch & Films/La Venta at www.laventa.it and www.naica.com.mx).

another version of the theorem states: take any two sets not extending to infinity and containing a solid sphere each; then it is always possible to dissect one into the other with a *finite* number of cuts. In particular it is possible to dissect a pea into the Earth, or vice versa. Size does not count!* In short, volume is thus not a useful concept at all!

The Banach–Tarski theorem raises two questions: first, can the result be applied to gold or bread? That would solve many problems. Second, can it be applied to empty space? In other words, are matter and empty space continuous? Both topics will be explored later in our walk; each issue will have its own, special consequences. For the moment, we eliminate this troubling issue by restricting our interest – again – to smoothly curved shapes (and cutting knives). With this restriction, volumes of matter and of empty space do behave nicely: they are additive and rigid, and show no paradoxes.** Indeed, the cuts required for the Banach–Tarski paradox are not smooth; it is not possible to perform them with an everyday knife, as they require (infinitely many) infinitely sharp bends performed with an infinitely sharp knife. Such a knife does not exist. Nevertheless, we keep in the back of our mind that the size of an object or of a piece of empty space is a tricky quantity – and that we need to be careful whenever we talk about it.

In summary,

> ▷ Volume only exists as an approximation. Volume is well defined only for regions with smooth surfaces. Volume does not exist in general, when infinitely sharp cuts are allowed.

We avoid strangely shaped volumes, surfaces and curves in the following, and we take special care when we talk about infinitely small entities. We can talk about length, area and volume *only* with this restriction. This avoidance is a central assumption in the first five volumes of this adventure. Again: we should never forget these restrictions, even

---

* The proof of the result does not need much mathematics; it is explained beautifully by Ian Stewart in *Paradox of the spheres*, New Scientist, 14 January 1995, pp. 28–31. The proof is based on the axiom of choice, which is presented later on. The Banach–Tarski paradox also exists in four dimensions, as it does in any higher dimension. More mathematical detail can be found in the beautiful book by Stan Wagon.
** Mathematicians say that a so-called *Lebesgue measure* is sufficient in physics. This countably additive isometrically invariant measure provides the most general way to define a volume.

FIGURE 28 A photograph of the Earth – seen from the direction of the Sun (NASA).

though they are not an issue in everyday life. We will come back to the assumptions at the end of our adventure.

What is straight?

When you see a solid object with a straight edge, it is a 99%-safe bet that it is man-made. Of course, there are exceptions, as shown in Figure 27.* The largest crystals ever found are 18 m in length. But in general, the contrast between the objects seen in a city – buildings, furniture, cars, electricity poles, boxes, books – and the objects seen in a forest – trees, plants, stones, clouds – is evident: in the forest no object is straight or flat, whereas in the city most objects are.

Any forest teaches us the origin of straightness; it presents tall tree trunks and rays of daylight entering from above through the leaves. For this reason we call a line *straight* if it touches either a plumb-line or a light ray along its whole length. In fact, the two definitions are equivalent. Can you confirm this? Can you find another definition? Obviously, we call a surface *flat* if for any chosen orientation and position the surface touches a plumb-line or a light ray along its whole extension.

In summary, the concept of straightness – and thus also of flatness – is defined with

---

* Why do crystals have straight edges? Another example of straight lines in nature, unrelated to atomic structures, is the well-known Irish geological formation called the Giant's Causeway. Other candidates that might come to mind, such as certain bacteria which have (almost) square or (almost) triangular shapes are not counter-examples, as the shapes are only approximate.

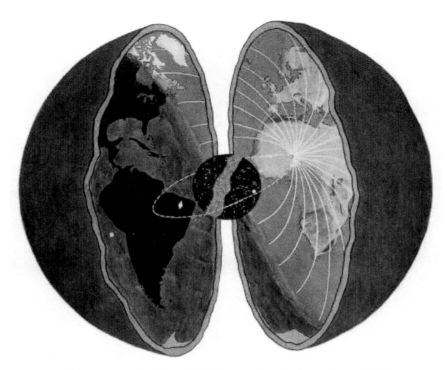

FIGURE 29 A model illustrating the hollow Earth theory, showing how day and night appear (© Helmut Diehl).

the help of bodies or radiation. In fact, all spatial concepts, like all temporal concepts, require motion for their definition.

A hollow Earth?

Space and straightness pose subtle challenges. Some strange people maintain that all humans live on the *inside* of a sphere; they call this the *hollow Earth model*. They claim that the Moon, the Sun and the stars are all near the centre of the hollow sphere, as illustrated in Figure 29. They also explain that light follows curved paths in the sky and that when conventional physicists talk about a distance r from the centre of the Earth, the real hollow Earth distance is $r_{he} = R_{\text{Earth}}^2/r$. Can you show that this model is wrong? Roman Sexl* used to ask this question to his students and fellow physicists.

The answer is simple: if you think you have an argument to show that the hollow Earth model is wrong, you are mistaken! There is *no way* of showing that such a view is wrong. It is possible to explain the horizon, the appearance of day and night, as well as the satellite photographs of the round Earth, such as Figure 28. To explain what happened during a flight to the Moon is also fun. A consistent hollow Earth view is fully *equivalent* to the usual picture of an infinitely extended space. We will come back to this problem in the section on general relativity.

---

* Roman Sexl, (1939–1986), important Austrian physicist, author of several influential textbooks on gravitation and relativity.

Curiosities and fun challenges about everyday space and time

How does one measure the speed of a gun bullet with a stop watch, in a space of 1 m^3, without electronics? Hint: the same method can also be used to measure the speed of light.

* *

For a striking and interactive way to zoom through all length scales in nature, from the Planck length to the size of the universe, see the website www.htwins.net/scale2/.

* *

What is faster: an arrow or a motorbike?

* *

Why are manholes always round?

* *

Can you show to a child that the sum of the angles in a triangle equals two right angles? What about a triangle on a sphere or on a saddle?

* *

Do you own a glass whose height is larger than its maximum circumference?

* *

A gardener wants to plant nine trees in such a way that they form ten straight lines with three trees each. How does he do it?

* *

How fast does the grim reaper walk? This question is the title of a publication in the British Medial Journal from the year 2011. Can you imagine how it is answered?

* *

Time measurements require periodic phenomena. Tree rings are traces of the seasons. Glaciers also have such traces, the *ogives*. Similar traces are found in teeth. Do you know more examples?

* *

A man wants to know how many stairs he would have to climb if the escalator in front of him, which is running upwards, were standing still. He walks up the escalator and counts 60 stairs; walking down the same escalator with the same speed he counts 90 stairs. What is the answer?

* *

You have two hourglasses: one needs 4 minutes and one needs 3 minutes. How can you use them to determine when 5 minutes are over?

FIGURE 30 At what height is a conical glass half full?

* *

You have two fuses of different length that each take one minute to burn. You are not allowed to bend them nor to use a ruler. How do you determine that 45 s have gone by? Now the tougher puzzle: How do you determine that 10 s have gone by with a single fuse?

* *

You have three wine containers: a full one of 8 litres, an empty one of 5 litres, and another empty one of 3 litres. How can you use them to divide the wine evenly into two?

* *

How can you make a hole in a postcard that allows you to step through it?

* *

What fraction of the height of a conical glass, shown in Figure 30, must be filled to make the glass half full?

* *

How many pencils are needed to draw a line as long as the Equator of the Earth?

* *

Can you place five equal coins so that each one touches the other four? Is the stacking of two layers of three coins, each layer in a triangle, a solution for six coins? Why?
What is the smallest number of coins that can be laid flat on a table so that every coin is touching exactly three other coins?

* *

Can you find three crossing points on a chessboard that lie on an equilateral triangle?

* *

The following bear puzzle is well known: A hunter leaves his home, walks 10 km to the South and 10 km to the West, shoots a bear, walks 10 km to the North, and is back home. What colour is the bear? You probably know the answer straight away. Now comes the harder question, useful for winning money in bets. The house could be on several *addi-*

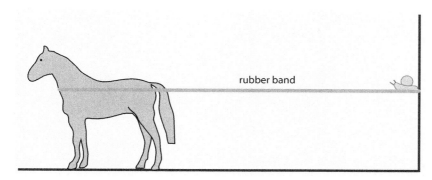

FIGURE 31 Can the snail reach the horse once it starts galloping away?

tional spots on the Earth; where are these less obvious spots from which a man can have *exactly* the same trip (forget the bear now) that was just described and be at home again?

* *

Imagine a rubber band that is attached to a wall on one end and is attached to a horse at the other end, as shown in Figure 31. On the rubber band, near the wall, there is a snail. Both the snail and the horse start moving, with typical speeds – with the rubber being infinitely stretchable. Can the snail reach the horse?

* *

For a mathematician, 1 km is the same as 1000 m. For a physicist the two are different! Indeed, for a physicist, 1 km is a measurement lying between 0.5 km and 1.5 km, whereas 1000 m is a measurement between 999.5 m and 1000.5 m. So be careful when you write down measurement values. The professional way is to write, for example, 1000(8) m to mean 1000 ± 8 m, i.e., a value that lies between 992 and 1008 m with a probability of 68.3 %.

* *

Imagine a black spot on a white surface. What is the colour of the line separating the spot from the background? This question is often called Peirce's puzzle.

* *

Also bread is an (approximate) fractal, though an irregular one. The fractal dimension of bread is around 2.7. Try to measure it!

* *

How do you find the centre of a beer mat using paper and pencil?

* *

How often in 24 hours do the hour and minute hands of a clock lie on top of each other? For clocks that also have a second hand, how often do all three hands lie on top of each other?

∗ ∗

Challenge 85 s   How often in 24 hours do the hour and minute hands of a clock form a right angle?

∗ ∗

Challenge 86 s   How many times in twelve hours can the two hands of a clock be *exchanged* with the result that the new situation shows a *valid* time? What happens for clocks that also have a third hand for seconds?

∗ ∗

Challenge 87 s   How many minutes does the Earth rotate in one minute?

∗ ∗

Challenge 88 s   What is the highest speed achieved by throwing (with and without a racket)? What was the projectile used?

∗ ∗

Challenge 89 s   A rope is put around the Earth, on the Equator, as tightly as possible. The rope is then lengthened by 1 m. Can a mouse slip under the rope? The original, tight rope is lengthened by 1 mm. Can a child slip under the rope?

∗ ∗

Challenge 90 s   Jack was rowing his boat on a river. When he was under a bridge, he dropped a ball into the river. Jack continued to row in the same direction for 10 minutes after he dropped the ball. He then turned around and rowed back. When he reached the ball, the ball had floated 600 m from the bridge. How fast was the river flowing?

∗ ∗

Challenge 91 e   Adam and Bert are brothers. Adam is 18 years old. Bert is twice as old as at the time when Adam was the age that Bert is now. How old is Bert?

∗ ∗

Challenge 92 s   'Where am I?' is a common question; 'When am I?' is almost never asked, not even in other languages. Why?

∗ ∗

Challenge 93 s   Is there a smallest time interval in nature? A smallest distance?

∗ ∗

Challenge 94 s   Given that you know what straightness is, how would you characterize or define the curvature of a curved line using numbers? And that of a surface?

∗ ∗

Challenge 95 s   What is the speed of your eyelid?

∗ ∗

The surface area of the human body is about 400 m^2. Can you say where this large num-

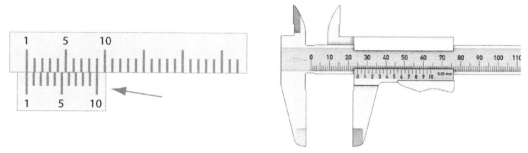

FIGURE 32 A 9-to-10 vernier/nonius/clavius and a 19-to-20 version (in fact, a 38-to-40 version) in a caliper (© www.medien-werkstatt.de).

ber comes from?

* *

How does a *vernier* work? It is called *nonius* in other languages. The first name is derived from a French military engineer* who did not invent it, the second is a play of words on the Latinized name of the Portuguese inventor of a more elaborate device** and the Latin word for 'nine'. In fact, the device as we know it today – shown in Figure 32 – was designed around 1600 by Christophonius Clavius,*** the same astronomer whose studies were the basis of the Gregorian calendar reform of 1582. Are you able to design a vernier/nonius/clavius that, instead of increasing the precision tenfold, does so by an arbitrary factor? Is there a limit to the attainable precision?

* *

Fractals in three dimensions bear many surprises. Let us generalize Figure 25 to three dimensions. Take a regular tetrahedron; then glue on every one of its triangular faces a smaller regular tetrahedron, so that the surface of the body is again made up of many equal regular triangles. Repeat the process, gluing still smaller tetrahedrons to these new (more numerous) triangular surfaces. What is the shape of the final fractal, after an infinite number of steps?

* *

Motoring poses many mathematical problems. A central one is the following *parallel parking* challenge: what is the shortest distance d from the car in front necessary to leave a parking spot without using reverse gear? (Assume that you know the geometry of your car, as shown in Figure 33, and its smallest outer turning radius R, which is known for every car.) Next question: what is the smallest gap required when you are allowed to manoeuvre back and forward as often as you like? Now a problem to which no solution seems to be available in the literature: How does the gap depend on the number, n, of

---

* Pierre Vernier (1580–1637), French military officer interested in cartography.
** Pedro Nuñes or Peter Nonnius (1502–1578), Portuguese mathematician and cartographer.
*** Christophonius Clavius or Schlüssel (1537–1612), Bavarian astronomer, one of the main astronomers of his time.

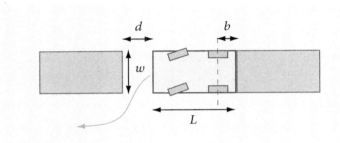

FIGURE 33 Leaving a parking space.

TABLE 12 The exponential notation: how to write small and large numbers.

Number	Exponential notation	Number	Exponential notation
1	10^0		
0.1	10^{-1}	10	10^1
0.2	$2 \cdot 10^{-1}$	20	$2 \cdot 10^1$
0.0324	$3.24 \cdot 10^{-2}$	32.4	$3.24 \cdot 10^1$
0.01	10^{-2}	100	10^2
0.001	10^{-3}	1000	10^3
0.0001	10^{-4}	10 000	10^4
0.000 056	$5.6 \cdot 10^{-5}$	56 000	$5.6 \cdot 10^4$
0.000 01	10^{-5} etc.	100 000	10^5 etc.

times you use reverse gear? (The author had offered 50 euro for the first well-explained solution; the winning solution by Daniel Hawkins is now found in the appendix.)

* *

Scientists use a special way to write large and small numbers, explained in Table 12.

* *

In 1996 the smallest experimentally probed distance was 10^{-19} m, achieved between quarks at Fermilab. (To savour the distance value, write it down without the exponent.) What does this measurement mean for the continuity of space?

* *

Zeno, the Greek philosopher, discussed in detail what happens to a moving object at a given instant of time. To discuss with him, you decide to build the fastest possible shutter for a photographic camera that you can imagine. You have all the money you want. What is the shortest shutter time you would achieve?

* *

Can you prove Pythagoras' theorem by geometrical means alone, without using

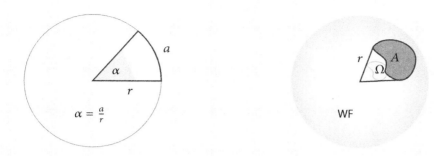

FIGURE 34 The definition of plane and solid angles.

coordinates? (There are more than 30 possibilities.)

* *

Why are most planets and moons, including ours, (almost) spherical (see, for example, Figure 28)?

* *

A rubber band connects the tips of the two hands of a clock. What is the path followed by the mid-point of the band?

* *

There are two important quantities connected to angles. As shown in Figure 34, what is usually called a *(plane) angle* is defined as the ratio between the lengths of the arc and the radius. A right angle is π/2 *radian* (or π/2 rad) or 90°.

The *solid angle* is the ratio between area and the square of the radius. An eighth of a sphere is π/2 *steradian* or π/2 sr. (Mathematicians, of course, would simply leave out the steradian unit.) As a result, a small solid angle shaped like a cone and the angle of the cone tip are *different*. Can you find the relationship?

* *

The definition of angle helps to determine the size of a firework display. Measure the time T, in seconds, between the moment that you see the rocket explode in the sky and the moment you hear the explosion, measure the (plane) angle α – pronounced 'alpha' – of the ball formed by the firework with your hand. The diameter D is

$$D \approx \frac{6\,\mathrm{m}}{\mathrm{s}\,°} T \alpha \,. \qquad (3)$$

Why? For more information about fireworks, see the cc.oulu.fi/~kempmp website. By the way, the angular distance between the knuckles of an extended fist are about 3°, 2° and 3°, the size of an extended hand 20°. Can you determine the other angles related to your hand?

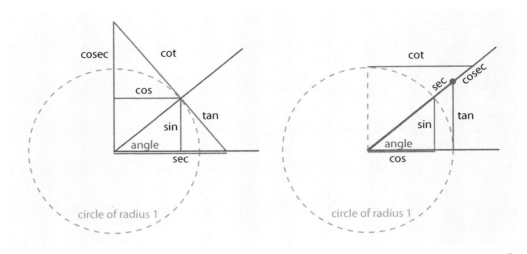

FIGURE 35 Two equivalent definitions of the sine, cosine, tangent, cotangent, secant and cosecant of an angle.

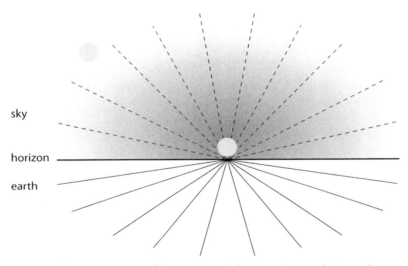

FIGURE 36 How the apparent size of the Moon and the Sun changes during a day.

* *

Using angles, the *sine*, *cosine*, *tangent*, *cotangent*, *secant* and *cosecant* can be defined, as shown in Figure 35. You should remember this from secondary school. Can you confirm that $\sin 15° = (\sqrt{6} - \sqrt{2})/4$, $\sin 18° = (-1 + \sqrt{5})/4$, $\sin 36° = \sqrt{10 - 2\sqrt{5}}/4$, $\sin 54° = (1 + \sqrt{5})/4$ and that $\sin 72° = \sqrt{10 + 2\sqrt{5}}/4$? Can you show also that

$$\frac{\sin x}{x} = \cos \frac{x}{2} \cos \frac{x}{4} \cos \frac{x}{8} ... \qquad (4)$$

is correct?

FIGURE 37 How the size of the Moon actually changes during its orbit (© Anthony Ayiomamitis).

* *

Measuring angular size with the eye only is tricky. For example, can you say whether the Moon is larger or smaller than the nail of your thumb at the end of your extended arm? Angular size is not an intuitive quantity; it requires measurement instruments.

A famous example, shown in Figure 36, illustrates the difficulty of estimating angles. Both the Sun and the Moon seem larger when they are on the horizon. In ancient times, Ptolemy explained this so-called *Moon illusion* by an unconscious apparent distance change induced by the human brain. Indeed, the Moon illusion disappears when you look at the Moon through your legs. In fact, the Moon is even *further away* from the observer when it is just above the horizon, and thus its image is *smaller* than it was a few hours earlier, when it was high in the sky. Can you confirm this?

The Moon's angular size changes even more due to another effect: the orbit of the Moon round the Earth is elliptical. An example of the consequence is shown in Figure 37.

* *

Galileo also made mistakes. In his famous book, the *Dialogues*, he says that the curve formed by a thin chain hanging between two nails is a parabola, i.e., the curve defined by $y = x^2$. That is not correct. What is the correct curve? You can observe the shape (approximately) in the shape of suspension bridges.

* *

Draw three circles, of different sizes, that touch each other, as shown in Figure 38. Now draw a fourth circle in the space between, touching the outer three. What simple relation do the inverse radii of the four circles obey?

* *

Take a tetrahedron OABC whose triangular sides OAB, OBC and OAC are rectangular in O, as shown in Figure 39. In other words, the edges OA, OB and OC are all perpendicular to each other. In the tetrahedron, the areas of the triangles OAB, OBC and OAC are

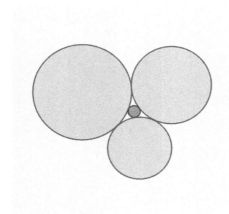

FIGURE 38 A famous puzzle: how are the four radii related?

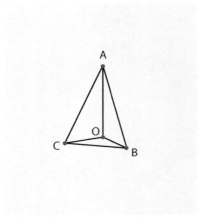

FIGURE 39 What is the area ABC, given the other three areas and three right angles at O?

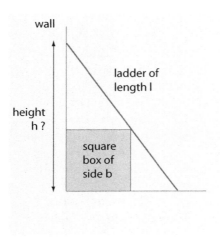
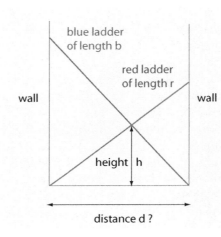

FIGURE 40 Two ladder puzzles: a moderately difficult (left) and a difficult one (right).

respectively 8, 4 and 1. What is the area of triangle ABC?

* *

There are many puzzles about ladders. Two are illustrated in Figure 40. If a 5 m ladder is put against a wall in such a way that it just touches a box with 1 m height and depth, how high does the ladder reach? If two ladders are put against two facing walls, and if the lengths of the ladders and the height of the crossing point are known, how distant are the walls?

* *

With two rulers, you can add and subtract numbers by lying them side by side. Are you able to design rulers that allow you to multiply and divide in the same manner?

FIGURE 41 Anticrepuscular rays - where is the Sun in this situation? (© Peggy Peterson)

* *

How many days would a year have if the Earth turned the other way with the same rotation frequency?

* *

The Sun is hidden in the spectacular situation shown in Figure 41 Where is it?

* *

A slightly different, but equally fascinating situation – and useful for getting used to perspective drawing – appears when you have a lighthouse in your back. Can you draw the rays you see in the sky up to the horizon?

* *

Two cylinders of equal radius intersect at a right angle. What is the value of the intersection volume? (First make a drawing.)

* *

Two sides of a hollow cube with side length 1 dm are removed, to yield a tunnel with square opening. Is it true that a cube with edge length 1.06 dm can be made to pass through the hollow cube with side length 1 dm?

* *

Could a two-dimensional universe exist? Alexander Dewdney imagined such a universe in great detail and wrote a well-known book about it. He describes houses, the transportation system, digestion, reproduction and much more. Can you explain why a two-dimensional universe is impossible?

* *

Ropes are wonderful structures. They are flexible, they are helically woven, but despite this, they do not unwind or twist, they are almost inextensible, and their geometry depends little on the material used in making them. What is the origin of all these properties?

FIGURE 42 Ideal configurations of ropes made of two, three and four strands. In the ideal configuration, the specific pitch angle relative to the equatorial plane – 39.4°, 42.8° and 43.8°, respectively – leads to zero-twist structures. In these ideal configurations, the rope will neither rotate in one nor in the other direction under vertical strain (© Jakob Bohr).

Laying rope is an old craft; it is based on a purely geometric result: among all possible helices of n strands of given length laid around a central structure of fixed radius, there is one helix for which the number of turns is *maximal*. For purely geometric reasons, ropes with that specific number of turns and the corresponding inner radius have the mentioned properties that make ropes so useful. The geometries of ideal ropes made of two, three and four strands are shown in Figure 42.

* *

Some researchers are investigating whether time could be two-dimensional. Can this be?

* *

Other researchers are investigating whether space could have more than three dimensions. Can this be?

* *

One way to compare speeds of animals and machines is to measure them in 'body lengths per second'. The click beetle achieves a value of around 2000 during its jump phase, certain Archaea (bacteria-like) cells a value of 500, and certain hummingbirds 380. These are the record-holders so far. Cars, aeroplanes, cheetahs, falcons, crabs, and all other motorized systems are much slower.

* *

Why is the cross section of a tube usually circular?

* *

What are the dimensions of an open rectangular water tank that contains $1\,\text{m}^3$ of water and uses the smallest amount of wall material?

FIGURE 43 An open research problem: What is the ropelength of a tight knot? (© Piotr Pieranski, from Ref. 54)

∗ ∗

Draw a square consisting of four equally long connecting line segments hinged at the vertices. Such a structure may be freely deformed into a rhombus if some force is applied. How many additional line interlinks of the same length must be supplemented to avoid this freedom and to prevent the square from being deformed? The extra line interlinks must be in the same plane as the square and each one may only be pegged to others at the endpoints.

∗ ∗

Area measurements can be difficult. In 2014 it became clear that the area of the gastro-intestinal tract of adult health humans is between 30 and 40 m^2. For many years, the mistaken estimate for the area was between 180 and 300 m^2.

∗ ∗

If you never explored plane geometry, do it once in your life. An excellent introduction is CLAUDI ALSINA & ROGER B. NELSEN, *Icons of Mathematics: An Exploration of Twenty Key Images*, MAA Press, 2011. This is a wonderful book with many simple and surprising facts about geometry that are never told in school or university. You will enjoy it.

∗ ∗

Triangles are full of surprises. Together, Leonhard Euler, Charles Julien Brianchon and Jean Victor Poncelet discovered that in any triangle, nine points lie on the same circle: the midpoints of the sides, the feet of the altitude lines, and the midpoints of the altitude segments connecting each vertex to the orthocenter. Euler also discovered that in every triangle, the orthocenter, the centroid, the circumcenter and the center of the nine-point-circle lie on the same line, now called the Euler line.

For the most recent research results on plane triangles, see the wonderful *Encyclopedia of Triangle Centers*, available at faculty.evansville.edu/ck6/encyclopedia/ETC.html.

∗ ∗

Here is a simple challenge on length that nobody has solved yet. Take a piece of ideal rope: of constant radius, ideally flexible, and completely slippery. Tie a tight knot into it, as shown in Figure 43. By how much did the two ends of the rope come closer together?

SUMMARY ABOUT EVERYDAY SPACE AND TIME

Motion defines speed, time and length. Observations of everyday life and precision experiments are conveniently and precisely described by describing velocity as a vector, space as three-dimensional set of points, and time as a one-dimensional real line, also made of points. These three definitions form the everyday, or *Galilean*, description of our environment.

Modelling velocity, time and space as *continuous* quantities is precise and convenient. The modelling works during most of the adventures that follows. However, this common model of space and time *cannot* be confirmed by experiment. For example, no experiment can check distances larger than 10^{25} m or smaller than 10^{-25} m; the continuum model is likely to be incorrect at smaller scales. We will find out in the last part of our adventure that this is indeed the case.

Chapter 3
HOW TO DESCRIBE MOTION – KINEMATICS

> La filosofia è scritta in questo grandissimo libro
> che continuamente ci sta aperto innanzi agli
> occhi (io dico l'universo) ... Egli è scritto in
> lingua matematica.**
> Galileo Galilei, *Il saggiatore* VI.

Experiments show that the properties of motion, time and space are extracted from the environment both by children and animals. This extraction has been confirmed for cats, dogs, rats, mice, ants and fish, among others. They all find the same results.

First of all, *motion is change of position with time*. This description is illustrated by rapidly flipping the lower left corners of this book, starting at page 242. Each page simulates an instant of time, and the only change that takes place during motion is in the position of the object, say a stone, represented by the dark spot. The other variations from one picture to the next, which are due to the imperfections of printing techniques, can be taken to simulate the inevitable measurement errors.

Stating that 'motion is the change of position with time' is *neither* an explanation *nor* a definition, since both the concepts of time and position are deduced from motion itself. It is only a *description* of motion. Still, the statement is useful, because it allows for high precision, as we will find out by exploring gravitation and electrodynamics. After all, precision is our guiding principle during this promenade. Therefore the detailed description of changes in position has a special name: it is called *kinematics*.

The idea of change of positions implies that the object can be *followed* during its motion. This is not obvious; in the section on quantum theory we will find examples where this is impossible. But in everyday life, objects can always be tracked. The set of all positions taken by an object over time forms its *path* or *trajectory*. The origin of this concept is evident when one watches fireworks or again the flip film in the lower left corners starting at page 242.

Ref. 55

In everyday life, animals and humans agree on the Euclidean properties of velocity, space and time. In particular, this implies that a trajectory can be described by specifying three numbers, three *coordinates* (x, y, z) – one for each dimension – as continuous

---

** Science is written in this huge book that is continuously open before our eyes (I mean the universe) ... It is written in mathematical language.

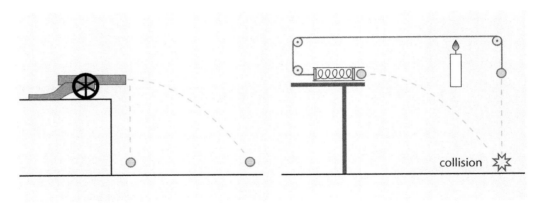

FIGURE 44 Two ways to test that the time of free fall does not depend on horizontal velocity.

functions of time t. (Functions are defined in detail later on.) This is usually written as

$$\boldsymbol{x} = \boldsymbol{x}(t) = (x(t), y(t), z(t)) \,. \tag{5}$$

For example, already Galileo found, using stopwatch and ruler, that the height z of any thrown or falling stone changes as

$$z(t) = z_0 + v_{z0}(t - t_0) - \tfrac{1}{2}g(t - t_0)^2 \tag{6}$$

where t_0 is the time the fall starts, z_0 is the initial height, v_{z0} is the initial velocity in the vertical direction and $g = 9.8\,\text{m/s}^2$ is a constant that is found to be the same, within about one part in 300, for all falling bodies on all points of the surface of the Earth. Where do the value $9.8\,\text{m/s}^2$ and its slight variations come from? A preliminary answer will be given shortly, but the complete elucidation will occupy us during the larger part of this hike.

The special case with no initial velocity is of great interest. Like a few people before him, Galileo made it clear that g is the same for all bodies, if air resistance can be neglected. He had many arguments for this conclusion; can you find one? And of course, his famous experiment at the leaning tower in Pisa confirmed the statement. (It is a *false* urban legend that Galileo never performed the experiment. He did it.)

Equation (6) therefore allows us to determine the depth of a well, given the time a stone takes to reach its bottom. The equation also gives the speed v with which one hits the ground after jumping from a tree, namely

$$v = \sqrt{2gh} \,. \tag{7}$$

A height of 3 m yields a velocity of 27 km/h. The velocity is thus proportional only to the square root of the height. Does this mean that one's strong fear of falling results from an overestimation of its actual effects?

Galileo was the first to state an important result about free fall: the motions in the horizontal and vertical directions are *independent*. He showed that the time it takes for

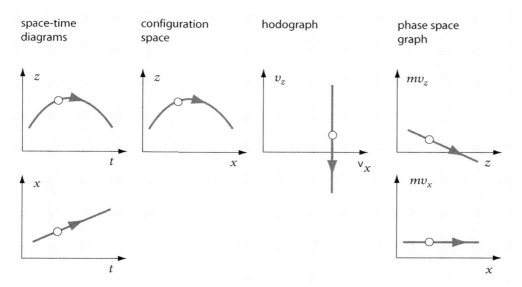

FIGURE 45 Various types of graphs describing the same path of a thrown stone.

a cannon ball that is shot exactly horizontally to fall is *independent* of the strength of the gunpowder, as shown in Figure 44. Many great thinkers did not agree with this statement even after his death: in 1658 the Academia del Cimento even organized an experiment to check this assertion, by comparing the flying cannon ball with one that simply fell vertically. Can you imagine how they checked the simultaneity? Figure 44 shows how you can check this at home. In this experiment, whatever the spring load of the cannon, the two bodies will always collide in mid-air (if the table is high enough), thus proving the assertion.

In other words, a flying cannon ball is not accelerated in the horizontal direction. Its horizontal motion is simply unchanging – as long as air resistance is negligible. By extending the description of equation (6) with the two expressions for the horizontal coordinates x and y, namely

$$x(t) = x_0 + v_{x0}(t - t_0)$$
$$y(t) = y_0 + v_{y0}(t - t_0) \,, \tag{8}$$

a *complete* description for the path followed by thrown stones results. A path of this shape is called a *parabola*; it is shown in Figures 18, 44 and 45. (A parabolic shape is also used for light reflectors inside pocket lamps or car headlights. Can you show why?)

Physicists enjoy generalizing the idea of a path. As Figure 45 shows, a path is a trace left in a diagram by a moving object. Depending on what diagram is used, these paths have different names. Space-time diagrams are useful to make the theory of relativity accessible. The *configuration space* is spanned by the coordinates of all particles of a system. For many particles, it has a high number of dimensions and plays an important role in self-organization. The difference between chaos and order can be described as a difference in the properties of paths in configuration space. *Hodographs*, the paths in 'velocity space', are used in weather forecasting. The phase space diagram is also called *state space diagram*. It plays an essential role in thermodynamics.

Throwing, jumping and shooting

The kinematic description of motion is useful for answering a whole range of questions.

∗ ∗

What is the upper limit for the long jump? The running peak speed world record in 2019 was over 12.5 m/s ≈ 45 km/h by Usain Bolt, and the 1997 women's record was 11 m/s ≈ 40 km/h. However, male long jumpers never run much faster than about 9.5 m/s. How much extra jump distance could they achieve if they could run full speed? How could they achieve that? In addition, long jumpers take off at angles of about 20°, as they are not able to achieve a higher angle at the speed they are running. How much would they gain if they could achieve 45°? Is 45° the optimal angle?

∗ ∗

Why was basketball player Dirk Nowitzki so successful? His trainer Holger Geschwindner explained him that a throw is most stable against mistakes when it falls into the basket at around 47 degrees from the horizontal. He further told Nowitzki that the ball flies in a plane, and that therefore the arms should also move in that plane only. And he explained that when the ball leaves the hand, it should roll over the last two fingers like a train moves on rails. Using these criteria to check and to improve Nowitzki's throws, he made him into one of the best basket ball throwers in the world.

∗ ∗

What do the athletes Usain Bolt and Michael Johnson, the last two world record holders on the 200 m race at time of this writing, have in common? They were tall, athletic, and had many fast twitch fibres in the muscles. These properties made them good sprinters. A last difference made them world class sprinters: they had a flattened spine, with almost no S-shape. This abnormal condition saves them a little bit of time at every step, because their spine is not as flexible as in usual people. This allows them to excel at short distance races.

∗ ∗

Athletes continuously improve speed records. Racing horses do not. Why? For racing horses, breathing rhythm is related to gait; for humans, it is not. As a result, racing horses cannot change or improve their technique, and the speed of racing horses is essentially the same since it is measured.

∗ ∗

What is the highest height achieved by a human throw of any object? What is the longest distance achieved by a human throw? How would you clarify the rules? Compare the results with the record distance with a crossbow, 1,871.8 m, achieved in 1988 by Harry Drake, the record distance with a footbow, 1854.4 m, achieved in 1971 also by Harry Drake, and with a hand-held bow, 1,222.0 m, achieved in 1987 by Don Brown.

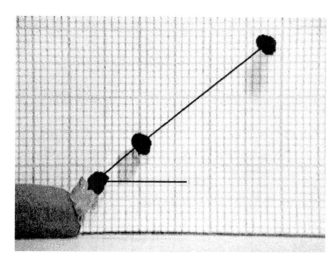

FIGURE 46 Three superimposed images of a frass pellet shot away by a caterpillar inside a rolled-up leaf (© Stanley Caveney).

∗ ∗

How can the speed of falling rain be measured using an umbrella? The answer is important: the same method can also be used to measure the speed of light, as we will find out later. (Can you guess how?)

∗ ∗

When a dancer jumps in the air, how many times can he or she rotate around his or her vertical axis before arriving back on earth?

∗ ∗

Numerous species of moth and butterfly caterpillars shoot away their frass – to put it more crudely: their faeces – so that its smell does not help predators to locate them. Stanley Caveney and his team took photographs of this process. Figure 46 shows a caterpillar (yellow) of the skipper *Calpodes ethlius* inside a rolled up green leaf caught in the act. Given that the record distance observed is 1.5 m (though by another species, *Epargyreus clarus*), what is the ejection speed? How do caterpillars achieve it?

∗ ∗

What is the horizontal distance one can reach by throwing a stone, given the speed and the angle from the horizontal at which it is thrown?

∗ ∗

What is the maximum numbers of balls that could be juggled at the same time?

∗ ∗

Is it true that rain drops would kill if it weren't for the air resistance of the atmosphere? What about hail?

∗ ∗

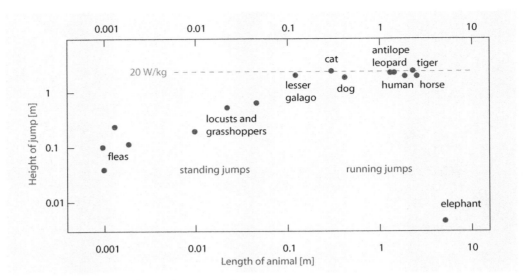

FIGURE 47 The height achieved by jumping land animals.

Are bullets, fired into the air from a gun, dangerous when they fall back down?

∗ ∗

Police finds a dead human body at the bottom of cliff with a height of 30 m, at a distance of 12 m from the cliff. Was it suicide or murder?

∗ ∗

All land animals, regardless of their size, achieve jumping heights of at most 2.2 m, as shown in Figure 47. The explanation of this fact takes only two lines. Can you find it?

The last two issues arise because the equation (6) describing free fall does not hold in all cases. For example, leaves or potato crisps do not follow it. As Galileo already knew, this is a consequence of air resistance; we will discuss it shortly. Because of air resistance, the path of a stone is not a parabola.

In fact, there are other situations where the path of a falling stone is not a parabola, even without air resistance. Can you find one?

Enjoying vectors

Physical quantities with a defined direction, such as speed, are described with three numbers, or three components, and are called *vectors*. Learning to calculate with such multi-component quantities is an important ability for many sciences. Here is a summary.

Vectors can be pictured by small arrows. Note that vectors do not have specified points at which they start: two arrows with same direction and the same length are the *same* vector, even if they start at different points in space. Since vectors behave like arrows, vectors can be added and they can be multiplied by numbers. For example, stretching an arrow $\boldsymbol{a} = (a_x, a_y, a_z)$ by a number c corresponds, in component notation, to the vector $c\boldsymbol{a} = (ca_x, ca_y, ca_z)$.

3 HOW TO DESCRIBE MOTION – KINEMATICS

In precise, mathematical language, a vector is an element of a set, called *vector space*, in which the following properties hold for all vectors a and b and for all numbers c and d:

$$c(a + b) = ca + cb \quad , \quad (c + d)a = ca + da \quad , \quad (cd)a = c(da) \quad \text{and} \quad 1a = a \, . \tag{9}$$

Examples of vector spaces are the set of all *positions* of an object, or the set of all its possible velocities. Does the set of all rotations form a vector space?

All vector spaces allow defining a unique *null vector* and a unique *negative vector* for each vector.

In most vector spaces of importance when describing nature the concept of *length* – specifying the 'magnitude' – of a vector can be introduced. This is done via an intermediate step, namely the introduction of the scalar product of two vectors. The product is called 'scalar' because its result is a scalar; a *scalar* is a number that is the same for all observers; for example, it is the same for observers with different orientations.* The *scalar product* between two vectors a and b is a number that satisfies

$$\begin{aligned} aa &\geq 0 \, , \\ ab &= ba \, , \\ (a + a')b &= ab + a'b \, , \\ a(b + b') &= ab + ab' \quad \text{and} \\ (ca)b &= a(cb) = c(ab) \, . \end{aligned} \tag{10}$$

This definition of a scalar product is not unique; however, there is a *standard* scalar product. In Cartesian coordinate notation, the standard scalar product is given by

$$ab = a_x b_x + a_y b_y + a_z b_z \, . \tag{11}$$

If the scalar product of two vectors vanishes the two vectors are *orthogonal*, at a right angle to each other. (Show it!) Note that one can write either ab or $a \cdot b$ with a central dot.

The *length* or *magnitude* or *norm* of a vector can then be defined as the square root of the scalar product of a vector with itself: $a = \sqrt{aa}$. Often, and also in this text, lengths are written in *italic* letters, whereas vectors are written in **bold** letters. The magnitude is often written as $a = \sqrt{a^2}$. A vector space with a scalar product is called an *Euclidean* vector space.

The scalar product is especially useful for specifying directions. Indeed, the scalar product between two vectors encodes the angle between them. Can you deduce this important relation?

---

* We mention that in mathematics, a scalar is a *number*; in physics, a scalar is an *invariant* number, i.e., a number that is the same for all observers. Likewise, in mathematics, a vector is an element of a vector space; in physics, a vector is an *invariant* element of a vector space, i.e., a quantity whose coordinates, when observed by different observers, change like the components of velocity.

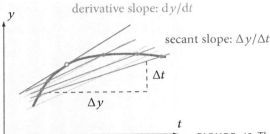

FIGURE 48 The derivative in a point as the limit of secants.

What is rest? What is velocity?

In the Galilean description of nature, motion and rest are opposites. In other words, a body is at rest when its position, i.e., its coordinates, do not change with time. In other words, (Galilean) *rest* is defined as

$$x(t) = \text{const} . \tag{12}$$

We recall that $x(t)$ is the abbreviation for the three coordinates $(x(t), y(t), z(t))$. Later we will see that this definition of rest, contrary to first impressions, is not much use and will have to be expanded. Nevertheless, any definition of rest implies that non-resting objects can be distinguished by comparing the rapidity of their displacement. Thus we can define the *velocity* v of an object as the change of its position x with time t. This is usually written as

$$v = \frac{dx}{dt} . \tag{13}$$

In this expression, valid for each coordinate separately, d/dt means 'change with time'. We can thus say that velocity is the *derivative* of position with respect to time. The *speed* v is the name given to the magnitude of the velocity v. Thus we have $v = \sqrt{vv}$. Derivatives are written as fractions in order to remind the reader that they are derived from the idea of slope. The expression

$$\frac{ds}{dt} \quad \text{is meant as an abbreviation of} \quad \lim_{\Delta t \to 0} \frac{\Delta s}{\Delta t} , \tag{14}$$

a shorthand for saying that the *derivative at a point* is the limit of the secant slopes in the neighbourhood of the point, as shown in Figure 48. This definition implies the working rules

$$\frac{d(s+r)}{dt} = \frac{ds}{dt} + \frac{dr}{dt} \quad , \quad \frac{d(cs)}{dt} = c\frac{ds}{dt} \quad , \quad \frac{d}{dt}\frac{ds}{dt} = \frac{d^2s}{dt^2} \quad , \quad \frac{d(sr)}{dt} = \frac{ds}{dt}r + s\frac{dr}{dt} , \tag{15}$$

c being any number. This is all one ever needs to know about derivatives in physics. Quantities such as dt and ds, sometimes useful by themselves, are called *differentials*.

FIGURE 49 Gottfried Wilhelm Leibniz (1646–1716).

These concepts are due to Gottfried Wilhelm Leibniz.* Derivatives lie at the basis of all calculations based on the continuity of space and time. Leibniz was the person who made it possible to describe and use velocity in physical formulae and, in particular, to use the idea of velocity at a given point in time or space for calculations.

The definition of velocity assumes that it makes sense to take the limit $\Delta t \to 0$. In other words, it is assumed that *infinitely small* time intervals do exist in nature. The definition of velocity with derivatives is possible only because both space and time are described by sets which are *continuous*, or in mathematical language, *connected and complete*. In the rest of our walk we shall not forget that from the beginning of classical physics, *infinities* are present in its description of nature. The infinitely small is part of our definition of velocity. Indeed, differential calculus can be defined as the study of infinity and its uses. We thus discover that the appearance of infinity does not automatically render a description impossible or imprecise. In order to remain precise, physicists use only the smallest two of the various possible types of infinities. Their precise definition and an overview of other types are introduced later on.

The appearance of infinity in the usual description of motion was first criticized in his famous ironical arguments by Zeno of Elea (around 445 BCE), a disciple of Parmenides. In his so-called third argument, Zeno explains that since at every instant a given object occupies a part of space corresponding to its size, the notion of velocity at a given instant makes no sense; he provokingly concludes that therefore motion does not exist. Nowadays we would not call this an argument against the *existence* of motion, but against its usual *description*, in particular against the use of infinitely divisible space and time. (Do you agree?) Nevertheless, the description criticized by Zeno actually works quite well in everyday life. The reason is simple but deep: in daily life, changes are indeed continuous.

Large changes in nature are made up of many small changes. This property of nature is not obvious. For example, we note that we have (again) tacitly assumed that the path of an object is not a fractal or some other badly behaved entity. In everyday life this is correct; in other domains of nature it is not. The doubts of Zeno will be partly rehabilitated later in our walk, and increasingly so the more we proceed. The rehabilitation is only partial, as the final solution will be different from that which he envisaged; on the other hand,

---

* Gottfried Wilhelm Leibniz (b. 1646 Leipzig, d. 1716 Hannover), lawyer, physicist, mathematician, philosopher, diplomat and historian. He was one of the great minds of mankind; he invented the differential calculus (before Newton) and published many influential and successful books in the various fields he explored, among them *De arte combinatoria*, *Hypothesis physica nova*, *Discours de métaphysique*, *Nouveaux essais sur l'entendement humain*, the *Théodicée* and the *Monadologia*.

TABLE 13 Some measured acceleration values.

Observation	Acceleration
What is the lowest you can find?	Challenge 154 s
Back-acceleration of the galaxy M82 by its ejected jet	$10\,\text{fm/s}^2$
Acceleration of a young star by an ejected jet	$10\,\text{pm/s}^2$
Fathoumi Acceleration of the Sun in its orbit around the Milky Way	$0.2\,\text{nm/s}^2$
Deceleration of the Pioneer satellites, due to heat radiation imbalance	$0.8\,\text{nm/s}^2$
Centrifugal acceleration at Equator due to Earth's rotation	$33\,\text{mm/s}^2$
Electron acceleration in household electricity wire due to alternating current	$50\,\text{mm/s}^2$
Acceleration of fast underground train	$1.3\,\text{m/s}^2$
Gravitational acceleration on the Moon	$1.6\,\text{m/s}^2$
Minimum deceleration of a car, by law, on modern dry asphalt	$5.5\,\text{m/s}^2$
Gravitational acceleration on the Earth's surface, depending on location	$9.8 \pm 0.3\,\text{m/s}^2$
Standard gravitational acceleration	$9.806\,65\,\text{m/s}^2$
Highest acceleration for a car or motorbike with engine-driven wheels	$15\,\text{m/s}^2$
Space rockets at take-off	20 to $90\,\text{m/s}^2$
Acceleration of cheetah	$32\,\text{m/s}^2$
Gravitational acceleration on Jupiter's surface	$25\,\text{m/s}^2$
Flying fly (*Musca domestica*)	$c.\ 100\,\text{m/s}^2$
Acceleration of thrown stone	$c.\ 120\,\text{m/s}^2$
Acceleration that triggers air bags in cars	$360\,\text{m/s}^2$
Fastest leg-powered acceleration (by the froghopper, *Philaenus spumarius*, an insect)	$4\,\text{km/s}^2$
Tennis ball against wall	$0.1\,\text{Mm/s}^2$
Bullet acceleration in rifle	$2\,\text{Mm/s}^2$
Fastest centrifuges	$0.1\,\text{Gm/s}^2$
Acceleration of protons in large accelerator	$90\,\text{Tm/s}^2$
Acceleration of protons inside nucleus	$10^{31}\,\text{m/s}^2$
Highest possible acceleration in nature	$10^{52}\,\text{m/s}^2$

the doubts about the idea of 'velocity at a point' will turn out to be well-founded. For the moment though, we have no choice: we continue with the basic assumption that in nature changes happen smoothly.

Why is velocity necessary as a concept? Aiming for precision in the description of motion, we need to find the complete list of aspects necessary to specify the state of an object. The concept of velocity is obviously on this list.

Acceleration

Continuing along the same line, we call *acceleration* **a** of a body the change of velocity **v** with time, or

$$\boldsymbol{a} = \frac{\mathrm{d}\boldsymbol{v}}{\mathrm{d}t} = \frac{\mathrm{d}^2\boldsymbol{x}}{\mathrm{d}t^2} \ . \tag{16}$$

Acceleration is what we feel when the Earth trembles, an aeroplane takes off, or a bicycle goes round a corner. More examples are given in Table 13. Acceleration is the time derivative of velocity. Like velocity, acceleration has both a magnitude and a direction. In short, acceleration, like velocity, is a vector quantity. As usual, this property is indicated by the use of a **bold** letter for its abbreviation.

In a usual car, or on a motorbike, we can *feel* being accelerated. (These accelerations are below $1g$ and are therefore harmless.) We feel acceleration because some part inside us is moved against some other part: acceleration deforms us. Such a moving part can be, for example, some small part inside our ear, or our stomach inside the belly, or simply our limbs against our trunk. All acceleration sensors, including those listed in Table 14 or those shown in Figure 50, whether biological or technical, work in this way.

Acceleration is felt. Our body is deformed and the sensors in our body detect it, for example in amusement parks. Higher accelerations can have stronger effects. For example, when accelerating a sitting person in the direction of the head at two or three times the value of usual gravitational acceleration, eyes stop working and the sight is greyed out, because the blood cannot reach the eye any more. Between 3 and $5g$ of continuous acceleration, or 7 to $9g$ of short time acceleration, consciousness is lost, because the brain does not receive enough blood, and blood may leak out of the feet or lower legs. High acceleration in the direction of the feet of a sitting person can lead to haemorrhagic strokes in the brain. The people most at risk are jet pilots; they have special clothes that send compressed air onto the pilot's bodies to avoid blood accumulating in the wrong places.

Can you think of a situation where you are accelerated but do *not* feel it?

Velocity is the time derivative of position. Acceleration is the second time derivative of position. Higher derivatives than acceleration can also be defined, in the same manner. They add little to the description of nature, because – as we will show shortly – neither these higher derivatives nor even acceleration itself are useful for the description of the state of motion of a system.

From objects to point particles

> Wenn ich den Gegenstand kenne, so kenne ich auch sämtliche Möglichkeiten seines Vorkommens in Sachverhalten.*
> Ludwig Wittgenstein, *Tractatus*, 2.0123

One aim of the study of motion is to find a complete and precise description of both states and objects. With the help of the concept of space, the description of objects can be refined considerably. In particular, we know from experience that all objects seen in daily life have an important property: they can be divided into *parts*. Often this observation is

---

* 'If I know an object, then I also know all the possibilities of its occurrence in atomic facts.'

TABLE 14 Some acceleration sensors.

Measurement	Sensor	Range
Direction of gravity in plants (roots, trunk, branches, leaves)	statoliths in cells	0 to 10 m/s^2
Direction and value of accelerations in mammals	the utricle and saccule in the inner ear (detecting linear accelerations), and the membranes in each semicircular canal (detecting rotational accelerations)	0 to 20 m/s^2
Direction and value of acceleration in modern step counters for hikers	piezoelectric sensors	0 to 20 m/s^2
Direction and value of acceleration in car crashes	airbag sensor using piezoelectric ceramics	0 to 2000 m/s^2

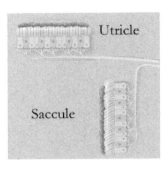
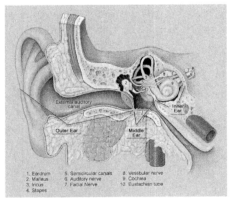
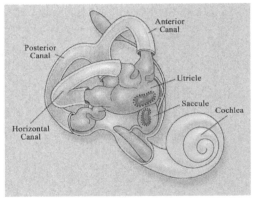

FIGURE 50 Three accelerometers: a one-axis piezoelectric airbag sensor, a three-axis capacitive accelerometer, and the utricle and saccule near the three semicircular canals inside the human ear (© Bosch, Rieker Electronics, Northwestern University).

expressed by saying that all objects, or bodies, have two properties. First, they are made out of *matter*,* defined as that aspect of an object responsible for its impenetrability, i.e., the property preventing two objects from being in the same place. Secondly, bodies

---

* Matter is a word derived from the Latin 'materia', which originally meant 'wood' and was derived via intermediate steps from 'mater', meaning 'mother'.

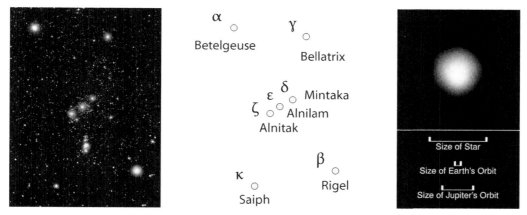

FIGURE 51 Orion in natural colours (© Matthew Spinelli) and Betelgeuse (ESA, NASA).

have a certain form or *shape*, defined as the precise way in which this impenetrability is distributed in space.

In order to describe motion as accurately as possible, it is convenient to start with those bodies that are as simple as possible. In general, the smaller a body, the simpler it is. A body that is so small that its parts no longer need to be taken into account is called a *particle*. (The older term *corpuscle* has fallen out of fashion.) Particles are thus idealized small stones. The extreme case, a particle whose size is *negligible* compared with the dimensions of its motion, so that its position is described completely by a *single* triplet of coordinates, is called a *point particle* or a *point mass* or a *mass point*. In equation (6), the stone was assumed to be such a point particle.

Do point-like objects, i.e., objects smaller than anything one can measure, exist in daily life? Yes and no. The most notable examples are the stars. At present, angular sizes as small as 2 μrad can be measured, a limit given by the fluctuations of the air in the atmosphere. In space, such as for the Hubble telescope orbiting the Earth, the angular limit is due to the diameter of the telescope and is of the order of 10 nrad. Practically all stars seen from Earth are smaller than that, and are thus effectively 'point-like', even when seen with the most powerful telescopes.

As an exception to the general rule, the size of a few large or nearby stars, mostly of red giant type, can be measured with special instruments.* Betelgeuse, the higher of the two shoulders of Orion shown in Figure 51, Mira in Cetus, Antares in Scorpio, Aldebaran in Taurus and Sirius in Canis Major are examples of stars whose size has been measured; they are all less than two thousand light years from Earth. For a comparison of dimensions, see Figure 52. Of course, like the Sun, also all other stars have a finite size, but one cannot prove this by measuring their dimension in photographs. (True?)

---

* The website stars.astro.illinois.edu/sow/sowlist.html gives an introduction to the different types of stars. The www.astro.wisc.edu/~dolan/constellations website provides detailed and interesting information about constellations.

For an overview of the planets, see the beautiful book by KENNETH R. LANG & CHARLES A. WHITNEY, *Vagabonds de l'espace – Exploration et découverte dans le système solaire*, Springer Verlag, 1993. Amazingly beautiful pictures of the stars can be found in DAVID MALIN, *A View of the Universe*, Sky Publishing and Cambridge University Press, 1993.

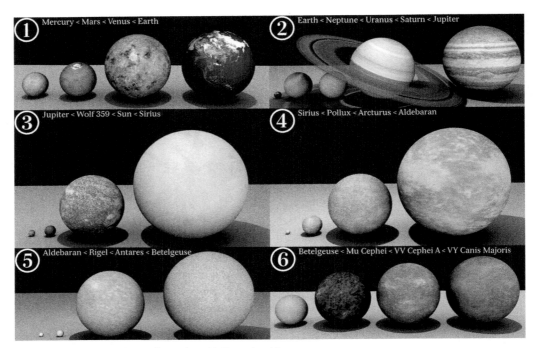

FIGURE 52 A comparison of star sizes (© Dave Jarvis).

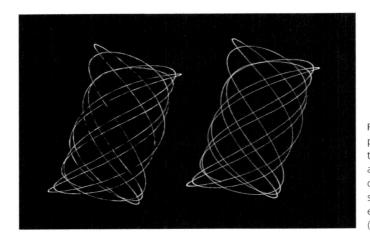

FIGURE 53 Regulus and Mars, photographed with an exposure time of 10 s on 4 June 2010 with a wobbling camera, show the difference between a point-like star that twinkles and an extended planet that does not (© Jürgen Michelberger).

The difference between 'point-like' and finite-size sources can be seen with the naked eye: at night, stars twinkle, but planets do not. (Check it!) A beautiful visualization is shown in Figure 53. This effect is due to the turbulence of air. Turbulence has an effect on the almost point-like stars because it deflects light rays by small amounts. On the other hand, air turbulence is too weak to lead to twinkling of sources of larger angular size, such as planets or artificial satellites,* because the deflection is averaged out in this

---

* A *satellite* is an object circling a planet, like the Moon; an *artificial satellite* is a system put into orbit by humans, like the Sputniks.

case.

An object is *point-like for the naked eye* if its angular size is smaller than about $2' = 0.6$ mrad. Can you estimate the size of a 'point-like' dust particle? By the way, an object is *invisible* to the naked eye if it is point-like *and* if its luminosity, i.e., the intensity of the light from the object reaching the eye, is below some critical value. Can you estimate whether there are any man-made objects visible from the Moon, or from the space shuttle?

The above definition of 'point-like' in everyday life is obviously misleading. Do proper, real point particles exist? In fact, is it at all possible to show that a particle has vanishing size? In the same way, we need to ask and check whether points in space do exist. Our walk will lead us to the astonishing result that all the answers to these questions are negative. Can you imagine why? Do not be disappointed if you find this issue difficult; many brilliant minds have had the same problem.

However, many particles, such as electrons, quarks or photons are point-like for all practical purposes. Once we know how to describe the motion of point particles, we can also describe the motion of extended bodies, rigid or deformable: we assume that they are made of parts. This is the same approach as describing the motion of an animal as a whole by combining the motion of its various body parts. The simplest description, the *continuum approximation*, describes extended bodies as an infinite collection of point particles. It allows us to understand and to predict the motion of milk and honey, the motion of the air in hurricanes and of perfume in rooms. The motion of fire and all other gaseous bodies, the bending of bamboo in the wind, the shape changes of chewing gum, and the growth of plants and animals can also be described in this way.

All observations so far have confirmed that the motion of large bodies can be described to full precision as the result of the motion of their parts. All machines that humans ever built are based on this idea. A description that is even more precise than the continuum approximation is given later on. Describing body motion with the motion of body parts will guide us through the first five volumes of our mountain ascent; for example, we will understand life in this way. Only in the final volume will we discover that, at a fundamental scale, this decomposition is impossible.

Legs and wheels

The parts of a body determine its shape. Shape is an important aspect of bodies: among other things, it tells us how to count them. In particular, living beings are always made of a single body. This is not an empty statement: from this fact we can deduce that animals cannot have large wheels or large propellers, but only legs, fins, or wings. Why?

Living beings have only one surface; simply put, they have only one piece of skin. Mathematically speaking, animals are *connected*. This is often assumed to be obvious, and it is often mentioned that the blood supply, the nerves and the lymphatic connections to a rotating part would get tangled up. However, this argument is not so simple, as Figure 54 shows. The figure proves that it is indeed possible to rotate a body continuously against a second one, without tangling up the connections. Three dimensions of space allow *tethered rotation*. Can you find an example for this kind of motion, often called *tethered rotation*, in your own body? Are you able to see how many cables may be attached to the rotating body of the figure without hindering the rotation?

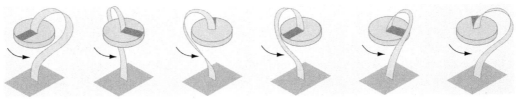

FIGURE 54 Tethered rotation: How an object can rotate continuously without tangling up the connection to a second object.

FIGURE 55 Tethered rotation: the continuous rotation of an object attached to its environment (QuickTime film © Jason Hise).

Despite the possibility of animals having rotating parts, the method of Figure 54 or Figure 55 still cannot be used to make a practical wheel or propeller. Can you see why? Therefore, evolution had no choice: it had to avoid animals with (large) parts rotating around axles. That is the reason that propellers and wheels do not exist in nature. Of course, this limitation does not rule out that living bodies move by rotation as a whole: tumbleweed, seeds from various trees, some insects, several spiders, certain other animals, children and dancers occasionally move by rolling or rotating as a whole.

Large single bodies, and thus all large living beings, can thus only move through *deformation* of their shape: therefore they are limited to walking, running, jumping, rolling, gliding, crawling, flapping fins, or flapping wings. Moving a leg is a common way to deform a body.

Extreme examples of leg use in nature are shown in Figure 56 and Figure 57. The most extreme example of rolling spiders – there are several species – are *Cebrennus villosus* and live in the sand in Morocco. They use their legs to accelerate the rolling, they can steer

3 HOW TO DESCRIBE MOTION – KINEMATICS

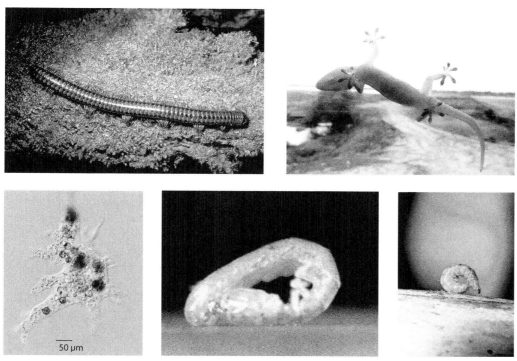

FIGURE 56 Legs and 'wheels' in living beings: the red millipede *Aphistogoniulus erythrocephalus* (15 cm body length), a gecko on a glass pane (15 cm body length), an amoeba *Amoeba proteus* (1 mm size), the rolling shrimp *Nannosquilla decemspinosa* (2 cm body length, 1.5 rotations per second, up to 2 m, can even roll slightly uphill slopes) and the rolling caterpillar *Pleurotya ruralis* (can only roll downhill, to escape predators), (© David Parks, Marcel Berendsen, Antonio Guillén Oterino, Robert Full, John Brackenbury / Science Photo Library).

FIGURE 57 Two of the rare lifeforms that are able to roll *uphill* also on steep slopes: the desert spider *Cebrennus villosus* and *Homo sapiens* (© Ingo Rechenberg, Karva Javi).

the rolling direction and can even roll uphill slopes of 30 % – a feat that humans are unable to perform. Films of the rolling motion can be found at www.bionik.tu-berlin. de.* Walking on water is shown in Figure 127 on page 170; examples of wings are given later on, as are the various types of deformations that allow swimming in water.

In contrast, *systems of several bodies*, such as bicycles, pedal boats or other machines, can move *without* any change of shape of their components, thus enabling the use of axles with wheels, propellers and other rotating devices.**

In short, whenever we observe a construction in which some part is turning continuously (and without the 'wiring' of Figure 54) we know immediately that it is an artefact: it is a machine, not a living being (but built by one). However, like so many statements about living creatures, this one also has exceptions.

The distinction between one and two bodies is poorly defined if the whole system is made of only a few molecules. This happens most clearly inside bacteria. Organisms such as *Escherichia coli*, the well-known bacterium found in the human gut, or bacteria from the *Salmonella* family, all swim using flagella. *Flagella* are thin filaments, similar to tiny hairs that stick out of the cell membrane. In the 1970s it was shown that each flagellum, made of one or a few long molecules with a diameter of a few tens of nanometres, does in fact turn about its axis.

Bacteria are able to rotate their flagella in both clockwise and anticlockwise directions, can achieve more than 1000 turns per second, and can turn all its flagella in perfect synchronization. These wheels are so tiny that they do not need a mechanical connection; Figure 58 shows a number of motor models found in bacteria. The motion and the construction of these amazing structures is shown in more details in the films Figure 59 and Figure 60.

In summary, wheels actually do exist in living beings, albeit only tiny ones. The growth and motion of these wheels are wonders of nature. Macroscopic wheels in living beings are not possible, though rolling motion is.

Curiosities and fun challenges about kinematics

What is the biggest wheel ever made?

* *

A football is shot, by a goalkeeper, with around 30 m/s. Use a video to calculate the distance it should fly and compare it with the distances realized in a soccer match. Where does the difference come from?

* *

---

* Rolling is also known for the Namibian wheel spiders of the *Carparachne* genus; films of their motion can be found on the internet.
** Despite the disadvantage of not being able to use rotating parts and of being restricted to one piece only, nature's moving constructions, usually called animals, often outperform human built machines. As an example, compare the size of the smallest flying systems built by evolution with those built by humans. (See, e.g., pixelito.reference.be.) There are two reasons for this discrepancy. First, nature's systems have integrated repair and maintenance systems. Second, nature can build large structures inside containers with small openings. In fact, nature is very good at what people do when they build sailing ships inside glass bottles. The human body is full of such examples; can you name a few?

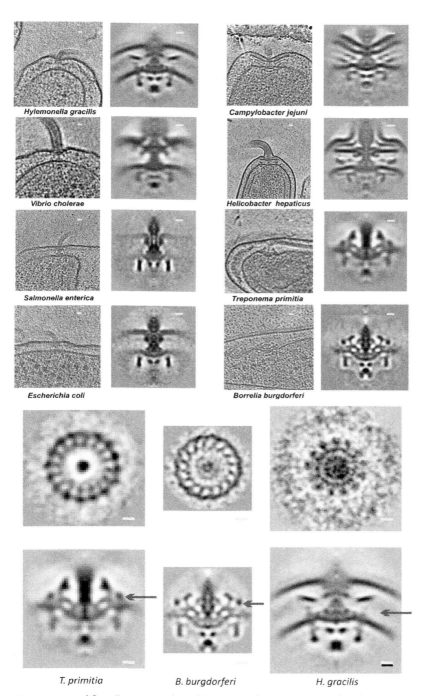

FIGURE 58 Some types of flagellar motors found in nature; the images are taken by cryotomography. All yellow scale bars are 10 nm long (© S. Chen & al., EMBO Journal, Wiley & Sons).

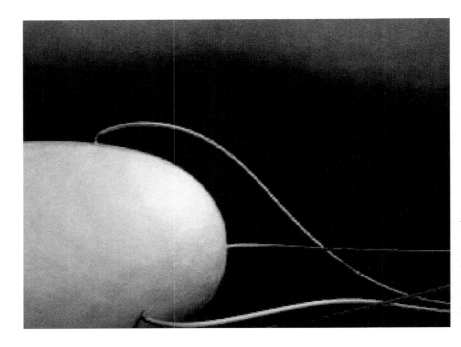

FIGURE 59 The rotational motion of a bacterial flagellum, and its reversal (QuickTime film © Osaka University).

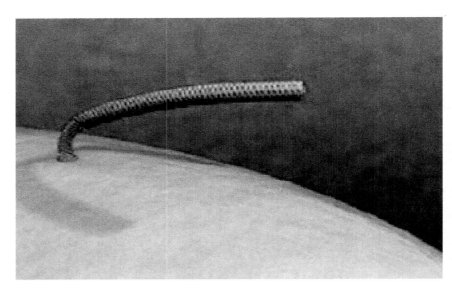

FIGURE 60 The growth of a bacterial flagellum, showing the molecular assembly (QuickTime film © Osaka University).

A train starts to travel at a constant speed of 10 m/s between two cities A and B, 36 km apart. The train will take one hour for the journey. At the same time as the train, a fast dove starts to fly from A to B, at 20 m/s. Being faster than the train, the dove arrives at B first. The dove then flies back towards A; when it meets the train, it turns back again, to city B. It goes on flying back and forward until the train reaches B. What distance did

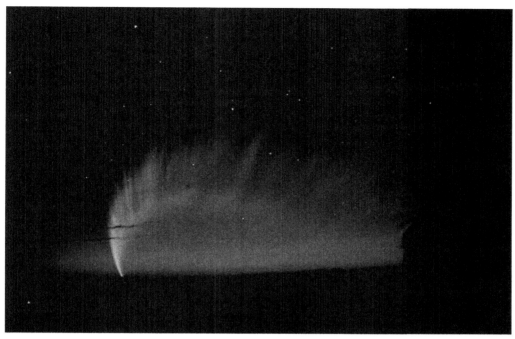

FIGURE 61 Are comets, such as the beautiful comet McNaught seen in 2007, images or bodies? How can you show it? (And why is the tail curved?) (© Robert McNaught)

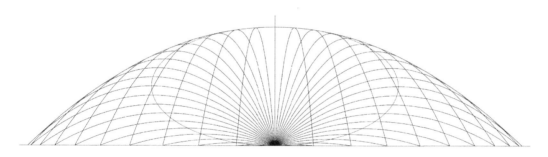

FIGURE 62 The parabola of safety around a cannon, shown in red. The highest points of all trajectories form an ellipse, shown in blue. (© Theon)

the dove cover?

* *

Figure 62 illustrates that around a cannon, there is a line outside which you cannot be hit. Already in the 17th century, Evangelista Torricelli showed, without algebra, that the line is a parabola, and called it the *parabola of safety*. Can you show this as well? Can you confirm that the highest points of all trajectories lie on an ellipse? The parabola of safety also appears in certain water fountains.

* *

FIGURE 63 Observation of sonoluminescence with a simple set-up that focuses ultrasound in water (© Detlef Lohse).

Balance a pencil vertically (tip upwards!) on a piece of paper near the edge of a table. How can you pull out the paper without letting the pencil fall?

* *

Is a return flight by aeroplane – from a point A to B and back to A – faster if the wind blows or if it does not?

* *

The level of acceleration that a human can survive depends on the duration over which one is subjected to it. For a tenth of a second, $30\,g = 300\,\text{m/s}^2$, as generated by an ejector seat in an aeroplane, is acceptable. (It seems that the record acceleration a human has survived is about $80\,g = 800\,\text{m/s}^2$.) But as a rule of thumb it is said that accelerations of $15\,g = 150\,\text{m/s}^2$ or more are fatal.

* *

The highest *microscopic* accelerations are observed in particle collisions, where values up to $10^{35}\,\text{m/s}^2$ are achieved. The highest *macroscopic* accelerations are probably found in the collapsing interiors of *supernovae*, exploding stars which can be so bright as to be visible in the sky even during the daytime. A candidate on Earth is the interior of collapsing bubbles in liquids, a process called *cavitation*. Cavitation often produces light, an effect discovered by Frenzel and Schultes in 1934 and called *sonoluminescence*. (See Figure 63.) It appears most prominently when air bubbles in water are expanded and contracted by underwater loudspeakers at around 30 kHz and allows precise measurements of bubble motion. At a certain threshold intensity, the bubble radius changes at $1500\,\text{m/s}$ in as little as a few µm, giving an acceleration of several $10^{11}\,\text{m/s}^2$.

* *

Legs are easy to build. Nature has even produced a millipede, *Illacme plenipes*, that has 750 legs. The animal is 3 to 4 cm long and about 0.5 mm wide. This seems to be the record so far. In contrast to its name, no millipede actually has a thousand legs.

Summary of kinematics

The description of everyday motion of mass points with three coordinates as $(x(t), y(t), z(t))$ is simple, precise and complete. This description of paths is the basis of kinematics. As a consequence, space is described as a three-dimensional Euclidean space and velocity and acceleration as Euclidean vectors.

The description of motion with paths assumes that the motion of objects can be *followed* along their paths. Therefore, the description often does not work for an important case: the motion of images.

Chapter 4
FROM OBJECTS AND IMAGES TO CONSERVATION

Walking through a forest we observe two rather different types of motion: we see the breeze move the leaves, and at the same time, on the ground, we see their shadows move. Shadows are a simple type of image. Both objects and images are able to move; both change position over time. Running tigers, falling snowflakes, and material ejected by volcanoes, but also the shadow following our body, the beam of light circling the tower of a lighthouse on a misty night, and the rainbow that constantly keeps the same apparent distance from us are examples of motion.

Both objects and images differ from their environment in that they have *boundaries* defining their size and shape. But everybody who has ever seen an animated cartoon knows that images can move in more surprising ways than objects. Images can change their size and shape, they can even change colour, a feat only few objects are able to perform.** Images can appear and disappear without trace, multiply, interpenetrate, go backwards in time and defy gravity or any other force. Images, even ordinary shadows, can move faster than light. Images can float in space and keep the same distance from approaching objects. Objects can do almost none of this. In general, the 'laws of cartoon physics' are rather different from those in nature. In fact, the motion of images does not seem to follow any rules, in contrast to the motion of objects. We feel the need for precise criteria allowing the two cases to be distinguished.

Making a clear distinction between images and objects is performed using the same method that children or animals use when they stand in front of a mirror for the first time: they try to *touch* what they see. Indeed,

> ▷ If we are able to touch what we see – or more precisely, if we are able to move it with a collision – we call it an *object*, otherwise an *image*.***

** Excluding very slow changes such as the change of colour of leaves in the Autumn, in nature only certain crystals, the octopus and other cephalopods, the chameleon and a few other animals achieve this. Of man-made objects, television, computer displays, heated objects and certain lasers can do it. Do you know more examples? An excellent source of information on the topic of colour is the book by K. Nassau, *The Physics and Chemistry of Colour – the fifteen causes of colour*, J. Wiley & Sons, 1983. In the popular science domain, the most beautiful book is the classic work by the Flemish astronomer Marcel G. J. Minnaert, *Light and Colour in the Outdoors*, Springer, 1993, an updated version based on his wonderful book series, *De natuurkunde van 't vrije veld*, Thieme & Cie, 1937. Reading it is a must for all natural scientists. On the web, there is also the – simpler, but excellent – webexhibits.org/causesofcolour website.

*** One could propose including the requirement that objects may be rotated; however, this requirement, surprisingly, gives difficulties in the case of atoms, as explained on page 85 in Volume IV.

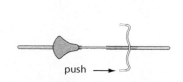

FIGURE 64 In which direction does the bicycle turn?

Images cannot be touched, but objects can. Images cannot hit each other, but objects can. And as everybody knows, touching something means feeling that it resists movement. Certain bodies, such as butterflies, pose little resistance and are moved with ease, others, such as ships, resist more, and are moved with more difficulty.

▷ The resistance to motion – more precisely, to change of motion – is called *inertia*, and the difficulty with which a body can be moved is called its *(inertial) mass*.

Images have neither inertia nor mass.

Summing up, for the description of motion we must distinguish bodies, which can be touched and are impenetrable, from images, which cannot and are not. Everything visible is either an object or an image; there is no third possibility. (Do you agree?) If the object is so far away that it cannot be touched, such as a star or a comet, it can be difficult to decide whether one is dealing with an image or an object; we will encounter this difficulty repeatedly. For example, how would you show that comets – such as the beautiful example of Figure 61 – are objects and not images, as Galileo (falsely) claimed?

In the same way that objects are made of *matter*, images are made of *radiation*. Images are the domain of shadow theatre, cinema, television, computer graphics, belief systems and drug experts. Photographs, motion pictures, ghosts, angels, dreams and many hallucinations are images (sometimes coupled with brain malfunction). To understand images, we need to study radiation (plus the eye and the brain). However, due to the importance of objects – after all we are objects ourselves – we study the latter first.

Motion and contact

“ Democritus affirms that there is only one type of movement: That resulting from collision.
Aetius, *Opinions*. ”

When a child rides a unicycle, she or he makes use of a general rule in our world: one body acting on another puts it in motion. Indeed, in about six hours, anybody can learn to ride and enjoy a unicycle. As in all of life's pleasures, such as toys, animals, women, machines, children, men, the sea, wind, cinema, juggling, rambling and loving, something pushes something else. Thus our first challenge is to describe the transfer of motion due to contact – and to collisions – in more precise terms.

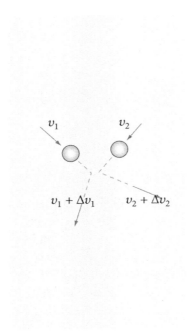

FIGURE 65 Collisions define mass.

FIGURE 66 The standard kilogram (until 2019) (© BIPM).

Contact is not the only way to put something into motion; a counter-example is an apple falling from a tree or one magnet pulling another. Non-contact influences are more fascinating: nothing is hidden, but nevertheless something mysterious happens. Contact motion seems easier to grasp, and that is why one usually starts with it. However, despite this choice, non-contact interactions cannot be avoided. Our choice to start with contact will lead us to a similar experience to that of riding a bicycle. (See Figure 64.) If we ride a bicycle at a sustained speed and try to turn left by pushing the right-hand steering bar, we will turn *right*. By the way, this surprising effect, also known to motor bike riders, obviously works only above a certain minimal speed. Can you determine what this speed is? Be careful! Too strong a push will make you fall.

Something similar will happen to us as well; despite our choice for contact motion, the rest of our walk will rapidly force us to study non-contact interactions.

What is mass?

> Δός μοί (φησι) ποῦ στῶ καὶ κινῶ τὴν γῆν.
> Da ubi consistam, et terram movebo.*
> Archimedes

When we push something we are unfamiliar with, such as when we kick an object on the street, we automatically pay attention to the same aspect that children explore when

---

* 'Give me a place to stand, and I'll move the Earth.' Archimedes (*c.* 283–212), Greek scientist and engineer. This phrase is attributed to him by Pappus. Already Archimedes knew that the distinction used by lawyers between movable and immovable objects made no sense.

FIGURE 67 Antoine Lavoisier (1743–1794) and his wife.

they stand in front of a mirror for the first time, or when they see a red laser spot for the first time. They check whether the unknown entity can be pushed or caught, and they pay attention to how the unknown object moves under their influence. All these are collision experiments. The high precision version of any collision experiment is illustrated in Figure 65. Repeating such experiments with various pairs of objects, we find:

▷ A *fixed* quantity m_i can be ascribed to every object i, determined by the relation
$$\frac{m_2}{m_1} = -\frac{\Delta v_1}{\Delta v_2} \qquad (17)$$
where Δv is the velocity change produced by the collision. The quantity m_i is called the *mass* of the object i.

The more difficult it is to move an object, the higher the mass value. In order to have mass values that are common to everybody, the mass value for one particular, selected object has to be fixed in advance. Until 2019, there really was one such special object in the world, shown in Figure 66; it was called the *standard kilogram*. It was kept with great care in a glass container in Sèvres near Paris. Until 2019, the standard kilogram determined the value of the mass of every other object in the world. The standard kilogram was touched only once every few years because otherwise dust, humidity, or scratches would change its mass. For example, the standard kilogram was *not* kept under vacuum, because this would lead to outgassing and thus to changes in its mass. All the care did not avoid the stability issues though, and in 2019, the kilogram unit has been redefined using the fundamental constants G (indirectly, via the caesium transition frequency), c and $\hbar$ that are shown in Figure 1. Since that change, everybody can produce his or her own standard kilogram in the laboratory – provided that sufficient care is used.

The *mass* thus measures *the difficulty of getting something moving*. High masses are harder to move than low masses. Obviously, only objects have mass; images don't. (By the way, the word 'mass' is derived, via Latin, from the Greek μαζα – bread – or the

Ref. 67

FIGURE 68 Christiaan Huygens (1629–1695).

Hebrew 'mazza' – unleavened bread. That is quite a change in meaning.)

Experiments with everyday life objects also show that throughout any collision, the sum of all masses is *conserved*:

$$\sum_i m_i = \text{const}. \qquad (18)$$

The principle of conservation of mass was first stated by Antoine-Laurent Lavoisier.* Conservation of mass also implies that the mass of a composite system is the sum of the mass of the components. In short, *mass is also a measure for the quantity of matter.*

In a famous experiment in the sixteenth century, for several weeks Santorio Santorio (Sanctorius) (1561–1636), friend of Galileo, lived with all his food and drink supply, and also his toilet, on a large balance. He wanted to test mass conservation. How did the measured weight change with time?

Various cult leaders pretended and still pretend that they can produce matter out of nothing. This would be an example of non-conservation of mass. How can you show that all such leaders are crooks?

Momentum and mass

The definition of mass can also be given in another way. We can ascribe a number m_i to every object i such that for collisions free of outside interference the following sum is unchanged *throughout* the collision:

$$\sum_i m_i v_i = \text{const}. \qquad (19)$$

The product of the velocity v_i and the mass m_i is called the (linear) *momentum* of the body. The sum, or *total momentum* of the system, is the same before and after the collision; momentum is a *conserved* quantity.

---

* Antoine-Laurent Lavoisier (b. 1743 Paris , d. 1794 Paris), chemist and genius. Lavoisier was the first to understand that combustion is a reaction with oxygen; he discovered the components of water and introduced mass measurements into chemistry. A famous story about his character: When he was (unjustly) sentenced to the guillotine during the French revolution, he decided to use the situations for a scientific experiment. He announced that he would try to blink his eyes as frequently as possible after his head was cut off, in order to show others how long it takes to lose consciousness. Lavoisier managed to blink eleven times. It is unclear whether the story is true or not. It is known, however, that it could be true. Indeed, after a decapitation without pain or shock, a person can remain conscious for up to half a minute.

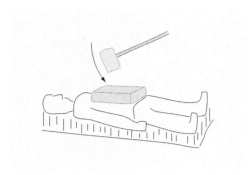

FIGURE 69 Is this dangerous?

▷ Momentum conservation defines mass.

The two conservation principles (18) and (19) were first stated in this way by the important physicist Christiaan Huygens:* *Momentum and mass are conserved in everyday motion of objects.* In particular, neither quantity can be defined for the motion of images. Some typical momentum values are given in Table 15.

Momentum conservation implies that when a moving sphere hits a resting one of the same mass and without loss of energy, a simple rule determines the angle between the directions the two spheres take after the collision. Can you find this rule? It is particularly useful when playing billiards. We will find out later that the rule is *not* valid for speeds near that of light.

Another consequence of momentum conservation is shown in Figure 69: a man is lying on a bed of nails with a large block of concrete on his stomach. Another man is hitting the concrete with a heavy sledgehammer. As the impact is mostly absorbed by the concrete, there is no pain and no danger – unless the concrete is missed. Why?

The above definition (17) of mass has been generalized by the physicist and philosopher Ernst Mach** in such a way that it is valid even if the two objects interact without contact, as long as they do so along the line connecting their positions.

▷ The mass ratio between two bodies is defined as a negative inverse acceleration ratio, thus as

$$\frac{m_2}{m_1} = -\frac{a_1}{a_2}, \qquad (20)$$

---

* Christiaan Huygens (b. 1629 's Gravenhage, d. 1695 Hofwyck) was one of the main physicists and mathematicians of his time. Huygens clarified the concepts of mechanics; he also was one of the first to show that light is a wave. He wrote influential books on probability theory, clock mechanisms, optics and astronomy. Among other achievements, Huygens showed that the Orion Nebula consists of stars, discovered Titan, the moon of Saturn, and showed that the rings of Saturn consist of rock. (This is in contrast to Saturn itself, whose density is lower than that of water.)
** Ernst Mach (1838 Chrlice–1916 Vaterstetten), Austrian physicist and philosopher. The *mach* unit for aeroplane speed as a multiple of the speed of sound in air (about 0.3 km/s) is named after him. He also studied the basis of mechanics. His thoughts about mass and inertia influenced the development of general relativity, and led to Mach's principle, which will appear later on. He was also proud to be the last scientist denying – humorously, and against all evidence – the existence of atoms.

TABLE 15 Some measured momentum values.

Observation	Momentum
Images	0
Momentum of a green photon	$1.2 \cdot 10^{-27}$ Ns
Average momentum of oxygen molecule in air	10^{-26} Ns
X-ray photon momentum	10^{-23} Ns
γ photon momentum	10^{-17} Ns
Highest particle momentum in accelerators	1 fNs
Highest possible momentum of a single elementary particle – the Planck momentum	6.5 Ns
Fast billiard ball	3 Ns
Flying rifle bullet	10 Ns
Box punch	15 to 50 Ns
Comfortably walking human	80 Ns
Lion paw strike	kNs
Whale tail blow	kNs
Car on highway	40 kNs
Impact of meteorite with 2 km diameter	100 TNs
Momentum of a galaxy in galaxy collision	up to 10^{46} Ns

where a is the acceleration of each body during the interaction.

This definition of mass has been explored in much detail in the physics community, mainly in the nineteenth century. A few points sum up the results:

— The definition of mass *implies* the conservation of total momentum $\sum mv$. Momentum conservation is *not* a separate principle. Conservation of momentum cannot be checked experimentally, because mass is defined in such a way that the momentum conservation holds.
— The definition of mass *implies* the equality of the products $m_1 a_1$ and $-m_2 a_2$. Such products are called *forces*. The equality of acting and reacting forces is not a separate principle; mass is defined in such a way that the principle holds.
— The definition of mass is *independent* of whether contact is involved or not, and whether the accelerations are due to electricity, gravitation, or other interactions.* Since the interaction does not enter the definition of mass, mass values defined with the help of the electric, nuclear or gravitational interaction all agree, as long as momentum is conserved. All known interactions conserve momentum. For some unfortunate historical reasons, the mass value measured with the electric or nuclear interactions is called the 'inertial' mass and the mass measured using gravity is called

---

* As mentioned above, only *central* forces obey the relation (20) used to define mass. Central forces act between the centre of mass of bodies. We give a precise definition later. However, since all fundamental forces are central, this is not a restriction. There seems to be one notable exception: magnetism. Is the definition of mass valid in this case?

4 FROM OBJECTS AND IMAGES TO CONSERVATION

TABLE 16 Some measured mass values.

Observation	Mass
Probably lightest known object: neutrino	$c.\ 2 \cdot 10^{-36}$ kg
Mass increase due to absorption of one green photon	$4.1 \cdot 10^{-36}$ kg
Lightest known charged object: electron	$9.109\,381\,88(72) \cdot 10^{-31}$ kg
Atom of argon	$39.962\,383\,123(3)$ u $= 66.359\,1(1)$ yg
Lightest object ever weighed (a gold particle)	0.39 ag
Human at early age (fertilized egg)	10^{-8} g
Water adsorbed on to a kilogram metal weight	10^{-5} g
Planck mass	$2.2 \cdot 10^{-5}$ g
Fingerprint	10^{-4} g
Typical ant	10^{-4} g
Water droplet	1 mg
Honey bee, *Apis mellifera*	0.1 g
Euro coin	7.5 g
Blue whale, *Balaenoptera musculus*	180 Mg
Heaviest living things, such as the fungus *Armillaria ostoyae* or a large Sequoia *Sequoiadendron giganteum*	10^6 kg
Heaviest train ever	$99.7 \cdot 10^6$ kg
Largest ocean-going ship	$400 \cdot 10^6$ kg
Largest object moved by man (Troll gas rig)	$687.5 \cdot 10^6$ kg
Large antarctic iceberg	10^{15} kg
Water on Earth	10^{21} kg
Earth's mass	$5.98 \cdot 10^{24}$ kg
Solar mass	$2.0 \cdot 10^{30}$ kg
Our galaxy's visible mass	$3 \cdot 10^{41}$ kg
Our galaxy's estimated total mass	$2 \cdot 10^{42}$ kg
virgo supercluster	$2 \cdot 10^{46}$ kg
Total mass visible in the universe	10^{54} kg

the 'gravitational' mass. As it turns out, this artificial distinction makes no sense; this becomes especially clear when we take an observation point that is *far away* from all the bodies concerned.
— The definition of mass requires observers at rest or in inertial motion.

By measuring the masses of bodies around us we can explore the science and art of experiments. An overview of mass measurement devices is given in Table 18 and Figure 71. Some measurement results are listed in Table 16.

Measuring mass vales around us we confirm the main properties of mass. First of all, mass is *additive* in everyday life, as the mass of two bodies combined is equal to the sum of the two separate masses. Furthermore, mass is *continuous*; it can seemingly take any positive value. Finally, mass is *conserved* in everyday life; the mass of a system, defined

TABLE 17 Properties of mass in everyday life.

MASSES	PHYSICAL PROPERTY	MATHEMATICAL NAME	DEFINITION
Can be distinguished	distinguishability	element of set	Vol. III, page 284
Can be ordered	sequence	order	Vol. IV, page 224
Can be compared	measurability	metricity	Vol. IV, page 236
Can change gradually	continuity	completeness	Vol. V, page 364
Can be added	quantity of matter	additivity	Page 81
Beat any limit	infinity	unboundedness, openness	Vol. III, page 285
Do not change	conservation	invariance	m = const
Do not disappear	impenetrability	positivity	$m \geq 0$

as the sum of the mass of all constituents, does not change over time if the system is kept isolated from the rest of the world. Mass is not only conserved in collisions but also during melting, evaporation, digestion and all other everyday processes.

All the properties of everyday mass are summarized in Table 17. Later we will find that several of the properties are only approximate. High-precision experiments show deviations.* However, the definition of mass remains unchanged throughout our adventure.

The definition of mass through momentum conservation implies that when an object falls, the Earth is accelerated upwards by a tiny amount. If we could measure this tiny amount, we could determine the mass of the Earth. Unfortunately, this measurement is impossible. Can you find a better way to determine the mass of the Earth?

The definition of mass and momentum allows to answer the question of Figure 70. A brick hangs from the ceiling; a second thread hangs down from the brick, and you can pull it. How can you tune your pulling method to make the upper thread break? The lower one?

Summarizing Table 17, the mass of a body is thus most precisely described by a *positive real number*, often abbreviated m or M. This is a direct consequence of the impenetrability of matter. Indeed, a *negative* (inertial) mass would mean that such a body would move in the opposite direction of any applied force or acceleration. Such a body could not be kept in a box; it would break through any wall trying to stop it. Strangely enough, negative mass bodies would still fall downwards in the field of a large positive mass (though more slowly than an equivalent positive mass). Are you able to confirm this? However, a small positive mass object would float away from a large negative-mass body, as you can easily deduce by comparing the various accelerations involved. A positive and a negative mass of the same value would stay at constant distance and spontaneously accelerate away along the line connecting the two masses. Note that both energy and momentum are conserved in all these situations.** Negative-mass bodies have never been observed. Antimatter, which will be discussed later, also has positive mass.

---

* For example, in order to define mass we must be able to *distinguish* bodies. This seems a trivial requirement, but we discover that this is not always possible in nature.

** For more curiosities, see R. H. PRICE, *Negative mass can be positively amusing*, American Journal of

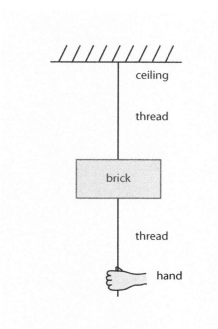

FIGURE 70 Depending on the way you pull, either the upper of the lower thread snaps. What are the options?

TABLE 18 Some mass sensors.

MEASUREMENT	SENSOR	RANGE
Precision scales	balance, pendulum, or spring	1 pg to 10^3 kg
Particle collision	speed	below 1 mg
Sense of touch	pressure sensitive cells	1 mg to 500 kg
Doppler effect on light reflected off the object	interferometer	1 mg to 100 g
Cosmonaut body mass measurement device	spring frequency	around 70 kg
Truck scales	hydraulic balance	10^3 to $60 \cdot 10^3$ kg
Ship weight	water volume measurement	up to $500 \cdot 10^6$ kg

---

Physics **61**, pp. 216–217, 1993. Negative mass particles in a box would heat up a box made of positive mass while traversing its walls, and accelerating, i.e., losing energy, at the same time. They would allow one to build a *perpetuum mobile* of the second kind, i.e., a device circumventing the second principle of thermodynamics. Moreover, such a system would have no thermodynamic equilibrium, because its energy could decrease forever. The more one thinks about negative mass, the more one finds strange properties contradicting observations. By the way, what is the range of possible mass values for tachyons?

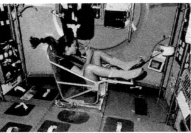

FIGURE 71 Mass measurement devices: a vacuum balance used in 1890 by Dmitriy Ivanovich Mendeleyev, a modern laboratory balance, a device to measure the mass of a cosmonaut in space and a truck scales (© Thinktank Trust, Mettler-Toledo, NASA, Anonymous).

Is motion eternal? – Conservation of momentum

> Every body continues in the state of rest or of uniform motion in a straight line except in so far as it doesn't.
>
> Arthur Eddington*

The product $\boldsymbol{p} = m\boldsymbol{v}$ of mass and velocity is called the *momentum* of a particle; it describes the tendency of an object to keep moving during collisions. The larger it is, the harder it is to stop the object. Like velocity, momentum has a direction and a magnitude: it is a vector. In French, momentum is called 'quantity of motion', a more appropriate term. In the old days, the term 'motion' was used instead of 'momentum', for example by Newton. The conservation of momentum, relation (19), therefore expresses the conservation of motion during interactions.

Momentum is an *extensive quantity*. That means that it can be said that it *flows* from one body to the other, and that it can be *accumulated* in bodies, in the same way that water flows and can be accumulated in containers. Imagining momentum as something

---

* Arthur Eddington (1882–1944), British astrophysicist.

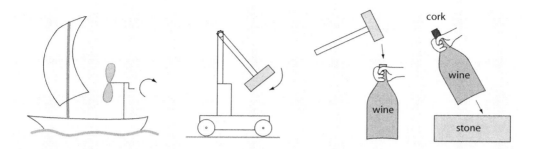

FIGURE 72 What happens in these four situations?

that can be *exchanged* between bodies in collisions is always useful when thinking about the description of moving objects.

Momentum is conserved. That explains the limitations you might experience when being on a perfectly frictionless surface, such as ice or a polished, oil covered marble: you cannot propel yourself forward by patting your own back. (Have you ever tried to put a cat on such a marble surface? It is not even able to stand on its four legs. Neither are humans. Can you imagine why?) Momentum conservation also answers the puzzles of Figure 72.

The conservation of momentum and mass also means that teleportation ('beam me up') is impossible in nature. Can you explain this to a non-physicist?

Momentum conservation implies that momentum can be imagined to be like an invisible *fluid*. In an interaction, the invisible fluid is transferred from one object to another. In such transfers, the amount of fluid is always constant.

Momentum conservation implies that motion never stops; it is only *exchanged*. On the other hand, motion often 'disappears' in our environment, as in the case of a stone dropped to the ground, or of a ball left rolling on grass. Moreover, in daily life we often observe the creation of motion, such as every time we open a hand. How do these examples fit with the conservation of momentum?

It turns out that apparent momentum disappearance is due to the microscopic aspects of the involved systems. A muscle only *transforms* one type of motion, namely that of the electrons in certain chemical compounds* into another, the motion of the fingers. The working of muscles is similar to that of a car engine transforming the motion of electrons in the fuel into motion of the wheels. Both systems need fuel and get warm in the process.

We must also study the microscopic behaviour when a ball rolls on grass until it stops. The apparent disappearance of motion is called *friction*. Studying the situation carefully, we find that the grass and the ball heat up a little during this process. *During friction, visible motion is transformed into heat.* A striking observation of this effect for a bicycle is shown below, in Figure 273. Later, when we discover the structure of matter, it will become clear that heat is the disorganized motion of the microscopic constituents of every material. When the microscopic constituents all move in the same direction, the object as a whole moves; when they oscillate randomly, the object is at rest, but is warm.

---

* The fuel of most processes in animals usually is adenosine triphosphate (ATP).

Heat is a form of motion. Friction thus only seems to be disappearance of motion; in fact it is a transformation of ordered into unordered motion.

Despite momentum conservation, *macroscopic* perpetual motion does not exist, since friction cannot be completely eliminated.* Motion is eternal only at the microscopic scale. In other words, the disappearance and also the spontaneous appearance of motion in everyday life is an illusion due to the limitations of our senses. For example, the motion proper of every living being exists before its birth, and stays after its death. The same happens with its energy. This result is probably the closest one can get to the idea of everlasting life from evidence collected by observation. It is perhaps less than a coincidence that energy used to be called *vis viva*, or 'living force', by Leibniz and many others.

Since motion is conserved, it has no origin. Therefore, at this stage of our walk we cannot answer the fundamental questions: Why does motion exist? What is its origin? The end of our adventure is nowhere near.

More conservation – energy

When collisions are studied in detail, a second conserved quantity turns up. Experiments show that in the case of perfect, or elastic collisions – collisions without friction – the following quantity, called the *kinetic energy* T of the system, is also conserved:

$$T = \sum_i \tfrac{1}{2} m_i v_i^2 = \text{const} . \tag{21}$$

Kinetic energy is the ability that a body has to induce change in bodies it hits. Kinetic energy thus depends on the mass and on the square of the speed v of a body. The full name 'kinetic energy' was introduced by Gustave-Gaspard Coriolis.** Some measured energy values are given in Table 19.

---

* Some funny examples of past attempts to built a *perpetual motion machine* are described in Stanislav Michel, *Perpetuum mobile*, VDI Verlag, 1976. Interestingly, the idea of eternal motion came to Europe from India, via the Islamic world, around the year 1200, and became popular as it opposed the then standard view that all motion on Earth disappears over time. See also the web.archive.org/web/20040812085618/http://www.geocities.com/mercutio78_99/pmm.html and the www.lhup.edu/~dsimanek/museum/unwork.htm websites. The conceptual mistake made by eccentrics and used by crooks is always the same: the hope of overcoming friction. (In fact, this applied only to the perpetual motion machines of the second kind; those of the first kind – which are even more in contrast with observation – even try to generate energy from nothing.)

If the machine is well constructed, i.e., with little friction, it can take the little energy it needs for the sustenance of its motion from very subtle environmental effects. For example, in the Victoria and Albert Museum in London one can admire a beautiful clock powered by the variations of air pressure over time.

Low friction means that motion takes a long time to stop. One immediately thinks of the motion of the planets. In fact, there *is* friction between the Earth and the Sun. (Can you guess one of the mechanisms?) But the value is so small that the Earth has already circled around the Sun for thousands of millions of years, and will do so for quite some time more.

** Gustave-Gaspard Coriolis (b. 1792 Paris, d. 1843 Paris) was engineer and mathematician. He introduced the modern concepts of 'work' and of 'kinetic energy', and explored the Coriolis effect discovered by Laplace. Coriolis also introduced the factor 1/2 in the kinetic energy T, in order that the relation $dT/dv = p$ would be obeyed. (Why?)

The experiments and ideas mentioned so far can be summarized in the following definition:

▷ (Physical) *energy* is the measure of the ability to generate motion.

A body has a lot of energy if it has the ability to move many other bodies. Energy is a number; energy, in contrast to momentum, has no direction. The total momentum of two equal masses moving with opposite velocities is zero; but their total energy is not, and it increases with velocity. Energy thus also measures motion, but in a different way than momentum. Energy measures motion in a more global way.

An equivalent definition is the following:

▷ Energy is the ability to perform work.

Here, the physical concept of work is just the precise version of what is meant by work in everyday life. As usual, (physical) *work* is the product of force and distance in direction of the force. In other words, work is the *scalar product* of force and distance. Physical work is a quantity that describes the effort of pushing of an object along a distance. As a result, in physics, work is a form of energy.

Another, equivalent definition of energy will become clear shortly:

▷ Energy is what can be transformed into heat.

Energy is a word taken from ancient Greek; originally it was used to describe character, and meant 'intellectual or moral vigour'. It was taken into physics by Thomas Young (1773–1829) in 1807 because its literal meaning is 'force within'. (The letters E, W, A and several others are also used to denote energy.)

Both energy and momentum measure how systems change. Momentum tells how systems change *over distance*: momentum is action (or change) divided by distance. Momentum is needed to compare motion here and there.

Energy measures how systems change *over time*: energy is action (or change) divided by time. Energy is needed to compare motion now and later.

Do not be surprised if you do not grasp the difference between momentum and energy straight away: physicists took about a century to figure it out! So you are allowed to take some time to get used to it. Indeed, for many decades, English physicists insisted on using the same term for both concepts; this was due to Newton's insistence that – no joke – the existence of god implied that energy was the same as momentum. Leibniz, instead, knew that energy increases with the square of the speed and proved Newton wrong. In 1722, Willem Jacob 's Gravesande even showed the difference between energy and momentum experimentally. He let metal balls of different masses fall into mud from different heights. By comparing the size of the imprints he confirmed that Newton was wrong both with his physical statements and his theological ones.

One way to explore the difference between energy and momentum is to think about the following challenges. Which running man is more difficult to stop? One of mass m running at speed v, or one with mass $m/2$ and speed $2v$, or one with mass $m/2$ and speed $\sqrt{2}\,v$? You may want to ask a rugby-playing friend for confirmation.

TABLE 19 Some measured energy values.

OBSERVATION	ENERGY
Average kinetic energy of oxygen molecule in air	6 zJ
Green photon energy	0.37 aJ
X-ray photon energy	1 fJ
γ photon energy	1 pJ
Highest particle energy in accelerators	0.1 μJ
Kinetic energy of a flying mosquito	0.2 μJ
Comfortably walking human	20 J
Flying arrow	50 J
Right hook in boxing	50 J
Energy in torch battery	1 kJ
Energy in explosion of 1 g TNT	4.1 kJ
Energy of 1 kcal	4.18 kJ
Flying rifle bullet	10 kJ
One gram of fat	38 kJ
One gram of gasoline	44 kJ
Apple digestion	0.2 MJ
Car on highway	0.3 to 1 MJ
Highest laser pulse energy	1.8 MJ
Lightning flash	up to 1 GJ
Planck energy	2.0 GJ
Small nuclear bomb (20 ktonne)	84 TJ
Earthquake of magnitude 7	2 PJ
Largest nuclear bomb (50 Mtonne)	210 PJ
Impact of meteorite with 2 km diameter	1 EJ
Yearly machine energy use	420 EJ
Rotation energy of Earth	$2 \cdot 10^{29}$ J
Supernova explosion	10^{44} J
Gamma-ray burst	up to 10^{47} J
Energy content $E = c^2 m$ of Sun's mass	$1.8 \cdot 10^{47}$ J
Energy content of Galaxy's central black hole	$4 \cdot 10^{53}$ J

Another distinction between energy and momentum is illustrated by athletics: the *real* long jump world record, almost 10 m, is still kept by an athlete who in the early twentieth century ran with two weights in his hands, and then threw the weights behind him at the moment he took off. Can you explain the feat?

When a car travelling at 100 m/s runs head-on into a parked car of the same kind and make, which car receives the greatest damage? What changes if the parked car has its brakes on?

To get a better feeling for energy, here is an additional aspect. The world consumption of energy by human machines (coming from solar, geothermal, biomass, wind, nuclear,

FIGURE 73 Robert Mayer (1814–1878).

hydro, gas, oil, coal, or animal sources) in the year 2000 was about 420 EJ,* for a world population of about 6000 million people. To see what this energy consumption means, we translate it into a personal power consumption; we get about 2.2 kW. The watt W is the unit of power, and is simply defined as 1 W = 1 J/s, reflecting the definition of *(physical) power* as energy used per unit time. The precise wording is: power is energy flowing per time through a defined closed surface. See Table 20 for some power values found in nature, and Table 21 for some measurement devices.

As a working person can produce mechanical work of about 100 W, the average human energy consumption corresponds to about 22 humans working 24 hours a day. In particular, if we look at the energy consumption in First World countries, the average inhabitant there has machines working for him or her that are equivalent to several hundred 'servants'. Machines do a lot of good. Can you point out some of these machines?

Kinetic energy is thus not conserved in everyday life. For example, in non-elastic collisions, such as that of a piece of chewing gum hitting a wall, kinetic energy is lost. *Friction* destroys kinetic energy. At the same time, friction produces heat. It was one of the important conceptual discoveries of physics that *total* energy is conserved if one includes the discovery that heat is a form of energy. Friction is thus a process transforming kinetic energy, i.e., the energy connected with the motion of a body, into heat. On a microscopic scale, *energy is always conserved*.

Any example of non-conservation of energy is only apparent. ** Indeed, without energy conservation, the concept of time would not be definable! We will show this important connection shortly.

In summary, in addition to mass and momentum, everyday linear motion also conserves energy. To discover the last conserved quantity, we explore another type of motion: rotation.

---

* For the explanation of the abbreviation E, see Appendix B.
** In fact, the conservation of energy was stated in its full generality in public only in 1842, by Julius Robert Mayer. He was a medical doctor by training, and the journal *Annalen der Physik* refused to publish his paper, as it supposedly contained 'fundamental errors'. What the editors called errors were in fact mostly – but not only – contradictions of their prejudices. Later on, Helmholtz, Thomson-Kelvin, Joule and many others acknowledged Mayer's genius. However, the first to have stated energy conservation in its modern form was the French physicist Sadi Carnot (1796–1832) in 1820. To him the issue was so clear that he did not publish the result. In fact he went on and discovered the second 'law' of thermodynamics. Today, energy conservation, also called the first 'law' of thermodynamics, is one of the pillars of physics, as it is valid in all its domains.

TABLE 20 Some measured power values.

Observation	Power
Radio signal from the Galileo space probe sending from Jupiter	10 zW
Power of flagellar motor in bacterium	0.1 pW
Power consumption of a typical cell	1 pW
sound power at the ear at hearing threshold	2.5 pW
CR-R laser, at 780 nm	40-80 mW
Sound output from a piano playing fortissimo	0.4 W
Dove (0.16 kg) basal metabolic rate	0.97 W
Rat (0.26 kg) basal metabolic rate	1.45 W
Pigeon (0.30 kg) basal metabolic rate	1.55 W
Hen (2.0 kg) basal metabolic rate	4.8 W
Incandescent light bulb light output	1 to 5 W
Dog (16 kg) basal metabolic rate	20 W
Sheep (45 kg) basal metabolic rate	50 W
Woman (60 kg) basal metabolic rate	68 W
Man (70 kg) basal metabolic rate	87 W
Incandescent light bulb electricity consumption	25 to 100 W
A human, during one work shift of eight hours	100 W
Cow (400 kg) basal metabolic rate	266 W
One horse, for one shift of eight hours	300 W
Steer (680 kg) basal metabolic rate	411 W
Eddy Merckx, the great bicycle athlete, during one hour	500 W
Metric horse power power unit (75 kg · 9.81 m/s² · 1 m/s)	735.5 W
British horse power power unit	745.7 W
Large motorbike	100 kW
Electrical power station output	0.1 to 6 GW
World's electrical power production in 2000 Ref. 90	450 GW
Power used by the geodynamo	200 to 500 GW
Limit on wind energy production Ref. 91	18 to 68 TW
Input on Earth surface: Sun's irradiation of Earth Ref. 92	0.17 EW
Input on Earth surface: thermal energy from inside of the Earth	32 TW
Input on Earth surface: power from tides (i.e., from Earth's rotation)	3 TW
Input on Earth surface: power generated by man from fossil fuels	8 to 11 TW
Lost from Earth surface: power stored by plants' photosynthesis	40 TW
World's record laser power	1 PW
Output of Earth surface: sunlight reflected into space	0.06 EW
Output of Earth surface: power radiated into space at 287 K	0.11 EW
Peak power of the largest nuclear bomb	5 YW
Sun's output	384.6 YW
Maximum power in nature, $c^5/4G$	$9.1 \cdot 10^{51}$ W

TABLE 21 Some power sensors.

Measurement	Sensor	Range
Heart beat as power meter	deformation sensor and clock	75 to 2 000 W
Fitness power meter	piezoelectric sensor	75 to 2 000 W
Electricity meter at home	rotating aluminium disc	20 to 10 000 W
Power meter for car engine	electromagnetic brake	up to 1 MW
Laser power meter	photoelectric effect in semiconductor	up to 10 GW
Calorimeter for chemical reactions	temperature sensor	up to 1 MW
Calorimeter for particles	light detector	up to a few µJ/ns

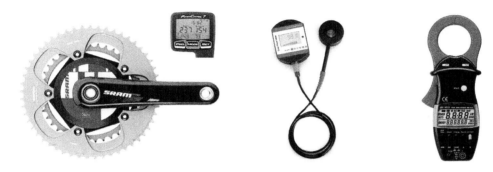

FIGURE 74 Some power measurement devices: a bicycle power meter, a laser power meter, and an electrical power meter (© SRAM, Laser Components, Wikimedia).

The cross product, or vector product

The discussion of rotation is easiest if we introduce an additional way to multiply vectors. This new product between two vectors a and b is called the *cross product* or *vector product* $a \times b$.

The result of the vector product is another *vector*; thus it differs from the *scalar* product, whose result is a scalar, i.e., a number. The result of the vector product is that vector
— that is orthogonal to both vectors to be multiplied,
— whose orientation is given by the *right-hand rule*, and
— whose length is given by the surface area of the parallelogram spanned by the two vectors, i.e., by $ab \sin \sphericalangle(a, b)$.

The definition implies that the cross product vanishes if and only if the vectors are parallel. From the definition you can also show that the vector product has the properties

$$\begin{aligned}
&\boldsymbol{a} \times \boldsymbol{b} = -\boldsymbol{b} \times \boldsymbol{a} \,, \quad \boldsymbol{a} \times (\boldsymbol{b} + \boldsymbol{c}) = \boldsymbol{a} \times \boldsymbol{b} + \boldsymbol{a} \times \boldsymbol{c} \,, \\
&\lambda \boldsymbol{a} \times \boldsymbol{b} = \lambda(\boldsymbol{a} \times \boldsymbol{b}) = \boldsymbol{a} \times \lambda \boldsymbol{b} \,, \quad \boldsymbol{a} \times \boldsymbol{a} = \boldsymbol{0} \,, \\
&\boldsymbol{a}(\boldsymbol{b} \times \boldsymbol{c}) = \boldsymbol{b}(\boldsymbol{c} \times \boldsymbol{a}) = \boldsymbol{c}(\boldsymbol{a} \times \boldsymbol{b}) \,, \quad \boldsymbol{a} \times (\boldsymbol{b} \times \boldsymbol{c}) = (\boldsymbol{a}\boldsymbol{c})\boldsymbol{b} - (\boldsymbol{a}\boldsymbol{b})\boldsymbol{c} \,, \\
&(\boldsymbol{a} \times \boldsymbol{b})(\boldsymbol{c} \times \boldsymbol{d}) = \boldsymbol{a}(\boldsymbol{b} \times (\boldsymbol{c} \times \boldsymbol{d})) = (\boldsymbol{a}\boldsymbol{c})(\boldsymbol{b}\boldsymbol{d}) - (\boldsymbol{b}\boldsymbol{c})(\boldsymbol{a}\boldsymbol{d}) \,, \\
&(\boldsymbol{a} \times \boldsymbol{b}) \times (\boldsymbol{c} \times \boldsymbol{d}) = ((\boldsymbol{a} \times \boldsymbol{b})\boldsymbol{d})\boldsymbol{c} - ((\boldsymbol{a} \times \boldsymbol{b})\boldsymbol{c})\boldsymbol{d} \,, \\
&\boldsymbol{a} \times (\boldsymbol{b} \times \boldsymbol{c}) + \boldsymbol{b} \times (\boldsymbol{c} \times \boldsymbol{a}) + \boldsymbol{c} \times (\boldsymbol{a} \times \boldsymbol{b}) = \boldsymbol{0} \,.
\end{aligned} \quad (22)$$

The vector product exists only in vector spaces with *three* dimensions. We will explore more details on this connection later on.

The vector product is useful to describe systems that *rotate* – and (thus) also systems with magnetic forces. The motion of an orbiting body is always perpendicular both to the axis and to the line that connects the body with the axis. In rotation, axis, radius and velocity form a right-handed set of mutually orthogonal vectors. This connection lies at the origin of the vector product.

Confirm that the best way to calculate the vector product $\boldsymbol{a} \times \boldsymbol{b}$ component by component is given by the symbolic determinant

$$\boldsymbol{a} \times \boldsymbol{b} = \begin{vmatrix} \boldsymbol{e}_x & a_x & b_x \\ \boldsymbol{e}_y & a_y & b_y \\ \boldsymbol{e}_z & a_z & b_z \end{vmatrix} \quad \text{or, sloppily} \quad \boldsymbol{a} \times \boldsymbol{b} = \begin{vmatrix} + & - & + \\ a_x & a_y & a_z \\ b_x & b_y & b_z \end{vmatrix} \,. \quad (23)$$

These symbolic determinants are easy to remember and easy to perform, both with letters and with numerical values. (Here, $\boldsymbol{e}_x$ is the unit basis vector in the x direction.) Written out, the symbolic determinants are equivalent to the relation

$$\boldsymbol{a} \times \boldsymbol{b} = (a_y b_z - b_y a_z, b_x a_z - a_x b_z, a_x b_y - b_x a_y) \quad (24)$$

which is harder to remember, though.

Show that the *parallelepiped* spanned by three arbitrary vectors $\boldsymbol{a}$, $\boldsymbol{b}$ and $\boldsymbol{c}$ has the volume $V = \boldsymbol{c}\,(\boldsymbol{a} \times \boldsymbol{b})$. Show that the *pyramid* or *tetrahedron* formed by the same three vectors has one sixth of that volume.

Rotation and angular momentum

Rotation keeps us alive. Without the change of day and night, we would be either fried or frozen to death, depending on our location on our planet. But rotation appears in many other settings, as Table 22 shows. A short exploration of rotation is thus appropriate.

All objects have the ability to rotate. We saw before that a body is described by its reluctance to move, which we called mass; similarly, a body also has a reluctance to turn. This quantity is called its *moment of inertia* and is often abbreviated Θ – pronounced 'theta'. The speed or rate of rotation is described by *angular velocity*, usually abbreviated ω – pronounced 'omega'. A few values found in nature are given in Table 22.

The observables that describe rotation are similar to those describing linear motion,

4 FROM OBJECTS AND IMAGES TO CONSERVATION

TABLE 22 Some measured rotation frequencies.

OBSERVATION	ANGULAR VELOCITY $\omega = 2\pi/T$
Galactic rotation	$2\pi \cdot 0.14 \cdot 10^{-15}/\text{s}$ = $2\pi/(220 \cdot 10^6 \text{ a})$
Average Sun rotation around its axis	$2\pi \cdot 3.8 \cdot 10^{-7}/\text{s} = 2\pi/30\,\text{d}$
Typical lighthouse	$2\pi \cdot 0.08/\text{s}$
Pirouetting ballet dancer	$2\pi \cdot 3/\text{s}$
Ship's diesel engine	$2\pi \cdot 5/\text{s}$
Helicopter rotor	$2\pi \cdot 5.3/\text{s}$
Washing machine	up to $2\pi \cdot 20/\text{s}$
Bacterial flagella	$2\pi \cdot 100/\text{s}$
Fast CD recorder	up to $2\pi \cdot 458/\text{s}$
Racing car engine	up to $2\pi \cdot 600/\text{s}$
Fastest turbine built	$2\pi \cdot 10^3/\text{s}$
Fastest pulsars (rotating stars)	up to at least $2\pi \cdot 716/\text{s}$
Ultracentrifuge	$> 2\pi \cdot 3 \cdot 10^3/\text{s}$
Dental drill	up to $2\pi \cdot 13 \cdot 10^3/\text{s}$
Technical record	$2\pi \cdot 333 \cdot 10^3/\text{s}$
Proton rotation	$2\pi \cdot 10^{20}/\text{s}$
Highest possible, Planck angular velocity	$2\pi \cdot 10^{35}/\text{s}$

as shown in Table 24. Like mass, the moment of inertia is defined in such a way that the sum of *angular momenta L* – the product of moment of inertia and angular velocity – is conserved in systems that do not interact with the outside world:

$$\sum_i \Theta_i \boldsymbol{\omega}_i = \sum_i \boldsymbol{L}_i = \text{const} . \tag{25}$$

In the same way that the conservation of linear momentum defines mass, the conservation of angular momentum defines the moment of inertia. Angular momentum is a concept introduced in the 1730s and 1740s by Leonhard Euler and Daniel Bernoulli.

The moment of inertia can be related to the mass and shape of a body. If the body is imagined to consist of small parts or mass elements, the resulting expression is

$$\Theta = \sum_n m_n r_n^2 , \tag{26}$$

where r_n is the distance from the mass element m_n to the axis of rotation. Can you confirm the expression? Therefore, the moment of inertia of a body depends on the chosen axis of rotation. Can you confirm that this is so for a brick?

In contrast to the case of mass, there is *no* conservation of the moment of inertia. In fact, the value of the moment of inertia depends both on the direction and on the

TABLE 23 Some measured angular momentum values.

Observation	Angular momentum
Smallest observed value in nature, $\hbar/2$, in elementary matter particles (fermions)	$0.53 \cdot 10^{-34}$ Js
Spinning top	$5 \cdot 10^{-6}$ Js
CD (compact disc) playing	c. 0.029 Js
Walking man (around body axis)	c. 4 Js
Dancer in a pirouette	5 Js
Typical car wheel at 30 m/s	10 Js
Typical wind generator at 12 m/s (6 Beaufort)	10^4 Js
Earth's atmosphere	1 to $2 \cdot 10^{26}$ Js
Earth's oceans	$5 \cdot 10^{24}$ Js
Earth around its axis	$7.1 \cdot 10^{33}$ Js
Moon around Earth	$2.9 \cdot 10^{34}$ Js
Earth around Sun	$2.7 \cdot 10^{40}$ Js
Sun around its axis	$1.1 \cdot 10^{42}$ Js
Jupiter around Sun	$1.9 \cdot 10^{43}$ Js
Solar System around Sun	$3.2 \cdot 10^{43}$ Js
Milky Way	10^{68} Js
All masses in the universe	0 (within measurement error)

TABLE 24 Correspondence between linear and rotational motion.

Quantity	Linear motion		Rotational motion	
State	time	t	time	t
	position	x	angle	φ
	momentum	$p = mv$	angular momentum	$L = \Theta\omega$
	energy	$mv^2/2$	energy	$\Theta\omega^2/2$
Motion	velocity	v	angular velocity	ω
	acceleration	a	angular acceleration	α
Reluctance to move	mass	m	moment of inertia	Θ
Motion change	force	ma	torque	$\Theta\alpha$

location of the axis used for its definition. For each axis direction, one distinguishes an *intrinsic* moment of inertia, when the axis passes through the centre of mass of the body, from an *extrinsic* moment of inertia, when it does not.* In the same way, we distinguish

---

* Extrinsic and intrinsic moment of inertia are related by

$$\Theta_{\text{ext}} = \Theta_{\text{int}} + md^2 \,, \tag{27}$$

where d is the distance between the centre of mass and the axis of extrinsic rotation. This relation is called *Steiner's parallel axis theorem*. Are you able to deduce it?

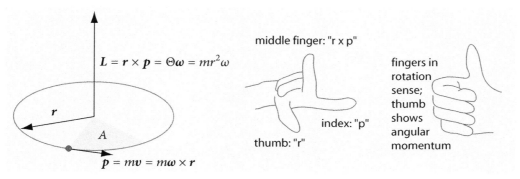

FIGURE 75 Angular momentum and other quantities for a point particle in circular motion, and the two versions of the right-hand rule.

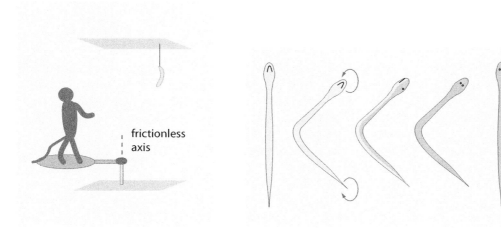

FIGURE 76 Can the ape reach the banana?

FIGURE 77 How a snake turns itself around its axis.

intrinsic and extrinsic angular momenta. (By the way, the *centre of mass* of a body is that imaginary point which moves straight during vertical fall, even if the body is rotating. Can you find a way to determine its location for a specific body?) Challenge 204 s

We now define the *rotational energy* as

$$E_{\text{rot}} = \tfrac{1}{2}\,\Theta\,\omega^2 = \frac{L^2}{2\Theta} \,. \tag{28}$$

The expression is similar to the expression for the kinetic energy of a particle. For rotating objects with fixed shape, rotational energy is conserved.

Can you guess how much larger the rotational energy of the Earth is compared with the yearly electricity usage of humanity? In fact, if you could find a way to harness the Earth's rotational energy, you would become famous. Challenge 205 s

Every object that has an orientation also has an intrinsic angular momentum. (What about a sphere?) Therefore, *point* particles do *not* have intrinsic angular momenta – at Challenge 206 s

least in classical physics. (This statement will change in quantum theory.) The *extrinsic* angular momentum L of a point particle is defined as

$$L = r \times p \qquad (29)$$

where p is the momentum of the particle and r the position vector. The angular momentum thus points along the rotation axis, following the right-hand rule, as shown in Figure 75. A few values observed in nature are given in Table 23. The definition implies that the angular momentum can also be determined using the expression

$$L = \frac{2A(t)m}{t}, \qquad (30)$$

where $A(t)$ is the area *swept* by the position vector r of the particle during time t. For example, by determining the swept area with the help of his telescope, Johannes Kepler discovered in the year 1609 that each planet orbiting the Sun has an angular momentum value that is *constant* over time.

A physical body can rotate simultaneously about *several* axes. The film of Figure 108 shows an example: The top rotates around its body axis and around the vertical at the same time. A detailed exploration shows that the exact rotation of the top is given by the *vector sum* of these two rotations. To find out, 'freeze' the changing rotation axis at a specific time. Rotations thus are a type of vectors.

As in the case of linear motion, rotational energy and angular momentum are not always conserved in the macroscopic world: rotational energy can change due to friction, and angular momentum can change due to external forces (torques). But for *closed* (undisturbed) systems, both angular momentum and rotational energy are always conserved. In particular, on a microscopic scale, most objects are undisturbed, so that conservation of rotational energy and angular momentum usually holds on microscopic scales.

Angular momentum is conserved. This statement is valid for any axis of a physical system, *provided* that external forces (torques) play no role. To make the point, Jean-Marc Lévy-Leblond poses the problem of Figure 76. Can the ape reach the banana without leaving the plate, assuming that the plate on which the ape rests can turn around the axis without any friction?

We note that many effects of rotation are the same as for acceleration: both acceleration and rotation of a car pushed us in our seat. Therefore, many sensors for rotation are the same as the acceleration sensors we explored above. But a few sensors for rotation are fundamentally new. In particular, we will meet the gyroscope shortly.

On a frictionless surface, as approximated by smooth ice or by a marble floor covered by a layer of oil, it is impossible to move forward. In order to move, we need to push *against* something. Is this also the case for rotation?

Surprisingly, it is possible to turn even *without* pushing against something. You can check this on a well-oiled rotating office chair: simply rotate an arm above the head. After each turn of the hand, the orientation of the chair has changed by a small amount. Indeed, conservation of angular momentum and of rotational energy do *not* prevent bodies from changing their orientation. Cats learn this in their youth. After they have learned

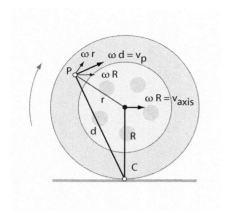

FIGURE 78 The velocities and unit vectors for a rolling wheel.

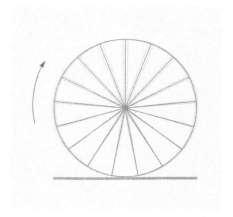

FIGURE 79 A simulated photograph of a rolling wheel with spokes.

the trick, if cats are dropped legs up, they can turn themselves in such a way that they always land feet first. Snakes also know how to rotate themselves, as Figure 77 shows. Also humans have the ability: during the Olympic Games you can watch board divers and gymnasts perform similar tricks. Rotation thus differs from translation in this important aspect. (Why?)

Rolling wheels

Rotation is an interesting phenomenon in many ways. A rolling wheel does *not* turn around its axis, but around its point of contact. Let us show this.

A wheel of radius R is *rolling* if the speed of the axis v_{axis} is related to the angular velocity ω by

$$\omega = \frac{v_{\text{axis}}}{R} \, . \tag{31}$$

For any point P on the wheel, with distance r from the axis, the velocity v_P is the sum of the motion of the axis and the motion around the axis. Figure 78 shows that v_P is orthogonal to d, the distance between the point P and the contact point of the wheel. The figure also shows that the length ratio between v_P and d is the same as between v_{axis} and R. As a result, we can write

$$\boldsymbol{v}_P = \boldsymbol{\omega} \times \boldsymbol{d} \, , \tag{32}$$

which shows that a rolling wheel does indeed rotate about its point of contact with the ground.

Surprisingly, when a wheel rolls, some points on it move *towards* the wheel's axis, some stay at a *fixed* distance and others move *away* from it. Can you determine where these various points are located? Together, they lead to an interesting pattern when a rolling wheel with spokes, such as a bicycle wheel, is photographed, as show in Figure 79.

With these results you can tackle the following beautiful challenge. When a turning bicycle wheel is deposed on a slippery surface, it will slip for a while, then slip and roll,

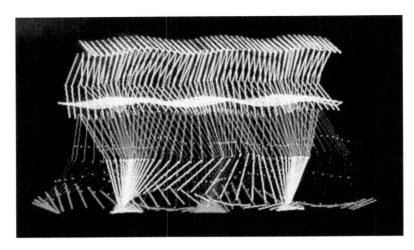

FIGURE 80 The measured motion of a walking human (© Ray McCoy).

and finally roll only. How does the final speed depend on the initial speed and on the friction?

How do we walk and run?

> Golf is a good walk spoiled.
> The Allens

Why do we move our arms when walking or running? To save energy or to be graceful? In fact, whenever a body movement is performed with as little energy as possible, it is both natural and graceful. This correspondence can indeed be taken as the actual definition of grace. The connection is common knowledge in the world of dance; it is also a central aspect of the methods used by actors to learn how to move their bodies as beautifully as possible.

To convince yourself about the energy savings, try walking or running with your arms fixed or moving in the opposite direction to usual: the effort required is considerably higher. In fact, when a leg is moved, it produces a torque around the body axis which has to be counterbalanced. The method using the least energy is the swinging of arms, as depicted in Figure 80. Since the arms are lighter than the legs, they must move further from the axis of the body, to compensate for the momentum; evolution has therefore moved the attachment of the arms, the shoulders, farther apart than those of the legs, the hips. Animals on two legs but without arms, such as penguins or pigeons, have more difficulty walking; they have to move their whole torso with every step.

Measurements show that all walking animals follow

$$v_{\text{max walking}} = (2.2 \pm 0.2 \text{ m/s}) \sqrt{l/\text{m}} . \tag{33}$$

Indeed, walking, the moving of one leg after the other, can be described as a concatenation of (inverted) pendulum swings. The pendulum length is given by the leg length l. The typical time scale of a pendulum is $t \sim \sqrt{l/g}$. The maximum speed of walking then becomes $v \sim l/t \sim \sqrt{gl}$, which is, up to a constant factor, the measured result.

Which muscles do most of the work when walking, the motion that experts call *gait*? In 1980, Serge Gracovetsky found that in human gait a large fraction of the power comes from the muscles along the *spine*, not from those of the legs. (Indeed, people without legs are also able to walk. However, a number of muscles in the legs must work in order to walk normally.) When you take a step, the lumbar muscles straighten the spine; this automatically makes it turn a bit to one side, so that the knee of the leg on that side automatically comes forward. When the foot is moved, the lumbar muscles can relax, and then straighten again for the next step. In fact, one can experience the increase in tension in the *back* muscles when walking without moving the arms, thus confirming where the human engine, the so-called *spinal engine* is located.

Human legs differ from those of apes in a fundamental aspect: humans are able to *run*. In fact the whole human body has been optimized for running, an ability that no other primate has. The human body has shed most of its hair to achieve better cooling, has evolved the ability to run while keeping the head stable, has evolved the right length of arms for proper balance when running, and even has a special ligament in the back that works as a shock absorber while running. In other words, running is the most human of all forms of motion.

CURIOSITIES AND FUN CHALLENGES ABOUT MASS, CONSERVATION AND ROTATION

> It is a mathematical fact that the casting of this pebble from my hand alters the centre of gravity of the universe.
> Thomas Carlyle,* *Sartor Resartus III.*

A cup with water is placed on a weighing scale, as shown in Figure 81. How does the mass result change if you let a piece of metal attached to a string hang into the water?

* *

Take ten coins of the same denomination. Put nine of them on a table and form a closed loop with them of any shape you like, a shown in Figure 82. (The nine coins thus look like a section of pearl necklace where the pearls touch each other.) Now take then tenth coin and let it roll around the loop, thus without ever sliding it. How many turns does this last coin make during one round?

* *

Conservation of momentum is best studied playing and exploring billiards, snooker or pool. The best introduction are the trickshot films found across the internet. Are you able to use momentum conservation to deduce ways for improving your billiards game?

Another way to explore momentum conservation is to explore the ball-chain, or ball collision pendulum, that was invented by Edme Mariotte. Decades later, Newton claimed it as his, as he often did with other people's results. Playing with the toy is fun – and explaining its behaviour even more. Indeed, if you lift and let go three balls on one side, you will see three balls departing on the other side; for the explanation of this behaviour

---

* Thomas Carlyle (1797–1881), Scottish essayist. Do you agree with the quotation?

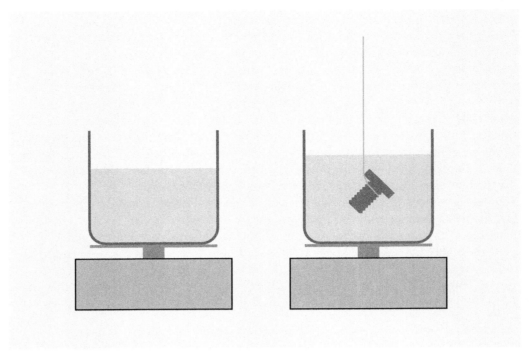

FIGURE 81 How does the displayed weight value change when an object hangs into the water?

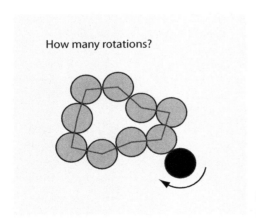

FIGURE 82 How many rotations does the tenth coin perform in one round?

the conservation of momentum and energy conservation are *not* sufficient, as you should be able to find out. Are you able to build a high-precision ball-chain?

Challenge 219 d

* *

Challenge 220 s There is a well-known way to experience 81 sunrises in just 80 days. How?

* *

Walking is a source of many physics problems. When climbing a mountain, the most

FIGURE 83 The ball-chain or cradle invented by Mariotte allows to explore momentum conservation, energy conservation, and the difficulties of precision manufacturing (© www.questacon.edu.au).

energy-effective way is not always to follow the steepest ascent; indeed, for steep slopes, zig-zagging is more energy efficient. Why? And can you estimate the slope angle at which this will happen?

* *

Asterix and his friends from the homonymous comic strip, fear only one thing: that the sky might fall down. Is the sky an object? An image?

* *

Death is a physical process and thus can be explored. In general, animals have a *lifespan* T that scales with fourth root of their mass M. In other terms, $T = M^{1/4}$. This is valid from bacteria to insects to blue whales. Animals also have a power consumption per mass, or *metabolic rate* per mass, that scales with the *inverse* fourth root. We conclude that death occurs for all animals when a certain fixed energy consumption per mass has been achieved. This is indeed the case; death occurs for most animals when they have consumed around 1 GJ/kg. (But quite a bit later for humans.) This surprisingly simple result is valid, *on average*, for all known animals.

Note that the argument is only valid when *different* species are compared. The dependence on mass is *not* valid when specimen of the same species are compared. (You cannot live longer by eating less.)

In short, animals die after they metabolized 1 GJ/kg. In other words, once we ate all the calories we were designed for, we die.

* *

A car at a certain speed uses 7 litres of gasoline per 100 km. What is the combined air and rolling resistance? (Assume that the engine has an efficiency of 25 %.)

* *

A cork is attached to a thin string a metre long. The string is passed over a long rod held horizontally, and a wine glass is attached at the other end. If you let go the cork in

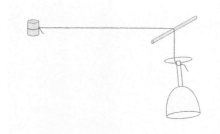

FIGURE 84 Is it safe to let the cork go?

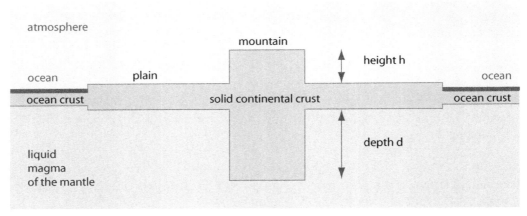

FIGURE 85 A simple model for continents and mountains.

Figure 84, nothing breaks. Why not? And what happens exactly?

* *

In 1907, Duncan MacDougalls, a medical doctor, measured the weight of dying people, in the hope to see whether death leads to a mass change. He found a sudden decrease between 10 and 20 g at the moment of death. He attributed it to the soul exiting the body. Can you find a more satisfying explanation?

* *

It is well known that the weight of a one-year old child depends on whether it wants to be carried or whether it wants to reach the floor. Does this contradict mass conservation?

* *

The Earth's crust is less dense (2.7 kg/l) than the Earth's mantle (3.1 kg/l) and floats on it. As a result, the lighter crust below a mountain ridge must be much deeper than below a plain. If a mountain rises 1 km above the plain, how much deeper must the crust be below it? The simple block model shown in Figure 85 works fairly well; first, it explains why, near mountains, measurements of the deviation of free fall from the vertical line lead to so much lower values than those expected without a deep crust. Later, sound measurements have confirmed directly that the continental crust is indeed thicker beneath mountains.

∗ ∗

All homogeneous cylinders roll down an inclined plane in the same way. True or false? And what about spheres? Can you show that spheres roll faster than cylinders?

∗ ∗

Which one rolls faster: a soda can filled with liquid or a soda can filled with ice? (And how do you make a can filled with ice?)

∗ ∗

Take two cans of the same size and weight, one full of ravioli and one full of peas. Which one rolls faster on an inclined plane?

∗ ∗

Another difference between matter and images: matter smells. In fact, the nose is a matter sensor. The same can be said of the tongue and its sense of taste.

∗ ∗

Take a pile of coins. You can push out the coins, starting with the one at the bottom, by shooting another coin over the table surface. The method also helps to visualize two-dimensional momentum conservation.

∗ ∗

In early 2004, two men and a woman earned £ 1.2 million in a single evening in a London casino. They did so by applying the formulae of Galilean mechanics. They used the method pioneered by various physicists in the 1950s who built various small computers that could predict the outcome of a roulette ball from the initial velocity imparted by the croupier. In the case in Britain, the group added a laser scanner to a smart phone that measured the path of a roulette ball and predicted the numbers where it would arrive. In this way, they increased the odds from 1 in 37 to about 1 in 6. After six months of investigations, Scotland Yard ruled that they could keep the money they won.

In fact around the same time, a few people earned around 400 000 euro over a few weeks by using the same method in Germany, but with no computer at all. In certain casinos, machines were throwing the roulette ball. By measuring the position of the zero to the incoming ball with the naked eye, these gamblers were able to increase the odds of the bets they placed during the last allowed seconds and thus win a considerable sum purely through fast reactions.

∗ ∗

Does the universe rotate?

∗ ∗

The toy of Figure 86 shows interesting behaviour: when a number of spheres are lifted and dropped to hit the resting ones, the same number of spheres detach on the other side, whereas the previously dropped spheres remain motionless. At first sight, all this seems

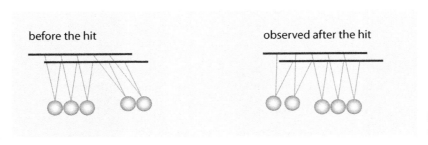

FIGURE 86 A well-known toy.

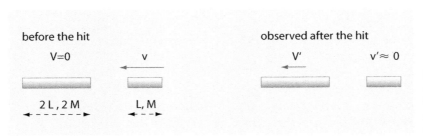

FIGURE 87 An elastic collision that seems not to obey energy conservation.

to follow from energy and momentum conservation. However, energy and momentum conservation provide only two equations, which are insufficient to explain or determine the behaviour of five spheres. Why then do the spheres behave in this way? And why do they all swing in phase when a longer time has passed?

* *

A surprising effect is used in home tools such as hammer drills. We remember that when a small ball elastically hits a large one at rest, both balls move after the hit, and the small one obviously moves faster than the large one. Despite this result, when a short cylinder hits a long one of the same diameter and material, but with a length that is some *integer* multiple of that of the short one, something strange happens. After the hit, the small cylinder remains almost at rest, whereas the large one moves, as shown in Figure 87. Even though the collision is elastic, conservation of energy seems not to hold in this case. (In fact this is the reason that demonstrations of elastic collisions in schools are always performed with spheres.) What happens to the energy?

* *

Is the structure shown in Figure 88 possible?

* *

Does a wall get a stronger jolt when it is hit by a ball rebounding from it or when it is hit by a ball that remains stuck to it?

* *

Housewives know how to extract a cork of a wine bottle using a cloth or a shoe. Can you imagine how? They also know how to extract the cork with the cloth if the cork has fallen

FIGURE 88 Is this possible?

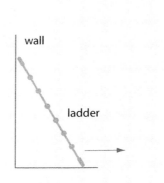

FIGURE 89 How does the ladder fall?

inside the bottle. How?

* *

The sliding ladder problem, shown schematically in Figure 89, asks for the detailed motion of the ladder over time. The problem is more difficult than it looks, even if friction is not taken into account. Can you say whether the lower end always touches the floor, or if is lifted into the air for a short time interval?

* *

A homogeneous ladder of length 5 m and mass 30 kg leans on a wall. The angle is 30°; the static friction coefficient on the wall is negligible, and on the floor it is 0.3. A person of mass 60 kg climbs the ladder. What is the maximum height the person can climb before the ladder starts sliding? This and many puzzles about ladders can be found on www.mathematische-basteleien.de/leiter.htm.

* *

A common fly on the stern of a 30 000 ton ship of 100 m length tilts it by less than the diameter of an atom. Today, distances that small are easily measured. Can you think of at least two methods, one of which should not cost more than 2000 euro?

* *

Is the image of three stacked spinning tops shown in Figure 90 a true photograph, showing a real observation, or is it the result of digital composition, showing an impossible situation?

* *

How does the kinetic energy of a rifle bullet compare to that of a running man?

* *

What happens to the size of an egg when one places it in a jar of vinegar for a few days?

FIGURE 90 Is this a possible situation or is it a fake photograph? (© Wikimedia)

* *

Challenge 242 s   What is the amplitude of a pendulum oscillating in such a way that the absolute value of its acceleration at the lowest point and at the return point are equal?

* *

Challenge 243 d   Can you confirm that the value of the acceleration of a drop of water falling through mist is $g/7$?

* *

Challenge 244 s   You have two hollow spheres: they have the same weight, the same size and are painted in the same colour. One is made of copper, the other of aluminium. Obviously, they fall with the same speed and acceleration. What happens if they both roll down a tilted plane?

* *

Challenge 245 s   What is the shape of a rope when rope jumping?

* *

Challenge 246 s   How can you determine the speed of a rifle bullet with only a scale and a metre stick?

* *

Challenge 247 e   Why does a gun make a hole in a door but cannot push it open, in exact contrast to what a finger can do?

* *

Challenge 248 s   What is the curve described by the mid point of a ladder sliding down a wall?

* *

Challenge 249 s   A high-tech company, see www.enocean.com, sells electric switches for room lights that have no cables and no power cell (battery). You can glue such a switch to the centre of a window pane. How is this possible?

4 FROM OBJECTS AND IMAGES TO CONSERVATION

FIGURE 91 A commercial clock that needs no special energy source, because it takes its energy from the environment (© Jaeger-LeCoultre).

* *

For over 50 years now, a famous Swiss clock maker is selling table clocks with a rotating pendulum that need no battery and no manual rewinding, as they take up energy from the environment. A specimen is shown in Figure 91. Can you imagine how this clock works?

* *

Ship lifts, such as the one shown in Figure 92, are impressive machines. How does the weight of the lift change when the ship enters?

* *

How do you measure the mass of a ship?

* *

All masses are measured by comparing them, directly or indirectly, to the *standard kilogram* in Sèvres near Paris. Since a few years, there is the serious doubt that the standard kilogram is losing weight, possibly through outgassing, with an estimated rate of around 0.5 μg/a. This is an awkward situation, and there is a vast, world-wide effort to find a better definition of the kilogram. Such an improved definition must be simple, precise, and make trips to Sèvres unnecessary. No such alternative has been defined yet.

FIGURE 92 The spectacular ship lift at Strépy-Thieux in Belgium. What engine power is needed to lift a ship, if the right and left lifts were connected by ropes or by a hydraulic system? (© Jean-Marie Hoornaert)

FIGURE 93 The famous Celtic wobble stone – above and right – and a version made by bending a spoon – bottom left (© Ed Keath).

* *

Which engine is more efficient: a moped or a human on a bicycle?

* *

Both mass and moment of inertia can be defined and measured both with and without contact. Can you do so?

* *

Figure 93 shows the so-called *Celtic wobble stone*, also called *anagyre* or *rattleback*, a stone that starts rotating on a plane surface when it is put into up-and-down oscillation. The size can vary between a few centimetres and a few metres. By simply bending a spoon one can realize a primitive form of this strange device, if the bend is not completely symmetrical. The rotation is always in the same direction. If the stone is put into rotation in the wrong direction, after a while it stops and starts rotating in the other sense! Can you explain the effect that seems to contradict the conservation of angular momentum?

* *

A beautiful effect, the *chain fountain*, was discovered in 2013 by Steve Mould. Certain chains, when flowing out of a container, first shoot up in the air. See the video at www.youtube.com/embed/_dQJBBklpQQ and the story of the discovery at stevemould.com. Can you explain the effect to your grandmother?

Summary on conservation in motion

> The gods are not as rich as one might think: what they give to one, they take away from the other.
>
> Antiquity

We have encountered four conservation principles that are valid for the motion of all closed systems in everyday life:

— conservation of total linear momentum,
— conservation of total angular momentum,
— conservation of total energy,
— conservation of total mass.

None of these conservation principles applies to the motion of images. These principles thus allow us to distinguish objects from images.

The conservation principles are among the great results in science. They limit the surprises that nature can offer: conservation means that linear momentum, angular momentum, and mass–energy can neither be created from nothing, nor can they disappear into nothing. Conservation limits creation. The quote below the section title expresses this idea.

Later on we will find out that these results could have been deduced from three simple observations: closed systems behave the same independently of where they are, in what direction they are oriented and of the time at which they are set up. In more abstract terms, physicists like to say that all conservation principles are consequences of the *invariances*, or *symmetries*, of nature.

Later on, the theory of special relativity will show that energy and mass are conserved only when taken together. Many adventures still await us.

Chapter 5
FROM THE ROTATION OF THE EARTH TO THE RELATIVITY OF MOTION

> "Eppur si muove!"
> Anonymous**

Is the Earth rotating? The search for definite answers to this question gives an interesting cross section of the history of classical physics. In the fourth century, in ancient Greece, Hicetas and Philolaus, already stated that the Earth rotates. Then, in the year 265 BCE, Aristarchus of Samos was the first to explore the issue in detail. He had measured the parallax of the Moon (today known to be up to 0.95°) and of the Sun (today known to be 8.8′).*** The *parallax* is an interesting effect; it is the angle describing the difference between the directions of a body in the sky when seen by an observer on the surface of the Earth and when seen by a hypothetical observer at the Earth's centre. (See Figure 94.) Aristarchus noticed that the Moon and the Sun *wobble* across the sky, and this wobble has a period of 24 hours. He concluded that the Earth rotates. It seems that Aristarchus received death threats for his conclusion.

Aristarchus' observation yields an even more powerful argument than the trails of the stars shown in Figure 95. Can you explain why? (And how do the trails look at the most populated places on Earth?)

Experiencing Figure 95 might be one reason that people dreamt and still dream about reaching the poles. Because the rotation and the motion of the Earth makes the poles extremely cold places, the adventure of reaching them is not easy. Many tried unsuccessfully. A famous crook, Robert Peary, claimed to have reached the North Pole in 1909. (In fact, Roald Amundsen reached both the South and the North Pole first.) Among others, Peary claimed to have taken a picture there, but that picture, which went round the world, turned out to be one of the proofs that he had not been there. Can you imagine how?

If the Earth rotates instead of being at rest, said the unconvinced, the speed at the equator has the substantial value of 0.46 km/s. How did Galileo explain why we do not feel or notice this speed?

Measurements of the aberration of light also show the rotation of the Earth; it can be detected with a telescope while looking at the stars. The *aberration* is a change of the

---

** 'And yet she moves' is the sentence about the Earth attributed, most probably incorrectly, to Galileo since the 1640s. It is true, however, that at his trial he was forced to publicly retract the statement of a moving Earth to save his life. For more details of this famous story, see the section on page 335.
*** For the definition of the concept of angle, see page 67, and for the definition of the measurement units for angle see Appendix B.

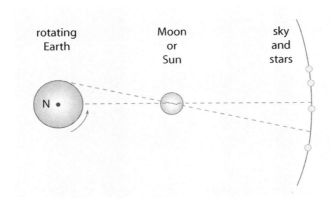

FIGURE 94 The parallax – not drawn to scale.

expected light direction, which we will discuss shortly. At the Equator, Earth rotation adds an angular deviation of 0.32 ′, changing sign every 12 hours, to the aberration due to the motion of the Earth around the Sun, about 20.5 ′. In modern times, astronomers have found a number of additional proofs for the rotation of the Earth, but none is accessible to the man on the street.

Also the measurements showing that the Earth is not a sphere, but is *flattened* at the poles, confirmed the rotation of the Earth. Figure 96 illustrates the situation. Again, however, this eighteenth century measurement by Maupertuis* is not accessible to everyday observation.

Then, in the years 1790 to 1792 in Bologna, Giovanni Battista Guglielmini (1763–1817) finally succeeded in measuring what Galileo and Newton had predicted to be the simplest proof for the Earth's rotation. On the rotating Earth, *objects do not fall vertically*, but are slightly deviated to the east. This deviation appears because an object keeps the larger horizontal velocity it had at the height from which it started falling, as shown in Figure 97. Guglielmini's result was the first non-astronomical proof of the Earth's rotation. The experiments were repeated in 1802 by Johann Friedrich Benzenberg (1777–1846). Using metal balls which he dropped from the Michaelis tower in Hamburg – a height of 76 m – Benzenberg found that the deviation to the east was 9.6 mm. Can you confirm that the value measured by Benzenberg almost agrees with the assumption that the Earth turns once every 24 hours? There is also a much smaller deviation towards the Equator, not measured by Guglielmini, Benzenberg or anybody after them up to this day; however, it completes the list of effects on free fall by the rotation of the Earth.

Both deviations from vertical fall are easily understood if we use the result (described below) that falling objects describe an ellipse around the centre of the rotating Earth. The elliptical shape shows that the path of a thrown stone does not lie on a plane for an

---

* Pierre Louis Moreau de Maupertuis (1698–1759), physicist and mathematician, was one of the key figures in the quest for the principle of least action, which he named in this way. He was also founding president of the Berlin Academy of Sciences. Maupertuis thought that the principle reflected the maximization of goodness in the universe. This idea was thoroughly ridiculed by Voltaire in his *Histoire du Docteur Akakia et du natif de Saint-Malo*, 1753. (Read it at www.voltaire-integral.com/Html/23/08DIAL.htm.) Maupertuis performed his measurement of the Earth to distinguish between the theory of gravitation of Newton and that of Descartes, who had predicted that the Earth is elongated at the poles, instead of flattened.

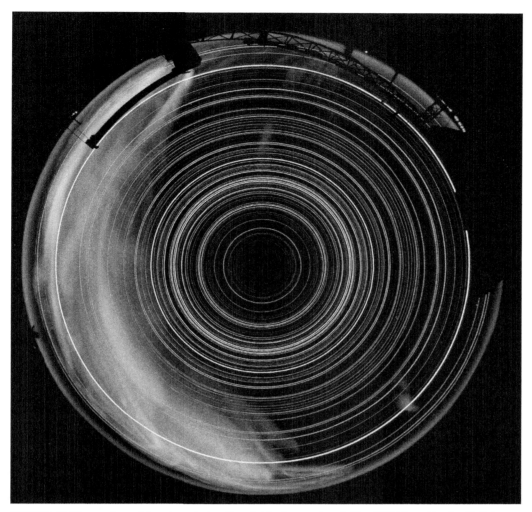

FIGURE 95 The motion of the stars during the night, observed on 1 May 2012 from the South Pole, together with the green light of an aurora australis (© Robert Schwartz).

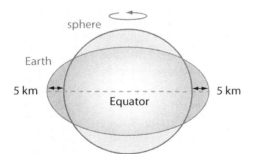

FIGURE 96 Earth's deviation from spherical shape due to its rotation (exaggerated).

observer standing on Earth; for such an observer, the exact path of a stone thus cannot

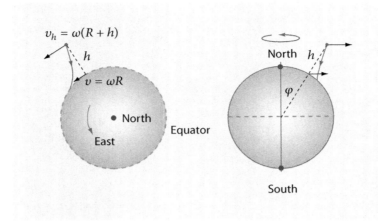

FIGURE 97 The deviations of free fall towards the east and towards the Equator due to the rotation of the Earth.

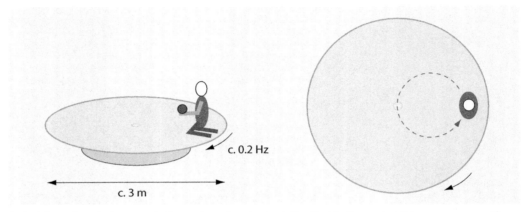

FIGURE 98 A typical carousel allows observing the Coriolis effect in its most striking appearance: if a person lets a ball roll with the proper speed and direction, the ball is deflected so strongly that it comes back to her.

be drawn on a flat piece of paper!

In 1798, Pierre Simon Laplace* explained how bodies move on the rotating Earth and showed that they feel an apparent force. In 1835, Gustave-Gaspard Coriolis then reformulated and simplified the description. Imagine a ball that rolls over a table. For a person on the floor, the ball rolls in a straight line. Now imagine that the table rotates. For the person on the floor, the ball still rolls in a straight line. But for a person on the rotating table, the ball traces a *curved* path. In short, any object that travels in a rotating background is subject to a transversal acceleration. The acceleration, discovered by Laplace, is nowadays called *Coriolis acceleration* or *Coriolis effect*. On a rotating background, travelling objects deviate from the straight line. The best way to understand the Coriolis effect is to exper-

---

* Pierre Simon Laplace (b. 1749 Beaumont-en-Auge, d. 1827 Paris), important mathematician. His famous treatise *Traité de mécanique céleste* appeared in five volumes between 1798 and 1825. He was the first to propose that the Solar System was formed from a rotating gas cloud, and one of the first people to imagine and explore black holes.

FIGURE 99 Cyclones, with their low pressure centre, differ in rotation sense between the southern hemisphere, here cyclone Larry in 2006, and the northern hemisphere, here hurricane Katrina in 2005. (Courtesy NOAA)

ience it yourself; this can be done on a carousel, as shown in Figure 98. Watching films on the internet on the topic is also helpful. You will notice that on a rotating carousel it is not easy to hit a target by throwing or rolling a ball.

Also the Earth is a rotating background. On the northern hemisphere, the rotation is anticlockwise. As the result, any moving object is slightly deviated to the right (while the magnitude of its velocity stays constant). On Earth, like on all rotating backgrounds, the *Coriolis acceleration* $\boldsymbol{a}_\text{C}$ results from the change of distance to the rotation axis. Can you deduce the analytical expression for the Coriolis effect, namely $\boldsymbol{a}_\text{C} = -2\boldsymbol{\omega} \times \boldsymbol{v}$?

On Earth, the Coriolis acceleration generally has a small value. Therefore it is best observed either in large-scale or high-speed phenomena. Indeed, the Coriolis acceleration determines the handedness of many large-scale phenomena with a spiral shape, such as the directions of cyclones and anticyclones in meteorology – as shown in Figure 99 – the general wind patterns on Earth and the deflection of ocean currents and tides. These phenomena have opposite handedness on the northern and the southern hemisphere. Most beautifully, the Coriolis acceleration explains why icebergs do not follow the direction of the wind as they drift away from the polar caps. The Coriolis acceleration also plays a role in the flight of cannon balls (that was the original interest of Coriolis), in satellite launches, in the motion of sunspots and even in the motion of electrons in molecules. All these Coriolis accelerations are of opposite sign on the northern and southern hemispheres and thus prove the rotation of the Earth. For example, in the First World War, many naval guns missed their targets in the southern hemisphere because the engineers had compensated them for the Coriolis effect in the northern hemisphere.

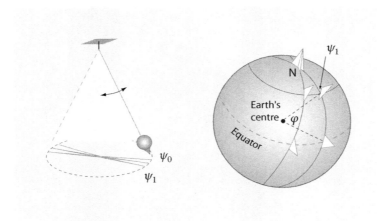

FIGURE 100 The turning motion of a pendulum showing the rotation of the Earth.

Only in 1962, after several earlier attempts by other researchers, Asher Shapiro was the first to verify that the Coriolis effect has a tiny influence on the direction of the vortex formed by the water flowing out of a bath-tub. Instead of a normal bath-tub, he had to use a carefully designed experimental set-up because, contrary to an often-heard assertion, no such effect can be seen in a real bath-tub. He succeeded only by carefully eliminating all disturbances from the system; for example, he waited 24 hours after the filling of the reservoir (and never actually stepped in or out of it!) in order to avoid any left-over motion of water that would disturb the effect, and built a carefully designed, completely rotationally-symmetric opening mechanism. Others have repeated the experiment in the southern hemisphere, finding opposite rotation direction and thus confirming the result. In other words, the handedness of usual bath-tub vortices is *not* caused by the rotation of the Earth, but results from the way the water starts to flow out. (A number of crooks in Quito, a city located on the Equator, show gullible tourists that the vortex in a sink changes when crossing the Equator line drawn on the road.) But let us go on with the story about the Earth's rotation.

In 1851, the physician-turned-physicist Jean Bernard Léon Foucault (b. 1819 Paris, d. 1868 Paris) performed an experiment that removed all doubts and rendered him world-famous practically overnight. He suspended a 67 m long pendulum* in the Panthéon in Paris and showed the astonished public that the direction of its swing changed over time, rotating slowly. To anybody with a few minutes of patience to watch the change of direction, the experiment proved that the Earth rotates. If the Earth did not rotate, the swing of the pendulum would always continue in the same direction. On a rotating Earth, in Paris, the direction changes to the right, in clockwise sense, as shown in Figure 100. The swing direction does not change if the pendulum is located at the Equator, and it changes to the left in the southern hemisphere.** A modern version of the

---

* Why was such a long pendulum necessary? Understanding the reasons allows one to repeat the experiment at home, using a pendulum as short as 70 cm, with the help of a few tricks. To observe Foucault's effect with a simple set-up, attach a pendulum to your office chair and rotate the chair slowly. Several pendulum animations, with exaggerated deviation, can be found at commons.wikimedia.org/wiki/Foucault_pendulum.
** The discovery also shows how precision and genius go together. In fact, the first person to observe the

pendulum can be observed via the web cam at pendelcam.kip.uni-heidelberg.de; high speed films of the pendulum's motion during day and night can also be downloaded at www.kip.uni-heidelberg.de/oeffwiss/pendel/zeitraffer/.

The time over which the orientation of the pendulum's swing performs a full turn – the *precession time* – can be calculated. Study a pendulum starting to swing in the North–South direction and you will find that the precession time T_{Foucault} is given by

$$T_{\text{Foucault}} = \frac{23 \text{ h } 56 \text{ min}}{\sin \varphi} \qquad (34)$$

where φ is the latitude of the location of the pendulum, e.g. 0° at the Equator and 90° at the North Pole. This formula is one of the most beautiful results of Galilean kinematics.*

Foucault was also the inventor and namer of the *gyroscope*. He built the device, shown in Figure 101 and Figure 102, in 1852, one year after his pendulum. With it, he again demonstrated the rotation of the Earth. Once a gyroscope rotates, the axis stays fixed in space – but only when seen from distant stars or galaxies. (By the way, this is not the same as talking about absolute space. Why?) For an observer on Earth, the axis direction changes regularly with a period of 24 hours. Gyroscopes are now routinely used in ships and in aeroplanes to give the direction of north, because they are more precise and more reliable than magnetic compasses. The most modern versions use laser light running in circles instead of rotating masses.**

In 1909, Roland von Eötvös measured a small but surprising effect: due to the rotation of the Earth, the weight of an object depends on the direction in which it moves. As a result, a balance in rotation around the vertical axis does not stay perfectly horizontal: the balance starts to oscillate slightly. Can you explain the origin of the effect?

In 1910, John Hagen published the results of an even simpler experiment, proposed by Louis Poinsot in 1851. Two masses are put on a horizontal bar that can turn around a vertical axis, a so-called *isotomeograph*. Its total mass was 260 kg. If the two masses are slowly moved towards the support, as shown in Figure 103, and if the friction is kept low enough, the bar rotates. Obviously, this would not happen if the Earth were not rotating. Can you explain the observation? This little-known effect is also useful for winning bets between physicists.

In 1913, Arthur Compton showed that a closed tube filled with water and some small floating particles (or bubbles) can be used to show the rotation of the Earth. The device is called a *Compton tube* or *Compton wheel*. Compton showed that when a horizontal tube filled with water is rotated by 180°, something happens that allows one to prove that the Earth rotates. The experiment, shown in Figure 104, even allows measuring the latitude of the point where the experiment is made. Can you guess what happens?

Another method to detect the rotation of the Earth using light was first realized in 1913 Georges Sagnac:*** he used an *interferometer* to produce bright and dark fringes of light

---

effect was Vincenzo Viviani, a student of Galileo, as early as 1661! Indeed, Foucault had read about Viviani's work in the publications of the Academia dei Lincei. But it took Foucault's genius to connect the effect to the rotation of the Earth; nobody had done so before him.

* The calculation of the period of Foucault's pendulum assumes that the precession rate is constant during a rotation. This is only an approximation (though usually a good one).
** Can you guess how rotation is detected in this case?
*** Georges Sagnac (b. 1869 Périgeux, d. 1928 Meudon-Bellevue) was a physicist in Lille and Paris, friend

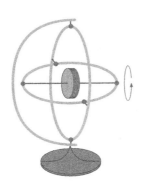

FIGURE 101 The gyroscope: the original system by Foucault with its freely movable spinning top, the mechanical device to bring it to speed, the optical device to detect its motion, the general construction principle, and a modern (triangular) ring laser gyroscope, based on colour change of rotating laser light instead of angular changes of a rotating mass (© CNAM, JAXA).

with two light beams, one circulating in clockwise direction, and the second circulating in anticlockwise direction. The interference fringes are *shifted* when the whole system rotates; the faster it rotates, the larger is the shift. A modern, high-precision version of the experiment, which uses lasers instead of lamps, is shown in Figure 105. (More details on interference and fringes are found in volume III.) Sagnac also determined the relation between the fringe shift and the details of the experiment. The rotation of a complete ring interferometer with angular frequency (vector) Ω produces a fringe shift of angular phase $\Delta\varphi$ given by

$$\Delta\varphi = \frac{8\pi\,\Omega\,a}{c\,\lambda} \qquad (35)$$

where a is the area (vector) enclosed by the two interfering light rays, λ their wavelength and c the speed of light. The effect is now called the *Sagnac effect* after its discoverer. It had

---

of the Curies, Langevin, Perrin, and Borel. Sagnac also deduced from his experiment that the speed of light was independent from the speed of its source, and thus confirmed a prediction of special relativity.

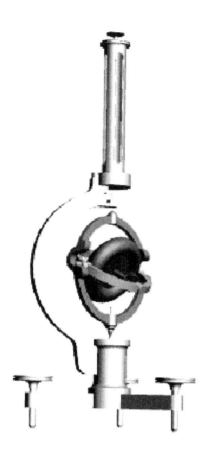

FIGURE 102 A three-dimensional model of Foucault's original gyroscope: in the pdf verion of this text, the model can be rotated and zoomed by moving the cursor over it (© Zach Joseph Espiritu).

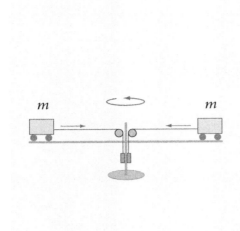
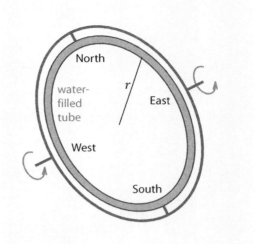

FIGURE 103 Showing the rotation of the Earth through the rotation of an axis.

FIGURE 104 Demonstrating the rotation of the Earth with water.

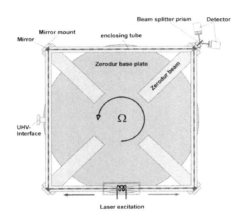

FIGURE 105 A modern precision ring laser interferometer (© Bundesamt für Kartographie und Geodäsie, Carl Zeiss).

already been predicted 20 years earlier by Oliver Lodge.* Today, Sagnac interferometers are the central part of laser gyroscopes – shown in Figure 101 – and are found in every passenger aeroplane, missile and submarine, in order to measure the changes of their motion and thus to determine their actual position.

A part of the fringe shift is due to the rotation of the Earth. Modern high-precision Sagnac interferometers use ring lasers with areas of a few square metres, as shown in

---

* Oliver Lodge (b. 1851, Stoke, d. on-Trent-1940, Wiltshire) was a physicist and spiritualist who studied electromagnetic waves and tried to communicate with the dead. A strange but influential figure, his ideas are often cited when fun needs to be made of physicists; for example, he was one of those (rare) physicists who believed that at the end of the nineteenth century physics was complete.

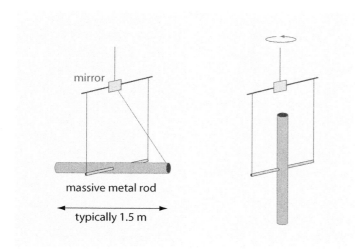

FIGURE 106 Observing the rotation of the Earth in two seconds.

Figure 105. Such a ring interferometer is able to measure variations of the rotation rates of the Earth of less than one part per million. Indeed, over the course of a year, the length of a day varies irregularly by a few milliseconds, mostly due to influences from the Sun or the Moon, due to weather changes and due to hot magma flows deep inside the Earth.* But also earthquakes, the El Niño effect in the climate and the filling of large water dams have effects on the rotation of the Earth. All these effects can be studied with such high-precision interferometers; they can also be used for research into the motion of the soil due to lunar tides or earthquakes, and for checks on the theory of special relativity.

Finally, in 1948, Hans Bucka developed the simplest experiment so far to show the rotation of the Earth. A metal rod allows anybody to detect the rotation of the Earth after only a few seconds of observation, using the set-up of Figure 106. The experiment can be easily be performed in class. Can you guess how it works?

In summary: all observations show that the Earth surface rotates at 464 m/s at the Equator, a larger value than that of the speed of sound in air, which is about 340 m/s at usual conditions. The rotation of the Earth also implies an acceleration, at the Equator, of 0.034 m/s^2. We are in fact *whirling* through the universe.

How does the Earth rotate?

Is the rotation of the Earth, the length of the day, *constant* over geological time scales? That is a hard question. If you find a method leading to an answer, publish it! (The same is true for the question whether the length of the year is constant.) Only a few methods are known, as we will find out shortly.

The rotation of the Earth is not even constant during a human lifespan. It varies by a few parts in 10^8. In particular, on a 'secular' time scale, the length of the day increases by about 1 to 2 ms per century, mainly because of the friction by the Moon and the melting of the polar ice caps. This was deduced by studying historical astronomical observations

---

* The growth of leaves on trees and the consequent change in the Earth's moment of inertia, already thought of in 1916 by Harold Jeffreys, is way too small to be seen, as it is hidden by larger effects.

of the ancient Babylonian and Arab astronomers. Additional 'decadic' changes have an amplitude of 4 or 5 ms and are due to the motion of the liquid part of the Earth's core. (The centre of the Earth's core is solid; this was discovered in 1936 by the Danish seismologist Inge Lehmann (1888–1993); her discovery was confirmed most impressively by two British seismologists in 2008, who detected shear waves of the inner core, thus confirming Lehmann's conclusion. There is a liquid core around the solid core.)

The seasonal and biannual changes of the length of the day – with an amplitude of 0.4 ms over six months, another 0.5 ms over the year, and 0.08 ms over 24 to 26 months – are mainly due to the effects of the *atmosphere*. In the 1950s the availability of precision measurements showed that there is even a 14 and 28 day period with an amplitude of 0.2 ms, due to the Moon. In the 1970s, when *wind oscillations* with a length scale of about 50 days were discovered, they were also found to alter the length of the day, with an amplitude of about 0.25 ms. However, these last variations are quite irregular.

Also the oceans influence the rotation of the Earth, due to the tides, the ocean currents, wind forcing, and atmospheric pressure forcing. Further effects are due to the ice sheet variations and due to water evaporation and rain falls. Last but not least, flows in the interior of the Earth, both in the mantle and in the core, change the rotation. For example, earthquakes, plate motion, post-glacial rebound and volcanic eruptions all influence the rotation.

But why does the Earth rotate at all? The rotation originated in the rotating gas cloud at the origin of the Solar System. This connection explains that the Sun and all planets, except two, turn around their axes in the same direction, and that they also all orbit the Sun in that same direction. But the complete story is outside the scope of this text.

The rotation around its axis is not the only motion of the Earth; it performs other motions as well. This was already known long ago. In 128 BCE, the Greek astronomer Hipparchos discovered what is today called the *(equinoctial) precession*. He compared a measurement he made himself with another made 169 years before. Hipparchos found that the Earth's axis points to different stars at different times. He concluded that the sky was moving. Today we prefer to say that the axis of the Earth is moving. (Why?) During a period of 25 800 years the axis draws a cone with an opening angle of 23.5°. This motion, shown in Figure 107, is generated by the tidal forces of the Moon and the Sun on the equatorial bulge of the Earth that results form its flattening. The Sun and the Moon try to align the axis of the Earth at right angles to the Earth's path; this torque leads to the precession of the Earth's axis.

Precession is a motion common to all rotating systems: it appears in planets, spinning tops and atoms. (Precession is also at the basis of the surprise related to the suspended wheel shown on page 244.) Precession is most easily seen in spinning tops, be they suspended or not. An example is shown in Figure 108; for atomic nuclei or planets, just imagine that the suspending wire is missing and the rotating body less flattened. On the Earth, precession leads to upwelling of deep water in the equatorial Atlantic Ocean and regularly changes the ecology of algae.

In addition, the axis of the Earth is not even fixed relative to the Earth's surface. In 1884, by measuring the exact angle above the horizon of the celestial North Pole, Friedrich Küstner (1856–1936) found that the axis of the Earth *moves* with respect to the Earth's crust, as Bessel had suggested 40 years earlier. As a consequence of Küstner's discovery, the International Latitude Service was created. The *polar motion* Küstner dis-

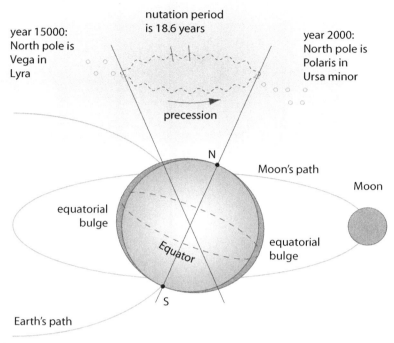

FIGURE 107 The precession and the nutation of the Earth's axis.

covered turned out to consist of three components: a small linear drift – not yet understood – a yearly elliptical motion due to seasonal changes of the air and water masses, and a circular motion* with a period of about 1.2 years due to fluctuations in the pressure at the bottom of the oceans. In practice, the North Pole moves with an amplitude of about 15 m around an average central position, as shown in Figure 109. Short term variations of the North Pole position, due to local variations in atmospheric pressure, to weather change and to the tides, have also been measured. The high precision of the GPS system is possible only with the help of the exact position of the Earth's axis; and only with this knowledge can artificial satellites be guided to Mars or other planets.

The details of the motion of the Earth have been studied in great detail. Table 25 gives an overview of the knowledge and the precision that is available today.

In 1912, the meteorologist and geophysicist Alfred Wegener (1880–1930) discovered an even larger effect. After studying the shapes of the continental shelves and the geological layers on both sides of the Atlantic, he conjectured that the continents *move*, and that they are all fragments of a single continent that broke up 200 million years ago.**

---

* The circular motion, a wobble, was predicted by the great Swiss mathematician Leonhard Euler (1707–1783). In a disgusting story, using Euler's and Bessel's predictions and Küstner's data, in 1891 Seth Chandler claimed to be the discoverer of the circular component.
** In this old continent, called Gondwanaland, there was a huge river that flowed westwards from the Chad

FIGURE 108 Precession of a suspended spinning top (mpg film © Lucas Barbosa)

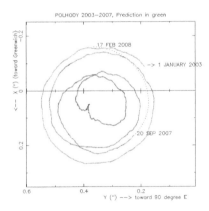
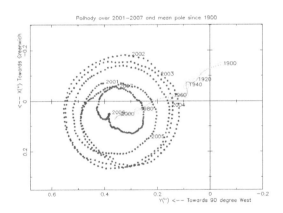

FIGURE 109 The motion of the North Pole – roughly speaking, the Earth's effective *polhode* – from 2003 to 2007, including the prediction until 2008 (left) and the average position since 1900 (right) – with 0.1 arcsecond being around 3.1 m on the surface of the Earth – not showing the diurnal and semidiurnal variations of a fraction of a millisecond of arc due to the tides (from hpiers.obspm.fr/eop-pc).

Even though at first derided across the world, Wegener's discoveries were correct. Modern satellite measurements, shown in Figure 110, confirm this model. For example, the American continent moves away from the European continent by about 23 mm every

---

to Guayaquil in Ecuador. After the continent split up, this river still flowed to the west. When the Andes appeared, the water was blocked, and many millions of years later, it reversed its flow. Today, the river still flows eastwards: it is called the Amazon River.

FIGURE 110 The continental plates are the objects of tectonic motion (HoloGlobe project, NASA).

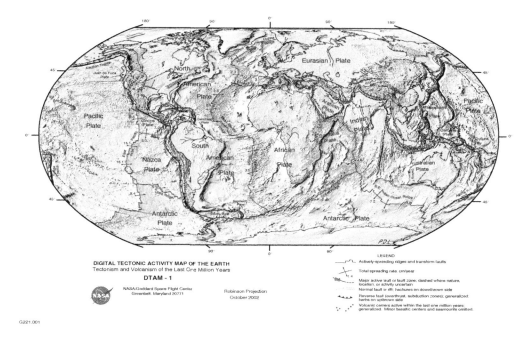

FIGURE 111 The tectonic plates of the Earth, with the relative speeds at the boundaries. (© NASA)

year, as shown in Figure 111. There are also speculations that this velocity may have been much higher at certain periods in the past. The way to check this is to look at the magnetization of sedimental rocks. At present, this is still a hot topic of research. Following the modern version of the model, called *plate tectonics*, the continents (with a density of

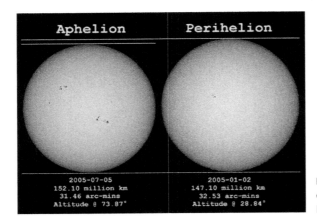

FIGURE 112 The angular size of the Sun changes due to the elliptical motion of the Earth (© Anthony Ayiomamitis).

FIGURE 113 Friedrich Wilhelm Bessel (1784–1846).

$2.7 \cdot 10^3$ kg/m^3) float on the fluid mantle of the Earth (with a density of $3.1 \cdot 10^3$ kg/m^3) like pieces of cork on water, and the convection inside the mantle provides the driving mechanism for the motion.

Does the Earth move?

Also the centre of the Earth is not at rest in the universe. In the third century BCE Aristarchus of Samos maintained that the Earth turns around the Sun. Experiments such as that of Figure 112 confirm that the orbit is an ellipse. However, a fundamental difficulty of the heliocentric system is that the stars look the same all year long. How can this be, if the Earth travels around the Sun? The distance between the Earth and the Sun has been known since the seventeenth century, but it was only in 1837 that Friedrich Wilhelm Bessel* became the first person to observe the *parallax* of a star. This was a result of extremely careful measurements and complex calculations: he discovered the *Bessel functions* in order to realize it. He was able to find a star, 61 Cygni, whose apparent position changed with the month of the year. Seen over the whole year, the star describes a small ellipse in the sky, with an opening of 0.588″ (this is the modern value). After care-

---

* Friedrich Wilhelm Bessel (1784–1846), Westphalian astronomer who left a successful business career to dedicate his life to the stars, and became the foremost astronomer of his time.

TABLE 25 Modern measurement data about the motion of the Earth (from hpiers.obspm.fr/eop-pc).

Observable	Symbol	Value
Mean angular velocity of Earth	Ω	72.921 150(1) μrad/s
Nominal angular velocity of Earth (epoch 1820)	Ω_N	72.921 151 467 064 μrad/s
Conventional mean solar day (epoch 1820)	d	86 400 s
Conventional sidereal day	d_{si}	86 164.090 530 832 88 s
Ratio conv. mean solar day to conv. sidereal day	$k = d/d_{si}$	1.002 737 909 350 795
Conventional duration of the stellar day	d_{st}	86 164.098 903 691 s
Ratio conv. mean solar day to conv. stellar day	$k' = d/d_{st}$	1.002 737 811 911 354 48
General precession in longitude	p	5.028 792(2) ″/a
Obliquity of the ecliptic (epoch 2000)	ε_0	23° 26′ 21.4119″
Küstner-Chandler period in terrestrial frame	T_{KC}	433.1(1.7) d
Quality factor of the Küstner-Chandler peak	Q_{KC}	170
Free core nutation period in celestial frame	T_F	430.2(3) d
Quality factor of the free core nutation	Q_F	$2 \cdot 10^4$
Astronomical unit	AU	149 597 870.691(6) km
Sidereal year (epoch 2000)	a_{si}	365.256 363 004 d = 365 d 6 h 9 min 9.76 s
Tropical year	a_{tr}	365.242 190 402 d = 365 d 5 h 48 min 45.25 s
Mean Moon period	T_M	27.321 661 55(1) d
Earth's equatorial radius	a	6 378 136.6(1) m
First equatorial moment of inertia	A	$8.0101(2) \cdot 10^{37}$ kg m²
Longitude of principal inertia axis A	λ_A	−14.9291(10)°
Second equatorial moment of inertia	B	$8.0103(2) \cdot 10^{37}$ kg m²
Axial moment of inertia	C	$8.0365(2) \cdot 10^{37}$ kg m²
Equatorial moment of inertia of mantle	A_m	$7.0165 \cdot 10^{37}$ kg m²
Axial moment of inertia of mantle	C_m	$7.0400 \cdot 10^{37}$ kg m²
Earth's flattening	f	1/298.25642(1)
Astronomical Earth's dynamical flattening	$h = (C - A)/C$	0.003 273 794 9(1)
Geophysical Earth's dynamical flattening	$e = (C - A)/A$	0.003 284 547 9(1)
Earth's core dynamical flattening	e_f	0.002 646(2)
Second degree term in Earth's gravity potential	$J_2 = -(A + B - 2C)/(2MR^2)$	$1.082 635 9(1) \cdot 10^{-3}$
Secular rate of J_2	dJ_2/dt	$-2.6(3) \cdot 10^{-11}$ /a
Love number (measures shape distortion by tides)	k_2	0.3
Secular Love number	k_s	0.9383
Mean equatorial gravity	g_{eq}	9.780 3278(10) m/s²
Geocentric constant of gravitation	GM	$3.986 004 418(8) \cdot 10^{14}$ m³/s²
Heliocentric constant of gravitation	$GM_\odot$	$1.327 124 420 76(50) \cdot 10^{20}$ m³/s²
Moon-to-Earth mass ratio	μ	0.012 300 038 3(5)

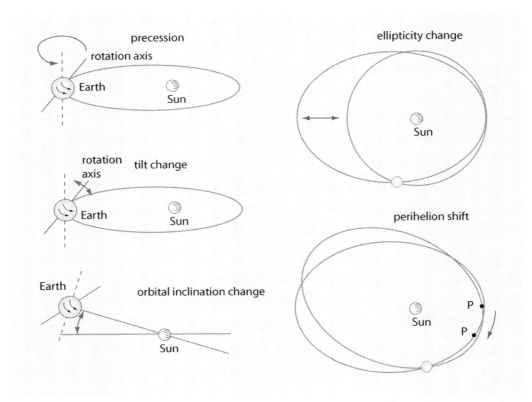

FIGURE 114 Changes in the Earth's motion around the Sun, as seen from different observers outside the orbital plane.

fully eliminating all other possible explanations, he deduced that the change of position was due to the motion of the Earth around the Sun, when seen from distant stars. From the size of the ellipse he determined the distance to the star to be 105 Pm, or 11.1 light years.

Bessel had thus managed, for the first time, to measure the distance of a star. By doing so he also proved that the Earth is not fixed with respect to the stars in the sky. The motion of the Earth was not a surprise. It confirmed the result of the mentioned aberration of light, discovered in 1728 by James Bradley* and to be discussed below. When seen from the sky, the Earth indeed revolves around the Sun.

With the improvement of telescopes, other motions of the Earth were discovered. In 1748, James Bradley announced that there is a small regular *change* of the precession, which he called *nutation*, with a period of 18.6 years and an angular amplitude of 19.2″. Nutation occurs because the plane of the Moon's orbit around the Earth is not exactly

---

* James Bradley (b. 1693 Sherborne, d. 1762 Chalford), was an important astronomer. He was one of the first astronomers to understand the value of precise measurement, and thoroughly modernized the Greenwich observatory. He discovered, independently of Eustachio Manfredi, the aberration of light, and showed with it that the Earth moves. In particular, the discovery allowed him to measure the speed of light and confirm the value of 0.3 Gm/s. He later discovered the nutation of the Earth's axis.

the same as the plane of the Earth's orbit around the Sun. Are you able to confirm that this situation produces nutation?

Astronomers also discovered that the 23.5° tilt – or *obliquity* – of the Earth's axis, the angle between its intrinsic and its orbital angular momentum, actually changes from 22.1° to 24.5° with a period of 41 000 years. This motion is due to the attraction of the Sun and the deviations of the Earth from a spherical shape. In 1941, during the Second World War, the Serbian astronomer Milutin Milankovitch (1879–1958) retreated into solitude and explored the consequences. In his studies he realized that this 41 000 year period of the obliquity, together with an average period of 22 000 years due to precession,* gives rise to the more than 20 *ice ages* in the last 2 million years. This happens through stronger or weaker irradiation of the poles by the Sun. The changing amounts of melted ice then lead to changes in average temperature. The last ice age had its peak about 20 000 years ago and ended around 11 800 years ago; the next is still far away. A spectacular confirmation of the relation between ice age cycles and astronomy came through measurements of oxygen isotope ratios in ice cores and sea sediments, which allow the average temperature over the past million years to be tracked. Figure 115 shows how closely the temperature follows the changes in irradiation due to changes in obliquity and precession.

The Earth's orbit also changes its *eccentricity* with time, from completely circular to slightly oval and back. However, this happens in very complex ways, not with periodic regularity, and is due to the influence of the large planets of the solar system on the Earth's orbit. The typical time scale is 100 000 to 125 000 years.

In addition, the Earth's orbit changes in *inclination* with respect to the orbits of the other planets; this seems to happen regularly every 100 000 years. In this period the inclination changes from +2.5° to −2.5° and back.

Even the direction in which the ellipse points changes with time. This so-called *perihelion shift* is due in large part to the influence of the other planets; a small remaining part will be important in the chapter on general relativity. The perihelion shift of Mercury was the first piece of data confirming Einstein's theory.

Obviously, the length of the year also changes with time. The measured variations are of the order of a few parts in 10^{11} or about 1 ms per year. However, knowledge of these changes and of their origins is much less detailed than for the changes in the Earth's rotation.

The next step is to ask whether the Sun itself moves. Indeed it does. Locally, it moves with a speed of 19.4 km/s towards the constellation of Hercules. This was shown by William Herschel in 1783. But globally, the motion is even more interesting. The diameter of the galaxy is at least 100 000 light years, and we are located 26 000 light years from the centre. (This has been known since 1918; the centre of the galaxy is located in the direction of Sagittarius.) At our position, the galaxy is 1 300 light years thick; presently, we are 68 light years 'above' the centre plane. The Sun, and with it the Solar System, takes about 225 million years to turn once around the galactic centre, its orbital velocity being around 220 km/s. It seems that the Sun will continue moving away from the galaxy plane until it is about 250 light years above the plane, and then move back, as shown in Figure 116. The oscillation period is estimated to be around 62 million years, and has been suggested as

---

* In fact, the 25 800 year precession leads to three insolation periods, of 23 700, 22 400 and 19 000 years, due to the interaction between precession and perihelion shift.

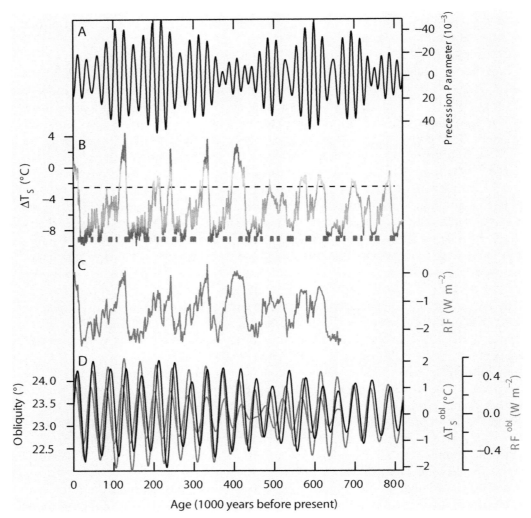

FIGURE 115 Modern measurements showing how Earth's precession parameter (black curve A) and obliquity (black curve D) influence the average temperature (coloured curve B) and the irradiation of the Earth (blue curve C) over the past 800 000 years: the obliquity deduced by Fourier analysis from the irradiation data RF (blue curve D) and the obliquity deduced by Fourier analysis from the temperature (red curve D) match the obliquity known from astronomical data (black curve D); sharp cooling events took place whenever the obliquity rose while the precession parameter was falling (marked red below the temperature curve) (© Jean Jouzel/Science from Ref. 124).

the mechanism for the mass extinctions of animal life on Earth, possibly because some gas cloud or some cosmic radiation source may be periodically encountered on the way. The issue is still a hot topic of research.

The motion of the Sun around the centre of the Milky Way implies that the planets of the Solar System can be seen as forming helices around the Sun. Figure 117 illustrates their helical path.

We turn around the galaxy centre because the formation of galaxies, like that of planetary systems, always happens in a whirl. By the way, can you confirm from your own

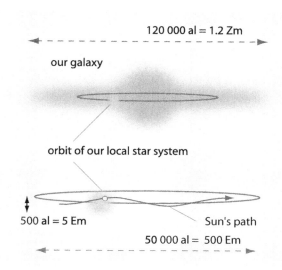

FIGURE 116 The motion of the Sun around the galaxy.

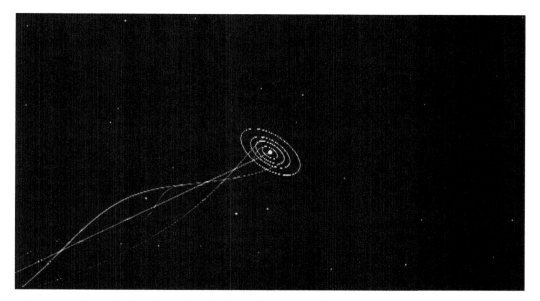

FIGURE 117 The helical motion of the first four planets around the path traced by the Sun during its travel around the centre of the Milky Way. Brown: Mercury, white: Venus, blue: Earth, red: Mars. (QuickTime film © Rhys Taylor at www.rhysy.net).

observation that our galaxy itself rotates?

Finally, we can ask whether the galaxy itself moves. Its motion can indeed be observed because it is possible to give a value for the motion of the Sun through the universe, defining it as the motion against the background radiation. This value has been measured to be 370 km/s. (The velocity of the *Earth* through the background radiation of course depends on the season.) This value is a combination of the motion of the Sun around the galaxy centre and of the motion of the galaxy itself. This latter motion is due to the

FIGURE 118 Driving through snowflakes shows the effects of relative motion in three dimensions. Similar effects are seen when the Earth speeds through the universe. (© Neil Provo at neilprovo.com).

gravitational attraction of the other, nearby galaxies in our local group of galaxies.*

In summary, the Earth really moves, and it does so in rather complex ways. As Henri Poincaré would say, if we are in a given spot today, say the Panthéon in Paris, and come back to the same spot tomorrow at the same time, we are in fact 31 million kilometres away. This state of affairs would make time travel extremely difficult even if it were possible (which it is not); whenever you went back to the past, you would have to get to the old spot exactly!

Is velocity absolute? – The theory of everyday relativity

Why don't we feel all the motions of the Earth? The two parts of the answer were already given in 1632. First of all, as Galileo explained, we do not feel the accelerations of the Earth because the effects they produce are too small to be detected by our senses. Indeed, many of the mentioned accelerations do induce measurable effects only in high-precision experiments, e.g. in atomic clocks.

But the second point made by Galileo is equally important: it is impossible to feel the high speed at which we are moving. We do not feel translational, unaccelerated motions because this is impossible *in principle*. Galileo discussed the issue by comparing the observations of two observers: one on the ground and another on the most modern means of unaccelerated transportation of the time, a ship. Galileo asked whether a man on the

---

* This is roughly the end of the ladder. Note that the expansion of the universe, to be studied later, produces no motion.

ground and a man in a ship moving at constant speed experience (or 'feel') anything different. Einstein used observers in trains. Later it became fashionable to use travellers in rockets. (What will come next?) Galileo explained that only *relative* velocities between bodies produce effects, not the absolute values of the velocities. For the senses and for all measurements we find:

> ▷ There is no difference between constant, undisturbed motion, however rapid it may be, and rest. This is called *Galileo's principle of relativity*.

Indeed, in everyday life we feel motion only if the means of transportation trembles – thus if it accelerates – or if we move against the air. Therefore Galileo concludes that two observers in straight and undisturbed motion against each other cannot say who is 'really' moving. Whatever their relative speed, neither of them 'feels' in motion.*

Rest is relative. Or more clearly: rest is an observer-dependent concept. This result of Galilean physics is so important that Poincaré introduced the expression 'theory of relativity' and Einstein repeated the principle explicitly when he published his famous theory of special relativity. However, these names are awkward. Galilean physics is also a theory of relativity! The relativity of rest is common to *all* of physics; it is an essential aspect of motion.

In summary, undisturbed or uniform motion has no observable effect; only *change* of motion does. Velocity cannot be felt; acceleration can. As a result, every physicist can deduce something simple about the following statement by Wittgenstein:

---

* In 1632, in his *Dialogo*, Galileo writes: 'Shut yourself up with some friend in the main cabin below decks on some large ship, and have with you there some flies, butterflies, and other small flying animals. Have a large bowl of water with some fish in it; hang up a bottle that empties drop by drop into a wide vessel beneath it. With the ship standing still, observe carefully how the little animals fly with equal speed to all sides of the cabin. The fish swim indifferently in all directions; the drops fall into the vessel beneath; and, in throwing something to your friend, you need throw it no more strongly in one direction than another, the distances being equal: jumping with your feet together, you pass equal spaces in every direction. When you have observed all these things carefully (though there is no doubt that when the ship is standing still everything must happen in this way), have the ship proceed with any speed you like, so long as the motion is uniform and not fluctuating this way and that, you will discover not the least change in all the effects named, nor could you tell from any of them whether the ship was moving or standing still. In jumping, you will pass on the floor the same spaces as before, nor will you make larger jumps toward the stern than toward the prow even though the ship is moving quite rapidly, despite the fact that during the time you are in the air the floor under you will be going in a direction opposite to your jump. In throwing something to your companion, you will need no more force to get it to him whether he is in the direction of the bow or the stern, with yourself situated opposite. The droplets will fall as before into the vessel beneath without dropping toward the stern, although while the drops are in the air the ship runs many spans. The fish in their water will swim toward the front of their bowl with no more effort than toward the back, and will go with equal ease to bait placed anywhere around the edges of the bowl. Finally the butterflies and flies will continue their flights indifferently toward every side, nor will it ever happen that they are concentrated toward the stern, as if tired out from keeping up with the course of the ship, from which they will have been separated during long intervals by keeping themselves in the air. And if smoke is made by burning some incense, it will be seen going up in the form of a little cloud, remaining still and moving no more toward one side than the other. The cause of all these correspondences of effects is the fact that the ship's motion is common to all the things contained in it, and to the air also. That is why I said you should be below decks; for if this took place above in the open air, which would not follow the course of the ship, more or less noticeable differences would be seen in some of the effects noted.' (Translation by Stillman Drake)

> Daß die Sonne morgen aufgehen wird, ist eine Hypothese; und das heißt:
> wir *wissen* nicht, ob sie aufgehen wird.*

The statement is *wrong*. Can you explain why Wittgenstein erred here, despite his strong desire not to?

Is rotation relative?

When we turn rapidly, our arms lift. Why does this happen? How can our body detect whether we are rotating or not? There are two possible answers. The first approach, promoted by Newton, is to say that there is an absolute space; whenever we rotate against this space, the system reacts. The other answer is to note that whenever the arms lift, the stars also rotate, and in exactly the same manner. In other words, our body detects rotation because we move against the average mass distribution in space.

The most cited discussion of this question is due to Newton. Instead of arms, he explored the water in a rotating bucket. In a rotating bucket, the water surface forms a concave shape, whereas the surface is flat for a non-rotating bucket. Newton asked why this is the case. As usual for philosophical issues, Newton's answer was guided by the mysticism triggered by his father's early death. Newton saw absolute space as a mystical and religious concept and was not even able to conceive an alternative. Newton thus saw rotation as an *absolute* type of motion. Most modern scientists have fewer personal problems and more common sense than Newton; as a result, today's consensus is that rotation effects are due to the mass distribution in the universe:

▷ Rotation is relative.

A number of high-precision experiments confirm this conclusion; thus it is also part of Einstein's theory of relativity.

Curiosities and fun challenges about rotation and relativity

When travelling in the train, you can test Galileo's statement about everyday relativity of motion. Close your eyes and ask somebody to turn you around several times: are you able to say in which direction the train is running?

* *

A good bathroom scales, used to determine the weight of objects, does not show a constant weight when you step on it and stay motionless. Why not?

* *

If a gun located at the Equator shoots a bullet vertically, where does the bullet fall?

* *

Why are most rocket launch sites as near as possible to the Equator?

---

* 'That the Sun will rise to-morrow, is an hypothesis; and that means that we do not *know* whether it will rise.' This well-known statement is found in Ludwig Wittgenstein, *Tractatus*, 6.36311.

∗ ∗

At the Equator, the speed of rotation of the Earth is 464 m/s, or about Mach 1.4; the latter number means that it is 1.4 times the speed of sound. This supersonic motion has two intriguing consequences.

First of all, the rotation speed determines the size of typical weather phenomena. This size, the so-called *Rossby radius*, is given by the speed of sound (or some other typical speed) divided by twice the local rotation speed, multiplied with the radius of the Earth. At moderate latitudes, the Rossby radius is about 2000 km. This is a sizeable fraction of the Earth's radius, so that only a few large weather systems are present on Earth at any specific time. If the Earth rotated more slowly, the weather would be determined by short-lived, local flows and have no general regularities. If the Earth rotated more rapidly, the weather would be much more violent – as on Jupiter – but the small Rossby radius implies that large weather structures have a huge lifetime, such as the red spot on Jupiter, which lasted for several centuries. In a sense, the rotation of the Earth has the speed that provides the most interesting weather.

The other consequence of the value of the Earth's rotation speed concerns the thickness of the atmosphere. Mach 1 is also, roughly speaking, the thermal speed of air molecules. This speed is sufficient for an air molecule to reach the characteristic height of the atmosphere, about 6 km. On the other hand, the speed of rotation Ω of the Earth determines its departure h from sphericity: the Earth is flattened, as we saw above. Roughly speaking, we have $gh = \Omega^2 R^2/2$, or about 12 km. (This is correct to within 50 %, the actual value is 21 km.) We thus find that the speed of rotation of the Earth implies that its flattening is comparable to the thickness of the atmosphere.

∗ ∗

The Coriolis effect influences rivers and their shores. This surprising connection was made in 1860 by Karl Ernst von Baer who found that in Russia, many rivers flowing north in lowlands had right shores that are steep and high, and left shores that are low and flat. (Can you explain the details?) He also found that rivers in the southern hemisphere show the opposite effect.

∗ ∗

The Coriolis effect saves lives and helps people. Indeed, it has an important application for navigation systems; the typical uses are shown in Figure 119. Insects use vibrating masses to stabilize their orientation, to determine their direction of travel and to find their way. Most two-winged insects, or diptera, use *vibrating halteres* for navigation: in particular, bees, house-flies, hover-flies and crane flies use them. Other insects, such as moths, use *vibrating antennae* for navigation. Cars, satellites, mobile phones, remote-controlled helicopter models, and computer games also use tiny vibrating masses as orientation and navigation sensors, in exactly the same way as insects do.

In all these navigation applications, one or a few tiny masses are made to vibrate; if the system to which they are attached turns, the change of orientation leads to a Coriolis effect. The effect is measured by detecting the ensuing change in geometry; the change, and thus the signal strength, depends on the angular velocity and its direction. Such orientation sensors are therefore called *vibrating Coriolis gyroscopes*. Their development

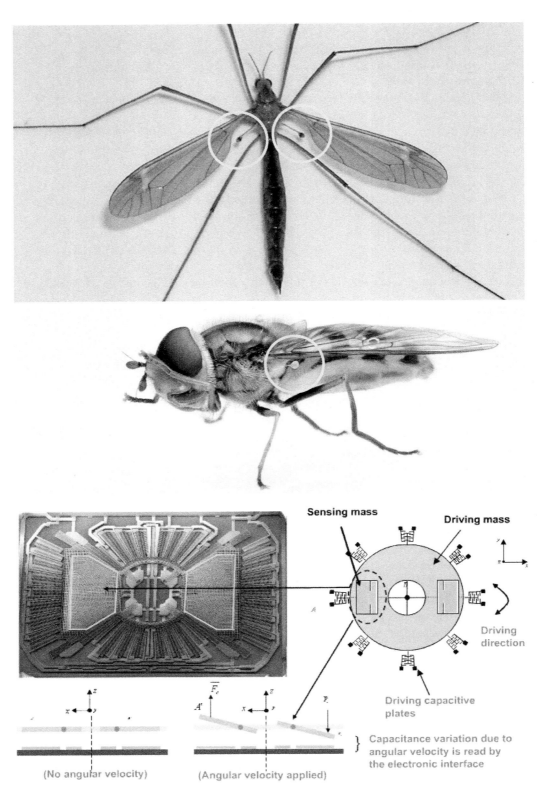

FIGURE 119 The use of the Coriolis effect in insects – here a crane fly and a hovering fly – and in micro-electromechanic systems (size about a few mm); all provide navigation signals to the systems to which they are attached (© Pinzo, Sean McCann, ST Microelectronics).

and production is a sizeable part of high-tech business – and of biological evolution.

∗ ∗

A wealthy and quirky customer asked his architect to plan and build a house whose four walls all faced south. How did the architect realize the request?

∗ ∗

Would travelling through interplanetary space be healthy? People often fantasize about long trips through the cosmos. Experiments have shown that on trips of long duration, cosmic radiation, bone weakening, muscle degeneration and psychological problems are the biggest dangers. Many medical experts question the viability of space travel lasting longer than a couple of years. Other dangers are rapid sunburn, at least near the Sun, and exposure to the vacuum. So far only one man has experienced vacuum without protection. He lost consciousness after 14 seconds, but survived unharmed.

∗ ∗

In which direction does a flame lean if it burns inside a jar on a rotating turntable?

∗ ∗

Galileo's principle of everyday relativity states that it is impossible to determine an absolute velocity. It is equally impossible to determine an absolute position, an absolute time and an absolute direction. Is this correct?

∗ ∗

Does *centrifugal acceleration* exist? Most university students go through the shock of meeting a teacher who says that it doesn't because it is a 'fictitious' quantity, in the face of what one experiences every day in a car when driving around a bend. Simply ask the teacher who denies it to define 'existence'. (The definition physicists usually use is given later on.) Then check whether the definition applies to the term and make up your own mind.

Whether you like the term 'centrifugal acceleration' or avoid it by using its negative, the so-called *centripetal acceleration*, you should know how it is calculated. We use a simple trick. For an object in circular motion of radius r, the magnitude v of the velocity $\boldsymbol{v} = \mathrm{d}\boldsymbol{x}/\mathrm{d}t$ is $v = 2\pi r/T$. The vector $\boldsymbol{v}$ behaves over time exactly like the position of the object: it rotates continuously. Therefore, the magnitude a of the centrifugal/centripetal acceleration $\boldsymbol{a} = \mathrm{d}\boldsymbol{v}/\mathrm{d}t$ is given by the corresponding expression, namely $a = 2\pi v/T$. Eliminating T, we find that the centrifugal/centripetal acceleration a of a body rotating at speed v at radius r is given by

$$a = \frac{v^2}{r} = \omega^2 r \ . \tag{36}$$

This is the acceleration we feel when sitting in a car that goes around a bend.

∗ ∗

Rotational motion holds a little surprise for anybody who studies it carefully. Angular momentum is a quantity with a magnitude and a direction. However, it is *not* a vector,

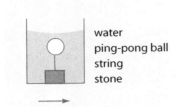

FIGURE 120 How does the ball move when the jar is accelerated in direction of the arrow?

as any mirror shows. The angular momentum of a body circling in a plane parallel to a mirror behaves in a different way from a usual arrow: its mirror image is not reflected if it points towards the mirror! You can easily check this for yourself. For this reason, angular momentum is called a *pseudo-vector*. (Are rotations pseudo-vectors?) The fact has no important consequences in classical physics; but we have to keep it in mind for later, when we explore nuclear physics.

* *

What is the best way to transport a number of full coffee or tea cups while at the same time avoiding spilling any precious liquid?

* *

A ping-pong ball is attached by a string to a stone, and the whole is put under water in a jar. The set-up is shown in Figure 120. Now the jar is accelerated horizontally, for example in a car. In which direction does the ball move? What do you deduce for a jar at rest?

* *

The Moon recedes from the Earth by 3.8 cm a year, due to friction. Can you find the mechanism responsible for the effect?

* *

What are earthquakes? *Earthquakes* are large examples of the same process that make a door squeak. The continental plates correspond to the metal surfaces in the joints of the door.

Earthquakes can be described as energy sources. The Richter scale is a direct measure of this energy. The *Richter magnitude* M_s of an earthquake, a pure number, is defined from its energy E in joule via

$$M_s = \frac{\log(E/1\,\text{J}) - 4.8}{1.5} \,. \tag{37}$$

The strange numbers in the expression have been chosen to put the earthquake values as near as possible to the older, qualitative Mercalli scale (now called EMS98) that classifies the intensity of earthquakes. However, this is not fully possible; the most sensitive instruments today detect earthquakes with magnitudes of −3. The highest value ever measured was a Richter magnitude of 10, in Chile in 1960. Magnitudes above 12 are probably im-

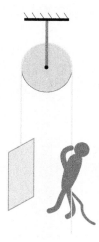

FIGURE 121 What happens when the ape climbs?

possible. Can you show why?

* *

What is the motion of the point on the surface of the Earth that has Sun in its zenith – i.e., vertically above it – when seen on a map of the Earth during one day? And day after day?

* *

Can it happen that a satellite dish for geostationary TV satellites focuses the sunshine onto the receiver?

* *

Why is it difficult to fire a rocket from an aeroplane in the direction opposite to the motion of the plane?

* *

An ape hangs on a rope. The rope hangs over a wheel and is attached to a mass of equal weight hanging down on the other side, as shown in Figure 121. The rope and the wheel are massless and frictionless. What happens when the ape climbs the rope?

* *

Can a water skier move with a higher speed than the boat pulling him?

* *

You might know the 'Dynabee', a hand-held gyroscopic device that can be accelerated to high speed by proper movements of the hand. How does it work?

* *

It is possible to make a spinning top with a metal paper clip. It is even possible to make

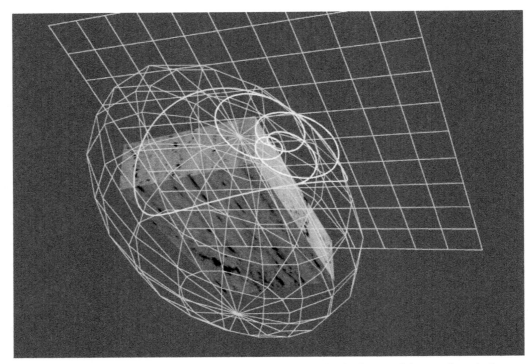

FIGURE 122 The motion of the angular velocity of a tumbling brick. The tip of the angular velocity vector moves along the yellow curve, the *polhode*. It moves together with the tumbling object, as does the elliptical mesh representing the energy ellipsoid; the *herpolhode* is the white curve and lies in the plane bright blue cross pattern that represents the invariable plane (see text). The animation behind the screenshot illustrates the well-known nerd statement: *the polhode rolls without slipping on the herpolhode lying in the invariable plane*. The full animation is available online at www.ialms.net/sim/3d-rigid-body-simulation/. Any rotating, irregular, free rigid body follows such a motion. For the Earth, which is rotating and irregular, but neither fully free nor fully rigid, this description is only an approximation. (© Svetoslav Zabunov)

one of those tops that turn onto their head when spinning. Can you find out how?

* *

The moment of inertia of a body depends on the shape of the body; usually, the angular momentum and the angular velocity do not point in the same direction. Can you confirm this with an example?

* *

What is the moment of inertia of a homogeneous sphere?

* *

The complete moment of inertia of a rigid body is determined by the values along its three principal axes. These values are all equal for a sphere and for a cube. Does it mean that it is impossible to distinguish a sphere from a cube by their inertial behaviour?

∗ ∗

Here is some mathematical fun about the rotation of a free rigid body. Even for a free rotating rigid body, such as a brick rotating in free space, the angular velocity is, in general, *not* constant: the brick *tumbles* while rotating. In this motion, both energy and angular momentum are constant, but the angular velocity is not. In particular, not even the direction of angular velocity is constant; in other words, the north pole changes with time. In fact, the north pole changes with time both for an observer on the body and for an observer in free space. How does the north pole, the end point of the angular velocity vector, move?

The moment of inertia is a tensor and can thus represented by an ellipsoid. In addition, the motion of a free rigid body is described, in the angular velocity space, by the kinetic energy ellipsoid – because its rotation energy is constant. When a free rigid body moves, the energy ellipsoid – not the moment of inertia ellipsoid – *rolls* on the invariable plane that is perpendicular to the initial (and constant) angular momentum of the body. This is the mathematical description of the tumbling motion. The curve traced by the angular velocity, or the extended north pole of the body, on the invariable plane is called the *herpolhode*. It is an involved curve that results from two superposed conical motions. For an observer on the rotating body, another curve is more interesting. The pole of the rotating body traces a curve on the energy ellipsoid – which is itself attached to the body. This curve is called the *polhode*. The polhode is a closed curve i three dimensions, the herpolhode is an open curve in two dimensions. The curves were named by Louis Poinsot and are especially useful for the description of the motion of rotating irregularly shaped bodies. For an complete visualization, see the excellent website www.ialms.net/sim/3d-rigid-body-simulation/.

The polhode is a circle only if the rigid body has rotational symmetry; the pole motion is then called *precession*; we encountered it above. As shown in Figure 109, the measured polhode of the Earth is not of the expected shape; this irregularity is due to several effects, among them to the non-rigidity of our planet.

Even though the angular momentum of a free tumbling brick is fixed in space, it is not fixed in the body frame. Can you confirm this? In the body frame, the end of the angular momentum of a tumbling brick moves along still another curve, given by the intersection of a sphere and an ellipsoid, as Jacques Binet pointed out.

∗ ∗

Is it true that the Moon in the first quarter in the northern hemisphere looks like the Moon in the last quarter in the southern hemisphere?

∗ ∗

An impressive confirmation that the Earth is a sphere can be seen at sunset, if we turn, against our usual habit, our back on the Sun. On the eastern sky we can then see the impressive rise of the Earth's shadow. We can admire the vast shadow rising over the whole horizon, clearly having the shape of a segment of a huge circle. Figure 124 shows an example. In fact, more precise investigations show that it is not the shadow of the Earth alone, but the shadow of its ionosphere.

∗ ∗

FIGURE 123 Long exposures of the stars at night – one when facing north, above the Gemini telescope in Hawaii, and one above the Alps that includes the celestial equator, with the geostationary satellites on it (© Gemini Observatory/AURA, Michael Kunze).

FIGURE 124 The shadow of the Earth – here a panoramic photograph taken at the South Pole – shows that the Earth is round (© Ian R. Rees).

How would Figure 123 look if taken at the Equator?

* *

Precision measurements show that not all planets move in exactly the same plane. Mercury shows the largest deviation. In fact, no planet moves exactly in an ellipse, nor even in a plane around the Sun. Almost all of these effects are too small and too complex to explain here.

* *

Since the Earth is round, there are many ways to drive from one point on the Earth to another along a circle segment. This freedom of choice has interesting consequences for volley balls and for men watching women. Take a volleyball and look at its air inlet. If you want to move the inlet to a different position with a simple rotation, you can choose the rotation axis in many different ways. Can you confirm this? In other words, when we look in a given direction and then want to look in another, the eye can accomplish this change in different ways. The option chosen by the human eye had already been studied by medical scientists in the eighteenth century. It is called *Listing's 'law'*.* It states that all axes that nature chooses lie in one plane. Can you imagine its position in space? Many men have a real interest that this mechanism is strictly followed; if not, looking at women on the beach could cause the muscles moving their eyes to get knotted up.

* *

Imagine to cut open a soft mattress, glue a steel ball into it, and glue the mattress together again. Now imagine that we use a magnetic field to rotate the steel ball glued inside. Intu-

---

* If you are interested in learning in more detail how nature and the eye cope with the complexities of three dimensions, see the schorlab.berkeley.edu/vilis/whatisLL.htm and www.physpharm.fmd.uwo.ca/undergrad/llconsequencesweb/ListingsLaw/perceptual1.htm websites.

FIGURE 125 A steel ball glued in a mattress can rotate forever. (QuickTime film © Jason Hise).

itively, we think that the ball can only be rotated by a finite angle, whose value would be limited by the elasticity of the mattress. But in reality, the steel ball can be rotated *infinitely often!* This surprising possibility is a consequence of the tethered rotation shown in Figure 54 and Figure 55. Such a continuous rotation in a mattress is shown in Figure 125. And despite its fascination, nobody has yet realized the feat. Can you?

Legs or wheels? – Again

The acceleration and deceleration of standard wheel-driven cars is never much greater than about 1 g = 9.8 m/s^2, the acceleration due to gravity on our planet. Higher accelerations are achieved by motorbikes and racing cars through the use of suspensions that divert weight to the axes and by the use of spoilers, so that the car is pushed downwards

FIGURE 126 A basilisk lizard (*Basiliscus basiliscus*) running on water, with a total length of about 25 cm, showing how the propulsing leg pushes into the water (© TERRA).

with more than its own weight. Modern spoilers are so efficient in pushing a car towards the track that racing cars could race on the roof of a tunnel without falling down.

Through the use of special tyres the downwards forces produced by aerodynamic effects are transformed into grip; modern racing tyres allow forward, backward and sideways accelerations (necessary for speed increase, for braking and for turning corners) of about 1.1 to 1.3 times the load. Engineers once believed that a factor 1 was the theoretical limit and this limit is still sometimes found in textbooks; but advances in tyre technology, mostly by making clever use of interlocking between the tyre and the road surface as in a gear mechanism, have allowed engineers to achieve these higher values. The highest accelerations, around 4 g, are achieved when part of the tyre melts and glues to the surface. Special tyres designed to make this happen are used for dragsters, but high performance radio-controlled model cars also achieve such values.

How do wheels compare to using legs? High jump athletes can achieve peak accelerations of about 2 to 4 g, cheetahs over 3 g, bushbabies up to 13 g, locusts about 18 g, and fleas have been measured to accelerate about 135 g. The maximum acceleration known for animals is that of click beetles, a small insect able to accelerate at over 2000 m/s^2 = 200 g, about the same as an airgun pellet when fired. Legs are thus definitively more efficient accelerating devices than wheels – a cheetah can easily beat any car or motorbike – and evolution developed legs, instead of wheels, to improve the chances of an animal in danger getting to safety.

In short, legs *outperform* wheels. But there are other reasons for using legs instead of wheels. (Can you name some?) For example, legs, unlike wheels, allow walking on water. Most famous for this ability is the *basilisk*,* a lizard living in Central America and shown in Figure 126. This reptile is up to 70 cm long and has a mass of up to 500 g. It looks like a miniature *Tyrannosaurus rex* and is able to run over water surfaces on its hind legs. The motion has been studied in detail with high-speed cameras and by measurements using aluminium models of the animal's feet. The experiments show that the feet slapping on the water provides only 25 % of the force necessary to run above water; the other 75 % is provided by a pocket of compressed air that the basilisks create between their feet and the water once the feet are inside the water. In fact, basilisks mainly walk on air. (Both

---

* In the Middle Ages, the term 'basilisk' referred to a mythical monster supposed to appear shortly before the end of the world. Today, it is a small reptile in the Americas.

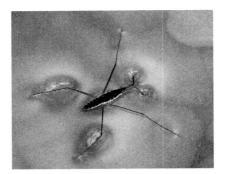

FIGURE 127 A water strider, total size about 10 mm (© Charles Lewallen).

FIGURE 128 A water walking robot, total size about 20 mm (© AIP).

effects used by basilisks are also found in fast canoeing.) It was calculated that humans are also able to walk on water, provided their feet hit the water with a speed of 100 km/h using the simultaneous physical power of 15 sprinters. Quite a feat for all those who ever did so.

There is a second method of walking and running on water; this second method even allows its users to remain immobile on top of the water surface. This is what water striders, insects of the family *Gerridae* with an overall length of up to 15 mm, are able to do (together with several species of spiders), as shown in Figure 127. Like all insects, the water strider has six legs (spiders have eight). The water strider uses the back and front legs to hover over the surface, helped by thousands of tiny hairs attached to its body. The hairs, together with the surface tension of water, prevent the strider from getting wet. If you put shampoo into the water, the water strider sinks and can no longer move.

The water strider uses its large middle legs as oars to advance over the surface, reaching speeds of up to 1 m/s doing so. In short, water striders actually row over water. The same mechanism is used by the small robots that can move over water and were developed by Metin Sitti and his group, as shown in Figure 128.

Robot design is still in its infancy. No robot can walk or even run as fast as the animal system it tries to copy. For two-legged robots, the most difficult kind, the speed record is around 3.5 leg lengths per second. In fact, there is a race going on in robotics departments: each department tries to build the first robot that is faster, either in metres per second or in leg lengths per second, than the original four-legged animal or two-legged human. The difficulties of realizing this development goal show how complicated walking motion is and how well nature has optimized living systems.

Legs pose many interesting problems. Engineers know that a staircase is comfortable to walk only if for each step the depth l plus *twice* the height h is a constant: $l + 2h = 0.63 \pm 0.02$ m. This is the so-called *staircase formula*. Why does it hold?

Most animals have an *even* number of legs. Do you know an exception? Why not? In fact, one can argue that no animal has less than four legs. Why is this the case?

On the other hand, all animals with two legs have the legs side by side, whereas most systems with two wheels have them one behind the other. Why is this not the other way round?

Legs are very efficient actuators. As Figure 129 shows, most small animals can run

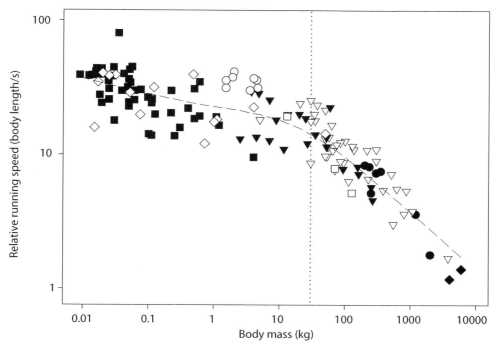

FIGURE 129 The graph shows how the relative running speed changes with the mass of terrestrial mammal species, for 142 different species. The graph also illustrates how the running performance changes above 30 kg. Filled squares show Rodentia; open squares show Primata; filled diamonds Proboscidae; open diamonds Marsupialia; filled triangles Carnivora; open triangles Artiodactyla; filled circles Perissodactyla; open circles Lagomorpha (© José Iriarte-Díaz/JEB).

with about 25 body lengths per second. For comparison, almost no car achieves such a speed. Only animals that weigh more than about 30 kg, including humans, are slower.

Legs also provide simple distance rulers: just count your steps. In 2006, it was discovered that this method is used by certain ant species, such as *Cataglyphis fortis*. They can count to at least 25 000, as shown by Matthias Wittlinger and his team. These ants use the ability to find the shortest way back to their home even in structureless desert terrain.

Why do 100 m sprinters run faster than ordinary people? A thorough investigation shows that the speed v of a sprinter is given by

$$v = f L_{\text{stride}} = f L_c \frac{F_c}{W}, \qquad (38)$$

where f is the frequency of the legs, L_{stride} is the stride length, L_c is the contact length – the length that the sprinter advances during the time the foot is in contact with the floor – W the weight of the sprinter, and F_c the average force the sprinter exerts on the floor during contact. It turns out that the frequency f is almost the same for all sprinters; the only way to be faster than the competition is to increase the stride length L_{stride}. Also the contact length L_c varies little between athletes. Increasing the stride length thus requires that the athlete hits the ground with strong strokes. This is what athletic training for sprinters has to achieve.

Summary on Galilean relativity

The Earth rotates. The acceleration is so small that we do not feel it. The speed of rotation is large, but we do not feel it, because there is *no way* to do so.

Undisturbed or inertial motion *cannot* be felt or measured. It is thus impossible to distinguish motion from rest. The distinction between rest and motion depends on the observer: *Motion of bodies is relative.* That is why the soil below our feet seems so stable to us, even though it moves with high speed across the universe.

Since motion is relative, speed values depend on the observer. Later on will we discover that one example of motion in nature has a speed value that is *not* relative: the motion of light. But we continue first with the study of motion transmitted over distance, without the use of any contact at all.

Chapter 6

MOTION DUE TO GRAVITATION

> "Caddi come corpo morto cade.
> Dante, *Inferno*, c. V, v. 142.**"

The first and main method to generate motion without any contact that we discover in our environment is *height*. Waterfalls, snow, rain, the ball of your favourite game and falling apples all rely on it. It was one of the fundamental discoveries of physics that height has this property because there is an interaction between every body and the Earth. *Gravitation* produces an *acceleration* along the line connecting the centres of gravity of the body and the Earth. Note that in order to make this statement, it is necessary to realize that the Earth is a body in the same way as a stone or the Moon, that this body is finite and that therefore it has a centre and a mass. Today, these statements are common knowledge, but they are by no means evident from everyday personal experience.

In several myths about the creation or the organization of the world, such as the biblical one or the Indian one, the Earth is not an object, but an imprecisely defined entity, such as an island floating or surrounded by water with unclear boundaries and unclear method of suspension. Are you able to convince a friend that the Earth is round and not flat? Can you find another argument apart from the roundness of the Earth's shadow when it is visible on the Moon, shown in Figure 134?

Gravitation as a limit to uniform motion

A productive way to define gravitation, or gravity for short, appears when we note that no object around us moves along a straight line. In nature, there is a *limit* to steady, or constant motion:

> ▷ Gravity prevents uniform motion, i.e., it prevents constant and straight motion.

In nature, we *never* observe bodies moving at constant speed along a straight line. Speaking with the vocabulary of kinematics: Gravity introduces an *acceleration* for every physical body. The gravitation of the objects in the environment curves the path of a body, changes its speed, or both. This limit has two aspects. First, gravity prevents unlimited uniform motion:

** 'I fell like dead bodies fall.' Dante Alighieri (1265, Firenze–1321, Ravenna), the powerful Italian poet.

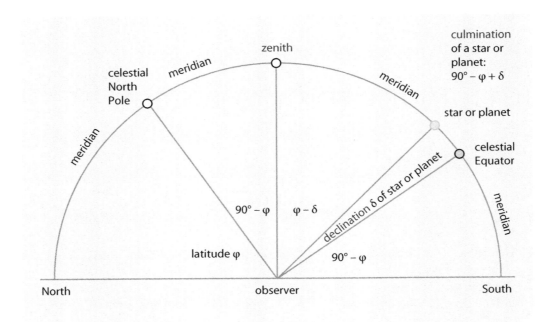

FIGURE 130 Some important concepts when observing the stars and at night.

▷ Motion cannot be straight for ever. Motion is not boundless.

We will learn later what this means for the universe as a whole. Secondly,

▷ Motion cannot be constant and straight even during short time intervals.

In other words, if we measure with sufficient precision, we will always find deviations from uniform motion. (Physicist also say that motion is never inertial.) These limits apply no matter how *powerful* the motion is and how *free* the moving body is from external influence. In nature, gravitation prevents the steady, uniform motion of atoms, billiard balls, planets, stars and even galaxies.

Gravitation is the first limitation to motion that we discover in nature. We will discover two additional limits to motion later on in our walk. These three fundamental limits are illustrated in Figure 1. To achieve a *precise* description of motion, we need to take each limit of motion into account. This is our main aim in the rest of our adventure.

Gravity affects *all* bodies, even if they are distant from each other. How exactly does gravitation affect two bodies that are far apart? We ask the experts for measuring distant objects: the astronomers.

GRAVITATION IN THE SKY

The gravitation of the Earth forces the Moon in an orbit around it. The gravitation of the Sun forces the Earth in an orbit around it and sets the length of the year. Similarly, the gravitation of the Sun determines the motion of all the other planets across the sky. We

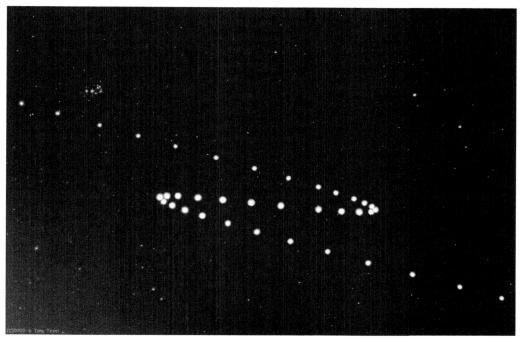

FIGURE 131 'Planet' means 'wanderer'. This composed image shows the retrograde motion of planet Mars across the sky – the Pleiades star cluster is at the top left – when the planet is on the other side of the Sun. The pictures were taken about a week apart and superimposed. The motion is one of the many examples that are fully explained by universal gravitation (© Tunc Tezel).

usually imagine to be located at the centre of the Sun and then say that the planets 'orbit the Sun'. The Sun thus prevents the planets from moving in straight lines and forces them into orbits. How can we check this?

First of all, looking at the sky at night, we can check that the planets always stay within the *zodiac*, a narrow stripe across the sky. The centre line of the zodiac gives the path of the Sun and is called the *ecliptic*, since the Moon must be located on it to produce an eclipse. This shows that planets move (approximately) in a single, common plane.*

To learn more about the motion in the sky, astronomers have performed numerous measurements of the movements of the Moon and the planets. The most industrious of all was Tycho Brahe,** who organized an industrial-scale search for astronomical facts sponsored by his king. His measurements were the basis for the research of his young assistant, the Swabian astronomer Johannes Kepler*** who found the first precise

---

* The apparent height of the ecliptic changes with the time of the year and is the reason for the changing seasons. Therefore seasons are a gravitational effect as well.
** Tycho Brahe (b. 1546 Scania, d. 1601 Prague), famous astronomer, builder of Uraniaborg, the astronomical castle. He consumed almost 10 % of the Danish gross national product for his research, which produced the first star catalogue and the first precise position measurements of planets.
*** Johannes Kepler (1571 Weil der Stadt–1630 Regensburg) studied Protestant theology and became a teacher of mathematics, astronomy and rhetoric. He helped his mother to defend herself successfully in a trial where she was accused of witchcraft. His first book on astronomy made him famous, and he became assistant to Tycho Brahe and then, at his teacher's death, the Imperial Mathematician. He was the first to

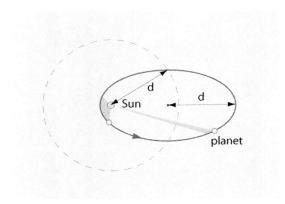

FIGURE 132 The motion of a planet around the Sun, showing its semimajor axis d, which is also the spatial average of its distance from the Sun.

description of planetary motion. This is not an easy task, as the observation of Figure 131 shows. In his painstaking research on the movements of the planets in the zodiac, Kepler discovered several 'laws', i.e., patterns or rules. The motion of all the planets follow the same rules, confirming that the Sun determines their orbits. The three main ones are as follows:

1. Planets move on ellipses with the Sun located at one focus (1609).
2. Planets sweep out equal areas in equal times (1609).
3. All planets have the same ratio T^2/d^3 between the orbit duration T and the semimajor axis d (1619).

Kepler's results are illustrated in Figure 132. The sheer work required to deduce the three 'laws' was enormous. Kepler had no calculating machine available. The calculation technology he used was the recently discovered logarithms. Anyone who has used tables of logarithms to perform calculations can get a feeling for the amount of work behind these three discoveries.

Finally, in 1684, all observations by Kepler about planets and stones were condensed into an astonishingly simple result by the English physicist Robert Hooke and a few others:*

▷ Every body of mass M attracts any other body towards its centre with an acceleration whose magnitude a is given by

$$a = G \frac{M}{r^2} \qquad (39)$$

where r is the centre-to-centre distance of the two bodies.

---

use mathematics in the description of astronomical observations, and introduced the concept and field of 'celestial physics'.

* Robert Hooke (1635–1703), important English physicist and secretary of the Royal Society. Apart from discovering the inverse square relation and many others, such as Hooke's 'law', he also wrote the *Micrographia*, a beautifully illustrated exploration of the world of the very small.

6 MOTION DUE TO GRAVITATION

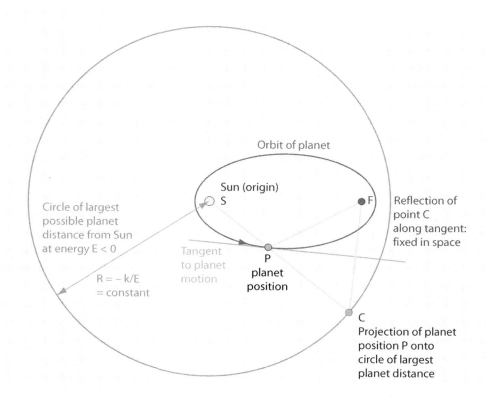

FIGURE 133 The proof that a planet moves in an ellipse (magenta) around the Sun, given an inverse square distance relation for gravitation. The proof – detailed in the text – is based on the relation SP+PF=R. Since R is constant, the orbit is an ellipse

This is called *universal gravitation*, or the *universal 'law' of gravitation*, because it is valid both in the sky and on Earth, as we will see shortly. The proportionality constant G is called the *gravitational constant*; it is one of the fundamental constants of nature, like the speed of light c or the quantum of action $\hbar$. More about G will be said shortly.

The effect of gravity thus decreases with increasing distance; the effect depends on the inverse distance squared of the bodies under consideration. If bodies are small compared with the distance r, or if they are spherical, expression (39) is correct as it stands; for non-spherical shapes the acceleration has to be calculated separately for each part of the bodies and then added together.

Why is the usual planetary orbit an ellipse? The simplest argument is given in Figure 133. We know that the acceleration due to gravity varies as $a = GM/r^2$. We also know that an orbiting body of mass m has a constant energy $E < 0$. We then can draw, around the Sun, the circle with radius $R = -GMm/E$, which gives the largest distance that a body with energy E can be from the Sun. We now project the planet position P onto this circle, thus constructing a position C. We then reflect C along the tangent to get a position F. This last position F is fixed in space and time, as a simple argument shows. (Can you find it?) As a result of the construction, the distance sum SP+PF is constant in time, and given by the radius $R = -GMm/E$. Since the distance sum is constant, the orbit is an ellipse, because an ellipse is precisely the curve that appears when this sum is constant. (Remember that an ellipse can be drawn with a piece of rope in this way.) Point *F*, like the Sun, is a focus of the ellipse. The construction thus shows that the motion of a planet defines two foci and follows an elliptical orbit defined by these two foci. In short, we have deduced the first of Kepler's 'laws' from the expression of universal gravitation.

The second of Kepler's 'laws', about equal swept areas, implies that planets move faster when they are near the Sun. It is a simple way to state the conservation of angular momentum. What does the third 'law' state?

Can you confirm that also the second and third of Kepler's 'laws' follow from Hooke's expression of universal gravity? Publishing this result – which was obvious to Hooke – was one of the achievements of Newton. Try to repeat this achievement; it will show you not only the difficulties, but also the possibilities of physics, and the joy that puzzles give.

Newton solved these puzzles with geometric drawings – though in quite a complex manner. It is well known that Newton was not able to write down, let alone handle, differential equations at the time he published his results on gravitation. In fact, Newton's notation and calculation methods were poor. (Much poorer than yours!) The English mathematician Godfrey Hardy* used to say that the insistence on using Newton's integral and differential notation, rather than the earlier and better method, still common today, due to his rival Leibniz – threw back English mathematics by 100 years.

To sum up, Kepler, Hooke and Newton became famous because they brought order to the description of planetary motion. They showed that all motion due to gravity follows from the same description, the inverse square distance. For this reason, the inverse square distance relation $a = GM/r^2$ is called the *universal* law of gravity. Achieving this unification of motion description, though of small practical significance, was widely publicized. The main reason were the age-old prejudices and fantasies linked with astrology.

In fact, the inverse square distance relation explains many additional phenomena. It explains the motion and shape of the Milky Way and of the other galaxies, the motion of many weather phenomena, and explains why the Earth has an atmosphere but the Moon does not. (Can you explain this?)

Gravitation on Earth

This inverse square dependence of gravitational acceleration is often, but incorrectly, called Newton's 'law' of gravitation. Indeed, the occultist and physicist Isaac Newton proved more elegantly than Hooke that the expression agreed with all astronomical and terrestrial observations. Above all, however, he organized a better public relations campaign, in which he falsely claimed to be the originator of the idea.

Newton published a simple proof showing that the description of astronomical gravitation also gives the correct description for stones thrown through the air, down here on 'father Earth'. To achieve this, he compared the acceleration a_m of the Moon with that of stones g. For the ratio between these two accelerations, the inverse square relation predicts a value $g/a_m = d_m^2/R^2$, where d_m the distance of the Moon and R is the radius of the Earth. The Moon's distance can be measured by triangulation, comparing

---

* Godfrey Harold Hardy (1877–1947) was an important number theorist, and the author of the well-known *A Mathematician's Apology*. He also 'discovered' the famous Indian mathematician Srinivasa Ramanujan, and brought him to Britain.

6 MOTION DUE TO GRAVITATION

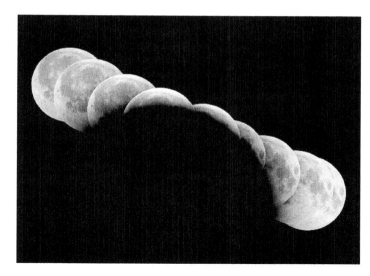

FIGURE 134 How to compare the radius of the Earth with that of the Moon during a partial lunar eclipse (© Anthony Ayiomamitis).

the position of the Moon against the starry background from two different points on Earth.* The result is $d_m/R = 60 \pm 3$, depending on the orbital position of the Moon, so that an average ratio $g/a_m = 3.6 \cdot 10^3$ is predicted from universal gravity. But both accelerations can also be measured directly. At the surface of the Earth, stones are subject to an acceleration due to gravitation with magnitude $g = 9.8 \text{ m/s}^2$, as determined by measuring the time that stones need to fall a given distance. For the Moon, the definition of acceleration, $a = dv/dt$, in the case of circular motion – roughly correct here – gives $a_m = d_m(2\pi/T)^2$, where $T = 2.4 \text{ Ms}$ is the time the Moon takes for one orbit around the Earth.** The measurement of the radius of the Earth*** yields $R = 6.4 \text{ Mm}$, so that

---

* The first precise – but not the first – measurement was achieved in 1752 by the French astronomers Lalande and La Caille, who simultaneously measured the position of the Moon seen from Berlin and from Le Cap.
** This expression for the centripetal acceleration is deduced easily by noting that for an object in circular motion, the magnitude v of the velocity $\boldsymbol{v} = d\boldsymbol{x}/dt$ is given as $v = 2\pi r/T$. The drawing of the vector $\boldsymbol{v}$ over time, the so-called *hodograph*, shows that it behaves exactly like the position of the object. Therefore the magnitude a of the acceleration $\boldsymbol{a} = d\boldsymbol{v}/dt$ is given by the corresponding expression, namely $a = 2\pi v/T$.
*** This is the hardest quantity to measure oneself. The most surprising way to determine the Earth's size is the following: watch a sunset in the garden of a house, with a stopwatch in hand, such as the one in your mobile phone. When the last ray of the Sun disappears, start the stopwatch and run upstairs. There, the Sun is still visible; stop the stopwatch when the Sun disappears again and note the time t. Measure the height difference h between the two eye positions where the Sun was observed. The Earth's radius R is then given by $R = k h/t^2$, with $k = 378 \cdot 10^6 \text{ s}^2$.

There is also a simple way to measure the distance to the Moon, once the size of the Earth is known. Take a photograph of the Moon when it is high in the sky, and call θ its zenith angle, i.e., its angle from the vertical above you. Make another photograph of the Moon a few hours later, when it is just above the horizon. On this picture, unlike the common optical illusion, the Moon is smaller, because it is further away. With a sketch the reason for this becomes immediately clear. If q is the ratio of the two angular diameters, the Earth–Moon distance d_m is given by the relation $d_m^2 = R^2 + (2Rq \cos\theta/(1 - q^2))^2$. Enjoy finding its derivation from the sketch.

Another possibility is to determine the size of the Moon by comparing it with the size of the shadow!Earth, during lunar eclipse, as shown in Figure 134. The distance to the Moon is then computed from its angular size, about 0.5°.

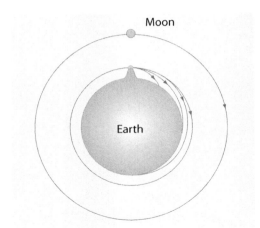 

FIGURE 135 A physicist's and an artist's view of the fall of the Moon: a diagram by Christiaan Huygens (not to scale) and a marble statue by Auguste Rodin.

the average Earth–Moon distance is $d_m = 0.38\,\text{Gm}$. One thus has $g/a_m = 3.6 \cdot 10^3$, in agreement with the above prediction. With this famous 'Moon calculation' we have thus shown that the inverse square property of gravitation indeed describes both the motion of the Moon and that of stones. You might want to deduce the value of the product GM for Earth.

Universal gravitation thus describes all motion due to gravity – both on Earth and in the sky. This was an important step towards the unification of physics. Before this discovery, from the observation that on the Earth all motion eventually comes to rest, whereas in the sky all motion is eternal, Aristotle and many others had concluded that motion in the *sublunar* world has *different* properties from motion in the *translunar* world. Several thinkers had criticized this distinction, notably the philosopher and rector of the University of Paris, Jean Buridan.* The Moon calculation was the most important result showing this distinction to be wrong. This is the reason for calling Hooke's expression (39) the *universal* gravitation.

Universal gravitation allows us to answer another old question. Why does the Moon not fall from the sky? Well, the preceding discussion showed that *fall* is motion due to gravitation. Therefore the Moon actually *is* falling, with the peculiarity that instead of falling *towards* the Earth, it is continuously falling *around* it. Figure 135 illustrates the idea. The Moon is continuously missing the Earth.**

The Moon is not the only object that falls around the Earth. Figure 137 shows another.

---

* Jean Buridan (*c.* 1295 to *c.* 1366) was also one of the first modern thinkers to discuss the rotation of the Earth about an axis.
** Another way to put it is to use the answer of the Dutch physicist Christiaan Huygens (1629–1695): the Moon does not fall from the sky because of the centrifugal acceleration. As explained on page 161, this explanation is often out of favour at universities.

There is a beautiful problem connected to the left side of the figure: Which points on the surface of the Earth can be hit by shooting from a mountain? And which points can be hit by shooting horizontally?

6 MOTION DUE TO GRAVITATION

FIGURE 136 A precision second pendulum, thus about 1 m in length; almost at the upper end, the vacuum chamber that compensates for changes in atmospheric pressure; towards the lower end, the wide construction that compensates for temperature variations of pendulum length; at the very bottom, the screw that compensates for local variations of the gravitational acceleration, giving a final precision of about 1 s per month (© Erwin Sattler OHG).

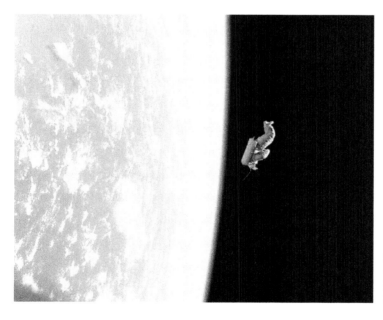

FIGURE 137 The man in orbit feels no weight, the blue atmosphere, which is not, does (NASA).

TABLE 26 Some measured values of the acceleration due to gravity.

PLACE	VALUE
Poles	9.83 m/s^2
Trondheim	9.8215243 m/s^2
Hamburg	9.8139443 m/s^2
Munich	9.8072914 m/s^2
Rome	9.8034755 m/s^2
Equator	9.78 m/s^2
Moon	1.6 m/s^2
Sun	273 m/s^2

Properties of gravitation: G and g

Gravitation implies that the path of a stone is not a parabola, as stated earlier, but actually an *ellipse* around the centre of the Earth. This happens for exactly the same reason that the planets move in ellipses around the Sun. Are you able to confirm this statement?

Universal gravitation allows us to understand the puzzling acceleration value $g = 9.8 \text{ m/s}^2$ we encountered in equation (6). The value is due to the relation

$$g = GM_{\text{Earth}}/R_{\text{Earth}}^2 \ . \tag{40}$$

The expression can be deduced from equation (39), universal gravity, by taking the Earth to be spherical. The everyday acceleration of gravity g thus results from the size of the Earth, its mass, and the universal constant of gravitation G. Obviously, the value for g is almost constant on the surface of the Earth, as shown in Table 26, because the Earth is almost a sphere. Expression (40) also explains why g gets smaller as one rises above the Earth, and the deviations of the shape of the Earth from sphericity explain why g is different at the poles and higher on a plateau. (What would g be on the Moon? On Mars? On Jupiter?)

By the way, it is possible to devise a simple machine, other than a yo-yo, that slows down the effective acceleration of gravity by a known amount, so that one can measure its value more easily. Can you imagine this machine?

Note that 9.8 is roughly π^2. This is *not* a coincidence: the metre has been chosen in such a way to make this (roughly) correct. The period T of a swinging pendulum, i.e., a

6 MOTION DUE TO GRAVITATION

back and forward swing, is given by*

$$T = 2\pi \sqrt{\frac{l}{g}}, \qquad (41)$$

where l is the length of the pendulum, and $g = 9.8 \text{ m/s}^2$ is the gravitational acceleration. (The pendulum is assumed to consist of a compact mass attached to a string of negligible mass.) The oscillation time of a pendulum depends only on the length of the string and on g, thus on the planet it is located on.

If the metre had been defined such that $T/2 = 1$ s, the value of the normal acceleration g would have been exactly $\pi^2 \text{ m/s}^2 = 9.869\,604\,401\,09 \text{ m/s}^2$. Indeed, this was the first proposal for the definition of the metre; it was made in 1673 by Huygens and repeated in 1790 by Talleyrand, but was rejected by the conference that defined the metre because variations in the value of g with geographical position, temperature-induced variations of the length of a pendulum and even air pressure variations induce errors that are too large to yield a definition of useful precision. (Indeed, all these effects must be corrected in pendulum clocks, as shown in Figure 136.)

Finally, the proposal was made to define the metre as $1/40\,000\,000$ of the circumference of the Earth through the poles, a so-called *meridian*. This proposal was almost identical to – but much more precise than – the pendulum proposal. The meridian definition of the metre was then adopted by the French national assembly on 26 March 1791, with the statement that 'a meridian passes under the feet of every human being, and all meridians are equal'. (Nevertheless, the distance from Equator to the poles is not exactly 10 Mm; that is a strange story. One of the two geographers who determined the size of the first metre stick was dishonest. The data he gave for his measurements – the general method of which is shown in Figure 138 – was fabricated. Thus the first official metre stick in Paris was shorter than it should be.)

Continuing our exploration of the gravitational acceleration g, we can still ask: Why does the Earth have the mass and size it has? And why does G have the value it has? The first question asks for a history of the Solar System; it is still unanswered and is topic of research. The second question is addressed in Appendix B.

If gravitation is indeed universal, and if all objects really attract each other, attraction should also occur between any two objects of everyday life. Gravity must also work *sideways*. This is indeed the case, even though the effects are extremely small. Indeed, the effects are so small that they were measured only long after universal gravity had predicted them. On the other hand, measuring this effect is the only way to determine the gravitational constant G. Let us see how to do it.

---

* Formula (41) is noteworthy mainly for all that is missing. The period of a pendulum does *not* depend on the mass of the swinging body. In addition, the period of a pendulum does *not* depend on the amplitude. (This is true as long as the oscillation angle is smaller than about 15°.) Galileo discovered this as a student, when observing a chandelier hanging on a long rope in the dome of Pisa. Using his heartbeat as a clock he found that even though the amplitude of the swing got smaller and smaller, the time for the swing stayed the same.

A leg also moves like a pendulum, when one walks normally. Why then do taller people tend to walk faster? Is the relation also true for animals of different size?

FIGURE 138 The measurements that lead to the definition of the metre (© Ken Alder).

We note that measuring the gravitational constant G is also the only way to determine the mass of the *Earth*. The first to do so, in 1798, was the English physicist Henry Cavendish; he used the machine, ideas and method of John Michell who died when attempting the experiment. Michell and Cavendish* called the aim and result of their experiments 'weighing the Earth'.

The idea of Michell was to suspended a horizontal handle, with two masses at the end, at the end of a long metal wire. He then approached two additional large masses at the two ends of the handle, avoiding any air currents, and measured how much the handle rotated. Figure 139 shows how to repeat this experiment in your basement, and Figure 140 how to perform it when you have a larger budget.

The value the gravitational constant G found in more elaborate versions of the Michell–Cavendish experiments is

$$G = 6.7 \cdot 10^{-11} \, \text{Nm}^2/\text{kg}^2 = 6.7 \cdot 10^{-11} \, \text{m}^3/\text{kg s}^2 \; . \tag{42}$$

Cavendish's experiment was thus the first to confirm that gravity also works sideways. The experiment also allows deducing the mass M of the Earth from its radius R and the relation $g = GM/R^2$. Therefore, the experiment also allows to deduce the average density of the Earth. Finally, as we will see later on, this experiment proves, if we keep in mind that the speed of light is finite and invariant, that space is curved. All this is achieved with this simple set-up!

---

* Henry Cavendish (b. 1731 Nice, d. 1810 London) was one of the great geniuses of physics; rich, autistic, misogynist, unmarried and solitary, he found many rules of nature, but never published them. Had he done so, his name would be much more well known. John Michell (1724–1793) was church minister, geologist and amateur astronomer.

6 MOTION DUE TO GRAVITATION

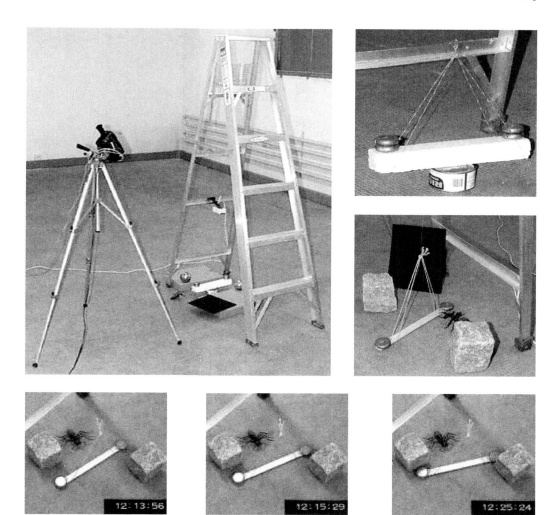

FIGURE 139 Home experiments that allow determining the gravitational constant G, weighing the Earth, proving that gravity also works sideways and showing that gravity curves space. Top left and right: a torsion balance made of foam and lead, with pétanque (boules) masses as fixed masses; centre right: a torsion balance made of wood and lead, with stones as fixed masses; bottom: a time sequence showing how the stones do attract the lead (© John Walker).

Cavendish found a mass density of the Earth of 5.5 times that of water. This was a surprising result, because rock only has 2.8 times the density of water. What is the origin of the large density value?

We note that G has a small value. Above, we mentioned that gravity limits motion. In fact, we can write the expression for universal gravitation in the following way:

$$\frac{ar^2}{M} = G > 0 \tag{43}$$

Gravity prevents uniform motion. In fact, we can say more: *Gravitation is the smallest*

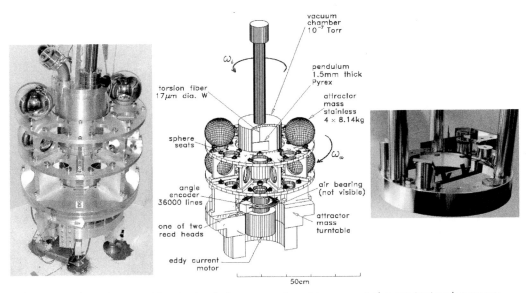

FIGURE 140 A modern precision torsion balance experiment to measure the gravitational constant, performed at the University of Washington (© Eöt-Wash Group).

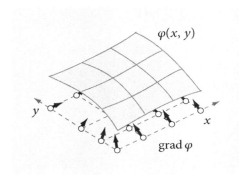

FIGURE 141 The potential and the gradient, visualized for two spatial dimensions.

possible effect of the environment on a moving body. All other effects come on top of gravity. However, it is not easy to put this statement in a simple formula.

Gravitation between everyday objects is weak. For example, two average people 1 m apart feel an acceleration towards each other that is less than that exerted by a common fly when landing on the skin. Therefore we usually do not notice the attraction to other people. When we notice it, it is much stronger than that. The measurement of G thus proves that gravitation cannot be the true cause of people falling in love, and also that erotic attraction is not of gravitational origin, but of a different source. The physical basis for love will be studied later in our walk: it is called *electromagnetism*.

The gravitational potential

Gravity has an important property: all effects of gravitation can also be described by another observable, namely the *(gravitational) potential* φ. We then have the simple relation

6 MOTION DUE TO GRAVITATION

that the acceleration is given by the *gradient* of the potential

$$\boldsymbol{a} = -\nabla\varphi \quad \text{or} \quad \boldsymbol{a} = -\text{grad}\,\varphi\,. \tag{44}$$

The gradient is just a learned term for 'slope along the steepest direction'. The gradient is defined for any point on a slope, is large for a steep one and small for a shallow one. The gradient points in the direction of steepest ascent, as shown in Figure 141. The gradient is abbreviated ∇, pronounced 'nabla', and is mathematically defined through the relation $\nabla\varphi = (\partial\varphi/\partial x, \partial\varphi/\partial y, \partial\varphi/\partial z) = \text{grad}\,\varphi$.* The minus sign in (44) is introduced by convention, in order to have higher potential values at larger heights. In everyday life, when the spherical shape of the Earth can be neglected, the gravitational potential is given by

$$\varphi = gh\,. \tag{45}$$

The potential φ is an interesting quantity; with a single number at every position in space we can describe the vector aspects of gravitational acceleration. It automatically gives that gravity in New Zealand acts in the opposite direction to gravity in Paris. In addition, the potential suggests the introduction of the so-called *potential energy U* by setting

$$U = m\varphi \tag{46}$$

and thus allowing us to determine the change of *kinetic* energy T of a body falling from a point 1 to a point 2 via

$$T_1 - T_2 = U_2 - U_1 \quad \text{or} \quad \tfrac{1}{2}m_1\boldsymbol{v}_1{}^2 - \tfrac{1}{2}m_2\boldsymbol{v}_2{}^2 = m\varphi_2 - m\varphi_1\,. \tag{47}$$

In other words, the *total energy*, defined as the sum of kinetic and potential energy, is *conserved* in motion due to gravity. This is a characteristic property of gravitation. Gravity conserves energy and momentum.

Not all accelerations can be derived from a potential; systems with this property are called *conservative*. Observation shows that accelerations due to friction are not conservative, but accelerations due to electromagnetism are. In short, we can either say that gravity can be described by a potential, or say that it conserves energy and momentum. Both mean the same. When the non-spherical shape of the Earth can be neglected, the potential energy of an object at height h is given by

$$U = mgh\,. \tag{48}$$

To get a feeling of how much energy this is, answer the following question. A car with mass 1 Mg falls down a cliff of 100 m. How much water can be heated from freezing point to boiling point with the energy of the car?

---

* In two or more dimensions slopes are written $\partial\varphi/\partial z$ – where ∂ is still pronounced 'd' – because in those cases the expression $d\varphi/dz$ has a slightly different meaning. The details lie outside the scope of this walk.

The shape of the Earth

Universal gravity also explains why the Earth and most planets are (almost) spherical. Since gravity increases with decreasing distance, a liquid body in space will always try to form a spherical shape. Seen on a large scale, the Earth is indeed liquid. We also know that the Earth is cooling down – that is how the crust and the continents formed. The sphericity of smaller solid objects encountered in space, such as the Moon, thus means that they used to be liquid in older times.

The Earth is thus not flat, but roughly spherical. Therefore, the top of two tall buildings is further apart than their base. Can this effect be measured?

Sphericity considerably simplifies the description of motion. For a spherical or a point-like body of mass M, the potential φ is

$$\varphi = -G\frac{M}{r}. \tag{49}$$

A potential considerably simplifies the description of motion, since a potential is additive: given the potential of a point particle, we can calculate the potential and then the motion around any other irregularly shaped object.* Interestingly, the number d of dimensions of space is coded into the potential φ of a spherical mass: the dependence of φ on the radius r is in fact $1/r^{d-2}$. The exponent $d-2$ has been checked experimentally to extremely high precision; no deviation of d from 3 has ever been found.

The concept of potential helps in understanding the *shape* of the Earth in more detail. Since most of the Earth is still liquid when seen on a large scale, its surface is always horizontal with respect to the direction determined by the combination of the accelerations of gravity and rotation. In short, the Earth is *not* a sphere. It is not an ellipsoid either. The mathematical shape defined by the equilibrium requirement is called a *geoid*. The geoid shape, illustrated in Figure 142, differs from a suitably chosen ellipsoid by at most 50 m. Can you describe the geoid mathematically? The geoid is an excellent approximation to the actual shape of the Earth; sea level differs from it by less than 20 metres. The differences can be measured with satellite radar and are of great interest to geologists and geographers. For example, it turns out that the South Pole is nearer to the equatorial

---

* Alternatively, for a general, extended body, the potential is found by requiring that the *divergence* of its gradient is given by the mass (or charge) density times some proportionality constant. More precisely, we have

$$\Delta\varphi = 4\pi G\rho \tag{50}$$

where $\rho = \rho(\boldsymbol{x}, t)$ is the mass volume density of the body and the so-called *Laplace operator* Δ, pronounced 'delta', is defined as $\Delta f = \nabla\nabla f = \partial^2 f/\partial x^2 + \partial^2 f/\partial y^2 + \partial^2 f/\partial z^2$. Equation (50) is called the *Poisson equation* for the potential φ. It is named after Siméon-Denis Poisson (1781–1840), eminent French mathematician and physicist. The positions at which ρ is not zero are called the *sources* of the potential. The so-called source term $\Delta\varphi$ of a function is a measure for how much the function $\varphi(x)$ at a point x differs from the average value in a region around that point. (Can you show this, by showing that $\Delta\varphi \sim \bar{\varphi} - \varphi(x)$?) In other words, the Poisson equation (50) implies that the actual value of the potential at a point is the same as the average value around that point minus the mass density multiplied by $4\pi G$. In particular, in the case of empty space the potential at a point is equal to the average of the potential around that point.

Often the concept of *gravitational field* is introduced, defined as $\boldsymbol{g} = -\nabla\varphi$. We avoid this in our walk, because we will discover that, following the theory of relativity, gravity is not due to a field at all; in fact even the concept of gravitational potential turns out to be only an approximation.

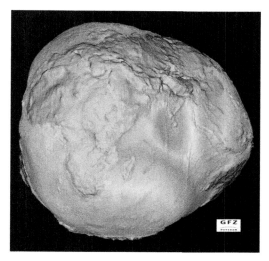

FIGURE 142 The shape of the Earth, with exaggerated height scale (© GeoForschungsZentrum Potsdam).

plane than the North Pole by about 30 m. This is probably due to the large land masses in the northern hemisphere.

Above we saw how the inertia of matter, through the so-called 'centrifugal force', increases the radius of the Earth at the Equator. In other words, the Earth is *flattened* at the poles. The Equator has a radius a of 6.38 Mm, whereas the distance b from the poles to the centre of the Earth is 6.36 Mm. The precise flattening $(a - b)/a$ has the value $1/298.3 = 0.0034$. As a result, the top of Mount Chimborazo in Ecuador, even though its height is only 6267 m above sea level, is about 20 km farther away from the centre of the Earth than the top of Mount Sagarmatha* in Nepal, whose height above sea level is 8850 m. The top of Mount Chimborazo is in fact the point on the surface most distant from the centre of the Earth.

The shape of the Earth has another important consequence. If the Earth stopped rotating (but kept its shape), the water of the oceans would flow from the Equator to the poles; all of Europe would be under water, except for the few mountains of the Alps that are higher than about 4 km. The northern parts of Europe would be covered by between 6 km and 10 km of water. Mount Sagarmatha would be over 11 km above sea level. We would also walk inclined. If we take into account the resulting change of shape of the Earth, the numbers come out somewhat smaller. In addition, the change in shape would produce extremely strong earthquakes and storms. As long as there are none of these effects, we can be *sure* that the Sun will indeed rise tomorrow, despite what some philosophers pretended.

Dynamics – how do things move in various dimensions?

The concept of potential is a powerful tool. If a body can move only along a – straight or curved – line, the concepts of kinetic and potential energy are sufficient to determine completely the way the body moves.

---

* Mount Sagarmatha is sometimes also called Mount Everest.

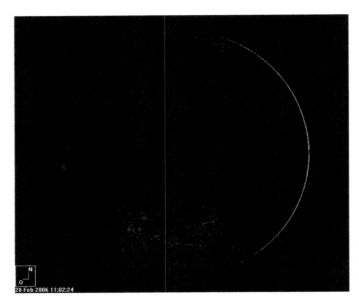

FIGURE 143 The change of the moon during the month, showing its libration (QuickTime film © Martin Elsässer)

In fact, motion in *one dimension* follows directly from energy conservation. For a body moving along a given curve, the speed at every instant is given by energy conservation.

If a body can move in *two dimensions* – i.e., on a flat or curved surface – *and* if the forces involved are *internal* (which is always the case in theory, but not in practice), the conservation of angular momentum can be used. The full motion in two dimensions thus follows from energy and angular momentum conservation. For example, all properties of free fall follow from energy and angular momentum conservation. (Are you able to show this?) Again, the potential is essential.

In the case of motion in *three dimensions*, a more general rule for determining motion is necessary. If more than *two spatial dimensions* are involved conservation is insufficient to determine how a body moves. It turns out that general motion follows from a simple principle: the time average of the difference between kinetic and potential energy must be as small as possible. This is called the *least action principle*. We will explain the details of this calculation method later. But again, the potential is the main ingredient in the calculation of change, and thus in the description of any example of motion.

For simple gravitational motions, motion is two-dimensional, in a plane. Most three-dimensional problems are outside the scope of this text; in fact, some of these problems are so hard that they still are subjects of research. In this adventure, we will explore three-dimensional motion only for selected cases that provide important insights.

The Moon

How long is a day on the Moon? The answer is roughly 29 Earth-days. That is the time that it takes for an observer on the Moon to see the Sun again in the same position in the sky.

One often hears that the Moon always shows the same side to the Earth. But this is

6 MOTION DUE TO GRAVITATION

FIGURE 144 High resolution maps (not photographs) of the near side (left) and far side (right) of the moon, showing how often the latter saved the Earth from meteorite impacts (courtesy USGS).

wrong. As one can check with the naked eye, a given feature in the centre of the face of the Moon at full Moon is not at the centre one week later. The various motions leading to this change are called *librations*; they are shown in the film in Figure 143. The motions appear mainly because the Moon does not describe a circular, but an elliptical orbit around the Earth and because the axis of the Moon is slightly inclined, when compared with that of its rotation around the Earth. As a result, only around 45 % of the Moon's surface is permanently hidden from Earth.

The first photographs of the hidden area of the Moon were taken in the 1960s by a Soviet artificial satellite; modern satellites provided exact maps, as shown in Figure 144. (Just zoom into the figure for fun.) The hidden surface is much more irregular than the visible one, as the hidden side is the one that intercepts most asteroids attracted by the Earth. Thus the gravitation of the Moon helps to deflect asteroids from the Earth. The number of animal life extinctions is thus reduced to a small, but not negligible number. In other words, the gravitational attraction of the Moon has saved the human race from extinction many times over.*

The trips to the Moon in the 1970s also showed that the Moon originated from the Earth itself: long ago, an object hit the Earth almost tangentially and threw a sizeable fraction of material up into the sky. This is the only mechanism able to explain the large size of the Moon, its low iron content, as well as its general material composition.

The Moon is receding from the Earth at 3.8 cm a year. This result confirms the old deduction that the tides slow down the Earth's rotation. Can you imagine how this measurement was performed? Since the Moon slows down the Earth, the Earth also changes shape due to this effect. (Remember that the shape of the Earth depends on its speed of

---

* The web pages www.minorplanetcenter.net/iau/lists/Closest.html and InnerPlot.html give an impression of the number of objects that almost hit the Earth every year. Without the Moon, we would have many additional catastrophes.

rotation.) These changes in shape influence the tectonic activity of the Earth, and maybe also the drift of the continents.

The Moon has many effects on animal life. A famous example is the midge *Clunio*, which lives on coasts with pronounced tides. Clunio spends between six and twelve weeks as a larva, sure then hatches and lives for only one or two hours as an adult flying insect, during which time it reproduces. The midges will only reproduce if they hatch during the low tide phase of a *spring tide*. Spring tides are the especially strong tides during the full and new moons, when the solar and lunar effects combine, and occur only every 14.8 days. In 1995, Dietrich Neumann showed that the larvae have two built-in clocks, a circadian and a circalunar one, which together control the hatching to precisely those few hours when the insect can reproduce. He also showed that the circalunar clock is synchronized by the brightness of the Moon at night. In other words, the larvae monitor the Moon at night and then decide when to hatch: they are the smallest known astronomers.

If insects can have circalunar cycles, it should come as no surprise that women also have such a cycle; however, in this case the precise origin of the cycle length is still unknown and a topic of research.

The Moon also helps to stabilize the tilt of the Earth's axis, keeping it more or less fixed relative to the plane of motion around the Sun. Without the Moon, the axis would change its direction irregularly, we would not have a regular day and night rhythm, we would have extremely large climate changes, and the evolution of life would have been impossible. Without the Moon, the Earth would also rotate much faster and we would have much less clement weather. The Moon's main remaining effect on the Earth, the precession of its axis, is responsible for the ice ages.

The orbit of the Moon is still a topic of research. It is still not clear why the Moon orbit is at a 5° to the ecliptic and how the orbit changed since the Moon formed. Possibly, the collision that led to the formation of the Moon tilted the rotation axis of the Earth and the original Moon; then over thousands of millions of years, both axes moved in complicated ways towards the ecliptic, one more than the other. During this evolution, the distance to the Moon is estimated to have increased by a factor of 15.

Orbits – conic sections and more

The path of a body continuously orbiting another under the influence of gravity is an *ellipse* with the central body at one focus. A circular orbit is also possible, a circle being a special case of an ellipse. Single encounters of two objects can also be *parabolas* or *hyperbolas*, as shown in Figure 145. Circles, ellipses, parabolas and hyperbolas are collectively known as *conic sections*. Indeed each of these curves can be produced by cutting a cone with a knife. Are you able to confirm this?

If orbits are mostly ellipses, it follows that comets *return*. The English astronomer Edmund Halley (1656–1742) was the first to draw this conclusion and to predict the return of a comet. It arrived at the predicted date in 1756, after his death, and is now named after him. The period of Halley's comet is between 74 and 80 years; the first recorded sighting was 22 centuries ago, and it has been seen at every one of its 30 passages since, the last time in 1986.

Depending on the initial energy and the initial angular momentum of the body with

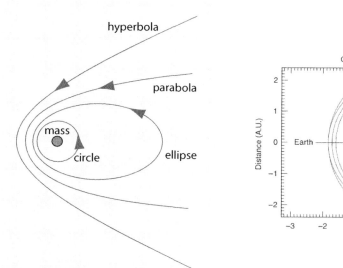
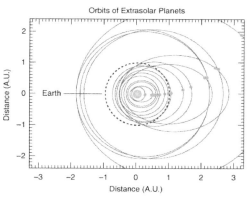

FIGURE 145 The possible orbits, due to universal gravity, of a small mass around a *single* large mass (left) and a few recent examples of measured orbits (right), namely those of some extrasolar planets and of the Earth, all drawn around their respective central star, with distances given in astronomical units (© Geoffrey Marcy).

respect to the central planet, paths are either *elliptic*, *parabolic* or *hyperbolic*. Can you determine the conditions for the energy and the angular momentum needed for these paths to appear?

In practice, parabolic orbits do not exist in nature. (Some comets seem to approach this case when moving around the Sun; but almost all comets follow elliptical paths – as long as they are far from other planets.). Hyperbolic paths do exist; artificial satellites follow them when they are shot towards a planet, usually with the aim of changing the direction of the satellite's journey across the Solar System.

Why does the inverse square 'law' lead to conic sections? First, for two bodies, the total angular momentum L is a constant:

$$L = mr^2\dot{\varphi} = mr^2\left(\frac{d\varphi}{dt}\right) \qquad (51)$$

and therefore the motion lies in a plane. Also the energy E is a constant

$$E = \tfrac{1}{2}m\left(\frac{dr}{dt}\right)^2 + \tfrac{1}{2}m\left(r\frac{d\varphi}{dt}\right)^2 - G\frac{mM}{r} \ . \qquad (52)$$

Together, the two equations imply that

$$r = \frac{L^2}{Gm^2 M} \frac{1}{1 + \sqrt{1 + \frac{2EL^2}{G^2 m^3 M^2}} \cos \varphi} \,. \tag{53}$$

Now, any curve defined by the general expression

$$r = \frac{C}{1 + e \cos \varphi} \quad \text{or} \quad r = \frac{C}{1 - e \cos \varphi} \tag{54}$$

is an ellipse for $0 < e < 1$, a parabola for $e = 1$ and a hyperbola for $e > 1$, one focus being at the origin. The quantity e, called the *eccentricity*, describes how squeezed the curve is. In other words, a body in orbit around a central mass follows a conic section.

In all orbits, also the heavy mass moves. In fact, both bodies orbit around the common centre of mass. Both bodies follow the same type of curve – ellipse, parabola or hyperbola – but the sizes of the two curves differ.

If more than two objects move under mutual gravitation, many additional possibilities for motions appear. The classification and the motions are quite complex. In fact, this so-called *many-body problem* is still a topic of research, both for astronomers and for mathematicians. Let us look at a few observations.

When several planets circle a star, they also attract each other. Planets thus do not move in perfect ellipses. The largest deviation is a perihelion shift, as shown in Figure 114. It is observed for Mercury and a few other planets, including the Earth. Other deviations from elliptical paths appear during a single orbit. In 1846, the observed deviations of the motion of the planet Uranus from the path predicted by universal gravity were used to predict the existence of another planet, Neptune, which was discovered shortly afterwards.

We have seen that mass is always positive and that gravitation is thus always attractive; there is *no antigravity*. Can gravity be used for *levitation* nevertheless, using more than two bodies? Yes; there are two examples.* The first are the geostationary satellites, which are used for easy transmission of television and other signals from and towards Earth.

The *Lagrangian libration points* are the second example. Named after their discoverer, these are points in space near a two-body system, such as Moon–Earth or Earth–Sun, in which small objects have a stable equilibrium position. An overview is given in Figure 147. Can you find their precise position, remembering to take rotation into account? There are three additional Lagrangian points on the Earth–Moon line (or Sun–planet line). How many of them are stable?

There are thousands of asteroids, called *Trojan asteroids*, at and around the Lagrangian points of the Sun–Jupiter system. In 1990, a Trojan asteroid for the Mars–Sun system was discovered. Finally, in 1997, an 'almost Trojan' asteroid was found that follows the Earth on its way around the Sun (it is only transitionary and follows a somewhat more complex orbit). This 'second companion' of the Earth has a diameter of 5 km. Similarly, on the

---

* Levitation is discussed in detail in the section on electrodynamics.

6 MOTION DUE TO GRAVITATION

FIGURE 146 Geostationary satellites, seen here in the upper left quadrant, move against the other stars and show the location of the celestial Equator. (MP4 film © Michael Kunze)

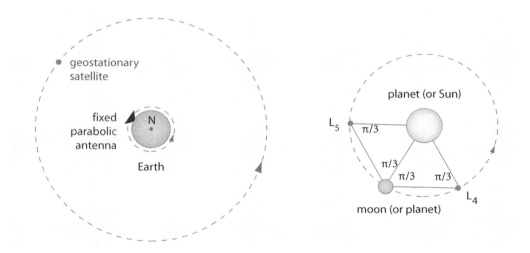

FIGURE 147 Geostationary satellites (left) and the main stable Lagrangian points (right).

main Lagrangian points of the Earth–Moon system a high concentration of dust has been observed.

Astronomers know that many other objects follow irregular orbits, especially asteroids. For example, asteroid 2003 YN107 followed an irregular orbit, shown in Figure 148, that accompanied the Earth for a number of years.

To sum up, the single equation $\boldsymbol{a} = -GM\boldsymbol{r}/r^3$ correctly describes a large number of phenomena in the sky. The first person to make clear that this expression describes

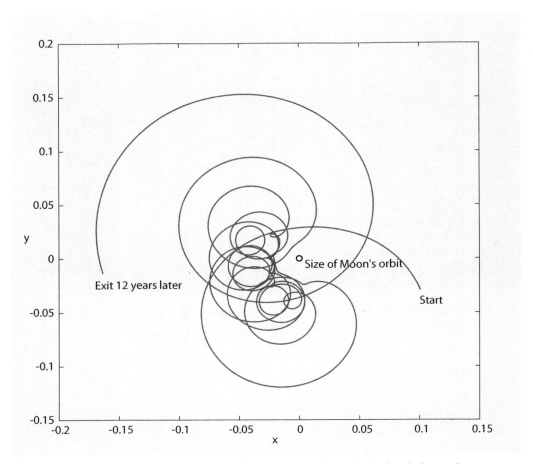

FIGURE 148 An example of irregular orbit, partly measured and partly calculated, due to the gravitational attraction of *several* masses: the orbit of the temporary Earth quasi-satellite 2003 YN107 in geocentric coordinates. This asteroid, with a diameter of 20(10) m, became orbitally trapped near the Earth around 1995 and remained so until 2006. The black circle represents the Moon's orbit around the Earth. (© Seppo Mikkola).

everything happening in the sky was Pierre Simon Laplace in his famous treatise *Traité de mécanique céleste*. When Napoleon told him that he found no mention about the creator in the book, Laplace gave a famous, one sentence summary of his book: *Je n'ai pas eu besoin de cette hypothèse.* 'I had no need for this hypothesis.' In particular, Laplace studied the stability of the Solar System, the eccentricity of the lunar orbit, and the eccentricities of the planetary orbits, always getting full agreement between calculation and measurement.

These results are quite a feat for the simple expression of universal gravitation; they also explain why it is called 'universal'. But how *accurate* is the formula? Since astronomy allows the most precise measurements of gravitational motion, it also provides the most stringent tests. In 1849, Urbain Le Verrier concluded after intensive study that there was only one known example of a discrepancy between observation and universal gravity, namely one observation for the planet Mercury. (Nowadays a few more are known.) The

FIGURE 149 Tides at Saint-Valéry en Caux on 20 September 2005 (© Gilles Régnier).

point of least distance to the Sun of the orbit of planet Mercury, its *perihelion*, rotates around the Sun at a rate that is slightly less than that predicted: he found a tiny difference, around 38 ″ per century. (This was corrected to 43 ″ per century in 1882 by Simon Newcomb.) Le Verrier thought that the difference was due to a planet between Mercury and the Sun, *Vulcan*, which he chased for many years without success. Indeed, Vulcan does not exist. The correct explanation of the difference had to wait for Albert Einstein.

Tides

Why do physics texts always talk about tides? Because, as general relativity will show, tides prove that space is curved! It is thus useful to study them in a bit more detail. Figure 149 how striking tides can be. Gravitation explains the sea tides as results of the attraction of the ocean water by the Moon and the Sun. Tides are interesting; even though the amplitude of the tides is only about 0.5 m on the open sea, it can be up to 20 m at special places near the coast. Can you imagine why? The *soil* is also lifted and lowered by the Sun and the Moon, by about 0.3 m, as satellite measurements show. Even the *atmosphere* is subject to tides, and the corresponding pressure variations can be filtered out from the weather pressure measurements.

Tides appear for any *extended* body moving in the gravitational field of another. To understand the origin of tides, picture a body in orbit, like the Earth, and imagine its components, such as the segments of Figure 150, as being held together by springs. Universal gravity implies that orbits are slower the more distant they are from a central body. As a result, the segment on the outside of the orbit would like to be slower than the cent-

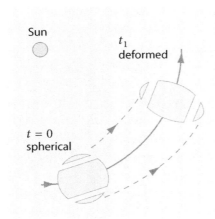

FIGURE 150 Tidal deformations due to gravity.

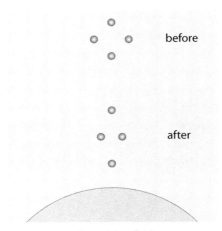

FIGURE 151 The origin of tides.

ral one; but it is *pulled* by the rest of the body through the springs. In contrast, the inside segment would like to orbit more rapidly but is *retained* by the others. Being slowed down, the inside segments want to fall towards the Sun. In sum, both segments feel a pull away from the centre of the body, limited by the springs that stop the deformation. Therefore, *extended bodies are deformed in the direction of the field inhomogeneity.*

For example, as a result of tidal forces, the Moon always has (roughly) the same face to the Earth. In addition, its radius in direction of the Earth is larger by about 5 m than the radius perpendicular to it. If the inner springs are too weak, the body is torn into pieces; in this way a *ring* of fragments can form, such as the asteroid ring between Mars and Jupiter or the rings around Saturn.

Let us return to the Earth. If a body is surrounded by water, it will form bulges in the direction of the applied gravitational field. In order to measure and compare the strength of the tides from the Sun and the Moon, we reduce tidal effects to their bare minimum. As shown in Figure 151, we can study the deformation of a body due to gravity by studying the arrangement of four bodies. We can study the free fall case, because orbital motion and free fall are equivalent. Now, gravity makes some of the pieces approach and others diverge, depending on their relative positions. The figure makes clear that the strength of the deformation – water has no built-in springs – depends on the change of gravitational acceleration with distance; in other words, the *relative* acceleration that leads to the tides is proportional to the derivative of the gravitational acceleration.

Using the numbers from Appendix B, the gravitational accelerations from the Sun and the Moon measured on Earth are

$$a_{\text{Sun}} = \frac{GM_{\text{Sun}}}{d_{\text{Sun}}^2} = 5.9 \, \text{mm/s}^2$$

$$a_{\text{Moon}} = \frac{GM_{\text{Moon}}}{d_{\text{Moon}}^2} = 0.033 \, \text{mm/s}^2 \tag{55}$$

and thus the attraction from the Moon is about 178 times weaker than that from the Sun.

6 MOTION DUE TO GRAVITATION

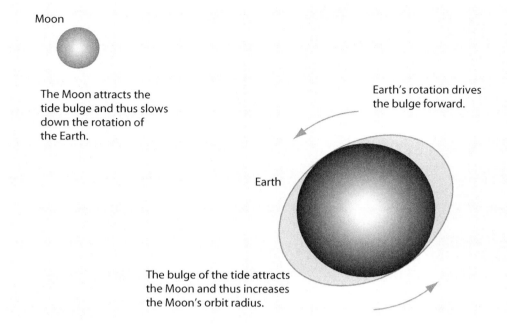

FIGURE 152 The Earth, the Moon and the friction effects of the tides (not to scale).

When two nearby bodies fall near a large mass, the relative acceleration is proportional to their distance, and follows $da = (da/dr)\, dr$. The proportionality factor $da/dr = \nabla a$, called the *tidal acceleration* (gradient), is the true measure of tidal effects. Near a large spherical mass M, it is given by

$$\frac{da}{dr} = -\frac{2GM}{r^3} \qquad (56)$$

which yields the values

$$\frac{da_{\text{Sun}}}{dr} = -\frac{2GM_{\text{Sun}}}{d_{\text{Sun}}^3} = -0.8 \cdot 10^{-13}/\text{s}^2$$
$$\frac{da_{\text{Moon}}}{dr} = -\frac{2GM_{\text{Moon}}}{d_{\text{Moon}}^3} = -1.7 \cdot 10^{-13}/\text{s}^2\ . \qquad (57)$$

In other words, despite the much weaker pull of the Moon, its tides are predicted to be over *twice as strong* as the tides from the Sun; this is indeed observed. When Sun, Moon and Earth are aligned, the two tides add up; these so-called *spring tides* are especially strong and happen every 14.8 days, at full and new moon.

Tides lead to a pretty puzzle. Moon tides are much stronger than Sun tides. This implies that the Moon is much denser than the Sun. Why?

Tides also produce *friction*, as shown in Figure 152. The friction leads to a slowing of the Earth's rotation. Nowadays, the slowdown can be measured by precise clocks (even

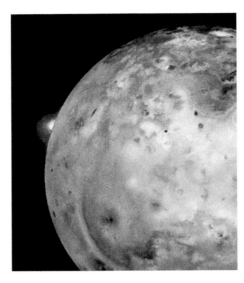

FIGURE 153 A spectacular result of tides: volcanism on Io (NASA).

though short time variations due to other effects, such as the weather, are often larger). The results fit well with fossil results showing that 400 million years ago, in the Devonian period, a year had 400 days, and a day about 22 hours. It is also estimated that 900 million years ago, each of the 481 days of a year were 18.2 hours long. The friction at the basis of this slowdown also results in an increase in the distance of the Moon from the Earth by about 3.8 cm per year. Are you able to explain why?

As mentioned above, the tidal motion of the soil is also responsible for the triggering of *earthquakes*. Thus the Moon can have also dangerous effects on Earth. (Unfortunately, knowing the mechanism does not allow predicting earthquakes.) The most fascinating example of tidal effects is seen on Jupiter's satellite Io. Its tides are so strong that they induce intense volcanic activity, as shown in Figure 153, with eruption plumes as high as 500 km. If tides are even stronger, they can destroy the body altogether, as happened to the body between Mars and Jupiter that formed the planetoids, or (possibly) to the moons that led to Saturn's rings.

In summary, tides are due to relative accelerations of nearby mass points. This has an important consequence. In the chapter on general relativity we will find that time multiplied by the speed of light plays the same role as length. Time then becomes an additional dimension, as shown in Figure 154. Using this similarity, two free particles moving in the same direction correspond to parallel lines in space-time. Two particles falling side-by-side also correspond to parallel lines. Tides show that such particles approach each other. In other words, tides imply that parallel lines approach each other. But parallel lines can approach each other *only* if space-time is curved. In short, tides imply *curved* space-time and space. This simple reasoning could have been performed in the eighteenth century; however, it took another 200 years and Albert Einstein's genius to uncover it.

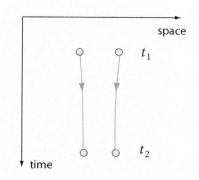

FIGURE 154 Particles falling side-by-side approach over time.

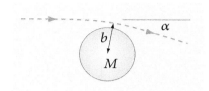

FIGURE 155 Masses bend light.

Can light fall?

> Die Maxime, jederzeit selbst zu denken, ist die Aufklärung.
> Immanuel Kant*

Towards the end of the seventeenth century people discovered that light has a finite velocity – a story which we will tell in detail later. An entity that moves with infinite velocity cannot be affected by gravity, as there is no time to produce an effect. An entity with a finite speed, however, should feel gravity and thus fall.

Does its speed increase when light reaches the surface of the Earth? For almost three centuries people had no means of detecting any such effect; so the question was not investigated. Then, in 1801, the Prussian astronomer Johann Soldner (1776–1833) was the first to put the question in a different way. Being an astronomer, he was used to measuring stars and their observation angles. He realized that light passing near a massive body would be *deflected* due to gravity.

Soldner studied a body on a hyperbolic path, moving with velocity c past a spherical mass M at distance b (measured from the centre), as shown in Figure 155. Soldner deduced the deflection angle

$$\alpha_{\text{univ. grav.}} = \frac{2}{b}\frac{GM}{c^2} \, . \qquad (58)$$

The value of the angle is largest when the motion is just grazing the mass M. For light deflected by the mass of the Sun, the angle turns out to be at most a tiny $0.88\,''= 4.3\,\mu\text{rad}$. In Soldner's time, this angle was too small to be measured. Thus the issue was forgotten. Had it been pursued, general relativity would have begun as an experimental science, and not as the theoretical effort of Albert Einstein! Why? The value just calculated is *different* from the measured value. The first measurement took place in 1919;** it found the correct dependence on the distance, but found a deflection of up to $1.75\,''$, exactly double that of expression (58). The reason is not easy to find; in fact, it is due to the curvature of space,

---

* The maxim to think at all times for oneself is the enlightenment.
** By the way, how would you measure the deflection of light near the bright Sun?

as we will see. In summary, light can fall, but the issue hides some surprises.

Mass: inertial and gravitational

Mass describes how an object interacts with others. In our walk, we have encountered two of its aspects. *Inertial mass* is the property that keeps objects moving and that offers resistance to a change in their motion. *Gravitational mass* is the property responsible for the acceleration of bodies nearby (the active aspect) or of being accelerated by objects nearby (the passive aspect). For example, the active aspect of the mass of the Earth determines the surface acceleration of bodies; the passive aspect of the bodies allows us to weigh them in order to measure their mass using distances only, e.g. on a scale or a balance. The gravitational mass is the basis of *weight*, the difficulty of lifting things.*

Is the gravitational mass of a body equal to its inertial mass? A rough answer is given by the experience that an object that is difficult to move is also difficult to lift. The simplest experiment is to take two bodies of different masses and let them fall. If the acceleration is the same for all bodies, inertial mass is equal to (passive) gravitational mass, because in the relation $ma = \nabla(GMm/r)$ the left-hand m is actually the inertial mass, and the right-hand m is actually the gravitational mass.

Already in the seventeenth century Galileo had made widely known an even older argument showing without a single experiment that the gravitational acceleration is indeed *the same* for all bodies. If larger masses fell more rapidly than smaller ones, then the following paradox would appear. Any body can be seen as being composed of a large fragment attached to a small fragment. If small bodies really fell less rapidly, the small fragment would slow the large fragment down, so that the complete body would have to fall *less* rapidly than the larger fragment (or break into pieces). At the same time, the body being larger than its fragment, it should fall *more* rapidly than that fragment. This is obviously impossible: all masses must fall with the same acceleration.

Many accurate experiments have been performed since Galileo's original discussion. In all of them the independence of the acceleration of free fall from mass and material composition has been confirmed with the precision they allowed. In other words, experiments confirm:

▷ Gravitational mass and inertial mass are *equal*.

What is the origin of this mysterious equality?

The equality of gravitational and inertial mass is not a mystery at all. Let us go back to the definition of mass as a negative inverse acceleration ratio. We mentioned that the physical origin of the accelerations does not play a role in the definition because the origin does not appear in the expression. In other words, the value of the mass is by definition independent of the interaction. That means in particular that inertial mass, based on and measured with the electromagnetic interaction, and gravitational mass are identical *by definition*.

The best proof of the equality of inertial and gravitational mass is illustrated in Figure 156: it shows that the two concepts only differ by the viewpoint of the observer. Inertial mass and gravitational mass describe the same observation.

---

* What are the weight values shown by a balance for a person of 85 kg juggling three balls of 0.3 kg each?

6 MOTION DUE TO GRAVITATION

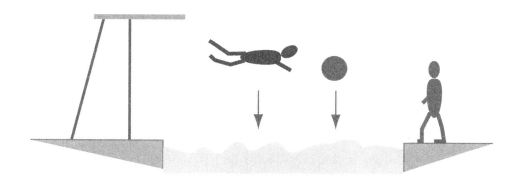

FIGURE 156 The falling ball is in inertial motion for a falling observer and in gravitational motion for an observer on the ground. Therefore, inertial mass is the same as gravitational mass.

We also note that we have not defined a separate concept of 'passive gravitational mass'. (This concept is sometimes found in research papers.) The mass being accelerated by gravitation is the inertial mass. Worse, there is no way to define a 'passive gravitational mass' that differs from inertial mass. Try it! All methods that measure a passive gravitational mass, such as weighing an object, cannot be distinguished from the methods that determine inertial mass from its reaction to acceleration. Indeed, all these methods use the same non-gravitational mechanisms. Bathroom scales are a typical example.

Indeed, if the 'passive gravitational mass' were different from the inertial mass, we would have strange consequences. Not only is it hard to distinguish the two in an experiment; for those bodies for which it were different we would get into trouble with energy conservation.

In fact, also assuming that ('active') 'gravitational mass' differs from inertial mass gets us into trouble. How could 'gravitational mass' differ from inertial mass? Would the difference depend on relative velocity, time, position, composition or on mass itself? No. Each of these possibilities contradicts either energy or momentum conservation.

In summary, it is no wonder that all measurements confirm the equality of all mass types: there is no other option – as Galileo pointed out. The lack of other options is due to the fundamental equivalence of all mass definitions:

▷ Mass ratios are acceleration ratios.

The topic is usually rehashed in general relativity, with no new results, because the definition of mass remains the same. Gravitational and inertial masses remain equal. In short:

▷ Mass is a unique property of each body.

Another, deeper issue remains, though. What is the *origin* of mass? Why does it exist? This simple but deep question cannot be answered by classical physics. We will need some patience to find out.

Curiosities and fun challenges about gravitation

> Fallen ist weder gefährlich noch eine Schande;
> Liegen bleiben ist beides.*
> Konrad Adenauer

Cosmonauts on the International Space Station face two challenges: cancer-inducing cosmic radiation and the lack of gravity. The lack of gravity often leads to orientation problems and nausea in the first days, the so-called *space sickness* and *motion sickness*. When these disappear, the muscles start to reduce in volume by a few % per month, bones get weaker every week, the immune systems is on permanent alarm state, blood gets pumped into the head more than usual and produces round 'baby' faces – easily seen on television – and strong headaches, legs loose blood and get thinner, body temperature permanently increases by over one degree Celsius, the brain gets compressed by the blood and spinal fluid, and the eyesight deteriorates, because also the eyes get compressed. When cosmonauts return to Earth after six months in space, they have weak bones and muscles, and they are unable to walk and stand. They need a day or two to learn to do so again. Later, they often get hernias, and because of the bone reduction, kidney stones. Other health issues are likely to exist; but they have been kept confidential by cosmonauts in order to maintain their image and their chances for subsequent missions.

* *

Gravity on the Moon is only one sixth of that on the Earth. Why does this imply that it is difficult to walk quickly and to run on the Moon (as can be seen in the TV images recorded there)?

* *

Understand and explain the following statement: a beam balance measures mass, a spring scale measures weight.

* *

Does the Earth have other satellites apart from the Moon and the artificial satellites shot into orbit up by rockets? Yes. The Earth has a number of mini-satellites and a large number of quasi-satellites. An especially long-lived quasi-satellite, an asteroid called 2016 HO3, has a size of about 60 m and was discovered in 2016. As shown in Figure 157, it orbits the Earth and will continue to do so for another few hundred years, at a distance from 40 to 100 times that of the Moon.

* *

---

* 'Falling is neither dangerous nor a shame; to keep lying is both.' Konrad Adenauer (b. 1876 Köln, d. 1967 Rhöndorf), West German Chancellor.

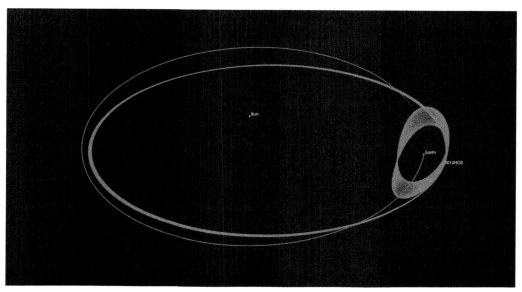

FIGURE 157 The calculated obit of the quasi-satellite 2016 HO3, a temporary companion of the Earth (courtesy NASA).

Show that a sphere bouncing – without energy loss – down an inclined plane, hits the plane in spots whose distances increase by a constant amount at every bounce.

* *

Is the acceleration due to gravity constant over time? Not really. Every day, it is estimated that 10^8 kg of material fall onto the Earth in the form of meteorites and asteroids. (Examples can be seen in Figure 158 and Figure 159.) Nevertheless, it is unknown whether the mass of the Earth increases with time (due to collection of meteorites and cosmic dust) or decreases (due to gas loss). If you find a way to settle the issue, publish it.

* *

Incidentally, discovering objects hitting the Earth is not at all easy. Astronomers like to point out that an asteroid as large as the one that led to the extinction of the dinosaurs could hit the Earth without any astronomer noticing in advance, if the direction is slightly unusual, such as from the south, where few telescopes are located.

* *

Several humans have survived free falls from aeroplanes for a thousand metres or more, even though they had no parachute. A minority of them even did so without any harm at all. How was this possible?

* *

Imagine that you have twelve coins of identical appearance, of which one is a forgery. The forged one has a different mass from the eleven genuine ones. How can you decide which is the forged one and whether it is lighter or heavier, using a simple balance only

FIGURE 158 A composite photograph of the Perseid meteor shower that is visible every year in mid August. In that month, the Earth crosses the cloud of debris stemming from comet Swift–Tuttle, and the source of the meteors appears to lie in the constellation of Perseus, because that is the direction in which the Earth is moving in mid August. The effect and the picture are thus similar to what is seen on the windscreen when driving by car while it is snowing. (© Brad Goldpaint at goldpaintphotography.com).

FIGURE 159 Two photographs, taken about a second apart, showing a meteor break-up (© Robert Mikaelyan).

three times?

You have nine identically-looking spheres, all of the same mass, except one, which is heavier. Can you determine which one, using the balance only two times?

∗ ∗

For a physicist, *antigravity* is repulsive gravity – it does not exist in nature. Nevertheless,

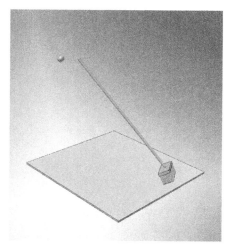

FIGURE 160 Brooms fall more rapidly than stones (© Luca Gastaldi).

the term 'antigravity' is used incorrectly by many people, as a short search on the internet shows. Some people call any effect that *overcomes* gravity, 'antigravity'. However, this definition implies that tables and chairs are antigravity devices. Following the definition, most of the wood, steel and concrete producers are in the antigravity business. The internet definition makes absolutely no sense.

* *

What is the cheapest way to switch gravity off for 25 seconds?

* *

Do all objects on Earth fall with the same acceleration of $9.8 \, \text{m/s}^2$, assuming that air resistance can be neglected? No; every housekeeper knows that. You can check this by yourself. As shown in Figure 160, a broom angled at around 35° hits the floor before a stone, as the sounds of impact confirm. Are you able to explain why?

* *

Also bungee jumpers are accelerated more strongly than g. For a bungee cord of mass m and a jumper of mass M, the maximum acceleration a is

$$a = g \left(1 + \frac{m}{8M} \left(4 + \frac{m}{M} \right) \right) . \tag{59}$$

Can you deduce the relation from Figure 161?

* *

Guess: What is the mass of a ball of cork with a radius of 1 m?

* *

Guess: One thousand 1 mm diameter steel balls are collected. What is the mass?

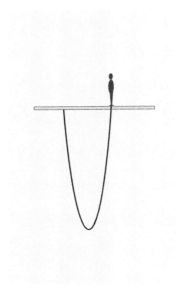

FIGURE 161 The starting situation for a bungee jumper.

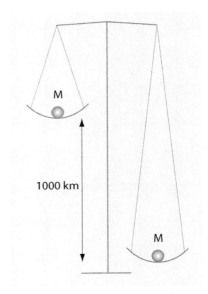

FIGURE 162 An honest balance?

∗ ∗

How can you use your observations made during your travels with a bathroom scale to show that the Earth is not flat?

∗ ∗

Both the Earth and the Moon attract bodies. The centre of mass of the Earth–Moon system is 4800 km away from the centre of the Earth, quite near its surface. Why do bodies on Earth still fall towards the centre of the Earth?

∗ ∗

Does every spherical body fall with the same acceleration? No. If the mass of the object is comparable to that of the Earth, the distance decreases in a different way. Can you confirm this statement? Figure 162 shows a related puzzle. What then is wrong about Galileo's argument about the constancy of acceleration of free fall?

∗ ∗

What is the fastest speed that a human can achieve making use of gravitational acceleration? There are various methods that try this; a few are shown in Figure 163. Terminal speed of free falling skydivers can be even higher, but no reliable record speed value exists. The last word is not spoken yet, as all these records will be surpassed in the coming years. It is important to require normal altitude; at stratospheric altitudes, speed values can be four times the speed values at low altitude.

∗ ∗

It is easy to put a mass of a kilogram onto a table. Twenty kilograms is harder. A thousand

FIGURE 163 Reducing air resistance increases the terminal speed: left, the 2007 speed skiing world record holder Simone Origone with 69.83 m/s and right, the 2007 speed world record holder for bicycles on snow Éric Barone with 61.73 m/s (© Simone Origone, Éric Barone).

is impossible. However, $6 \cdot 10^{24}$ kg is easy. Why?

* *

The friction between the Earth and the Moon slows down the rotation of both. The Moon stopped rotating millions of years ago, and the Earth is on its way to doing so as well. When the Earth stops rotating, the Moon will stop moving away from Earth. How far will the Moon be from the Earth at that time? Afterwards however, even further in the future, the Moon will move back towards the Earth, due to the friction between the Earth–Moon system and the Sun. Even though this effect would only take place if the Sun burned for ever, which is known to be false, can you explain it?

* *

When you run towards the east, you *lose weight*. There are two different reasons for this: the 'centrifugal' acceleration increases so that the force with which you are pulled down diminishes, and the Coriolis force appears, with a similar result. Can you estimate the size of the two effects?

* *

Laboratories use two types of ultracentrifuges: *preparative* ultracentrifuges isolate viruses, organelles and biomolecules, whereas *analytical* ultracentrifuges measure shape and mass of macromolecules. The fastest commercially available models achieve 200 000 rpm, or 3.3 kHz, and a centrifugal acceleration of $10^6 \cdot g$.

* *

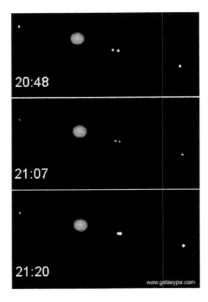

FIGURE 164 The four satellites of Jupiter discovered by Galileo and their motion (© Robin Scagell).

What is the relation between the time a stone takes falling through a distance l and the time a pendulum takes swinging though half a circle of radius l? (This problem is due to Galileo.) How many digits of the number π can one expect to determine in this way?

* *

Why can a spacecraft accelerate through the *slingshot effect* when going round a planet, despite momentum conservation? It is speculated that the same effect is also the reason for the few exceptionally fast stars that are observed in the galaxy. For example, the star HE0457-5439 moves with 720 km/s, which is much higher than the 100 to 200 km/s of most stars in the Milky Way. It seems that the role of the accelerating centre was taken by a black hole.

* *

The orbit of a planet around the Sun has many interesting properties. What is the hodograph of the orbit? What is the hodograph for parabolic and hyperbolic orbits?

* *

The *Galilean satellites* of Jupiter, shown in Figure 164, can be seen with small amateur telescopes. Galileo discovered them in 1610 and called them the *Medicean satellites*. (Today, they are named, in order of increasing distance from Jupiter, as Io, Europa, Ganymede and Callisto.) They are almost mythical objects. They were the first bodies found that obviously did not orbit the Earth; thus Galileo used them to deduce that the Earth is not at the centre of the universe. The satellites have also been candidates to be the first *standard clock*, as their motion can be predicted to high accuracy, so that the 'standard time' could be read off from their position. Finally, due to this high accuracy, in 1676, the speed of light was first measured with their help, as told in the section on special relativity.

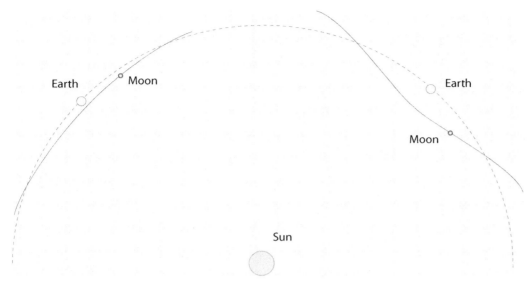

FIGURE 165 Which of the two Moon paths is correct?

* *

A simple, but difficult question: if all bodies attract each other, why don't or didn't all stars fall towards each other? Indeed, the inverse square expression of universal gravity has a limitation: it does not allow one to make sensible statements about the matter in the universe. Universal gravity predicts that a homogeneous mass distribution is unstable; indeed, an inhomogeneous distribution is observed. However, universal gravity does not predict the average mass density, the darkness at night, the observed speeds of the distant galaxies, etc. In fact, 'universal' gravity does not explain or predict a single property of the universe. To do this, we need general relativity.

* *

The acceleration g due to gravity at a depth of 3000 km is 10.05 m/s^2, over 2 % more than at the surface of the Earth. How is this possible? Also, on the Tibetan plateau, g is influenced by the material below it.

* *

When the Moon circles the Sun, does its path have sections *concave* towards the Sun, as shown at the right of Figure 165, or not, as shown on the left? (Independent of this issue, both paths in the diagram disguise that the Moon path does *not* lie in the same plane as the path of the Earth around the Sun.)

* *

You can prove that objects *attract each other* (and that they are not only attracted by the Earth) with a simple experiment that anybody can perform at home, as described on the www.fourmilab.ch/gravitation/foobar website.

* *

It is instructive to calculate the *escape velocity* from the Earth, i.e., that velocity with which a body must be thrown so that it never falls back. It turns out to be around 11 km/s. (This was called the *second cosmic velocity* in the past; the *first cosmic velocity* was the name given to the lowest speed for an orbit, 7.9 km/s.) The exact value of the escape velocity depends on the latitude of the thrower, and on the direction of the throw. (Why?)

What is the escape velocity from the Solar System? (It was once called the *third cosmic velocity*.) By the way, the escape velocity from our galaxy is over 500 km/s. What would happen if a planet or a system were so heavy that the escape velocity from it would be larger than the speed of light?

* *

What is the largest asteroid one can escape from by jumping?

* *

For bodies of irregular shape, the centre of gravity of a body is *not* the same as the centre of mass. Are you able to confirm this? (Hint: Find and use the simplest example possible.)

* *

Can gravity produce repulsion? What happens to a small test body on the inside of a large C-shaped mass? Is it pushed towards the centre of mass?

* *

A heavily disputed argument for the equality of inertial and gravitational mass was given by Chubykalo, and Vlaev. The total kinetic energy T of two bodies circling around their common centre of mass, like the Earth and the Moon, is given by $T = GmM/2R$, where the two quantities m and M are the *gravitational* masses of the two bodies and R their distance. From this expression, in which the inertial masses do *not* appear on the right side, they deduce that the inertial and gravitational mass must be proportional to each other. Can you see how? Is the reasoning correct?

* *

The *shape* of the Earth is not a sphere. As a consequence, a plumb line usually does not point to the centre of the Earth. What is the largest deviation in degrees?

* *

Owing to the slightly flattened shape of the Earth, the source of the Mississippi is about 20 km nearer to the centre of the Earth than its mouth; the water effectively runs uphill. How can this be?

* *

If you look at the sky every day at 6 a.m., the Sun's position varies during the year. The result of photographing the Sun on the same film is shown in Figure 166. The curve, called the analemma, is due to two combined effects: the inclination of the Earth's axis and the elliptical shape of the Earth's orbit around the Sun. The top and the (hidden) bottom points of the analemma correspond to the solstices. How does the analemma look if photographed every day at local noon? Why is it not a straight line pointing exactly

FIGURE 166 The analemma over Delphi, taken between January and December 2002 (© Anthony Ayiomamitis).

south?

* *

The constellation in which the Sun stands at noon (at the centre of the time zone) is supposedly called the 'zodiacal sign' of that day. Astrologers say there are twelve of them, namely Aries, Taurus, Gemini, Cancer, Leo, Virgo, Libra, Scorpius, Sagittarius, Capricornus, Aquarius and Pisces and that each takes (quite precisely) a twelfth of a year or a twelfth of the ecliptic. Any check with a calendar shows that at present, the midday Sun is never in the zodiacal sign during the days usually connected to it. The relation has shifted by about a month since it was defined, due to the precession of the Earth's axis. A check with a map of the star sky shows that the twelve constellations do not have the same length and that on the ecliptic there are fourteen of them, not twelve. There is *Ophiuchus* or *Serpentarius*, the serpent bearer constellation, between Scorpius and Sagittarius, and *Cetus*, the whale, between Aquarius and Pisces. In fact, not a single astronomical statement about zodiacal signs is correct. To put it clearly, astrology, in contrast to its name, is *not* about stars. (In German, the word 'Strolch', meaning 'rogue' or 'scoundrel', is derived from the word 'astrologer'.)

* *

For a long time, it was thought that there is no additional planet in our Solar System outside Neptune and Pluto, because their orbits show no disturbances from another body. Today, the view has changed. It is known that there are only eight planets: Pluto is not a planet, but the first of a set of smaller objects in the so-called *Kuiper belt*. Kuiper belt

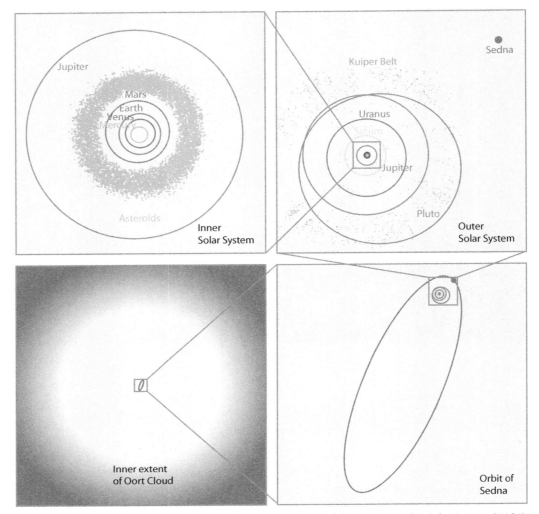

FIGURE 167 The orbit of Sedna in comparison with the orbits of the planets in the Solar System (NASA).

objects are regularly discovered; over 1000 are known today.

In 2003, two major Kuiper objects were discovered; one, called *Sedna*, is almost as large as Pluto, the other, called *Eris*, is even larger than Pluto and has a moon. Both have strongly elliptical orbits (see Figure 167). Since Pluto and Eris, like the asteroid Ceres, have cleaned their orbit from debris, these three objects are now classified as *dwarf planets*.

Outside the Kuiper belt, the Solar System is surrounded by the so-called *Oort cloud*. In contrast to the flattened Kuiper belt, the Oort cloud is spherical in shape and has a radius of up to 50 000 AU, as shown in Figure 167 and Figure 168. The Oort cloud consists of a huge number of icy objects consisting of mainly of water, and to a lesser degree, of methane and ammonia. Objects from the Oort cloud that enter the inner Solar System become comets; in the distant past, such objects have brought water onto the Earth.

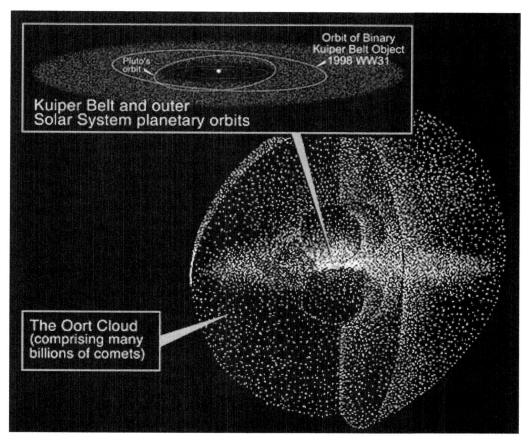

FIGURE 168 The Kuiper belt, containing mainly planetoids, and the Oort cloud orbit, containing comets, around the Solar System (NASA, JPL, Donald Yeoman).

* *

In astronomy new examples of motion are regularly discovered even in the present century. Sometimes there are also false alarms. One example was the alleged fall of *mini comets* on the Earth. They were supposedly made of a few dozen kilograms of ice, hitting the Earth every few seconds. It is now known not to happen.

* *

Universal gravity allows only elliptical, parabolic or hyperbolic orbits. It is impossible for a small object approaching a large one to be captured. At least, that is what we have learned so far. Nevertheless, all astronomy books tell stories of capture in our Solar System; for example, several outer satellites of Saturn have been captured. How is this possible?

* *

How would a tunnel have to be shaped in order that a stone would fall through it without touching the walls? (Assume constant density.) If the Earth did not rotate, the tunnel would be a straight line through its centre, and the stone would fall down and up again, in a oscillating motion. For a rotating Earth, the problem is much more difficult. What is the shape when the tunnel starts at the Equator?

* *

The International Space Station circles the Earth every 90 minutes at an altitude of about 380 km. You can see where it is from the website www.heavens-above.com. By the way, whenever it is just above the horizon, the station is the third brightest object in the night sky, superseded only by the Moon and Venus. Have a look at it.

* *

Is it true that the centre of mass of the Solar System, its barycentre, is always inside the Sun? Even though the Sun or a star move very little when planets move around them, this motion can be detected with precision measurements making use of the Doppler effect for light or radio waves. Jupiter, for example, produces a speed change of 13 m/s in the Sun, the Earth 1 m/s. The first planets outside the Solar System, around the pulsar PSR1257+12 and around the normal G-type star Pegasi 51, were discovered in this way, in 1992 and 1995. In the meantime, several thousand so-called exoplanets have been discovered with this and other methods. Some have even masses comparable to that of the Earth. This research also showed that exoplanets are more numerous than stars, and that earth-like planets are rare.

* *

Not all points on the Earth receive the same number of daylight hours during a year. The effects are difficult to spot, though. Can you find one?

* *

Can the phase of the Moon have a measurable effect on the human body, for example through tidal effects?

* *

There is an important difference between the heliocentric system and the old idea that all planets turn around the Earth. The heliocentric system states that certain planets, such as Mercury and Venus, can be *between* the Earth and the Sun at certain times, and *behind* the Sun at other times. In contrast, the geocentric system states that they are always in between. Why did such an important difference not immediately invalidate the geocentric system? And how did the observation of phases, shown in Figure 169 and Figure 170, invalidate the geocentric system?

* *

The strangest reformulation of the description of motion given by $m\boldsymbol{a} = \boldsymbol{\nabla} U$ is the almost absurd looking equation

$$\boldsymbol{\nabla} v = \mathrm{d}\boldsymbol{v}/\mathrm{d}s \tag{60}$$

where s is the motion path length. It is called the *ray form* of the equation of motion. Can you find an example of its application?

FIGURE 169 The phases of the Moon and of Venus, as observed from Athens in summer 2007 (© Anthony Ayiomamitis).

FIGURE 170 Universal gravitation also explains the observations of Venus, the evening and morning star. In particular, universal gravitation, and the elliptical orbits it implies, explains its phases and its change of angular size. The pictures shown here were taken in 2004 and 2005. The observations can easily be made with a binocular or a small telescope (© Wah!; film available at apod.nasa.gov/apod/ap060110.html).

* *

Seen from Neptune, the size of the Sun is the same as that of Jupiter seen from the Earth at the time of its closest approach. True?

* *

The gravitational acceleration for a particle inside a spherical shell is zero. The vanishing of gravity in this case is independent of the particle shape and its position, and independent of the thickness of the shell. Can you find the argument using Figure 171? This works only because of the $1/r^2$ dependence of gravity. Can you show that the result does not

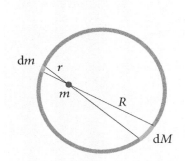

FIGURE 171 The vanishing of gravitational force inside a spherical shell of matter.

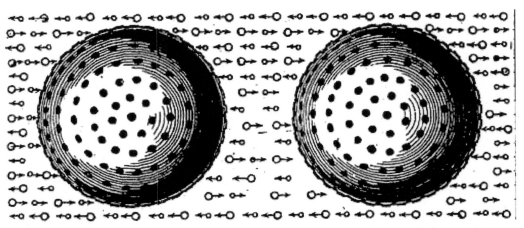

FIGURE 172 Le Sage's own illustration of his model, showing the smaller density of 'ultramondane corpuscules' between the attracting bodies and the higher density outside them (© Wikimedia)

hold for non-spherical shells? Note that the vanishing of gravity inside a spherical shell usually does not hold if other matter is found outside the shell. How could one eliminate the effects of outside matter?

* *

What is gravity? This simple question has a long history. In 1690, Nicolas Fatio de Duillier and in 1747, Georges-Louis Le Sage proposed an explanation for the $1/r^2$ dependence. Le Sage argued that the world is full of small particles – he called them 'corpuscules ultra-mondains' – flying around randomly and hitting all objects. Single objects do not feel the hits, since they are hit continuously and randomly from all directions. But when two objects are near to each other, they produce shadows for part of the flux to the other body, resulting in an attraction, as shown in Figure 172. Can you show that such an attraction has a $1/r^2$ dependence?

However, Le Sage's proposal has a number of problems. First, the argument only works if the collisions are inelastic. (Why?) However, that would mean that all bodies would heat up with time, as Jean-Marc Lévy-Leblond explains. Secondly, a moving body in free space would be hit by more or faster particles in the front than in the back; as a result, the body should be decelerated. Finally, gravity would depend on size, but in a strange way. In particular, three bodies lying on a line should *not* produce shadows, as no such shadows are observed; but the naive model predicts such shadows.

Despite all criticisms, the idea that gravity is due to particles has regularly resurfaced in physics research ever since. In the most recent version, the hypothetical particles are called *gravitons*. On the other hand, no such particles have never been observed. We will understand the origin of gravitation in the final part of our mountain ascent.

* *

For which bodies does gravity decrease as you approach them?

* *

Could one put a satellite into orbit using a cannon? Does the answer depend on the direction in which one shoots?

* *

Two old computer users share experiences. 'I threw my Pentium III and Pentium IV out of the window.' 'And?' 'The Pentium III was faster.'

* *

How often does the Earth rise and fall when seen from the Moon? Does the Earth show phases?

* *

What is the weight of the Moon? How does it compare with the weight of the Alps?

* *

If a star is made of high density material, the speed of a planet orbiting near to it could be greater than the speed of light. How does nature avoid this strange possibility?

* *

What will happen to the Solar System in the future? This question is surprisingly hard to answer. The main expert of this topic, French planetary scientist Jacques Laskar, simulated a few hundred million years of evolution using computer-aided calculus. He found that the planetary orbits are stable, but that there is clear evidence of chaos in the evolution of the Solar System, at a small level. The various planets influence each other in subtle and still poorly understood ways. Effects in the past are also being studied, such as the energy change of Jupiter due to its ejection of smaller asteroids from the Solar System, or energy gains of Neptune. There is still a lot of research to be done in this field.

* *

One of the open problems of the Solar System is the description of planet distances discovered in 1766 by Johann Daniel Titius (1729–1796) and publicized by Johann Elert Bode (1747–1826). Titius discovered that planetary distances d from the Sun can be ap-

TABLE 27 An unexplained property of nature: planet distances from the Sun and the values resulting from the Titius–Bode rule.

Planet	n	Predicted	Measured
		\multicolumn{2}{c}{distance in AU}	
Mercury	$-\infty$	0.4	0.4
Venus	0	0.7	0.7
Earth	1	1.0	1.0
Mars	2	1.6	1.5
Planetoids	3	2.8	2.2 to 3.2
Jupiter	4	5.2	5.2
Saturn	5	10.0	9.5
Uranus	6	19.6	19.2
Neptune	7	38.8	30.1
Pluto	8	77.2	39.5

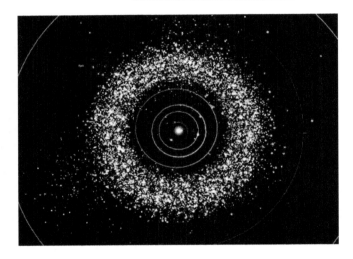

FIGURE 173 The motion of the planetoids compared to that of the planets (Shockwave animation © Hans-Christian Greier)

proximated by
$$d = a + 2^n b \quad \text{with} \quad a = 0.4\,\text{AU}\,,\ b = 0.3\,\text{AU} \tag{61}$$

where distances are measured in astronomical units and n is the number of the planet. The resulting approximation is compared with observations in Table 27.

Interestingly, the last three planets, as well as the planetoids, were discovered *after* Bode's and Titius' deaths; the rule had successfully predicted Uranus' distance, as well as that of the planetoids. Despite these successes – and the failure for the last two planets – nobody has yet found a model for the formation of the planets that explains Titius' rule. The large satellites of Jupiter and of Uranus have regular spacing, but not according to the Titius–Bode rule.

TABLE 28 The orbital periods known to the Babylonians.

BODY	PERIOD
Saturn	29 a
Jupiter	12 a
Mars	687 d
Sun	365 d
Venus	224 d
Mercury	88 d
Moon	29 d

Explaining or disproving the rule is one of the challenges that remains in classical mechanics. Some researchers maintain that the rule is a consequence of scale invariance, others maintain that it is an accident or even a red herring. The last interpretation is also suggested by the non-Titius–Bode behaviour of practically all extrasolar planets. The issue is not closed.

∗ ∗

Around 3000 years ago, the Babylonians had measured the orbital times of the seven celestial bodies that move across the sky. Ordered from longest to shortest, they wrote them down in Table 28. Six of the celestial bodies are visible in the beautiful Figure 174.

The Babylonians also introduced the week and the division of the day into 24 hours. They dedicated every one of the 168 hours of the week to a celestial body, following the order of Table 28. They also dedicated the whole day to that celestial body that corresponds to the first hour of that day. The first day of the week was dedicated to Saturn; the present ordering of the other days of the week then follows from Table 28. This story was told by Cassius Dio (*c.* 160 to *c.* 230). Towards the end of Antiquity, the ordering was taken up by the Roman empire. In Germanic languages, including English, the Latin names of the celestial bodies were replaced by the corresponding Germanic gods. The order Saturday, Sunday, Monday, Tuesday, Wednesday, Thursday and Friday is thus a consequence of both the astronomical measurements and the astrological superstitions of the ancients.

∗ ∗

In 1722, the great mathematician Leonhard Euler made a mistake in his calculation that led him to conclude that if a tunnel, or better, a deep hole were built from one pole of the Earth to the other, a stone falling into it would arrive at the Earth's centre and then immediately turn and go back up. Voltaire made fun of this conclusion for many years. Can you correct Euler and show that the real motion is an oscillation from one pole to the other, and can you calculate the time a pole-to-pole fall would take (assuming homogeneous density)?

What would be the oscillation time for an arbitrary straight surface-to-surface tunnel of length l, thus *not* going from pole to pole?

FIGURE 174 These are the six celestial bodies that are visible at night with the naked eye and whose positions vary over the course of the year. The nearly vertical line connecting them is the *ecliptic*, the narrow stripe around it the *zodiac*. Together with the Sun, the seven celestial bodies were used to name the days of the week. (© Alex Cherney)

6 MOTION DUE TO GRAVITATION   223

FIGURE 175 The solar eclipse of 11 August 1999, photographed by Jean-Pierre Haigneré, member of the Mir 27 crew, and the (enhanced) solar eclipse of 29 March 2006 (© CNES and Laurent Laveder/PixHeaven.net).

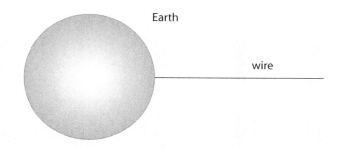

FIGURE 176 A wire attached to the Earth's Equator.

The previous challenges circumvented the effects of the Earth's rotation. The topic becomes much more interesting if rotation is included. What would be the shape of a tunnel so that a stone falling through it never touches the wall?

* *

Figure 175 shows a photograph of a solar eclipse taken from the Russian space station *Mir* and a photograph taken at the centre of the shadow from the Earth. Indeed, a global view of a phenomenon can be quite different from a local one. What is the speed of the shadow?

* *

In 2005, satellite measurements have shown that the water in the Amazon river presses down the land up to 75 mm more in the season when it is full of water than in the season when it is almost empty.

* *

Imagine that wires existed that do not break. How long would such a wire have to be so that, when attached to the Equator, it would stand upright in the air, as shown in Figure 176? Could one build an elevator into space in this way?

* *

Usually there are roughly two tides per day. But there are places, such as on the coast of Vietnam, where there is only *one* tide per day. See www.jason.oceanobs.com/html/applications/marees/marees_m2k1_fr.html. Why?

* *

It is sufficient to use the concept of centrifugal force to show that the rings of Saturn cannot be made of massive material, but must be made of separate pieces. Can you find out how?

* *

Why did Mars lose its atmosphere? Nobody knows. It has recently been shown that the solar wind is too weak for this to happen. This is one of the many open riddles of the solar system.

* *

All bodies in the Solar System orbit the Sun in the same direction. All? No; there are exceptions. One intriguing asteroid that orbits the Sun near Jupiter in the wrong direction was discovered in 2015: it has a size of 3 km. For an animation of its astonishing orbit, opposite to all Trojan asteroids, see www.astro.uwo.ca/~wiegert.

* *

The observed motion due to gravity can be shown to be the *simplest* possible, in the following sense. If we measure change of a falling object with the expression $\int mv^2/2 - mgh \, dt$, then a constant acceleration due to gravity *minimizes* the change in every example of fall. Can you confirm this?

* *

Motion due to gravity is fun: think about roller coasters. If you want to know more at how they are built, visit www.vekoma.com.

> " The scientific theory I like best is that the rings of Saturn are made of lost airline luggage. "
> Mark Russel

Summary on gravitation

Spherical bodies of mass M attract other bodies at a distance r by inducing an acceleration towards them given by $a = GM/r^2$. This expression, *universal gravitation*, describes snowboarders, skiers, paragliders, athletes, couch potatoes, pendula, stones, canons, rockets, tides, eclipses, planet shapes, planet motion and much more. Universal gravitation is the first example of a unified description: it describes *how everything falls*. By the acceleration it produces, gravitation limits the appearance of uniform motion in nature.

Chapter 7
CLASSICAL MECHANICS, FORCE AND THE PREDICTABILITY OF MOTION

All those types of motion in which the only permanent property of a body is mass define the field of *mechanics*. The same name is given also to the experts studying the field. We can think of mechanics as the athletic part of physics.** Both in athletics and in mechanics only lengths, times and masses are measured – and of interest at all.

More specifically, our topic of investigation so far is called *classical* mechanics, to distinguish it from *quantum* mechanics. The main difference is that in classical physics arbitrary small values are assumed to exist, whereas this is not the case in quantum physics. Classical mechanics is often also called *Galilean physics* or *Newtonian physics*.***

Classical mechanics states that motion is *predictable*: it thus states that there are no surprises in motion. Is this correct in all cases? Is predictability valid in the presence of friction? Of free will? Are there really no surprises in nature? These issues merit a discussion; they will accompany us for a stretch of our adventure.

We know that there is more to the world than gravity. Simple observations make this point: *floors* and *friction*. Neither can be due to gravity. Floors do not fall, and thus are not described by gravity; and friction is not observed in the skies, where motion is purely due to gravity.**** Also on Earth, friction is unrelated to gravity, as you might want to check yourself. There must be another interaction responsible for friction. We shall study it in the third volume. But a few issues merit a discussion right away.

** This is in contrast to the actual origin of the term 'mechanics', which means 'machine science'. It derives from the Greek μηκανή, which means 'machine' and even lies at the origin of the English word 'machine' itself. Sometimes the term 'mechanics' is used for the study of motion of *solid* bodies only, excluding, e.g., hydrodynamics. This use fell out of favour in physics in the twentieth century.

*** The basis of classical mechanics, the description of motion using only space and time, is called *kinematics*. An example is the description of free fall by $z(t) = z_0 + v_0(t - t_0) - \frac{1}{2}g(t - t_0)^2$. The other, main part of classical mechanics is the description of motion as a consequence of interactions between bodies; it is called *dynamics*. An example of dynamics is the formula of universal gravity. The distinction between kinematics and dynamics can also be made in relativity, thermodynamics and electrodynamics.

**** This is not completely correct: in the 1980s, the first case of gravitational friction was discovered: the emission of gravity waves. We discuss it in detail in the chapter on general relativity. The discovery does not change the main point, however.

FIGURE 177 The parabola shapes formed by accelerated water beams show that motion in everyday life is predictable (© Oase GmbH).

TABLE 29 Some force values in nature.

Observation	Force
Value measured in a magnetic resonance force microscope	820 zN
Force needed to rip a DNA molecule apart by pulling at its two ends	600 pN
Maximum force exerted by human bite	2.1 kN
Typical peak force exerted by sledgehammer	2 kN
Force exerted by quadriceps	up to 3 kN
Force sustained by 1 cm^2 of a good adhesive	up to 10 kN
Force needed to tear a good rope used in rock climbing	30 kN
Maximum force measurable in nature	$3.0 \cdot 10^{43}$ N

Should one use force? Power?

> " The direct use of physical force is so poor a
> solution [...] that it is commonly employed only
> by small children and great nations. "
> David Friedman

Everybody has to take a stand on this question, even students of physics. Indeed, many types of forces are used and observed in daily life. One speaks of muscular, gravitational, psychic, sexual, satanic, supernatural, social, political, economic and many others. Physicists see things in a simpler way. They call the different types of forces observed between objects *interactions*. The study of the details of all these interactions will show that, in everyday life, they are of electrical or gravitational origin.

For physicists, all change is due to motion. The term force then also takes on a more restrictive definition. *(Physical) force* is defined as the *change of momentum with time*, i.e., as

$$F = \frac{d\mathbf{p}}{dt} \, . \tag{62}$$

A few measured values are listed in Figure 29.

Ref. 86
Challenge 414 s

A horse is running so fast that the hooves touch the ground only 20 % of the time. What is the load carried by its legs during contact?

Force is the *change* of momentum. Since momentum is conserved, we can also say that force measures the *flow* of momentum. As we will see in detail shortly, whenever a force accelerates a body, momentum flows into it. Indeed, momentum can be imagined to be some invisible and intangible substance. Force measures how much of this substance flows into or out of a body per unit time.

Ref. 86

▷ Force is momentum flow.

The conservation of momentum is due to the conservation of this liquid. Like any liquid, momentum flows through a surface.

Using the Galilean definition of linear momentum $\mathbf{p} = m\mathbf{v}$, we can rewrite the definition of force (for constant mass) as

$$\mathbf{F} = m\mathbf{a} \, , \tag{63}$$

where $\mathbf{F} = \mathbf{F}(t, \mathbf{x})$ is the force acting on an object of mass m and where $\mathbf{a} = \mathbf{a}(t, \mathbf{x}) = d\mathbf{v}/dt = d^2\mathbf{x}/dt^2$ is the acceleration of the same object, that is to say its change of velocity.* The expression states in precise terms that force is what changes the *velocity* of masses. The quantity is called 'force' because it corresponds in many, but *not* all aspects to everyday muscular force. For example, the more force is used, the further a stone can be thrown. Equivalently, the more momentum is pumped into a stone, the further it can be thrown. As another example, the concept of *weight* describes the flow of momentum due to gravity.

▷ Gravitation constantly pumps momentum into massive bodies.

Challenge 415 s

Sand in an hourglass is running, and the hourglass is on a scale. Is the weight shown on the scale larger, smaller or equal to the weight when the sand has stopped falling?

Forces are measured with the help of deformations of bodies. Everyday force values can be measured by measuring the extension of a spring. Small force values, of the order of 1 nN, can be detected by measuring the deflection of small levers with the help of a reflected laser beam.

Ref. 28
Vol. II, page 83

* This equation was first written down by the mathematician and physicist Leonhard Euler (b. 1707 Basel, d. 1783 St. Petersburg) in 1747, 20 years after the death of Newton, to whom it is usually and falsely ascribed. It was Euler, one of the greatest mathematicians of all time (and not Newton), who first understood that this definition of force is useful in *every* case of motion, whatever the appearance, be it for point particles or extended objects, and be it rigid, deformable or fluid bodies. Surprisingly and in contrast to frequently-made statements, equation (63) is even correct in relativity.

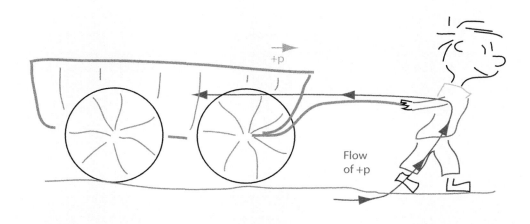

FIGURE 178 The pulling child pumps momentum into the chariot. In fact, some momentum flows back to the ground due to dynamic friction (not drawn).

However, whenever the concept of force is used, it should be remembered that *physical force is different from everyday force or everyday effort*. Effort is probably best approximated by the concept of *(physical) power*, usually abbreviated P, and defined (for constant force) as

$$P = \frac{dW}{dt} = \boldsymbol{F v} \qquad (64)$$

in which *(physical) work* W is defined as $W = \boldsymbol{Fs}$, where $\boldsymbol{s}$ is the distance along which the force acts. Physical work is a form of energy, as you might want to check. Work, as a form of energy, has to be taken into account when the conservation of energy is checked.

With the definition of work just given you can solve the following puzzles. What happens to the electricity consumption of an escalator if you walk on it instead of standing still? What is the effect of the definition of power for the salary of scientists? A man who walks carrying a heavy rucksack is hardly doing any work; why then does he get tired?

When students in exams say that the force acting on a thrown stone is least at the highest point of the trajectory, it is customary to say that they are using an incorrect view, namely the so-called *Aristotelian view*, in which force is proportional to velocity. Sometimes it is even said that they are using a different concept of *state* of motion. Critics then add, with a tone of superiority, how wrong all this is. This is an example of intellectual disinformation. Every student knows from riding a bicycle, from throwing a stone or from pulling an object that increased *effort* results in increased speed. The student is right; those theoreticians who deduce that the student has a mistaken concept of *force* are wrong. In fact, instead of the *physical* concept of force, the student is just using the *everyday* version, namely effort. Indeed, the effort exerted by gravity on a flying stone is least at the highest point of the trajectory. Understanding the difference between physical force and everyday effort is the main hurdle in learning mechanics.*

---

* This stepping stone is so high that many professional physicists do not really take it themselves; this is

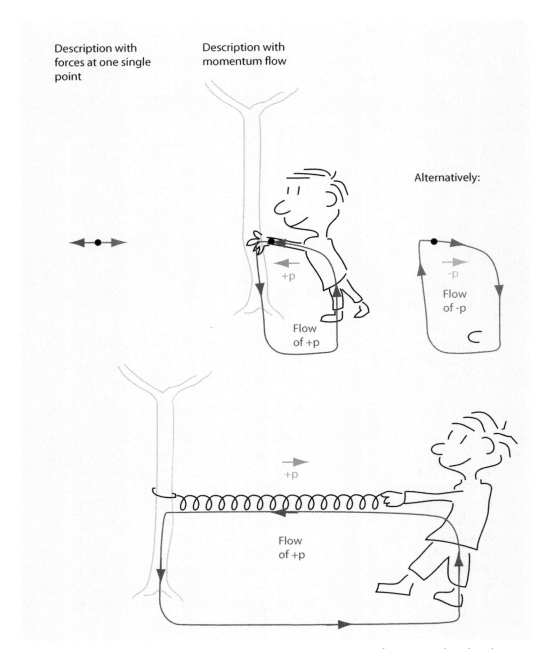

FIGURE 179 The two equivalent descriptions of situations with zero net force, i.e., with a closed momentum flow. Compression occurs when momentum flow and momentum point in the same direction; extension occurs when momentum flow and momentum point in opposite directions.

---

confirmed by the innumerable comments in papers that state that physical force is defined using mass, and, at the same time, that mass is defined using force (the latter part of the sentence being a fundamental mistake).

7 CLASSICAL MECHANICS, FORCE AND THE PREDICTABILITY OF MOTION

Often the flow of momentum, equation (62), is not recognized as the definition of force. This is mainly due to an everyday observation: there seem to be forces without any associated acceleration or change in momentum, such as in a string under tension or in water at high pressure. When one pushes against a tree, as shown in Figure 179, there is no motion, yet a force is applied. If force is momentum flow, where does the momentum go? It flows into the slight deformations of the arms and the tree. In fact, when one starts pushing and thus deforming, the associated momentum change of the molecules, the atoms, or the electrons of the two bodies can be observed. After the deformation is established a continuous and equal flow of momentum is going on in both directions.

Because force is net momentum flow, the concept of force is not really needed in the description of motion. But sometimes the concept is practical. This is the case in everyday life, where it is useful in situations where net momentum values are small or negligible. For example, it is useful to define pressure as force per area, even though it is actually a momentum flow per area. At the microscopic level, momentum alone suffices for the description of motion.

In the section title we asked about on the usefulness of force and power. Before we can answer conclusively, we need more arguments. Through its definition, the concepts of force and power are distinguished clearly from 'mass', 'momentum', 'energy' and from each other. But where do forces originate? In other words, which effects in nature have the capacity to accelerate bodies by pumping momentum into objects? Table 30 gives an overview.

FORCES, SURFACES AND CONSERVATION

We saw that force is the change of momentum. We also saw that momentum is conserved. How do these statements come together? The answer is the same for all conserved quantities. We imagine a closed surface that is the boundary of a volume in space. Conservation implies that the conserved quantity enclosed *inside* the surface can only change by *flowing through* that surface.*

All conserved quantities in nature – such as energy, linear momentum, electric charge, angular momentum – can only change by flowing through surfaces. In particular, whenever the momentum of a body changes, this happens through a surface. Momentum change is due to momentum flow. In other words, the concept of force always implies a surface through which momentum flows.

---

* Mathematically, the conservation of a quantity q is expressed with the help of the volume density $\rho = q/V$, the current $I = q/t$, and the flow or flux $\boldsymbol{j} = \rho \boldsymbol{v}$, so that $j = q/At$. Conservation then implies

$$\frac{dq}{dt} = \int_V \frac{\partial \rho}{\partial t} dV = -\int_{A=\partial V} \boldsymbol{j} d\boldsymbol{A} = -I \qquad (65)$$

or, equivalently,

$$\frac{\partial \rho}{\partial t} + \nabla \boldsymbol{j} = 0 . \qquad (66)$$

This is the *continuity equation* for the quantity q. All this only states that a conserved quantity in a closed volume V can only change by flowing through the surface A. This is a typical example of how complex mathematical expressions can obfuscate the simple physical content.

FIGURE 180 Friction-based processes (courtesy Wikimedia).

▷ Force is the flow of momentum through a surface.

This point is essential in understanding physical force. Every force requires a surface for its definition.

To refine your own concept of force, you can search for the relevant surface when a rope pulls a chariot, or when an arm pushes a tree, or when a car accelerates. It is also helpful to compare the definition of force with the definition of power: both quantities are flows through surfaces. As a result, we can say:

▷ A *motor* is a momentum pump.

Friction and motion

Every example of motion, from the motion that lets us choose the direction of our gaze to the motion that carries a butterfly through the landscape, can be put into one of the two left-most columns of Table 30. Physically, those two columns are separated by the following criterion: in the first class, the acceleration of a body can be in a different direction from its velocity. The second class of examples produces only accelerations that are exactly *opposed* to the velocity of the moving body, as seen from the frame of reference

TABLE 30 Selected processes and devices changing the motion of bodies.

Situations that can lead to acceleration	Situations that only lead to deceleration	Motors and actuators
piezoelectricity		
quartz under applied voltage	thermoluminescence	walking piezo tripod
collisions		
satellite in planet encounter	car crash	rocket motor
growth of mountains	meteorite crash	swimming of larvae
magnetic effects		
compass needle near magnet	electromagnetic braking	electromagnetic gun
magnetostriction	transformer losses	linear motor
current in wire near magnet	electric heating	galvanometer
electric effects		
rubbed comb near hair	friction between solids	electrostatic motor
bombs	fire	muscles, sperm flagella
cathode ray tube	electron microscope	Brownian motor
light		
levitating objects by light	light bath stopping atoms	(true) light mill
solar sail for satellites	light pressure inside stars	solar cell
elasticity		
bow and arrow	trouser suspenders	ultrasound motor
bent trees standing up again	pillow, air bag	bimorphs
osmosis		
water rising in trees	salt conservation of food	osmotic pendulum
electro-osmosis		tunable X-ray screening
heat & pressure		
freezing champagne bottle	surfboard water resistance	hydraulic engines
tea kettle	quicksand	steam engine
barometer	parachute	air gun, sail
earthquakes	sliding resistance	seismometer
attraction of passing trains	shock absorbers	water turbine
nuclei		
radioactivity	plunging into the Sun	supernova explosion
biology		
bamboo growth	decreasing blood vessel diameter	molecular motors
gravitation		
falling	emission of gravity waves	pulley

of the braking medium. Such a resisting force is called *friction*, *drag* or a *damping*. All examples in the second class are types of friction. Just check. Some examples of processes based on friction are given in Figure 180.

Here is a puzzle on cycling: does side wind brake – and why?

Friction can be so strong that all motion of a body against its environment is made impossible. This type of friction, called *static friction* or *sticking friction*, is common and important: without it, turning the wheels of bicycles, trains or cars would have no effect. Without static friction, wheels driven by a motor would have no grip. Similarly, not a single screw would stay tightened and no hair clip would work. We could neither run nor walk in a forest, as the soil would be more slippery than polished ice. In fact not only our own motion, but all *voluntary motion* of living beings is *based* on friction. The same is the case for all self-moving machines. Without static friction, the propellers in ships, aeroplanes and helicopters would not have any effect and the wings of aeroplanes would produce no lift to keep them in the air. (Why?)

In short, static friction is necessary whenever we or an engine want to move against the environment.

Friction, sport, machines and predictability

Once an object moves through its environment, it is hindered by another type of friction; it is called *dynamic friction* and acts between all bodies in relative motion.* Without dynamic friction, falling bodies would always rebound to the same height, without ever coming to a stop; neither parachutes nor brakes would work; and even worse, we would have no memory, as we will see later.

All motion examples in the second column of Table 30 include friction. In these examples, macroscopic energy is not conserved: the systems are *dissipative*. In the first column, macroscopic energy is constant: the systems are *conservative*.

The first two columns can also be distinguished using a more abstract, mathematical criterion: on the left are accelerations that can be derived from a potential, on the right, decelerations that can not. As in the case of gravitation, the description of any kind of motion is much simplified by the use of a potential: at every position in space, one needs only the single value of the potential to calculate the trajectory of an object, instead of the three values of the acceleration or the force. Moreover, the magnitude of the velocity of an object at any point can be calculated directly from energy conservation.

The processes from the second column *cannot* be described by a potential. These are the cases where it is best to use force if we want to describe the motion of the system. For example, the friction or drag force F due to *wind resistance* of a body is *roughly* given by

$$F = \frac{1}{2} c_w \rho A v^2 \qquad (67)$$

where A is the area of its cross-section and v its velocity relative to the air, ρ is the density of air. The *drag coefficient* c_w is a pure number that depends on the shape of the moving

---

* There might be one exception. Recent research suggest that maybe in certain crystalline systems, such as tungsten bodies on silicon, under ideal conditions gliding friction can be extremely small and possibly even vanish in certain directions of motion. This so-called *superlubrication* is presently a topic of research.

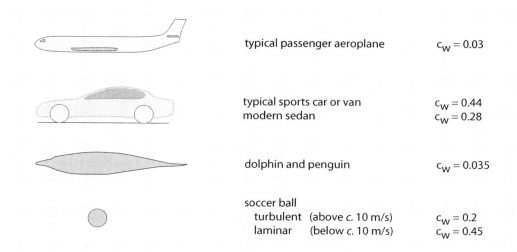

FIGURE 181 Shapes and air/water resistance.

object. A few examples are given in Figure 181. The formula is valid for all fluids, not only for air, below the speed of sound, as long as the drag is due to turbulence. This is usually the case in air and in water. (At very low velocities, when the fluid motion is not turbulent but laminar, drag is called *viscous* and follows an (almost) linear relation with speed.) You may check that drag, or aerodynamic resistance *cannot* be derived from a potential.*

The drag coefficient c_w is a measured quantity. Calculating drag coefficients with computers, given the shape of the body and the properties of the fluid, is one of the most difficult tasks of science; the problem is still not solved. An aerodynamic car has a value between 0.25 and 0.3; many sports cars share with vans values of 0.44 and higher, and racing car values can be as high as 1, depending on the amount of the force that is used to keep the car fastened to the ground. The lowest known values are for dolphins and penguins.**

Wind resistance is also of importance to humans, in particular in athletics. It is estimated that 100 m sprinters spend between 3 % and 6 % of their power overcoming drag. This leads to varying sprint times t_w when wind of speed w is involved, related by the expression

$$\frac{t_0}{t_w} = 1.03 - 0.03 \left(1 - \frac{wt_w}{100\,\mathrm{m}}\right)^2 , \qquad (68)$$

---

* Such a statement about friction is correct only in three dimensions, as is the case in nature; in the case of a single dimension, a potential can *always* be found.
** It is unclear whether there is, in nature, a smallest possible value for the drag coefficient.

The topic of aerodynamic shapes is also interesting for fluid bodies. They are kept together by *surface tension*. For example, surface tension keeps the wet hairs of a soaked brush together. Surface tension also determines the shape of rain drops. Experiments show that their shape is spherical for drops smaller than 2 mm diameter, and that larger rain drops are *lens* shaped, with the flat part towards the bottom. The usual tear shape is *not* encountered in nature; something vaguely similar to it appears during drop detachment, but *never* during drop fall.

where the more conservative estimate of 3 % is used. An opposing wind speed of −2 m/s gives an increase in time of 0.13 s, enough to change a potential world record into an 'only' excellent result. (Are you able to deduce the c_w value for running humans from the formula?)

Likewise, parachuting exists due to wind resistance. Can you determine how the speed of a falling body, with or without parachute, changes with time, assuming *constant* shape and drag coefficient?

In contrast, static friction has different properties. It is proportional to the force pressing the two bodies together. Why? Studying the situation in more detail, sticking friction is found to be proportional to the actual contact area. It turns out that putting two solids into contact is rather like turning Switzerland upside down and putting it onto Austria; the area of contact is much smaller than that estimated macroscopically. The important point is that the area of actual contact is proportional to the *normal* force, i.e., the force component that is perpendicular to the surface. The study of what happens in that contact area is still a topic of research; researchers are investigating the issues using instruments such as atomic force microscopes, lateral force microscopes and triboscopes. These efforts resulted in computer hard discs which last longer, as the friction between disc and the reading head is a central quantity in determining the lifetime.

All forms of friction are accompanied by an increase in the temperature of the moving body. The reason became clear after the discovery of atoms. Friction is not observed in few – e.g. 2, 3, or 4 – particle systems. Friction only appears in systems with *many* particles, usually millions or more. Such systems are called *dissipative*. Both the temperature changes and friction itself are due to the motion of large numbers of microscopic particles against each other. This motion is not included in the Galilean description. When it is included, friction and energy loss disappear, and potentials can then be used throughout. Positive accelerations – of microscopic magnitude – then also appear, and motion is found to be conserved.

In short, all motion is conservative on a microscopic scale. On a microscopic scale it is thus possible and most practical to describe *all* motion without the concept of force.*

The moral of the story is twofold: First, one should use force and power only in one situation: in the case of friction, and only when one does not want to go into the details.** Secondly, friction is not an obstacle to predictability. Motion remains predictable.

---

* The first scientist who eliminated force from the description of nature was Heinrich Rudolf Hertz (b. 1857 Hamburg, d. 1894 Bonn), the famous discoverer of electromagnetic waves, in his textbook on mechanics, *Die Prinzipien der Mechanik*, Barth, 1894, republished by Wissenschaftliche Buchgesellschaft, 1963. His idea was strongly criticized at that time; only a generation later, when quantum mechanics quietly got rid of the concept for good, did the idea become commonly accepted. (Many have speculated about the role Hertz would have played in the development of quantum mechanics and general relativity, had he not died so young.) In his book, Hertz also formulated the principle of the straightest path: particles follow geodesics. This same description is one of the pillars of general relativity, as we will see later on.

** But the cost is high; in the case of human relations the evaluation should be somewhat more discerning, as research on violence has shown.

7 CLASSICAL MECHANICS, FORCE AND THE PREDICTABILITY OF MOTION

> " Et qu'avons-nous besoin de ce moteur, quand
> l'étude réfléchie de la nature nous prouve que le
> mouvement perpétuel est la première de ses
> lois ?*
> Donatien de Sade *Justine, ou les malheurs de la vertu.*

Complete states – initial conditions

> " Quid sit futurum cras, fuge quaerere ...**
> Horace, *Odi*, lib. I, ode 9, v. 13.

Let us continue our exploration of the predictability of motion. We often describe the motion of a body by specifying the time dependence of its position, for example as

$$x(t) = x_0 + v_0(t - t_0) + \tfrac{1}{2}a_0(t - t_0)^2 + \tfrac{1}{6}j_0(t - t_0)^3 + \ldots \;. \tag{69}$$

The quantities with an index 0, such as the starting position x_0, the starting velocity v_0, etc., are called *initial conditions*. Initial conditions are necessary for any description of motion. Different physical systems have different initial conditions. Initial conditions thus specify the *individuality* of a given system. Initial conditions also allow us to distinguish the present situation of a system from that at any previous time: initial conditions specify the *changing aspects* of a system. Equivalently, they summarize the *past* of a system.

Initial conditions are thus precisely the properties we have been seeking for a description of the *state* of a system. To find a complete description of states we thus need only a complete description of initial conditions, which we can thus rightly call also *initial states*. It turns out that for gravitation, as for all other microscopic interactions, there is *no* need for initial acceleration a_0, initial jerk j_0, or higher-order initial quantities. In nature, acceleration and jerk depend only on the properties of objects and their environment; they do not depend on the past. For example, the expression $a = GM/r^2$ of universal gravity, giving the acceleration of a small body near a large one, does not depend on the past, but only on the environment. The same happens for the other fundamental interactions, as we will find out shortly.

The *complete state* of a moving mass point is thus described by specifying its position and its momentum at all instants of time. Thus we have now achieved a *complete* description of the *intrinsic properties* of point objects, namely by their mass, and of their *states of motion*, namely by their momentum, energy, position and time. For *extended rigid* objects we also need orientation and angular momentum. This is the full list for rigid objects; no other state observables are needed.

Can you specify the necessary state observables in the cases of extended elastic bodies and of fluids? Can you give an example of an intrinsic property that we have so far missed?

---

* 'And whatfor do we need this motor, when the reasoned study of nature proves to us that perpetual motion is the first of its laws?'
** 'What future will be tomorrow, never ask ...' Horace is Quintus Horatius Flaccus (65–8 BCE), the great Roman poet.

The set of all possible states of a system is given a special name: it is called the *phase space*. We will use the concept repeatedly. Like any space, it has a number of dimensions. Can you specify this number for a system consisting of N point particles?

It is interesting to recall an older challenge and ask again: does the universe have initial conditions? Does it have a phase space?

Given that we now have a description of both properties and states for point objects, extended rigid objects and deformable bodies, can we predict all motion? Not yet. There are situations in nature where the motion of an object depends on characteristics other than its mass; motion can depend on its colour (can you find an example?), on its temperature, and on a few other properties that we will soon discover. And for each intrinsic property there are state observables to discover. Each additional intrinsic property is the basis of a field of physical enquiry. Speed was the basis for mechanics, temperature is the basis for thermodynamics, charge is the basis for electrodynamics, etc. We must therefore conclude that as yet we do not have a complete description of motion.

> An optimist is somebody who thinks that the future is uncertain.
> Anonymous

Do surprises exist? Is the future determined?

> Die Ereignisse der Zukunft *können* wir nicht aus den gegenwärtigen erschließen. Der Glaube an den Kausalnexus ist ein Aberglaube.*
> Ludwig Wittgenstein, *Tractatus*, 5.1361

> Freedom is the recognition of necessity.
> Friedrich Engels (1820–1895)

If, after climbing a tree, we jump down, we cannot halt the jump in the middle of the trajectory; once the jump has begun, it is unavoidable and determined, like all passive motion. However, when we begin to move an arm, we can stop or change its motion from a hit to a caress. Voluntary motion does not seem unavoidable or predetermined. Which of these two cases is the general one?

Let us start with the example that we can describe most precisely so far: the fall of a body. Once the gravitational potential φ acting on a particle is given and taken into account, we can use the expression

$$\boldsymbol{a}(x) = -\nabla\varphi = -GM\boldsymbol{r}/r^3 , \tag{70}$$

and we can use the state at a given time, given by initial conditions such as

$$\boldsymbol{x}(t_0) \quad \text{and} \quad \boldsymbol{v}(t_0) , \tag{71}$$

to determine the motion of the particle in advance. Indeed, with these two pieces of information, we can calculate the complete trajectory $\boldsymbol{x}(t)$.

---

* 'We cannot infer the events of the future from those of the present. Belief in the causal nexus is superstition.' Our adventure, however, will confirm the everyday observation that this statement is wrong.

An equation that has the potential to predict the course of events is called an *evolution equation*. Equation (70), for example, is an evolution equation for the fall of the object. (Note that the term 'evolution' has different meanings in physics and in biology.) An evolution equation embraces the observation that not all types of change are observed in nature, but only certain specific cases. Not all imaginable sequences of events are observed, but only a limited number of them. In particular, equation (70) embraces the idea that from one instant to the next, falling objects change their motion based on the gravitational potential acting on them.

Evolution equations do not exist only for motion due to gravity, but for motion due to all forces in nature. Given an evolution equation and initial state, the whole motion of a system is thus *uniquely fixed*, a property of motion often called *determinism*. For example, astronomers can calculate the position of planets with high precision for thousands of years in advance.

Let us carefully distinguish determinism from several similar concepts, to avoid misunderstandings. Motion can be deterministic and at the same time be *unpredictable in practice*. The unpredictability of motion can have four origins:

1. an impracticably large number of particles involved, including situations with friction,
2. insufficient information about initial conditions, and
3. the mathematical complexity of the evolution equations,
4. strange shapes of space-time.

For example, in case of the *weather* the first three conditions are fulfilled at the same time. It is hard to predict the weather over periods longer than about a week or two. (In 1942, Hitler made once again a fool of himself across Germany by requesting a precise weather forecast for the following twelve months.) Despite the difficulty of prediction, weather change is still deterministic. As another example, near *black holes* all four origins apply together. We will discuss black holes in the section on general relativity. Despite being unpredictable, motion is deterministic near black holes.

Motion can be both deterministic and time *random*, i.e., with different outcomes in similar experiments. A roulette ball's motion is deterministic, but it is also random.* As we will see later, quantum systems fall into this category, as do all examples of irreversible motion, such as a drop of ink spreading out in clear water. Also the fall of a die is both deterministic and random. In fact, studies on how to predict the result of a die throw with the help of a computer are making rapid progress; these studies also show how to throw a die in order to increase the odds to get a desired result. In all such cases the randomness and the irreproducibility are only apparent; they disappear when the description of states and initial conditions in the microscopic domain are included. In short, determinism does not contradict *(macroscopic) irreversibility*. However, on the microscopic scale, deterministic motion is always reversible.

A final concept to be distinguished from determinism is *acausality*. Causality is the requirement that a cause must precede the effect. This is trivial in Galilean physics, but

---

* Mathematicians have developed a large number of tests to determine whether a collection of numbers may be called *random*; roulette results pass all these tests – in honest casinos only, however. Such tests typically check the equal distribution of numbers, of pairs of numbers, of triples of numbers, etc. Other tests are the χ^2 test, the Monte Carlo test(s), and the gorilla test.

becomes of importance in special relativity, where causality implies that the speed of light is a limit for the spreading of effects. Indeed, it seems impossible to have deterministic motion (of matter and energy) which is *acausal*, in other words, faster than light. Can you confirm this? This topic will be looked at more deeply in the section on special relativity.

Saying that motion is 'deterministic' means that it is fixed in the future *and also in the past*. It is sometimes stated that predictions of *future* observations are the crucial test for a successful description of nature. Owing to our often impressive ability to influence the future, this is not necessarily a good test. Any theory must, first of all, describe *past* observations correctly. It is our lack of freedom to change the past that results in our lack of choice in the description of nature that is so central to physics. In this sense, the term 'initial condition' is an unfortunate choice, because in fact, initial conditions summarize the *past* of a system.* The central ingredient of a deterministic description is that all motion can be reduced to an evolution equation plus one specific state. This state can be either initial, intermediate, or final. Deterministic motion is uniquely specified into the past and into the future.

To get a clear concept of determinism, it is useful to remind ourselves why the concept of 'time' is introduced in our description of the world. We introduce time because we observe first that we are able to define sequences in observations, and second, that unrestricted change is impossible. This is in contrast to films, where one person can walk through a door and exit into another continent or another century. In nature we do not observe metamorphoses, such as people changing into toasters or dogs into toothbrushes. We are able to introduce 'time' only because the sequential changes we observe are extremely restricted. If nature were not reproducible, time could not be used. In short, determinism expresses the observation that *sequential changes are restricted to a single possibility*.

Since determinism is connected to the use of the concept of time, new questions arise whenever the concept of time changes, as happens in special relativity, in general relativity and in theoretical high energy physics. There is a lot of fun ahead.

In summary, every description of nature that uses the concept of time, such as that of everyday life, that of classical physics and that of quantum mechanics, is intrinsically and inescapably deterministic, since it connects observations of the past and the future, *eliminating* alternatives. In short,

▷ The use of time implies determinism, and vice versa.

When drawing metaphysical conclusions, as is so popular nowadays when discussing quantum theory, one should never forget this connection. Whoever uses clocks but denies determinism is nurturing a split personality!** The future is determined.

---

* The problems with the term 'initial conditions' become clear near the big bang: at the big bang, the universe has no past, but it is often said that it has initial conditions. This contradiction will be explored later in our adventure.
** That can be a lot of fun though.

7 CLASSICAL MECHANICS, FORCE AND THE PREDICTABILITY OF MOTION

FREE WILL

> "You do have the ability to surprise yourself."
> Richard Bandler and John Grinder

The idea that motion is determined often produces fear, because we are taught to associate determinism with lack of freedom. On the other hand, we do experience freedom in our actions and call it *free will*. We know that it is necessary for our creativity and for our happiness. Therefore it seems that determinism is opposed to happiness.

But what precisely is free will? Much ink has been consumed trying to find a precise definition. One can try to define free will as the arbitrariness of the choice of initial conditions. However, initial conditions must themselves result from the evolution equations, so that there is in fact no freedom in their choice. One can try to define free will from the idea of unpredictability, or from similar properties, such as uncomputability. But these definitions face the same simple problem: whatever the definition, there is *no way* to prove experimentally that an action was performed freely. The possible definitions are useless. In short, because free will cannot be defined, it *cannot* be observed. (Psychologists also have a lot of additional data to support this conclusion, but that is another topic.)

No process that is *gradual* – in contrast to *sudden* – can be due to free will; gradual processes are described by time and are deterministic. In this sense, the question about free will becomes one about the existence of sudden changes in nature. This will be a recurring topic in the rest of this walk. Can nature surprise us? In everyday life, nature does not. Sudden changes are not observed. Of course, we still have to investigate this question in other domains, in the very small and in the very large. Indeed, we will change our opinion several times during our adventure, but the conclusion remains.

We note that the lack of surprises in everyday life is built deep into our nature: evolution has developed curiosity because everything that we discover is useful afterwards. If nature continually surprised us, curiosity would make no sense.

Many observations contradict the existence of surprises: in the beginning of our walk we defined time using the continuity of motion; later on we expressed this by saying that time is a consequence of the conservation of energy. Conservation is the opposite of surprise. By the way, a challenge remains: can you show that time would not be definable even if surprises existed only *rarely*?

In summary, so far we have no evidence that surprises exist in nature. Time exists because nature is deterministic. Free will cannot be defined with the precision required by physics. Given that there are no sudden changes, there is only one consistent conclusion: free will is a *feeling*, in particular of independence of others, of independence from fear and of accepting the consequences of one's actions.* Free will is a strange name for a

---

* That free will is a feeling can also be confirmed by careful introspection. Indeed, the idea of free will always arises *after* an action has been started. It is a beautiful experiment to sit down in a quiet environment, with the intention to make, within an unspecified number of minutes, a small gesture, such as closing a hand. If you carefully observe, in all detail, what happens inside yourself around the very moment of decision, you find either a mechanism that led to the decision, or a diffuse, unclear mist. You never find free will. Such an experiment is a beautiful way to experience deeply the wonders of the self. Experiences of this kind might also be one of the origins of human spirituality, as they show the connection everybody has with the rest of nature.

feeling of satisfaction. This solves the apparent paradox; free will, being a feeling, exists as a human experience, even though all objects move without any possibility of choice. There is no contradiction.

Even if human action is determined, it is still authentic. So why is determinism so frightening? That is a question everybody has to ask themselves. What difference does determinism imply for your life, for the actions, the choices, the responsibilities and the pleasures you encounter?* If you conclude that being determined is different from being free, you should change your life! Fear of determinism usually stems from refusal to take the world the way it is. Paradoxically, it is precisely the person who insists on the existence of free will who is running away from responsibility.

Summary on predictability

Despite difficulties to predict specific cases, all motion we encountered so far is both deterministic and predictable. Even friction is predictable, in principle, if we take into account the microscopic details of matter.

In short, classical mechanics states that the future is determined. In fact, we will discover that *all* motion in nature, even in the domains of quantum theory and general relativity, is predictable.

Motion is predictable. This is not a surprising result. If motion were not predictable, we could not have introduced the concepts of 'motion' and 'time' in the first place. We can only talk and think about motion because it is predictable.

From predictability to global descriptions of motion

> Πλεῖν ἀνάγκε, ζῆν οὐκ ἀνάγκη.**
> Pompeius

Physicists aim to talk about motion with the highest precision possible. Predictability is an aspect of precision. The highest predictability – and thus the highest precision – is possible when motion is described as globally as possible.

All over the Earth – even in Australia – people observe that stones fall 'down'. This ancient observation led to the discovery of universal gravity. To find it, all that was necessary was to look for a description of gravity that was valid *globally*. The only additional observation that needs to be recognized in order to deduce the result $a = GM/r^2$ is the variation of gravity with height.

In short, thinking *globally* helps us to make our description of motion more precise and our predictions more useful. How can we describe motion as globally as possible? It turns out that there are six approaches to this question, each of which will be helpful on our way to the top of Motion Mountain. We first give an overview; then we explore each of them.

1. *Action principles* or *variational principles*, the first global approach to motion, arise

---

* If nature's 'laws' are deterministic, are they in contrast with moral or ethical 'laws'? Can people still be held responsible for their actions?
** Navigare necesse, vivere non necesse. 'To navigate is necessary, to live is not.' Gnaeus Pompeius Magnus (106–48 BCE) is cited in this way by Plutarchus (c. 45 to c. 125).

7 CLASSICAL MECHANICS, FORCE AND THE PREDICTABILITY OF MOTION

FIGURE 182 What shape of rail allows the black stone to glide most rapidly from point A to the lower point B?

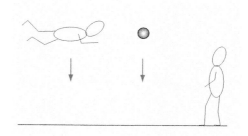

FIGURE 183 Can motion be described in a manner common to all observers?

when we overcome a fundamental limitation of what we have learned so far. When we predict the motion of a particle from its current acceleration with an evolution equation, we are using the most *local* description of motion possible. We use the acceleration of a particle at a certain place and time to determine its position and motion *just after* that moment and *in the immediate neighbourhood* of that place. Evolution equations thus have a mental 'horizon' of radius zero.

The contrast to evolution equations are variational principles. A famous example is illustrated in Figure 182. The challenge is to find the path that allows the fastest possible gliding motion from a high point to a distant low point. The sought path is the *brachistochrone*, from ancient Greek for 'shortest time'. This puzzle asks about a property of motion as a whole, for *all times* and *all positions*. The global approach required by questions such as this one will lead us to a description of motion which is simple, precise and fascinating: the so-called principle of *cosmic laziness*, also known as the principle of *least action*.

2. *Relativity*, the second global approach to motion, emerges when we compare the various descriptions of the same system produced by *all possible observers*. For example, the observations by somebody falling from a cliff – as shown in Figure 183 – a passenger in a roller coaster, and an observer on the ground will usually differ. The relationships between these observations, the so-called *symmetry transformations*, lead us to a global description of motion, valid for everybody. Later, this approach will lead us to Einstein's special and general theory of relativity.

3. *Mechanics of extended and rigid bodies,* rather than mass points, is required to understand the objects, plants and animals of everyday life. For such bodies, we want to understand how *all parts* of them move. As an example, the counter-intuitive result of the experiment in Figure 184 shows why this topic is worthwhile. The rapidly rotating wheel suspended on only one end of the axis remains almost horizontal, but slowly rotates around the rope.

In order to design machines, it is essential to understand how a group of rigid bodies interact with one another. For example, take the Peaucellier-Lipkin linkage shown in Figure 185. A joint F is fixed on a wall. Two movable rods lead to two opposite corners of a movable rhombus, whose rods connect to the other two corners C and P. This mechanism has several astonishing properties. First of all, it implicitly defines a

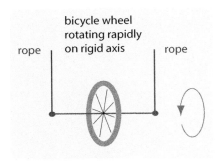

FIGURE 184 What happens when one rope is cut?

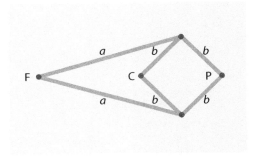

FIGURE 185 A famous mechanism, the *Peaucellier-Lipkin linkage*, consists of (grey) rods and (red) joints and allows drawing a straight line with a compass: fix point F, put a pencil into joint P, and then move C with a compass along a circle.

circle of radius R so that one always has the relation $r_C = R^2/r_P$ between the distances of joints C and P from the centre of this circle. This is called an *inversion at a circle*. Can you find this special circle? Secondly, if you put a pencil in joint P, and let joint C follow a certain circle, the pencil P draws a straight line. Can you find that circle? The mechanism thus allows drawing a *straight* line with the help of a compass.

A famous machine challenge is to devise a wooden carriage, with gearwheels that connect the wheels to an arrow, with the property that, whatever path the carriage takes, the arrow always points south (see Figure 187). The solution to this puzzle will even be useful in helping us to understand general relativity, as we will see. Such a wagon allows measuring the curvature of a surface and of space.

Another important machine part is the *differential gearbox*. Without it, cars could not follow bends on the road. Can you explain it to your friends?

Also nature uses machine parts. In 2011, screws and nuts were found in a joint of a weevil beetle, *Trigonopterus oblongus*. In 2013, the first example of biological *gears* have been discovered: in young plant hoppers of the species *Issus coleoptratus*, toothed gears ensure that the two back legs jump synchronously. Figure 186 shows some details. You might enjoy the video on this discovery available at www.youtube.com/watch?v=Q8fyUOxD2EA.

Another interesting example of rigid motion is the way that human movements, such as the general motions of an arm, are composed from a small number of basic motions. All these examples are from the fascinating field of engineering; unfortunately, we will have little time to explore this topic in our hike.

4. The next global approach to motion is the description of *non-rigid extended bodies*. For example, *fluid mechanics* studies the flow of fluids (like honey, water or air) around solid bodies (like spoons, ships, sails or wings). The aim is to understand how *all parts* of the fluid move. Fluid mechanics thus describes how insects, birds and aeroplanes fly,* why sailing-boats can sail against the wind, what happens when

---

* The mechanisms of insect flight are still a subject of active research. Traditionally, fluid dynamics has

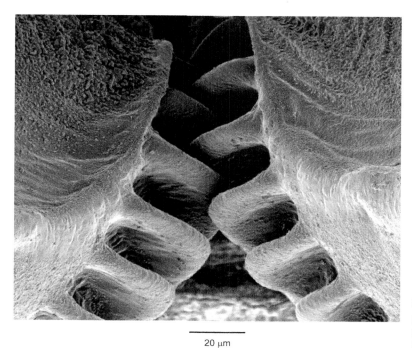

FIGURE 186 The gears found in young plant hoppers (© Malcolm Burrows).

20 μm

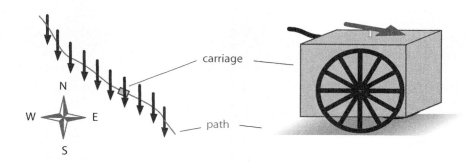

FIGURE 187 A south-pointing carriage: whatever the path it follows, the arrow on it always points south.

a hard-boiled egg is made to spin on a thin layer of water, or how a bottle full of wine can be emptied in the fastest way possible.

As well as fluids, we can study the behaviour of *deformable solids*. This area of research is called *continuum mechanics*. It deals with deformations and oscillations of

---

concentrated on large systems, like boats, ships and aeroplanes. Indeed, the smallest human-made object that can fly in a controlled way – say, a radio-controlled plane or helicopter – is much larger and heavier than many flying objects that evolution has engineered. It turns out that controlling the flight of small things requires more knowledge and more tricks than controlling the flight of large things. There is more about this topic on page 278 in Volume V.

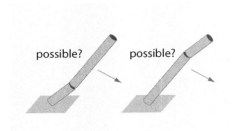

FIGURE 188 How and where does a falling brick chimney break?

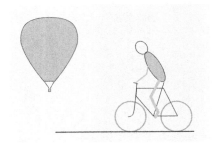

FIGURE 189 Why do hot-air balloons stay inflated? How can you measure the weight of a bicycle rider using only a ruler?

FIGURE 190 Why do marguerites – or ox-eye daisies, *Leucanthemum vulgare* – usually have around 21 (left and centre) or around 34 (right) petals? (© Anonymous, Giorgio Di Iorio and Thomas Lüthi)

extended structures. It seeks to explain, for example, why bells are made in particular shapes; how large bodies – such as the falling chimneys shown in Figure 188 – or small bodies – such as diamonds – break when under stress; and how cats can turn themselves the right way up as they fall. During the course of our journey we will repeatedly encounter issues from this field, which impinges even upon general relativity and the world of elementary particles.

5. *Statistical mechanics* is the study of the motion of *huge numbers of particles*. Statistical mechanics is yet another global approach to the study of motion. The concepts needed to describe gases, such as temperature, entropy and pressure (see Figure 189), are essential tools of this discipline. In particular, the concepts of statistical physics help us to understand why some processes in nature do not occur backwards. These concepts will also help us take our first steps towards the understanding of black holes.

6. The last global approach to motion, *self-organization*, involves all of the above-mentioned viewpoints *at the same time*. Such an approach is needed to understand everyday experience, and *life* itself. Why does a flower form a specific number of petals, as shown in Figure 190? How does an embryo differentiate in the womb? What makes our hearts beat? How do mountains ridges and cloud patterns emerge? How do stars and galaxies evolve? How are sea waves formed by the wind?

All these phenomena are examples of *self-organization* processes; life scientists speak of *growth* processes. Whatever we call them, all these processes are characterized by the spontaneous appearance of patterns, shapes and cycles. Self-organized processes are a common research theme across many disciplines, including biology, chemistry, medicine, geology and engineering.

We will now explore the six global approaches to motion. We will begin with the first approach, namely, the description of motion using a variational principle. This beautiful method for describing, understanding and predicting motion was the result of several centuries of collective effort, and is the highlight of Galilean physics. Variational principles also provide the basis for all the other global approaches just mentioned. They are also needed for all the further descriptions of motion that we will explore afterwards.

Chapter 8
MEASURING CHANGE WITH ACTION

Motion can be described by numbers. Take a single particle that moves. The expression $(x(t), y(t), z(t))$ describes how, during its motion, position changes with time. The description of particle motion is completed by stating how the velocity $(v_x(t), v_y(t), v_z(t))$ changes over time. Realizing that these two expressions fully describe the behaviour of a moving point particle was a milestone in the development of modern physics.

The next milestone of modern physics is achieved by answering a short but hard question. If motion is a type of *change*, as the Greek already said,

▷ How can we *measure* the amount of change?

Physicists took almost two centuries of attempts to uncover the way to measure change. In fact, change can be measured by a single number. Due to the long search, the quantity that measures change has a strange name: it is called *(physical) action*,** usually abbreviated S. To remember the connection of 'action' with change, just think about a Hollywood film: a lot of action means a large amount of change.

Introducing physical action as a measure of change is important, because it provides the first and also the most useful *global* description of motion. In fact, we already know enough to define action straight away.

Imagine taking two snapshots of a system at different times. How could you define the amount of change that occurred in between? When do things change a lot, and when do they change only a little? First of all, a system with *many* moving parts shows *a lot* of

---

** Note that this 'action' is not the same as the 'action' appearing in statements such as 'every action has an equal and opposite reaction'. This other usage, coined by Newton for certain forces, has not stuck; therefore the term has been recycled. After Newton, the term 'action' was first used with an intermediate meaning, before it was finally given the modern meaning used here. This modern meaning is the only meaning used in this text.

Another term that has been recycled is the 'principle of least action'. In old books it used to have a different meaning from the one in this chapter. Nowadays, it refers to what used to be called *Hamilton's principle* in the Anglo-Saxon world, even though it is (mostly) due to others, especially Leibniz. The old names and meanings are falling into disuse and are not continued here.

Behind these shifts in terminology is the story of an intense two-centuries-long attempt to describe motion with so-called *extremal* or *variational principles*: the objective was to complete and improve the work initiated by Leibniz. These principles are only of historical interest today, because all are special cases of the principle of least action described here.

8 MEASURING CHANGE WITH ACTION

FIGURE 191 Giuseppe Lagrangia/Joseph Lagrange (1736–1813).

FIGURE 192 Physical action measures change: an example of process with large action value (© Christophe Blanc).

change. So it makes sense that the action of a system composed of independent subsystems should be the *sum* of the actions of these subsystems.

Secondly, systems with *high energy* or speed, such as the explosions shown in Figure 192, show larger change than systems at lower energy or speed. Indeed, we already introduced energy as the quantity that measures how much a system changes over time.

Thirdly, change often – but not always – *builds up over time*; in other cases, recent change can compensate for previous change, as in a pendulum, when the system can return back to the original state. Change can thus increase or decrease with time.

Finally, for a system in which motion is *stored, transformed* or *shifted* from one subsystem to another, especially when kinetic energy is stored or changed to potential energy, change *diminishes over time*.

TABLE 31 Some action values for changes and processes either observed or imagined.

System and process	Approximate action value
Smallest measurable action	$1.1 \cdot 10^{-34}$ Js
Light	
Smallest blackening of photographic film	$< 10^{-33}$ Js
Photographic flash	$c.\ 10^{-17}$ Js
Electricity	
Electron ejected from atom or molecule	$c.\ 10^{-33}$ Js
Current flow in lightning bolt	$c.\ 10^{4}$ Js
Mechanics and materials	
Tearing apart two neighbouring iron atoms	$c.\ 10^{-33}$ Js
Breaking a steel bar	$c.\ 10^{1}$ Js
Tree bent by the wind from one side to the other	$c.$ 500 Js
Making a white rabbit vanish by 'real' magic	$c.$ 100 PJs
Hiding a white rabbit	$c.$ 0.1 Js
Car crash	$c.$ 2 kJs
Driving car stops within the blink of an eye	$c.$ 20 kJs
Levitating yourself within a minute by 1 m	$c.$ 40 kJs
Large earthquake	$c.$ 1 PJs
Driving car disappears within the blink of an eye	$c.$ 1 ZJs
Sunrise	$c.$ 0.1 ZJs
Chemistry	
Atom collision in liquid at room temperature	$c.\ 10^{-33}$ Js
Smelling one molecule	$c.\ 10^{-31}$ Js
Burning fuel in a cylinder in an average car engine explosion	$c.\ 10^{4}$ Js
Held versus dropped glass	$c.$ 0.8 Js
Life	
Air molecule hitting eardrum	$c.\ 10^{-32}$ Js
Ovule fertilization	$c.\ 10^{-20}$ Js
Cell division	$c.\ 10^{-15}$ Js
Fruit fly's wing beat	$c.\ 10^{-10}$ Js
Flower opening in the morning	$c.$ 1 nJs
Getting a red face	$c.$ 10 mJs
Maximum brain change in a minute	$c.$ 5 Js
Person walking one body length	$c.\ 10^{2}$ Js
Birth	$c.$ 2 kJs
Change due to a human life	$c.$ 1 EJs
Nuclei, stars and more	
Single nuclear fusion reaction in star	$c.\ 10^{-15}$ Js
Explosion of gamma-ray burster	$c.\ 10^{46}$ Js
Universe after one second has elapsed	undefinable

8 MEASURING CHANGE WITH ACTION

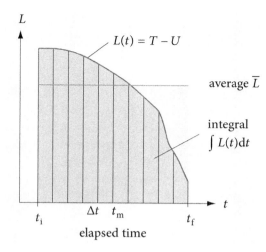

FIGURE 193 Defining a total change or action as an accumulation (addition, or integral) of small changes or actions over time (simplified for clarity).

All the mentioned properties, taken together, imply:

▷ The natural measure of change is the average difference between kinetic and potential energy multiplied by the elapsed time.

This quantity has all the right properties: it is the sum of the corresponding quantities for all subsystems if these are independent; it generally increases with time; and the quantity decreases if the system transforms motion into potential energy.

Thus the (physical) *action* S, measuring the change in a (physical) system, is defined as

$$S = \overline{L}\, \Delta t = \overline{T - U}\, (t_f - t_i) = \int_{t_i}^{t_f} (T - U)\, dt = \int_{t_i}^{t_f} L\, dt \;, \tag{72}$$

where T is the kinetic energy, U the potential energy we already know, L is the difference between these, and the overbar indicates a time average. The quantity L is called the *Lagrangian (function)* of the system,* describes what is being added over time, whenever things change. The sign $\int$ is a stretched 'S', for 'sum', and is pronounced 'integral of'. In intuitive terms it designates the operation – called *integration* – of adding up the values of a varying quantity in infinitesimal time steps dt. The initial and the final times are written below and above the integration sign, respectively. Figure 193 illustrates the idea: the integral is simply the size of the dark area below the curve $L(t)$.

---

* It is named for Giuseppe Lodovico Lagrangia (b. 1736 Torino, d. 1813 Paris), better known as Joseph Louis Lagrange. He was the most important mathematician of his time; he started his career in Turin, then worked for 20 years in Berlin, and finally for 26 years in Paris. Among other things he worked on number theory and analytical mechanics, where he developed most of the mathematical tools used nowadays for calculations in classical mechanics and classical gravitation. He applied them successfully to many motions in the solar system.

Mathematically, the integral of the Lagrangian, i.e., of the curve $L(t)$, is defined as

$$\int_{t_i}^{t_f} L(t)\, dt = \lim_{\Delta t \to 0} \sum_{m=i}^{f} L(t_m)\Delta t = \overline{L} \cdot (t_f - t_i)\,. \tag{73}$$

In other words, the *integral* is the limit, as the time slices get smaller, of the sum of the areas of the individual rectangular strips that approximate the function. Since the $\sum$ sign also means a sum, and since an infinitesimal Δt is written dt, we can understand the notation used for integration. Integration is a sum over slices. The notation was developed by Gottfried Wilhelm Leibniz to make exactly this point. Physically speaking, the integral of the Lagrangian measures the total *effect* that L builds up over time. Indeed, action is called 'effect' in some languages, such as German. The effect that builds up is the *total change* in the system. In short,

> ▷ The integral of the Lagrangian, the action, measures the total change that occurs in a system.

Physical action is total change. Action, or change, is the integral of the Lagrangian over time. The unit of action, and thus of change, is the unit of energy, the Joule, times the unit of time, the second.

> ▷ Change is measured in Js.

A large value means a big change. Table 31 shows some action values observed in nature. To understand the definition of action in more detail, we start with the simplest possible case: a system for which the potential energy vanishes, such as a particle moving freely. Obviously, the higher the kinetic energy is, the more change there is in a given time. Also, if we observe the particle at two instants, the more distant they are the larger the change. The change of a free particle accumulates with time. This is as expected.

Next, we explore a single particle moving in a potential. For example, a falling stone loses potential energy in exchange for a gain in kinetic energy. The more kinetic energy is stored into potential energy, the less change there is. Hence the minus sign in the definition of L. If we explore a particle that is first thrown up in the air and then falls, the curve for $L(t)$ first is below the times axis, then above. We note that the definition of integration makes us count the grey surface *below* the time axis *negatively*. Change can thus be negative, and be compensated by subsequent change, as expected.

To measure change for a system made of several independent components, we simply add all the kinetic energies and subtract all the potential energies. This technique allows us to define action values for gases, liquids and solid matter. In short, action is an *additive* quantity. Even if the components interact, we still get a sensible result.

In summary, physical action measures, in a single number, the change observed in a system between two instants of time. Action, or change, is measured in Js. Physical action quantifies the change due to a physical process. This is valid for *all* observations, i.e., for all processes and all systems: an explosion of a firecracker, a caress of a loved one or a colour change of computer display. We will discover later that describing change with a

8 MEASURING CHANGE WITH ACTION

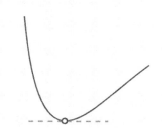

FIGURE 194 The minimum of a curve has vanishing slope.

single number is also possible in relativity and quantum theory: *any* change going on in any system of nature, be it transport, transformation or growth, can be measured with a single number.

THE PRINCIPLE OF LEAST ACTION

> The optimist thinks this is the best of all possible worlds, and the pessimist knows it.
> Robert Oppenheimer

We now have a precise measure of change. This, as it turns out, allows a simple, global and powerful description of motion. In nature, the change happening between two instants is always the *smallest* possible.

▷ *In nature, action is minimal.*

This is the essence of the famous *principle of least action*. It is valid for every example of motion.* Of all possible motions, nature always chooses for which the change is *minimal*. Let us study a few examples.

The simple case of a free particle, when no potentials are involved, the principle of least action implies that the particle moves in a *straight* line with *constant* velocity. All other paths would lead to larger actions. Can you verify this?

When gravity is present, a thrown stone flies along a parabola – or more precisely, along an ellipse – because any other path, say one in which the stone makes a loop in the air, would imply a *larger* action. Again you might want to verify this for yourself.

All observations support the simple and basic statement: things always move in a way that produces the smallest possible value for the action. This statement applies to the full path and to any of its segments. Bertrand Russell called it the 'law of cosmic laziness'. We could also call it the principle of *maximal efficiency* of nature.

It is customary to express the idea of minimal change in a different way. The action varies when the path is varied. The actual path is the one with the smallest action. You will

---

* In fact, in some macroscopic situations the action can be a saddle point, so that the snobbish form of the principle is that the action is 'stationary'. In contrast to what is often heard, the action is *never* a maximum. Moreover, for motion on small (infinitesimal) scales, the action is always a minimum. The mathematical condition of vanishing variation, given below, encompasses all these details.

recall from school that at a minimum the derivative of a quantity vanishes: a minimum has a horizontal slope. This relation is shown in Figure 194. In the present case, we do not vary a quantity, but a complete path; hence we do not speak of a derivative or slope, but of a variation. It is customary to write the variation of action as δS. The *principle of least action* thus states:

▷ The actual trajectory between specified end points satisfies $\delta S = 0$.

Mathematicians call this a *variational principle*. Note that the end points have to be specified: we have to compare motions with the *same* initial and final situations.

Before discussing the principle of least action further, we check that it is indeed equivalent to the evolution equation.* To do this, we can use a standard procedure, part of

---

* For those interested, here are a few comments on the equivalence of Lagrangians and evolution equations. First of all, Lagrangians do not exist for non-conservative, or *dissipative* systems. We saw that there is no potential for any motion involving *friction* (and more than one dimension); therefore there is no action in these cases. One approach to overcome this limitation is to use a generalized formulation of the principle of least action. Whenever there is no potential, we can express the *work* variation δW between different trajectories x_i as

$$\delta W = \sum_i m_i \ddot{x}_i \delta x_i \ . \tag{74}$$

Motion is then described in the following way:

▷ *The actual trajectory satifies* $\int_{t_i}^{t_f} (\delta T + \delta W) \mathrm{d}t = 0 \quad provided \quad \delta x(t_i) = \delta x(t_f) = 0$. (75)

The quantity being varied has no name; it represents a generalized notion of change. You might want to check that it leads to the correct evolution equations. Thus, although *proper* Lagrangian descriptions exist only for *conservative* systems, for dissipative systems the principle can be generalized and remains useful.

Many physicists will prefer another approach. What a mathematician calls a generalization is a special case for a physicist: the principle (75) hides the fact that *all* results from the usual principle of minimal action, if we include the complete microscopic details. There is no friction in the microscopic domain. Friction is an approximate, macroscopic concept.

Nevertheless, more mathematical viewpoints are useful. For example, they lead to interesting limitations for the use of Lagrangians. These limitations, which apply only if the world is viewed as purely classical – which it isn't – were discovered about a hundred years ago. In those times, computers were not available, and the exploration of new calculation techniques was important. Here is a summary.

The coordinates used in connection with Lagrangians are not necessarily the Cartesian ones. *Generalized* coordinates are especially useful when there are *constraints* on the motion. This is the case for a pendulum, where the weight always has to be at the same distance from the suspension, or for an ice skater, where the skate has to move in the direction in which it is pointing. Generalized coordinates may even be mixtures of positions and momenta. They can be divided into a few general types.

Generalized coordinates are called *holonomic–scleronomic* if they are related to Cartesian coordinates in a fixed way, independently of time: physical systems described by such coordinates include the pendulum and a particle in a potential. Coordinates are called *holonomic–rheonomic* if the dependence involves time. An example of a rheonomic systems would be a pendulum whose length depends on time. The two terms rheonomic and scleronomic are due to Ludwig Boltzmann. These two cases, which concern systems that are only described by their geometry, are grouped together as *holonomic systems*. The term is due to Heinrich Hertz.

The more general situation is called *anholonomic*, or *nonholonomic*. Lagrangians work well only for holonomic systems. Unfortunately, the meaning of the term 'nonholonomic' has changed. Nowadays, the term is also used for certain rheonomic systems. The modern use calls nonholonomic any system which involves

8 MEASURING CHANGE WITH ACTION

the so-called *calculus of variations*. The condition $\delta S = 0$ implies that the action, i.e., the area under the curve in Figure 193, is a minimum. A little bit of thinking shows that if the Lagrangian is of the form $L(x_n, v_n) = T(v_n) - U(x_n)$, then the minimum area is achieved when

$$\frac{d}{dt}\left(\frac{\partial T}{\partial v_n}\right) = -\frac{\partial U}{\partial x_n} \tag{76}$$

where n counts all coordinates of all particles.* For a single particle, these *Lagrange's equations of motion* reduce to

$$m\boldsymbol{a} = -\nabla U . \tag{78}$$

This is the evolution equation: it says that the acceleration times the mass of a particle is the gradient of the potential energy U. The principle of least action thus implies the equation of motion. (Can you show the converse, which is also correct?)

In other words, *all systems evolve in such a way that the change or action is as small as possible*. Nature is economical. Nature is maximally efficient. Or: Nature is lazy. Nature is thus the opposite of a Hollywood thriller, in which the action is maximized; nature is more like a wise old man who keeps his actions to a minimum.

The principle of minimal action states that the actual trajectory is the one for which the *average* of the Lagrangian over the whole trajectory is minimal (see Figure 193). Nature is a Dr. Dolittle. Can you verify this? This viewpoint allows one to deduce Lagrange's equations (76) directly.

The principle of least action distinguishes the actual trajectory from all other imaginable ones. This observation lead Leibniz to his famous interpretation that the actual world is the 'best of all possible worlds.'** We may dismiss this as metaphysical speculation, but we should still be able to feel the fascination of the issue. Leibniz was so excited about the principle of least action because it was the first time that actual observations were distinguished from all other imaginable possibilities. For the first time, the search for reasons why things are the way they are became a part of physical investigation. Could the world be different from what it is? In the principle of least action, we have a hint of

---

velocities. Therefore, an ice skater or a rolling disc is often called a nonholonomic system. Care is thus necessary to decide what is meant by nonholonomic in any particular context.

Even though the use of Lagrangians, and of action, has its limitations, these need not bother us at microscopic level, because microscopic systems are always conservative, holonomic and scleronomic. At the fundamental level, evolution equations and Lagrangians are indeed equivalent.

* The most general form for a Lagrangian $L(q_n, \dot{q}_n, t)$, using generalized holonomic coordinates q_n, leads to Lagrange equations of the form

$$\frac{d}{dt}\left(\frac{\partial L}{\partial \dot{q}_n}\right) = \frac{\partial L}{\partial q_n} . \tag{77}$$

In order to deduce these equations, we also need the relation $\delta \dot{q} = d/dt(\delta q)$. This relation is valid only for *holonomic* coordinates introduced in the previous footnote and explains their importance.

We remark that the Lagrangian for a moving system is not unique; however, the study of how the various Lagrangians for a given moving system are related is not part of this walk.

By the way, the letter q for position and p for momentum were introduced in physics by the mathematician Carl Jacobi (b. 1804 Potsdam, d. 1851 Berlin).

** This idea was ridiculed by the influential philosopher Voltaire (b. 1694 Paris, d. 1778 Paris) in his lucid writings, notably in the brilliant book *Candide*, written in 1759, and still widely available.

a negative answer. Leibniz also deduced from the result that gods cannot choose their actions. (What do you think?)

Lagrangians and motion

> Never confuse movement with action.
> Ernest Hemingway

Systems evolve by minimizing change. Change, or action, is the time integral of the Lagrangian. As a way to describe motion, the Lagrangian has several advantages over the evolution equation. First of all, the Lagrangian is usually more *compact* than writing the corresponding evolution equations. For example, only *one* Lagrangian is needed for one system, however many particles it includes. One makes fewer mistakes, especially sign mistakes, as one rapidly learns when performing calculations. Just try to write down the evolution equations for a chain of masses connected by springs; then compare the effort with a derivation using a Lagrangian. (The system is often studied because it behaves in many aspects like a chain of atoms.) We will encounter another example shortly: David Hilbert took only a few weeks to deduce the equations of motion of general relativity using a Lagrangian, whereas Albert Einstein had worked for ten years searching for them directly.

In addition, the description with a Lagrangian is valid with *any* set of coordinates describing the objects of investigation. The coordinates do not have to be Cartesian; they can be chosen as we prefer: cylindrical, spherical, hyperbolic, etc. These so-called *generalized coordinates* allow one to rapidly calculate the behaviour of many mechanical systems that are in practice too complicated to be described with Cartesian coordinates. For example, for programming the motion of robot arms, the angles of the joints provide a clearer description than Cartesian coordinates of the ends of the arms. Angles are non-Cartesian coordinates. They simplify calculations considerably: the task of finding the most economical way to move the hand of a robot from one point to another is solved much more easily with angular variables.

More importantly, the Lagrangian allows us to quickly deduce the essential properties of a system, namely, its *symmetries* and its *conserved quantities*. We will develop this important idea shortly, and use it regularly throughout our walk.

Finally, the Lagrangian formulation can be generalized to encompass *all types of interactions*. Since the concepts of kinetic and potential energy are general, the principle of least action can be used in electricity, magnetism and optics as well as mechanics. The principle of least action is central to general relativity and to quantum theory, and allows us to easily relate both fields to classical mechanics.

As the principle of least action became well known, people applied it to an ever-increasing number of problems. Today, Lagrangians are used in everything from the study of elementary particle collisions to the programming of robot motion in artificial intelligence. (Table 32 shows a few examples.) However, we should not forget that despite its remarkable simplicity and usefulness, the Lagrangian formulation is *equivalent* to the evolution equations. It is neither more general nor more specific. In particular, it is *not an explanation* for any type of motion, but only a different view of it. In fact, the search for a new physical 'law' of motion is *just* the search for a new Lagrangian. This makes

TABLE 32 Some Lagrangians.

System	Lagrangian	Quantities
Free, non-relativistic mass point	$L = \frac{1}{2}mv^2$	mass m, speed $v = dx/dt$
Particle in potential	$L = \frac{1}{2}mv^2 - m\varphi(x)$	gravitational potential φ
Mass on spring	$L = \frac{1}{2}mv^2 - \frac{1}{2}kx^2$	elongation x, spring constant k
Mass on frictionless table attached to spring	$L = \frac{1}{2}mv^2 - k(x^2 + y^2)$	spring constant k, coordinates x, y
Chain of masses and springs (simple model of atoms in a linear crystal)	$L = \frac{1}{2}m \sum v_i^2 - \frac{1}{2}m\omega^2 \sum_{i,j}(x_i - x_j)^2$	coordinates x_i, lattice frequency ω
Free, relativistic mass point	$L = -c^2 m \sqrt{1 - v^2/c^2}$	mass m, speed v, speed of light c

sense, as the description of nature always requires the description of change. Change in nature is always described by actions and Lagrangians.

The principle of least action states that the action is minimal when the end points of the motion, and in particular the time between them, are fixed. It is less well known that the reciprocal principle also holds: if the action value – the change value – is kept fixed, the elapsed time for the actual motion is *maximal*. Can you show this?

Even though the principle of least action is not an explanation of motion, the principle somehow calls for such an explanation. We need some patience, though. *Why* nature follows the principle of least action, and *how* it does so, will become clear when we explore quantum theory.

Why is motion so often bounded?

Looking around ourselves on Earth or in the sky, we find that matter is not evenly distributed. Matter tends to be near other matter: it is lumped together in *aggregates*. Figure 195 shows a typical example. Some major examples of aggregates are listed in Figure 196 and Table 33. All aggregates have mass and size. In the mass–size diagram of Figure 196, both scales are logarithmic. We note three straight lines: a line $m \sim l$ extending from the Planck mass* upwards, via black holes, to the universe itself; a line $m \sim 1/l$ extending from the Planck mass downwards, to the lightest possible aggregate; and the usual matter line with $m \sim l^3$, extending from atoms upwards, via everyday objects, the Earth to the Sun. The first of the lines, the black hole limit, is explained by general relativity; the last two, the aggregate limit and the common matter line, by quantum theory.**

The aggregates outside the common matter line also show that the stronger the interaction that keeps the components together, the smaller the aggregate. But why is matter mainly found in lumps?

---

* The Planck mass is given by $m_{Pl} = \sqrt{\hbar c/G} = 21.767(16)$ μg.
** Figure 196 suggests that domains beyond physics exist; we will discover later on that this is not the case, as mass and size are not definable in those domains.

FIGURE 195 Motion in the universe is bounded. (© Mike Hankey)

First of all, aggregates form because of the existence of *attractive* interactions between objects. Secondly, they form because of *friction*: when two components approach, an aggregate can only be formed if the released energy can be changed into heat. Thirdly, aggregates have a finite size because of *repulsive* effects that prevent the components from collapsing completely. Together, these three factors ensure that in the universe, bound motion is much more common than unbound, 'free' motion.

Only three types of attraction lead to aggregates: gravity, the attraction of electric charges, and the strong nuclear interaction. Similarly, only three types of repulsion are observed: rotation, pressure, and the Pauli exclusion principle (which we will encounter later on). Of the nine possible combinations of attraction and repulsion, not all appear in nature. Can you find out which ones are missing from Figure 196 and Table 33, and why?

Together, attraction, friction and repulsion imply that change and action are minim-

8 MEASURING CHANGE WITH ACTION

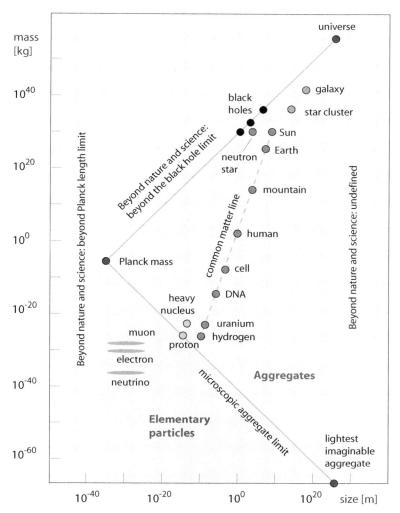

FIGURE 196 Elementary particles and aggregates found in nature.

ized when objects come and stay together. The principle of least action thus implies the stability of aggregates. By the way, the formation history also explains why so many aggregates *rotate*. Can you tell why?

But why does friction exist at all? And why do attractive and repulsive interactions exist? And why is it – as it would appear from the above – that in some distant past matter was *not* found in lumps? In order to answer these questions, we must first study another global property of motion: symmetry.

Challenge 461 s

TABLE 33 Some major aggregates observed in nature.

Aggregate	Size (diameter)	Obs. num.	Constituents

Gravitationally bound aggregates

Aggregate	Size (diameter)	Obs. num.	Constituents
Matter across universe	c. 100 Ym	1	superclusters of galaxies, hydrogen and helium atoms
Quasar	10^{12} to 10^{14} m	$20 \cdot 10^6$	baryons and leptons
Supercluster of galaxies	c. 3 Ym	10^7	galaxy groups and clusters
Galaxy cluster	c. 60 Zm	$25 \cdot 10^9$	10 to 50 galaxies
Galaxy group or cluster	c. 240 Zm		50 to over 2000 galaxies
Our local galaxy group	50 Zm	1	c. 40 galaxies
General galaxy	0.5 to 2 Zm	$3.5 \cdot 10^{12}$	10^{10} to $3 \cdot 10^{11}$ stars, dust and gas clouds, probably star systems
Our galaxy	1.0(0.1) Zm	1	10^{11} stars, dust and gas clouds, solar systems
Interstellar clouds	up to 15 Em	$\gg 10^5$	hydrogen, ice and dust
Solar System[a]	unknown	> 400	star, planets
Our Solar System	30 Pm	1	Sun, planets (Pluto's orbit's diameter: 11.8 Tm), moons, planetoids, comets, asteroids, dust, gas
Oort cloud	6 to 30 Pm	1	comets, dust
Kuiper belt	60 Tm	1	planetoids, comets, dust
Star[b]	10 km to 100 Gm	$10^{22\pm1}$	ionized gas: protons, neutrons, electrons, neutrinos, photons
Our star, the Sun	1.39 Gm		
Planet[a] (Jupiter, Earth)	143 Mm, 12.8 Mm	8+ > 400	solids, liquids, gases; in particular, heavy atoms
Planetoids (Varuna, etc)	50 to 1 000 km	> 100 (est. 10^9)	solids
Moons	10 to 1 000 km	> 50	solids
Neutron stars	10 km	> 1000	mainly neutrons
Electromagnetically bound aggregates[c]			
Dwarf planets, minor planets, asteroids[d]	1 m to 2400 km	> 10^6	(10^9 estimated) solids, usually monolithic
Comets	10 cm to 50 km	> 10^9	(10^{12} possible) ice and dust
Mountains, solids, liquids, gases, cheese	1 nm to > 100 km	n.a.	molecules, atoms
Animals, plants, kefir	5 μm to 1 km	$10^{26\pm2}$	organs, cells
brain, human	0.2 m	10^{10}	neurons and other cell types
Cells:		$10^{31\pm1}$	organelles, membranes, molecules
smallest (*Nanoarchaeum equitans*)	c. 400 nm		molecules
amoeba	c. 600 μm		molecules

Aggregate	Size (diameter)	Obs. num.	Constituents
largest (whale nerve, single-celled plants)	c. 30 m		molecules
Molecules:		$10^{78\pm2}$	atoms
H$_2$	c. 50 pm	$10^{72\pm2}$	atoms
DNA (human)	2 m (total per cell)	10^{21}	atoms
Atoms, ions	30 pm to 300 pm	$10^{80\pm2}$	electrons and nuclei
Aggregates bound by the weak interaction[c]			
None			
Aggregates bound by the strong interaction[c]			
Nucleus	0.9 to > 7 fm	$10^{79\pm2}$	nucleons
Nucleon (proton, neutron)	0.9 fm	$10^{80\pm2}$	quarks
Mesons	c. 1 fm	n.a.	quarks
Neutron stars: see further up			

a. Only in 1994 was the first evidence found for objects circling stars other than our Sun; of over 1000 *extrasolar planets* found so far, most are found around F, G and K stars, including neutron stars. For example, three objects circle the pulsar PSR 1257+12, and a matter ring circles the star β Pictoris. The objects seem to be dark stars, brown dwarfs or large gas planets like Jupiter. Due to the limitations of observation systems, none of the systems found so far form solar systems of the type we live in. In fact, only a few Earth-like planets have been found so far.

b. The Sun is among the brightest 7 % of stars. Of all stars, 80 %, are red M dwarfs, 8 % are orange K dwarfs, and 5 % are white D dwarfs: these are all faint. Almost all stars visible in the night sky belong to the bright 7 %. Some of these are from the rare blue O class or blue B class (such as Spica, Regulus and Rigel); 0.7 % consist of the bright, white A class (such as Sirius, Vega and Altair); 2 % are of the yellow–white F class (such as Canopus, Procyon and Polaris); 3.5 % are of the yellow G class (like Alpha Centauri, Capella or the Sun). Exceptions include the few visible K giants, such as Arcturus and Aldebaran, and the rare M supergiants, such as Betelgeuse and Antares. More on stars later on.

c. For more details on *microscopic* aggregates, see the table of composites.

d. It is estimated that there are up to 10^{20} *small Solar System bodies* (asteroids, meteoroids, planetoids or minor planets) that are heavier than 100 kg. Incidentally, no asteroids between Mercury and the Sun – the hypothetical *Vulcanoids* – have been found so far.

Curiosities and fun challenges about Lagrangians

The principle of least action as a mathematical description is due to Leibniz. He understood its validity in 1707. It was then rediscovered and named by Maupertuis in 1746, who wrote:

> Lorsqu'il arrive quelque changement dans la Nature, la quantité d'action nécessaire pour ce changement est la plus petite qu'il soit possible.*

---

* 'When some change occurs in Nature, the quantity of action necessary for this change is the smallest that is possible.'

Samuel König, the first scientist to state publicly and correctly in 1751 that the principle was due to Leibniz, and not to Maupertuis, was expelled from the Prussian Academy of Sciences for stating so. This was due to an intrigue of Maupertuis, who was the president of the academy at the time. The intrigue also made sure that the bizarre term 'action' was retained. Despite this disgraceful story, Leibniz' principle quickly caught on, and was then used and popularized by Euler, Lagrange and finally by Hamilton.

∗ ∗

The basic idea of the principle of least action, that nature is as *lazy* as possible, is also called *lex parismoniae*. This general idea is already expressed by Ptolemy, and later by Fermat, Malebranche, and 's Gravesande. But Leibniz was the first to understand its validity and mathematical usefulness for the description of *all* motion.

∗ ∗

When Lagrange published his book *Mécanique analytique*, in 1788, it formed one of the high points in the history of mechanics and established the use of variational principles. He was proud of having written a systematic exposition of mechanics without a single figure. Obviously the book was difficult to read and was not a sales success at all. Therefore his methods took another generation to come into general use.

∗ ∗

Given that action is the basic quantity describing motion, we can define energy as action per unit time, and momentum as action per unit distance. The *energy* of a system thus describes how much it changes over time, and the *momentum* describes how much it changes over distance. What are angular momentum and rotational energy?

∗ ∗

In Galilean physics, the Lagrangian is the difference between kinetic and potential energy. Later on, this definition will be generalized in a way that sharpens our understanding of this distinction: the Lagrangian becomes the difference between a term for free particles and a term due to their interactions. In other words, every particle motion is a continuous compromise between what the particle would do if it were free and what other particles want it to do. In this respect, particles behave a lot like humans beings.

∗ ∗

'In nature, effects of telekinesis or prayer are impossible, as in most cases the change inside the brain is much smaller than the change claimed in the outside world.' Is this argument correct?

∗ ∗

How is action measured? What is the best device or method to measure action?

∗ ∗

Explain: why is $T + U$ constant, whereas $T - U$ is minimal?

∗ ∗

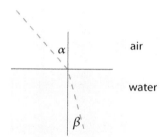

FIGURE 197 Refraction of light is due to travel-time optimization.

In nature, the sum $T + U$ of kinetic and potential energy is *constant* during motion (for closed systems), whereas the action is *minimal*. Is it possible to deduce, by combining these two facts, that systems tend to a state with minimum potential energy?

* *

Another minimization principle can be used to understand the construction of animal bodies, especially their size and the proportions of their inner structures. For example, the heart pulse and breathing frequency both vary with animal mass m as $m^{-1/4}$, and the dissipated power varies as $m^{3/4}$. It turns out that such exponents result from three properties of living beings. First, they transport energy and material through the organism via a branched network of vessels: a few large ones, and increasingly many smaller ones. Secondly, the vessels all have the same minimum size. And thirdly, the networks are optimized in order to minimize the energy needed for transport. Together, these relations explain many additional scaling rules; they might also explain why animal lifespan scales as $m^{-1/4}$, or why most mammals have roughly the same number of heart beats in a lifetime.

A competing explanation, using a different minimization principle, states that quarter powers arise in any network built in order that the flow arrives to the destination by the most direct path.

* *

The minimization principle for the motion of light is even more beautiful: light always takes the path that requires the shortest travel time. It was known long ago that this idea describes exactly how light changes direction when it moves from air to water, and effect illustrated in Figure 197. In water, light moves more slowly; the speed ratio between air and water is called the *refractive index* of water. The refractive index, usually abbreviated n, is material-dependent. The value for water is about 1.3. This speed ratio, together with the minimum-time principle, leads to the 'law' of refraction, a simple relation between the sines of the two angles. Can you deduce it?

* *

Can you confirm that all the mentioned minimization principles – that for the growth of trees, that for the networks inside animals, that for the motion of light – are special cases of the principle of least action? In fact, this is the case for *all* known minimization principles in nature. Each of them, like the principle of least action, is a principle of least change.

∗ ∗

In Galilean physics, the value of the action depends on the speed of the observer, but not on his position or orientation. But the action, when properly defined, should *not* depend on the observer. All observers should agree on the value of the observed change. Only special relativity will fulfil the requirement that action be independent of the observer's speed. How will the relativistic action be defined?

∗ ∗

What is the amount of change accumulated in the universe since the big bang? Measuring all the change that is going on in the universe presupposes that the universe is a physical system. Is this the case?

∗ ∗

One motion for which action is particularly well minimized in nature is dear to us: walking. Extensive research efforts try to design robots which copy the energy saving functioning and control of human legs. For an example, see the website by Tao Geng at cswww.essex.ac.uk/tgeng/research.html.

∗ ∗

Can you prove the following integration challenge?

$$\int_0^\varphi \sec t \, dt = \ln \tan\left(\frac{\pi}{4} + \frac{\varphi}{2}\right) \tag{79}$$

∗ ∗

What is the shape of the ideal halfpipe for skateboarding? What does 'ideal' imply? Which requirement leads to a *cycloid*? Which requirement speaks against a *cycloid*?

∗ ∗

As mentioned above, animal death is a physical process and occurs when an animal has consumed or metabolized around 1 GJ/kg. Show that the total action of an animal scales as $M^{5/4}$.

Summary on action

Systems move by minimizing change. Change, or action, is the time average of kinetic energy minus potential energy. The statement 'motion minimizes change' expresses motion's predictability and its continuity. The statement also implies that all motion is as simple as possible.

Systems move by minimizing change. Equivalently, systems move by maximizing the elapsed time between two situations. Both statements show that nature is lazy.

Systems move by minimizing change. In the next chapters we show that this statement implies the observer-invariance, conservation, mirror-invariance, reversibility and relativity of everyday motion.

Chapter 9
MOTION AND SYMMETRY

> "Am Anfang war die Symmetrie.**"
> Werner Heisenberg

The second way to describe motion globally is to describe it in such a way that *all* observers agree. Now, whenever an observation stays exactly the *same* when switching from one observer to another, we call the observation *invariant* or *absolute* or *symmetric*. Whenever an observation *changes* when switching from one observer to another, we call it *relative*. To explore relativity thus means to explore invariance and symmetry.

▷ *Symmetry* is invariance after change.

Change of observer, or change of point of view, is one such possible change; another possibility can be some change operated on the system under observation itself. For example, a forget-me-not flower, shown in Figure 198, is symmetrical because it looks the same after turning around it, or after turning it, by 72 degrees; many fruit tree flowers have the same symmetry. One also says that under certain changes of viewpoint the flower has an *invariant property*, namely its shape. If many such viewpoints are possible, one talks about a *high* symmetry, otherwise a *low* symmetry. For example, a four-leaf clover has a higher symmetry than a usual, three-leaf one. In physics, the viewpoints are often called *frames of reference*.

Whenever we speak about symmetry in flowers, in everyday life, in architecture or in the arts we usually mean mirror symmetry, rotational symmetry or some combination. These are *geometric symmetries*. Like all symmetries, geometric symmetries imply invariance under specific change operations. The complete list of geometric symmetries is known for a long time. Table 34 gives an overview of the basic types. Figure 199 and Figure 200 give some important examples. Additional geometric symmetries include *colour symmetries*, where colours are exchanged, and *spin groups*, where symmetrical objects do not contain only points but also spins, with their special behaviour under rotations. Also combinations with scale symmetry, as they appear in fractals, and variations on curved backgrounds are extension of the basic table.

** 'In the beginning, there was symmetry.' Do you agree with this statement? It has led many researchers astray during the search for the unification of physics. Probably, Heisenberg meant to say that in the beginning, there was *simplicity*. However, there are many conceptual and mathematical differences between symmetry and simplicity.

9 MOTION AND SYMMETRY

FIGURE 198 Forget-me-not, also called *Myosotis* (Boraginaceae), has five-fold symmetry (© Markku Savela).

TABLE 34 The classification and the number of simple geometric symmetries.

Dimension	Repetition Types	Translations			
		0	1	2	3
		POINT GROUPS	LINE GROUPS	PLANE GROUPS	SPACE GROUPS
1	1 row	2	2	n.a.	n.a.
2	5 nets or plane lattice types (square, oblique, hexagonal, rectangular, centred rectangular)	2 (cyclic, dihedral) or 10 rosette groups (C_1, C_2, C_3, C_4, C_6, D_1, D_2, D_3, D_4, D_6)	7 friezes	17 wall-papers	n.a.
3	14 (Bravais) lattices (3 cubic, 2 tetragonal, 4 orthorhombic, 1 hexagonal, 1 trigonal, 2 monoclinic, 1 triclinic type)	32 crystal groups, also called crystallographic point groups	75 rods	80 layers	230 crystal structures, also called space groups, Fedorov groups or crystallographic groups

A *high* symmetry means that *many* possible changes leave an observation invariant. At first sight, not many objects or observations in nature seem to be symmetrical: after all, in the nature around us, geometric symmetry is more the exception than the rule. But this is a fallacy. On the contrary, we can deduce that nature as a whole is symmetric from the simple fact that we have the ability to talk about it! Moreover, the symmetry of nature is considerably higher than that of a forget-me-not or of any other symmetry from

The 17 wallpaper patterns and a way to identify them quickly.

Is the maximum rotation order 1, 2, 3, 4 or 6?
Is there a mirror (m)? Is there an indecomposable glide reflection (g)?
Is there a rotation axis on a mirror? Is there a rotation axis not on a mirror?

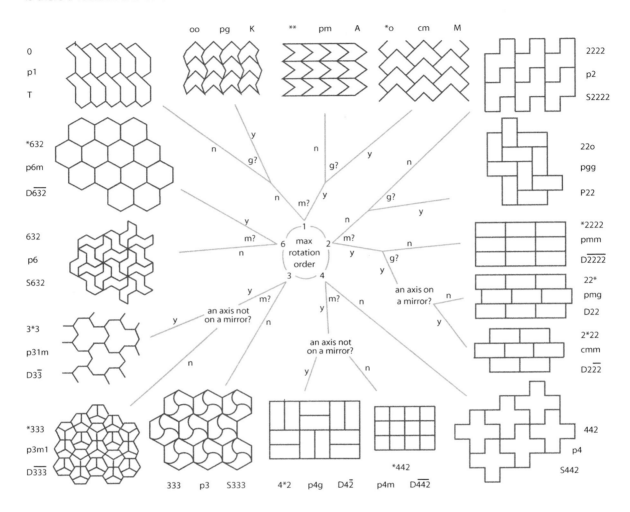

Every pattern is identified according to three systems of notation:

442	The Conway-Thurston notation.
p4	The International Union of Crystallography notation.
S442	The Montesinos notation, as in his book "Classical Tesselations and Three Manifolds"

FIGURE 199 The full list of possible symmetries of wallpaper patterns, the so-called *wallpaper groups*, their usual names, and a way to distinguish them (© Dror Bar-Natan).

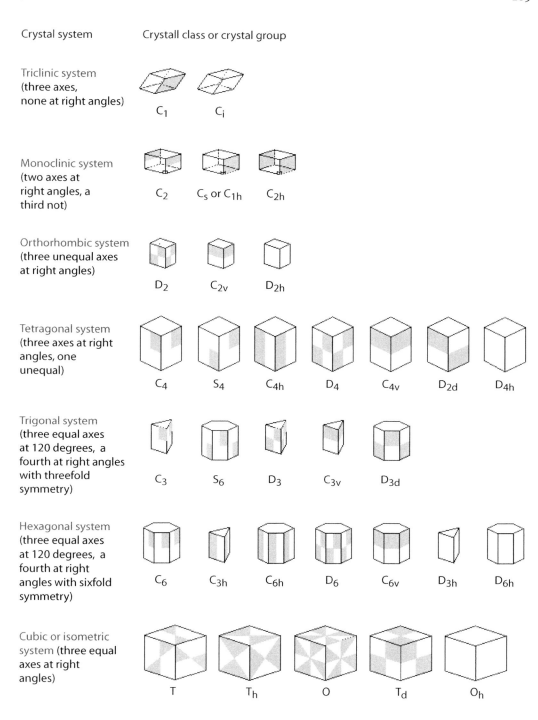

FIGURE 200 The full list of possible symmetries of units cells in crystals, the *crystallographic point groups* or *crystal groups* or *crystal classes* (© Jonathan Goss, after Neil Ashcroft and David Mermin).

Table 34. A consequence of this high symmetry is, among others, the famous expression $E_0 = c^2 m$.

Why can we think and talk about the world?

> "The hidden harmony is stronger than the apparent."
> Heraclitus of Ephesus, about 500 BCE

Why can we understand somebody when he is talking about the world, even though we are not in his shoes? We can for two reasons: because most things look *similar* from different viewpoints, and because most of us have already had similar experiences *beforehand*.

'Similar' means that what *we* and what *others* observe somehow correspond. In other words, many aspects of observations do not depend on viewpoint. For example, the number of petals of a flower has the same value for all observers. We can therefore say that this quantity has the highest possible symmetry. We will see below that mass is another such example. Observables with the highest possible symmetry are called *scalars* in physics. Other aspects change from observer to observer. For example, the apparent size varies with the distance of observation. However, the actual size is observer-independent. In general terms,

> ▷ Any type of *viewpoint-independence* is a form of symmetry.

The observation that two people looking at the same thing from different viewpoints can understand each other proves that nature is symmetric. We start to explore the details of this symmetry in this section and we will continue during most of the rest of our hike.

In the world around us, we note another general property: not only does the same phenomenon look similar to different observers, but *different* phenomena look similar to the *same* observer. For example, we know that if fire burns the finger in the kitchen, it will do so outside the house as well, and also in other places and at other times. Nature shows *reproducibility*. Nature shows no surprises. In fact, our memory and our thinking are only possible because of this basic property of nature. (Can you confirm this?) As we will see, reproducibility leads to additional strong restrictions on the description of nature.

Without viewpoint-independence and reproducibility, talking to others or to oneself would be impossible. Even more importantly, we will discover that viewpoint-independence and reproducibility do more than determine the possibility of talking to each other: they also fix much (but not all) of the *content* of what we can say to each other. In other words, we will see that most of our description of nature follows logically, almost without choice, from the simple fact that we can talk about nature to our friends.

9 MOTION AND SYMMETRY

Viewpoints

> Toleranz ... ist der Verdacht der andere könnte Recht haben.*
> Kurt Tucholsky (b. 1890 Berlin, d. 1935 Göteborg), German writer

> Toleranz – eine Stärke, die man vor allem dem politischen Gegner wünscht.**
> Wolfram Weidner (b. 1925) German journalist

When a young human starts to meet other people in childhood, he quickly finds out that certain experiences are shared, while others, such as dreams, are not. Learning to make this distinction is one of the adventures of human life. In these pages, we concentrate on a section of the first type of experiences: *physical* observations. However, even between these, distinctions are to be made. In daily life we are used to assuming that weights, volumes, lengths and time intervals are independent of the viewpoint of the observer. We can talk about these observed quantities to anybody, and there are no disagreements over their values, provided they have been measured correctly. However, other quantities do depend on the observer. Imagine talking to a friend after he jumped from one of the trees along our path, while he is still falling downwards. He will say that the forest floor is approaching with high speed, whereas the observer below will maintain that the floor is stationary. Obviously, the difference between the statements is due to their different viewpoints. The velocity of an object (in this example that of the forest floor or of the friend himself) is thus a less symmetric property than weight or size. Not all observers agree on the value. of velocity, nor even on its direction.

In the case of viewpoint-dependent observations, understanding each other is still possible with the help of a little effort: each observer can *imagine* observing from the point of view of the other, and *check* whether the imagined result agrees with the statement of the other.*** If the statement thus imagined and the actual statement of the other observer agree, the observations are consistent, and the difference in statements is due only to the different viewpoints; otherwise, the difference is fundamental, and they cannot agree or talk. Using this approach, you can even argue whether human feelings, judgements, or tastes arise from fundamental differences or not.

The distinction between viewpoint-independent – or invariant – quantities and viewpoint-dependent – or relative –quantities is an essential one. Invariant quantities, such as mass or shape, describe *intrinsic* properties, and relative quantities, depending on the observer, make up the *state* of the system. Therefore, in order to find a *complete* description of the state of a physical system, we must answer the following questions:

— Which viewpoints are possible?
— How are descriptions transformed from one viewpoint to another?
— Which observables do these symmetries admit?

---

* 'Tolerance ... is the suspicion that the other might be right.'
** 'Tolerance – a strength one mainly wishes to political opponents.'
*** Humans develop the ability to imagine that others can be in situations *different* from their own at the age of about four years. Therefore, before the age of four, humans are unable to conceive special relativity; afterwards, they can.

— What do these results tell us about motion?

So far, in our exploration of motion we have first of all studied viewpoints that differ in location, in orientation, in time and, most importantly, in motion. With respect to each other, observers can be at rest, can be rotated, can move with constant speed or can even accelerate. These 'concrete' changes of viewpoint are those we will study first. In this case the requirement of consistency of observations made by different observers is called the *principle of relativity*. The symmetries associated with this type of invariance are also called *external* symmetries. They are listed in Table 36.

A second class of fundamental changes of viewpoint concerns 'abstract' changes. Viewpoints can differ by the mathematical description used: such changes are called *changes of gauge*. They will be introduced first in the section on electrodynamics. Again, it is required that all statements be consistent across different mathematical descriptions. This requirement of consistency is called the *principle of gauge invariance*. The associated symmetries are called *internal* symmetries.

The third class of changes, whose importance may not be evident from everyday life, is that of the behaviour of a system under exchange of its parts. The associated invariance is called *permutation symmetry*. It is a *discrete* symmetry, and we will encounter it as a fundamental principle when we explore quantum theory.

The three consistency requirements just described are called 'principles' because these basic statements are so strong that they almost completely determine the 'laws' of physics – i.e., the description of motion – as we will see shortly. Later on we will discover that looking for a complete description of the *state* of objects will also yield a complete description of their *intrinsic properties*. But enough of introduction: let us come to the heart of the topic.

Symmetries and groups

Because we are looking for a description of motion that is complete, we need to understand and describe the *full set of symmetries* of nature. But what is symmetry?

A system is said to be *symmetric* or to possess a *symmetry* if it appears identical when observed from different viewpoints. We also say that the system possesses an *invariance* under change from one viewpoint to the other. Viewpoint changes are equivalent to *symmetry operations* or *transformations* of a system. A symmetry is thus a set of transformations that leaves a system invariant. However, a symmetry is more than a set: the successive application of two symmetry operations is another symmetry operation. In other terms, a symmetry is a set $G = \{a, b, c, ...\}$ of elements, the transformations, together with a binary operation $\circ$ called *concatenation* or *multiplication* and pronounced 'after' or 'times', in which the following properties hold for all elements a, b and c:

$$\begin{aligned}
&\textit{associativity}, \text{i.e.,} \quad (a \circ b) \circ c = a \circ (b \circ c) \\
&\text{a \textit{neutral element} } e \text{ exists such that} \quad e \circ a = a \circ e = a \\
&\text{an \textit{inverse element} } a^{-1} \text{ exists such that} \quad a^{-1} \circ a = a \circ a^{-1} = e \quad .
\end{aligned} \qquad (80)$$

Any set that fulfils these three defining properties, or axioms, is called a *(mathematical) group*. Historically, the notion of group was the first example of a mathematical struc-

9 MOTION AND SYMMETRY

FIGURE 201 A flower of *Crassula ovata* showing three fivefold multiplets: petals, stems and buds (© J.J. Harrison)

ture which was defined in a completely abstract manner.* Can you give an example of a group taken from daily life? Groups appear frequently in physics and mathematics, because symmetries are almost everywhere, as we will see.** Can you list the symmetry operations of the pattern of Figure 202?

Multiplets

Looking at a symmetric and composed system such as the ones shown in Figure 201 or Figure 202, we notice that each of its parts, for example each red patch, belongs to a set of similar objects, called a *multiplet*.

> ▷ Each part or component of a symmetric system can be classified according to what type of multiplet it belongs to.

---

* The term 'group' is due to Evariste Galois (b. 1811 Bourg-la-Reine, d. 1832 Paris), its structure to Augustin-Louis Cauchy (b. 1789 Paris, d. 1857 Sceaux) and the axiomatic definition to Arthur Cayley (b. 1821 Richmond upon Thames, d. 1895 Cambridge).
** In principle, mathematical groups need not be symmetry groups; but it can be proven that all groups can be seen as transformation groups on some suitably defined mathematical space, so that in mathematics we can use the terms 'symmetry group' and 'group' interchangeably.

A group is called *Abelian* if its concatenation operation is commutative, i.e., if $a \circ b = b \circ a$ for all pairs of elements a and b. In this case the concatenation is sometimes called *addition*. Do rotations form an Abelian group?

A subset $G_1 \subset G$ of a group G can itself be a group; one then calls it a *subgroup* and often says sloppily that G is *larger* than G_1 or that G is a *higher* symmetry group than G_1.

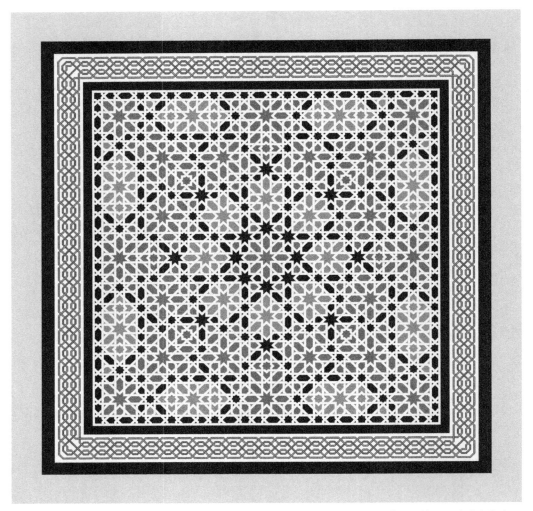

FIGURE 202 A Hispano–Arabic ornament from the Governor's Palace in Sevilla (© Christoph Schiller).

For some of the coloured patches in Figure 202 we need *four* objects to make up a full multiplet, whereas for others we need *two*, or only *one*, as in the case of the central star. Taken as a whole, each multiplet has (at least) the symmetry properties of the whole system.

Therefore we have two challenges to solve. First of all, we need to find all symmetries of nature. Secondly, throughout our adventure, we need to determine the full multiplet for every part of nature that we observe. Above all, we will need to determine the multiplets for the smallest parts found in nature, the elementary particles.

> ▷ A *multiplet* is a set of parts or components that transform into each other under all symmetry transformations.

Representations

Mathematicians often call abstract multiplets *representations*. By specifying to which multiplet or representation a part or component belongs, we describe in which way the component is part of the whole system. Let us see how this classification is achieved.

In mathematical language, symmetry transformations are often described by matrices. For example, in the plane, a reflection along the first diagonal is represented by the matrix

$$D(\text{refl}) = \begin{pmatrix} 0 & 1 \\ 1 & 0 \end{pmatrix} , \tag{81}$$

since every point (x, y) becomes transformed to (y, x) when multiplied by the matrix $D(\text{refl})$. Therefore, for a mathematician a *representation* of a symmetry group G is an assignment of a matrix $D(a)$ to each group element a such that the representation of the concatenation of two elements a and b is the product of the representations D of the elements:

$$D(a \circ b) = D(a)D(b) . \tag{82}$$

For example, the matrix of equation (81), together with the corresponding matrices for all the other symmetry operations, have this property.*

For every symmetry group, the construction and classification of all possible representations is an important task. It corresponds to the classification of all possible multiplets a symmetric system can be made of. Therefore, if we understand the classification of all multiplets and parts which can appear in Figure 202, we will also understand how to classify all possible parts of which an object or an example of motion can be composed!

A representation D is called *unitary* if all matrices $D(a)$ are unitary.** All representations appearing in physics, with only a handful of exceptions, are unitary: this term is

---

* There are some obvious, but important, side conditions for a representation: the matrices $D(a)$ must be invertible, or non-singular, and the identity operation of G must be mapped to the unit matrix. In even more compact language one says that a representation is a *homomorphism* from G into the group of non-singular or invertible matrices. A matrix D is invertible if its determinant $\det D$ is not zero.

In general, if a mapping f from a group G to another G' satisfies

$$f(a \circ_G b) = f(a) \circ_{G'} f(b) , \tag{83}$$

the mapping f is called an *homomorphism*. A homomorphism f that is one-to-one (injective) and onto (surjective) is called an *isomorphism*. If a representation is also injective, it is called *faithful*, *true* or *proper*.

In the same way as groups, more complex mathematical structures such as rings, fields and associative algebras may also be represented by suitable classes of matrices. A representation of the field of complex numbers is given later on.

** The *transpose* A^T of a matrix A is defined element-by-element by $(A^T)_{ik} = A_{ki}$. The *complex conjugate* A^* of a matrix A is defined by $(A^*)_{ik} = (A_{ik})^*$. The *adjoint* $A^\dagger$ of a matrix A is defined by $A^\dagger = (A^T)^*$. A matrix is called *symmetric* if $A^T = A$, *orthogonal* if $A^T = A^{-1}$, *Hermitean* or *self-adjoint* (the two are synonymous in all physical applications) if $A^\dagger = A$ (Hermitean matrices have real eigenvalues), and *unitary* if $A^\dagger = A^{-1}$. Unitary matrices have eigenvalues of norm one. Multiplication by a unitary matrix is a one-to-one mapping; since the time evolution of physical systems is a mapping from one time to another, evolution is always described by a unitary matrix.

An *antisymmetric* or *skew-symmetric* matrix is defined by $A^T = -A$, an *anti-Hermitean* matrix by $A^\dagger = -A$ and an *anti-unitary* matrix by $A^\dagger = -A^{-1}$. All the corresponding mappings are one-to-one.

A matrix is *singular*, and the corresponding vector transformation is not one-to-one, if $\det A = 0$.

the most restrictive, since it specifies that the corresponding transformations are one-to-one and invertible, which means that one observer never sees more or less than another. Obviously, if an observer can talk to a second one, the second one can also talk to the first. Unitarity is a natural property of representations in natural systems.

The final important property of a multiplet, or representation, concerns its structure. If a multiplet can be seen as composed of sub-multiplets, it is called *reducible*, else *irreducible*; the same is said about representations. The irreducible representations obviously cannot be decomposed any further. For example, the (almost perfect) symmetry group of Figure 202, commonly called D_4, has eight elements. It has the general, faithful, unitary and irreducible matrix representation

$$\begin{pmatrix} \cos n\pi/2 & -\sin n\pi/2 \\ \sin n\pi/2 & \cos n\pi/2 \end{pmatrix} n = 0..3, \begin{pmatrix} -1 & 0 \\ 0 & 1 \end{pmatrix}, \begin{pmatrix} 1 & 0 \\ 0 & -1 \end{pmatrix}, \begin{pmatrix} 0 & 1 \\ 1 & 0 \end{pmatrix}, \begin{pmatrix} 0 & -1 \\ -1 & 0 \end{pmatrix}. \quad (84)$$

The representation is an *octet*. The complete list of possible irreducible representations of the group D_4 also includes *singlets*, *doublets* and *quartets*. Can you find them all? These representations allow the classification of all the white and black ribbons that appear in the figure, as well as all the coloured patches. The most symmetric elements are singlets, the least symmetric ones are members of the quartets. The complete system is always a singlet as well.

With these concepts we are now ready to talk about motion and moving systems with improved precision.

The symmetries and vocabulary of motion

Every day we experience that we are able to talk to each other about motion. It must therefore be possible to find an *invariant* quantity describing it. We already know it: it is the *action*, the measure of change. For example, lighting a match is a change. The magnitude of the change is the same whether the match is lit here or there, in one direction or another, today or tomorrow. Indeed, the (Galilean) action is a number whose value is the same for each observer *at rest*, independent of his orientation or the time at which he makes his observation.

In the case of the Arabic pattern of Figure 202, the symmetry allows us to deduce the list of multiplets, or representations, that can be its building blocks. This approach must be possible for a moving system as well. Table 35 shows how. In the case of the Arabic pattern, from the various possible observation viewpoints, we deduced the classification of the ribbons into singlets, doublets, etc. For a moving system, the building blocks, corresponding to the ribbons, are the *(physical) observables*. Since we observe that nature is symmetric under many different changes of viewpoint, we can classify all observables. To do so, we first need to take the list of all viewpoint transformations and then deduce the list of all their representations.

Our everyday life shows that the world stays unchanged after changes in position, orientation and instant of observation. We also speak of space translation invariance, rotation invariance and time translation invariance. These transformations are different from those of the Arabic pattern in two respects: they are *continuous* and they are *unbounded*. As a result, their representations will generally be continuously variable and

9 MOTION AND SYMMETRY

TABLE 35 Correspondences between the symmetries of an ornament, a flower and motion.

System	Hispano-Arabic pattern	Flower	Motion
Structure and components	set of ribbons and patches	set of petals, stem	motion path and observables
System symmetry	pattern symmetry	flower symmetry	symmetry of Lagrangian
Mathematical description of the symmetry group	D_4	C_5	in Galilean relativity: position, orientation, instant and velocity changes
Invariants	number of multiplet elements	petal number	number of coordinates, magnitude of scalars, vectors and tensors
Representations of the components	multiplet types of elements	multiplet types of components	tensors, including scalars and vectors
Most symmetric representation	singlet	part with circular symmetry	scalar
Simplest faithful representation	quartet	quintet	vector
Least symmetric representation	quartet	quintet	no limit (tensor of infinite rank)

without bounds: they will be *quantities*, or *magnitudes*. In other words,

> ▷ Because the continuity of observation change, observables must be constructed with (real) *numbers*.

In this way we have deduced why numbers are *necessary* for any description of motion.*

Since observers can differ in orientation, representations will be mathematical objects possessing a direction. To cut a long story short, the symmetry under change of observation position, orientation or instant leads to the result that all observables are either 'scalars', 'vectors' or higher-order 'tensors.'**

> ▷ A *scalar* is an observable quantity which stays the same for all observers.

A scalar corresponds to a singlet. Examples are the mass or the charge of an object, the distance between two points, the distance of the horizon, and many others. The possible

---

* Only scalars, in contrast to vectors and higher-order tensors, may also be quantities that only take a discrete set of values, such as +1 or −1 only. In short, only scalars may be *discrete* observables.
** Later on, *spinors* will be added to, and complete, this list.

values of a scaler are (usually) continuous, unbounded and without direction. Other examples of scalars are the potential at a point and the temperature at a point. Velocity is obviously not a scalar; nor is the coordinate of a point. Can you find more examples and counter-examples?

Energy is a puzzling observable. It is a scalar if only changes of place, orientation and instant of observation are considered. But energy is not a scalar if changes of observer speed are included. Nobody ever searched for a generalization of energy that is a scalar also for moving observers. Only Albert Einstein discovered it, completely by accident. More about this issue will be told shortly.

▷ Any quantity which has a magnitude and a direction and which 'stays the same' with respect to the environment when changing viewpoint is a *vector*.

For example, the arrow between two fixed points on the floor is a vector. Its length is the same for all observers; its direction changes from observer to observer, but not with respect to its environment. On the other hand, the arrow between a tree and the place where a rainbow touches the Earth is *not* a vector, since that place does not stay fixed when the observer changes.

Mathematicians say that vectors are directed entities staying invariant under coordinate transformations. Velocities of objects, accelerations and field strength are examples of vectors. (Can you confirm this?) The magnitude of a vector is a scalar: it is the same for any observer. By the way, a famous and baffling result of nineteenth-century experiments is that the velocity of a light beam is *not* a vector like the velocity of a car; the velocity of a light beam is not a vector for Galilean transformations.* This mystery will be solved shortly.

Tensors are generalized vectors. As an example, take the moment of inertia of an object. It specifies the dependence of the angular momentum on the angular velocity. For any object, doubling the magnitude of angular velocity doubles the magnitude of angular momentum; however, the two vectors are not parallel to each other if the object is not a sphere. In general, if any two vector quantities are proportional, in the sense that doubling the magnitude of one vector doubles the magnitude of the other, but without the two vectors being parallel to each other, then the proportionality 'factor' is a (second order) *tensor*. Like all proportionality factors, tensors have a magnitude. In addition, tensors have a direction and a *shape*: they describe the connection between the vectors they relate. Just as vectors are the simplest quantities with a magnitude and a direction, so tensors are the simplest quantities with a magnitude, a direction and a shape, i.e., a direction depending on a second, chosen direction. Just as vectors can be visualized as *oriented arrows*, symmetric tensors – but not non-symmetric ones – can be visualized as oriented *ellipsoids*.** Can you name another example of tensor?

---

* *Galilean transformations* are changes of viewpoints from one observer to a second one, moving with respect to the first. 'Galilean transformation' is just a term for what happens in everyday life, where velocities add and time is the same for everybody. The term, introduced in 1908 by Philipp Frank, is mostly used as a contrast to the Lorentz transformation that is so common in special relativity.

** A rank-n tensor is the proportionality factor between a rank-1 tensor – i.e., a vector – and an rank-$(n-1)$ tensor. Vectors and scalars are rank 1 and rank 0 tensors. Scalars can be pictured as spheres, vectors as arrows, and symmetric rank-2 tensors as ellipsoids. A general, non-symmetric rank-2 tensor can be split

9 MOTION AND SYMMETRY

Let us get back to the description of motion. Table 35 shows that in physical systems – like in a Hispano-Arabic ornament – we always have to distinguish between the symmetry of the whole Lagrangian – corresponding to the symmetry of the complete ornament – and the representation of the observables – corresponding to the ribbon multiplets. Since the action must be a scalar, and since all observables must be tensors, Lagrangians contain sums and products of tensors only in combinations forming scalars. Lagrangians thus contain only scalar products or generalizations thereof. In short, Lagrangians always look like

$$L = \alpha\, a_i b^i + \beta\, c_{jk} d^{jk} + \gamma\, e_{lmn} f^{lmn} + ... \qquad (85)$$

where the indices attached to the variables a, b, c etc. always come in matching pairs to be summed over. (Therefore summation signs are usually simply left out.) The Greek letters represent constants. For example, the action of a free point particle in Galilean physics was given as

$$S = \int L\, dt = \frac{m}{2} \int v^2\, dt \qquad (86)$$

which is indeed of the form just mentioned. We will encounter many other cases during our study of motion.*

Galileo already understood that motion is also invariant under change of viewpoints

---

uniquely into a symmetric and an antisymmetric tensor. An antisymmetric rank-2 tensor corresponds to a polar vector. Tensors of higher rank correspond to more and more complex shapes.

A vector has the same length and direction for every observer; a tensor (of rank 2) has the same determinant, the same trace, and the same sum of diagonal subdeterminants for all observers.

A vector is described mathematically by a *list* of components; a tensor (of rank 2) is described by a *matrix* of components. The rank or order of a tensor thus gives the number of indices the observable has. Can you show this?

* By the way, is the usual list of possible observation viewpoints – namely different positions, different observation instants, different orientations, and different velocities – also *complete* for the action (86)? Surprisingly, the answer is no. One of the first who noted this fact was Niederer, in 1972. Studying the quantum theory of point particles, he found that even the action of a Galilean free point particle is invariant under some additional transformations. If the two observers use the coordinates (t, x) and (τ, ξ), the action (86) is invariant under the transformations

$$\xi = \frac{rx + x_0 + vt}{\gamma t + \delta} \quad \text{and} \quad \tau = \frac{\alpha t + \beta}{\gamma t + \delta} \quad \text{with} \quad r^T r = 1 \quad \text{and} \quad \alpha\delta - \beta\gamma = 1 . \qquad (87)$$

where r describes the rotation from the orientation of one observer to the other, v the velocity between the two observers, and x_0 the vector between the two origins at time zero. This group contains two important special cases of transformations:

$$\text{The connected, static Galilei group } \xi = rx + x_0 + vt \quad \text{and} \quad \tau = t$$

$$\text{The transformation group SL(2,R) } \xi = \frac{x}{\gamma t + \delta} \quad \text{and} \quad \tau = \frac{\alpha t + \beta}{\gamma t + \delta} \qquad (88)$$

The latter, three-parameter group includes *spatial inversion, dilations, time translation* and a set of time-dependent transformations such as $\xi = x/t, \tau = 1/t$ called *expansions*. Dilations and expansions are rarely mentioned, as they are symmetries of point particles only, and do not apply to everyday objects and systems. They will return to be of importance later on, however.

with different velocity. However, the action just given does not reflect this. It took some years to find out the correct generalization: it is given by the theory of special relativity. But before we study it, we need to finish the present topic.

Reproducibility, conservation and Noether's theorem

> I will leave my mass, charge and momentum to science.
> Graffito

The reproducibility of observations, i.e., the symmetry under change of instant of time or 'time translation invariance', is a case of viewpoint-independence. (That is not obvious; can you find its irreducible representations?) The connection has several important consequences. We have seen that symmetry implies invariance. It turns out that for *continuous* symmetries, such as time translation symmetry, this statement can be made more precise:

> ▷ For any continuous symmetry of the Lagrangian there is an associated *conserved constant of motion* and vice versa.

The exact formulation of this connection is the theorem of Emmy Noether.* She found the result in 1915 when helping Albert Einstein and David Hilbert, who were both struggling and competing at constructing general relativity. However, the result applies to any type of Lagrangian.

Noether investigated continuous symmetries depending on a continuous parameter b. A viewpoint transformation is a symmetry if the action S does not depend on the value of b. For example, changing position as

$$x \mapsto x + b \tag{89}$$

leaves the action

$$S_0 = \int T(v) - U(x) \, dt \tag{90}$$

invariant, since $S(b) = S_0$. This situation implies that

$$\frac{\partial T}{\partial v} = p = \text{const}. \tag{91}$$

In short, symmetry under change of position implies conservation of momentum. The converse is also true.

In the case of symmetry under shift of observation instant, we find

$$T + U = \text{const}. \tag{92}$$

---

* Emmy Noether (b. 1882 Erlangen, d. 1935 Bryn Mawr), mathematician. The theorem is only a sideline in her career which she dedicated mostly to number theory. The theorem also applies to gauge symmetries, where it states that to every gauge symmetry corresponds an identity of the equation of motion, and vice versa.

9 MOTION AND SYMMETRY

In other words, time translation invariance implies constant energy. Again, the converse is also correct.

The conserved quantity for a continuous symmetry is sometimes called the *Noether charge*, because the term *charge* is used in theoretical physics to designate conserved extensive observables. So, energy and momentum are Noether charges. 'Electric charge', 'gravitational charge' (i.e., mass) and 'topological charge' are other common examples. What is the conserved charge for rotation invariance?

We note that the expression 'energy is conserved' has several meanings. First of all, it means that the energy of a *single* free particle is constant in time. Secondly, it means that the total energy of any number of independent particles is constant. Finally, it means that the energy of a *system* of particles, i.e., including their interactions, is constant in time. Collisions are examples of the latter case. Noether's theorem makes all of these points at the same time, as you can verify using the corresponding Lagrangians.

But Noether's theorem also makes, or rather repeats, an even stronger statement: if energy were not conserved, time could not be defined. The whole description of nature requires the existence of conserved quantities, as we noticed when we introduced the concepts of object, state and environment. For example, we defined objects as *permanent* entities, that is, as entities characterized by conserved quantities. We also saw that the introduction of time is possible only because in nature there are 'no surprises'. Noether's theorem describes exactly what such a 'surprise' would have to be: the non-conservation of energy. However, energy jumps have never been observed – not even at the quantum level.

Since symmetries are so important for the description of nature, Table 36 gives an overview of all the symmetries of nature that we will encounter. Their main properties are also listed. Except for those marked as 'approximate', an experimental proof of incorrectness of any of them would be a big surprise indeed – and guarantee eternal fame. Various speculations about additional symmetries exist; so far, all these speculations and quests for even more eternal fame have turned out to be mistaken. The list of symmetries is also the full list of universal statements, i.e., of statements about *all* observations, that scientists make. For example, when it is said that "all stones fall down" the statement implies the existence of time and space translation invariance. For philosophers interested in logical *induction*, the list is thus important also from this point of view.

TABLE 36 The known symmetries of nature, with their properties; also the complete list of logical *inductions* used in physics.

SYMMETRY	TYPE [NUMBER OF PARAMETERS]	SPACE OF ACTION	GROUP TOPOLOGY	POSSIBLE REPRESENTATIONS	CONSERVED QUANTITY/CHARGE	VACUUM/MATTER IS SYMMETRIC	MAIN EFFECT
Geometric or space-time, external, symmetries							
Time and space translation	$R \times R^3$ [4 par.]	space, time	not compact	scalars, vectors,	momentum and energy	yes/yes	allow everyday

TABLE 36 (Continued) The known symmetries of nature, with their properties; also the complete list of logical *inductions* used in physics.

SYMMETRY	TYPE [NUMBER OF PARAMETERS]	SPACE OF ACTION	GROUP TOPOLOGY	POSSIBLE REPRESENTATIONS	CONSERVED QUANTITY/ CHARGE	VACUUM/ MATTER IS SYMMETRIC	MAIN EFFECT
Rotation	SO(3) [3 par.]	space	S^2	tensors	angular momentum	yes/yes	communication
Galilei boost	R^3 [3 par.]	space, time	not compact	scalars, vectors, tensors	velocity of centre of mass	yes/for low speeds	relativity of motion
Lorentz	homogeneous Lie SO(3,1) [6 par.]	space-time	not compact	tensors, spinors	energy-momentum $T^{\mu\nu}$	yes/yes	constant light speed
Poincaré ISL(2,C)	inhomogeneous Lie [10 par.]	space-time	not compact	tensors, spinors	energy-momentum $T^{\mu\nu}$	yes/yes	
Dilation invariance	R^+ [1 par.]	space-time	ray	n-dimen. continuum	none	yes/no	massless particles
Special conformal invariance	R^4 [4 par.]	space-time	R^4	n-dimen. continuum	none	yes/no	massless particles
Conformal invariance	[15 par.]	space-time	involved	massless tensors, spinors	none	yes/no	light cone invariance

Dynamic, interaction-dependent symmetries: gravity

$1/r^2$ gravity	SO(4) [6 par.]	config. space	as SO(4)	vector pair	perihelion direction	yes/yes	closed orbits
Diffeomorphism invariance	[∞ par.]	space-time	involved	space-times	local energy-momentum	yes/no	perihelion shift

Dynamic, classical and quantum-mechanical motion symmetries

Parity ('spatial') inversion P	discrete	Hilbert or phase space	discrete	even, odd	P-parity	yes/no	mirror world exists
Motion ('time') inversion T	discrete	Hilbert or phase space	discrete	even, odd	T-parity	yes/no	reversibility

9 MOTION AND SYMMETRY

TABLE 36 (Continued) The known symmetries of nature, with their properties; also the complete list of logical *inductions* used in physics.

SYMMETRY	TYPE [NUMBER OF PARAMETERS]	SPACE OF ACTION	GROUP TOPOLOGY	POSSIBLE REPRESENTATIONS	CONSERVED QUANTITY/ CHARGE	VACUUM/ MATTER IS SYMMETRIC	MAIN EFFECT
Charge conjugation C	global, antilinear, anti-Hermitean	Hilbert or phase space	discrete	even, odd	C-parity	yes/no	antiparticles exist
CPT	discrete	Hilbert or phase space	discrete	even	CPT-parity	yes/yes	makes field theory possible

Dynamic, interaction-dependent, gauge symmetries

Electromagnetic classical gauge invariance	[∞ par.]	space of fields	unimportant	unimportant	electric charge	yes/yes	massless light
Electromagnetic q.m. gauge inv.	Abelian Lie U(1) [1 par.]	Hilbert space	circle S^1	fields	electric charge	yes/yes	massless photon
Electromagnetic duality	Abelian Lie U(1) [1 par.]	space of fields	circle S^1	abstract	abstract	yes/no	none
Weak gauge	non-Abelian Lie SU(2) [3 par.]	Hilbert space	as $SU(3)$	particles	weak charge	no/ approx.	
Colour gauge	non-Abelian Lie SU(3) [8 par.]	Hilbert space	as $SU(3)$	coloured quarks	colour	yes/yes	massless gluons
Chiral symmetry	discrete	fermions	discrete	left, right	helicity	approximately	'massless' fermions[a]

Permutation symmetries

Particle exchange	discrete	Fock space etc.	discrete	fermions and bosons	none	n.a./yes	Gibbs' paradox

For details about the connection between symmetry and induction, see later on. The explanation of the terms in the table will be completed in the rest of the walk. The real numbers are denoted as R.

a. Only approximate; 'massless' means that $m \ll m_{Pl}$, i.e., that $m \ll 22$ μg.

Parity inversion and motion reversal

The symmetries in Table 36 include two so-called *discrete* symmetries that are important for the discussion of motion.

The first symmetry is *parity invariance* for objects or processes under spatial inversion. The symmetry is also called *mirror invariance* or *right-left symmetry*. Both objects and processes can be mirror symmetric. A single glove or a pair of scissors are not mirror-symmetric. How far can you throw a stone with your other hand? Most people have a preferred hand, and the differences are quite pronounced. Does nature have such a right-left preference? In everyday life, the answer is clear: everything that exists or happens in one way can also exist or happen in its mirrored way.

Numerous precision experiments have tested mirror invariance; they show that

> Every process due to gravitation, electricity or magnetism can also happen in a mirrored way.

There are *no* exceptions. For example, there are people with the heart on the right side; there are snails with left-handed houses; there are planets that rotate the other way. Astronomy and everyday life – which are governed by gravity and electromagnetic processes – are *mirror-invariant*. We also say that gravitation and electromagnetism are *parity invariant*. Later we will discover that certain rare processes not due to gravity or electromagnetism, but to the weak nuclear interaction, violate parity.

Mirror symmetry has two representations: '+ or singlet', such as mirror-symmetric objects, and '– or doublet', such as handed objects. Because of mirror symmetry, scalar quantities can thus be divided into *true* scalars, like temperature, and *pseudo-scalars*, like magnetic flux or magnetic charge. True scalars do not change sign under mirror reflection, whereas pseudo-scalars do. In the same way, one distinguishes *true* vectors, or *polar* vectors, such as velocity, from *pseudo-vectors*, or axial vectors, like angular velocity, angular momentum, torque, vorticity and the magnetic field. (Can you find an example of a pseudo-tensor?)

The other discrete symmetry is *motion reversal*. It is sometimes also called, falsely, 'time reversal'. Can phenomena happen backwards? Does reverse motion trace out the forward path? This question is not easy. Exploring motion due to gravitation shows that such motion can always also happen in the reverse direction. (Also for motion reversal, observables belong either to a + or to a – representation.) In case of motion due to electricity and magnetism, such as the behaviour of atoms in gases, solids and liquids, the question is more involved. Can broken objects be made to heal? We will discuss the issue in the section of thermodynamics, but we will reach the same conclusion, despite the appearance to the contrary:

> Every motion due to gravitation, electricity or magnetism can also happen in the reverse direction.

Motion reversion is a symmetry for all processes due to gravitation and the electromagnetic interaction. Everyday motion is *reversible*. And again, certain even rarer nuclear processes will provide exceptions.

9 MOTION AND SYMMETRY

INTERACTION SYMMETRIES

In nature, when we observe a system, we can often neglect the environment. Many processes occur independently of what happens around them. This independence is a physical symmetry. Given the independence of observations from the details occurring in the environment, we deduce that interactions between systems and the environment *decrease with distance*. In particular, we deduce that gravitational attraction, electric attraction and repulsion, and also magnetic attraction and repulsion must vanish for distant bodies. Can you confirm this?

Gauge symmetry is also an interaction symmetry. We will encounter them in our exploration of quantum physics. In a sense, these symmetries are more specific cases of the general decrease of interactions with distance.

CURIOSITIES AND FUN CHALLENGES ABOUT SYMMETRY

Right-left symmetry is an important property in everyday life; for example, humans prefer faces with a high degree of right-left symmetry. Humans also prefer that objects on the walls have shapes that are right-left symmetric. In turns out that the eye and the brain has symmetry detectors built in. They detect deviations from perfect right-left symmetry.

* *

What is the path followed by four turtles starting on the four angles of a square, if each of them continuously walks, at constant speed, towards the next one? How long is the distance they travel?

* *

What is the symmetry of a simple oscillation? And of a wave?

* *

For what systems is motion reversal a symmetry transformation?

* *

What is the symmetry of a continuous rotation?

* *

A sphere has a tensor for the moment of inertia that is diagonal with three equal numbers. The same is true for a cube. Can you distinguish spheres and cubes by their rotation behaviour?

* *

Is there a motion in nature whose symmetry is perfect?

* *

Can you show that in *two* dimensions, *finite* objects can have only rotation and reflection symmetry, in contrast to *infinite* objects, which can have also translation and glide-reflection symmetry? Can you prove that for finite objects in two dimensions, if no rotation symmetry is present, there is only one reflection symmetry? And that all possible rotations are always about the same centre? Can you deduce from this that at least one point is unchanged in all symmetrical finite two-dimensional objects?

* *

Which object of everyday life, common in the 20th century, had sevenfold symmetry?

* *

Here is little puzzle about the *lack* of symmetry. A *general* acute triangle is defined as a triangle whose angles differ from a right angle and from each other by at least 15 degrees. Show that there is only one such general triangle and find its angles.

* *

Can you show that in *three* dimensions, *finite* objects can have only rotation, reflection, inversion and rotatory inversion symmetry, in contrast to *infinite* objects, which can have also translation, glide-reflection, and screw rotation symmetry? Can you prove that for finite objects in three dimensions, if no rotation symmetry is present, there is only one reflection plane? And that for all inversions or rotatory inversions the centre must lie on a rotation axis or on a reflection plane? Can you deduce from this that at least one point is unchanged in all symmetrical finite three-dimensional objects?

Summary on symmetry

Symmetry is invariance to change. The simplest symmetries are geometrical: the point symmetries of flowers or the translation symmetries of infinite crystals are examples. All the possible changes that leave a system invariant – i.e., all possible symmetry transformations of a system – form a mathematical group. Apart from the *geometrical* symmetry groups, several additional symmetry groups appear for motion itself.

Motion is universal. Any universality statement implies a symmetry. The reproducibility and predictability of nature implies a number of fundamental continuous symmetries: since we can *talk* about nature we can deduce that above all, nature is symmetrical under *time and space translations* and *rotations*. Space-time symmetries form a group. More precisely, they form a *continuous* symmetry group.

From nature's continuous symmetries, using Noether's theorem, we can deduce conserved 'charges'. These are energy, linear momentum and angular momentum. They are described by real numbers. In other words, the definition of mass, space and time, together with their symmetry properties, is *equivalent* to the conservation of energy and momenta. Conservation and symmetry are two ways to express the same property of nature. To put it simply, our ability to talk about nature means that energy, linear momentum and angular momentum are conserved and described by numbers.

Additionally, there are two fundamental discrete symmetries about motion: first, everyday observations are found to be *mirror symmetric*; secondly, many simple motion examples are found to be symmetric under *motion reversal*.

Finally, the isolability of systems from their surroundings implies that interactions must have *no effect at large distances*. The full list of nature's symmetries also includes gauge symmetry, particle exchange symmetry and certain vacuum symmetries.

All aspects of motion, like all components of a symmetric system, can be classified by their symmetry behaviour, i.e., by the multiplet or the representation to which they belong. As a result, observables are either scalars, vectors, spinors or tensors.

An fruitful way to formulate the patterns and 'laws' of nature – i.e., the Lagrangian of a physical system – has been the search for the complete set of nature's symmetries first. For example, this is helpful for oscillations, waves, relativity, quantum physics and quantum electrodynamics. We will use the method throughout our walk; in the last part of our adventure we will discover some symmetries which are even more mind-boggling than those of relativity and those of interactions. In the next section, though, we will move on to the next approach for a global description of motion.

Chapter 10

SIMPLE MOTIONS OF EXTENDED BODIES – OSCILLATIONS AND WAVES

The observation of change is a fundamental aspect of nature. Among all these observations, *periodic* change is frequent around us. Indeed, throughout everyday life be observe oscillations and waves: Talking, singing, hearing and seeing would be impossible without them. Exploring oscillations and waves, the next global approach to motion in our adventure, is both useful and beautiful.

Oscillations

Oscillations are recurring changes, i.e., cyclic or periodic changes. Above, we defined action, and thus change, as the integral of the Lagrangian, and we defined the Lagrangian as the difference between kinetic and potential energy. All oscillating systems periodically exchange one kind of energy with the other. One of the simplest oscillating systems in nature is a mass m attached to a (linear) *spring*. The Lagrangian for the mass position x is given by

$$L = \tfrac{1}{2}mv^2 - \tfrac{1}{2}kx^2 \, , \qquad (93)$$

where k is a quantity characterizing the spring, the so-called spring constant. The Lagrangian is due to Robert Hooke, in the seventeenth century. Can you confirm the expression?

The motion that results from this Lagrangian is periodic, and illustrated in Figure 203. The Lagrangian (93) thus describes the *oscillation* of the spring length over time. The motion is exactly the same as that of a long pendulum at small amplitude. The motion is called *harmonic motion*, because an object vibrating rapidly in this way produces a completely pure – or harmonic – musical sound. (The musical instrument producing the purest harmonic waves is the transverse flute. This instrument thus gives the best idea of how harmonic motion 'sounds'.)

The graph of this harmonic oscillation, also called *linear oscillation*, shown in Figure 203, is called a *sine curve*; it can be seen as the basic building block of all oscillations. All other, anharmonic oscillations in nature can be composed from harmonic ones, i.e., from sine curves, as we shall see shortly. Any quantity $x(t)$ that oscillates harmonically is described by its amplitude A, its angular frequency ω and its phase φ:

$$x(t) = A\sin(\omega t + \varphi) \, . \qquad (94)$$

The amplitude and the phase depend on the way the oscillation is started. In contrast,

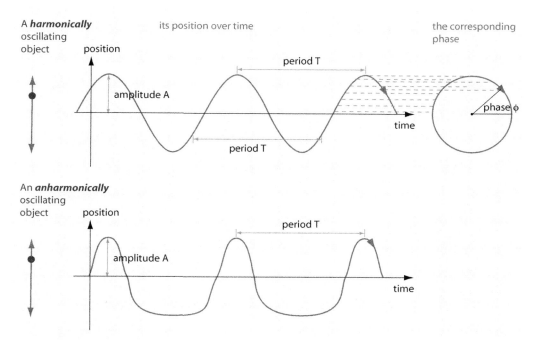

FIGURE 203 Above: the simplest oscillation, the *linear* or *harmonic* oscillation: how position changes over time, and how it is related to rotation. Below: an example of anharmonic oscillation.

the angular frequency ω is an intrinsic property of the system. Can you show that for the mass attached to the spring, we have $\omega = 2\pi f = 2\pi/T = \sqrt{k/m}$?

Every harmonic oscillation is thus described by just three quantities: the *amplitude*, the *period* (the inverse of the frequency) and the *phase*. The phase, illustrated in Figure 205, distinguishes oscillations of the same amplitude and period T; the phase defines at what time the oscillation starts.

Some observed oscillation frequencies are listed in Table 37. Figure 203 shows how a harmonic oscillation is related to an imaginary rotation. As a result, the phase is best described by an angle value between 0 and 2π.

Damping

Every oscillating motion continuously transforms kinetic energy into potential energy and vice versa. This is the case for the tides, the pendulum, or any radio receiver. But many oscillations also diminish in time: they are *damped*. Systems with large damping, such as the shock absorbers in cars, are used to avoid oscillations. Systems with *small* damping are useful for making precise and long-running clocks. The simplest measure of damping is the number of oscillations a system takes to reduce its amplitude to $1/e \approx 1/2.718$ times the original value. This characteristic number is the so-called *Q-factor*, named after the abbreviation of 'quality factor'. A poor Q-factor is 1 or less, an extremely good one is 100 000 or more. (Can you write down a simple Lagrangian for a damped oscillation with a given Q-factor?) In nature, damped oscillations do not usually keep

TABLE 37 Some sound frequency values found in nature.

Observation	Frequency
Sound frequencies in gas, emitted by black holes	c. 1 fHz
Precision in measured vibration frequencies of the Sun	down to 2 nHz
Vibration frequencies of the Sun	down to c. 300 nHz
Vibration frequencies that disturb gravitational radiation detection	down to 3 μHz
Lowest vibration frequency of the Earth Ref. 217	309 μHz
Resonance frequency of stomach and internal organs (giving the 'sound in the belly' experience)	1 to 10 Hz
Common music tempo	2 Hz
Frequency used for communication by farting fish	c. 10 Hz
Sound produced by loudspeaker sets (horn, electro-magetic, piezoelectric, electret, plasma, laser)	c. 18 Hz to over 150 kHz
Sound audible to young humans	20 Hz to 20 kHz
Hum of electrical appliances, depending on country	50 or 60 Hz
Fundamental voice frequency of speaking adult human male	85 Hz to 180 Hz
Fundamental voice frequency of speaking adult human female	165 Hz to 255 Hz
Official value, or *standard pitch*, of musical note 'A' or 'la', following ISO 16 (and of the telephone line signal in many countries)	440 Hz
Common values of musical note 'A' or 'la' used by orchestras	442 to 451 Hz
Wing beat of tiniest flying insects	c. 1000 Hz
Fundamental sound frequency produced by the feathers of the club-winged manakin, *Machaeropterus deliciosus*	1 to 1.5 kHz
Fundamental sound frequency of crickets	2 kHz to 9 kHz
Quartz oscillator frequencies	20 kHz up to 350 MHz
Sonar used by bats	up to over 100 kHz
Sonar used by dolphins	up to 150 kHz
Sound frequency used in ultrasound imaging	2 to 20 MHz
Phonon (sound) frequencies measured in single crystals	up to 20 THz and more

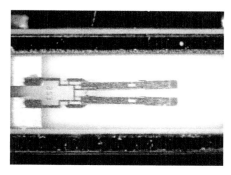
FIGURE 204 The interior of a commercial quartz oscillator, a few millimetres in size, driven at high amplitude. (QuickTime film © Microcrystal)

constant frequency; however, for the simple pendulum this remains the case to a high degree of accuracy. The reason is that for a pendulum, the frequency does not depend significantly on the amplitude (as long as the amplitude is smaller than about 20°). This is one reason why pendulums are used as oscillators in mechanical clocks.

Obviously, for a good clock, the driving oscillation must not only show small damping, but must also be independent of temperature and be insensitive to other external influences. An important development of the twentieth century was the introduction of quartz crystals as oscillators. Technical quartzes are crystals of the size of a few grains of sand; they can be made to oscillate by applying an electric signal. They have little temperature dependence and a large Q-factor, and therefore low energy consumption, so that precise clocks can now run on small batteries. The inside of a quartz oscillator is shown in Figure 204.

Resonance

In most physical systems that are brought to oscillate by an external source, the resulting amplitude depends on the frequency. The selected frequencies for which the amplitude is maximal are called *resonance frequencies* or simply *resonances*. For example, the quartz oscillator of Figure 204, or the usual vibration frequencies of guitar strings or bells – shown in Figure 206 – are resonance frequencies.

Usually, the oscillations at which a system will oscillate when triggered by a short hit will occur at resonance frequencies. Most musical instruments are examples. Almost all systems have several resonance frequencies; flutes, strings and bells are well-known examples.

In contrast to music or electronics, in many other situations resonance needs to be avoided. In buildings, earthquakes can trigger resonances; in bridges, the wind can trigger resonant oscillations; similarly, in many machines resonances need to be dampened or blocked in order to avoid that the large amplitude of a resonance destroys the system. In modern high-quality cars, the resonances of each part and of each structure are calculated and, if necessary, adjusted in such a way that no annoying vibrations disturb the driver or the passenger.

All systems that oscillate also emit *waves*. In fact, resonance only occurs because all oscillations are in fact *localized waves*. Indeed, oscillations only appear in *extended* sys-

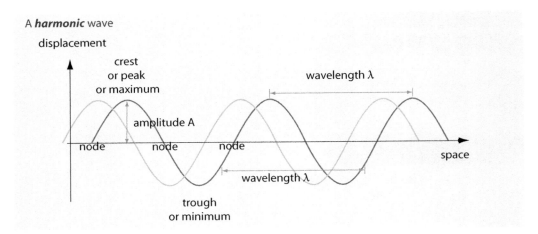
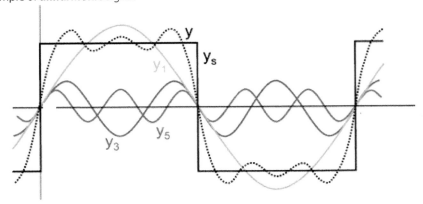

FIGURE 205 Top: the main properties of a harmonic wave, or sine wave. Bottom: A general periodic signal, or anharmonic wave – here a black square wave – can be decomposed uniquely into simplest, or *harmonic* waves. The first three components (green, blue and red) and also their intermediate sum (black dotted line) are shown. This is called a *Fourier decomposition* and the general method to do this *Fourier analysis*. (© Wikimedia) Not shown: the unique decomposition into harmonic waves is even possible for *non-periodic* signals.

FIGURE 206 The measured fundamental vibration patterns of a bell. Bells – like every other source of oscillations, be it an atom, a molecule, a music instrument or the human voice – show that all oscillations in nature are due to waves. (© H. Spiess & al.).

FIGURE 207 The centre of the grooves in an old vinyl record show the amplitude of the sound pressure, averaged over the two stereo channels (scanning electron microscope image by © Chris Supranowitz/University of Rochester).

tems; and oscillations are only simplified descriptions of the repetitive motion of any extended system. The complete and general repetitive motion of an extended system is the *wave*.

Waves: general and harmonic

Waves are travelling imbalances, or, equivalently, travelling oscillations of a substrate. Waves travel, though the substrate does not. Every wave can be seen as a superposition of *harmonic* waves. Every sound effect can be thought of as being composed of harmonic waves. Harmonic waves, also called *sine waves* or *linear waves*, are the building blocks of which all internal motions of an extended body are constructed, as shown in Figure 205. Can you describe the difference in wave shape between a pure harmonic tone, a musical sound, a noise and an explosion?

Every harmonic wave is characterized by an oscillation *frequency* f, a propagation or phase *velocity* c, a *wavelength* λ, an amplitude A and a *phase* φ, as can be deduced from Figure 205. Low-amplitude water waves are examples of harmonic waves – in contrast to water waves of large amplitude. In a harmonic wave, every position by itself performs a harmonic oscillation. The phase of a wave specifies the position of the wave (or a crest) at a given time. It is an angle between 0 and 2π.

TABLE 38 Some wave velocities.

Wave	Velocity
Tsunami	around 0.2 km/s
Sound in most gases	0.3 ± 0.1 km/s
Sound in air at 273 K	0.331 km/s
Sound in air at 293 K	0.343 km/s
Sound in helium at 293 K	0.983 km/s
Sound in most liquids	1.2 ± 0.2 km/s
Seismic waves	1 to 14 km/s
Sound in water at 273 K	1.402 km/s
Sound in water at 293 K	1.482 km/s
Sound in sea water at 298 K	1.531 km/s
Sound in gold	4.5 km/s
Sound in steel	5.8 to 5.960 km/s
Sound in granite	5.8 km/s
Sound in glass (longitudinal)	4 to 5.9 km/s
Sound in beryllium (longitudinal)	12.8 km/s
Sound in boron	up to 15 km/s
Sound in diamond	up to 18 km/s
Sound in fullerene (C_{60})	up to 26 km/s
Plasma wave velocity in InGaAs	600 km/s
Light in vacuum	$2.998 \cdot 10^8$ m/s

The phase velocity c is the speed with which a wave maximum moves. A few examples are listed in Table 38. Can you show that frequency and wavelength in a wave are related by $f\lambda = c$?

Waves appear inside all *extended* bodies, be they solids, liquids, gases or plasmas. Inside fluid bodies, waves are *longitudinal*, meaning that the wave motion is in the same direction as the wave oscillation. Sound in air is an example of a longitudinal wave. Inside solid bodies, waves can also be *transverse*; in that case the wave oscillation is perpendicular to the travelling direction.

Waves appear also on *interfaces* between bodies: water–air interfaces are a well-known case. Even a saltwater–freshwater interface, so-called *dead water*, shows waves: they can appear even if the upper surface of the water is immobile. Any flight in an aeroplane provides an opportunity to study the regular cloud arrangements on the interface between warm and cold air layers in the atmosphere. Seismic waves travelling along the boundary between the sea floor and the sea water are also well-known. Low-amplitude water waves are transverse; however, general surface waves are usually neither longitudinal nor transverse, but of a mixed type.

To get a first idea about waves, we have a look at water waves.

Water surface:

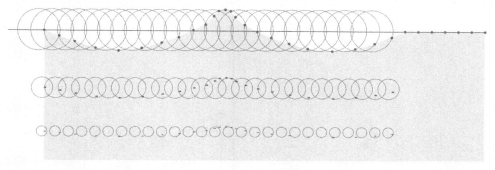

At a depth of *half the wavelength*, the amplitude is negligible

FIGURE 208 The formation of the shape of deep gravity waves, on and under water, from the circular motion of the water particles. Note the non-sinusoidal shape of the wave.

Water waves

Water waves on water surfaces show a large range of fascinating phenomena. First of all, there are two different types of surface water waves. In the first type, the force that restores the plane surface is the surface tension of the wave. These so-called *surface tension waves* play a role on scales up to a few centimetres. In the second, larger type of waves, the restoring force is gravity, and one speaks of *gravity waves*.* The difference is easily noted by watching them: surface tension waves have a sinusoidal shape, whereas gravity waves have a shape with sharper maxima and broader troughs. This occurs because of the special way the water moves in such a wave. As shown in Figure 208, the surface water for a (short) gravity water wave moves in circles; this leads to the typical wave shape with short sharp crests and long shallow troughs: the waves are not up–down symmetric. Under the crests, the water particles move *in* the direction of the wave motion; under the troughs, the water particles move *against* the wave motion. As long as there is no wind and the floor below the water is horizontal, gravity waves are symmetric under front-to-back reflection. If the amplitude is very high, or if the wind is too strong, waves break, because a cusp angle of more than 120° is not possible. Such waves have no front-to-back symmetry.

In addition, water waves need to be distinguished according to the depth of the water, when compared to their wavelength. One speaks of *short* or *deep water* waves, when the depth of the water is so high that the floor below plays no role. In the opposite case one speaks of *long* or *shallow water* waves. The transitional region between the two cases are waves whose wavelength is between twice and twenty times the water depth. It turns out that all deep water waves and all ripples are *dispersive*, i.e., their speed depends on their frequency; only shallow gravity water waves are *non-dispersive*.

Water waves can be generated by wind and storms, by earthquakes, by the Sun and

---

* Meteorologists also know of a third type of water wave: there are large wavelength waves whose restoring force is the Coriolis force.

FIGURE 209 Three of the four main types of water waves. Top: a shallow water gravity wave, non-sinusoidal. Bottom left: a deep water ripple – a sinusoidal surface tension wave. The not-shown shallow water ripples look the same. Bottom right: a deep water gravity wave, here a boat wake, again non-sinusoidal. (© Eric Willis, Wikimedia, allyhook)

Moon, and by any other effect that deplaces water. The spectrum of water waves reaches from periods shorter than 100 ms to periods longer than 24 h. An overview is given in Table 39. The table also includes the lesser known infra-gravity waves, ultra-gravity waves,, tides and trans-tidal waves.

The classification of periodic water waves according to their restoring force and to the influence of the floor give *four* limit cases; they are shown in Figure 210. Each of the four limit cases is interesting.

Experiments and theory show that the phase speed of *gravity waves*, the lower two cases in Figure 210, depends on the wavelength λ and on the depth of the water d in the following way:

$$c = \sqrt{\frac{g\lambda}{2\pi} \tanh \frac{2\pi d}{\lambda}}, \qquad (95)$$

where g is the acceleration due to gravity (and an amplitude much smaller than the

TABLE 39 Spectrum of water waves.

TYPE	PERIOD	PROPERTIES	GENERATION
Surface tension waves/capillary waves/ripples	< 0.1 s	wavelength below a few cm	wind with more than 1 m/s, other disturbances, temperature
Ultra-gravity waves	0.1 to 1 s	restoring forces are surface tension and gravity	wind, other disturbances
Ordinary gravity waves	1 to 30 s	amplitude up to many meters, restoring force is gravity	wind
Infra-gravity waves/surf beat	30 s to 5 min	amplitude up to 30 cm, related to seiches, restoring force is gravity	wind, gravity waves
Long-period waves	5 min to 12 h	amplitude typically below 10 cm in deep water, up to 40 m in shallow water, restoring force is gravity and Coriolis effect	storms, earthquakes, air pressure changes
Ordinary tides	12 h to 24 h	amplitude depends on location, restoring force is gravity and Coriolis effect	Moon, Sun
Trans-tidal waves	above 24 h	amplitude depends on location, restoring force is gravity and Coriolis effect	Moon, Sun, storms, seasons, climate change

wavelength is assumed*). The formula shows two limiting regimes.

First, so-called *deep water* or *short* gravity waves appear when the water depth is larger than about half the wavelength. The usual sea wave is a deep water gravity wave, and so are the wakes generated by ships. Deep water gravity waves generated by wind are called *sea* they are generated by local winds, and *swell* if they are generated by distant winds. The typical phase speed of a gravity wave is of the order of the wind speed that generates it. For deep water waves, the phase velocity is related to the wavelength by $c \approx \sqrt{g\lambda/2\pi}$; the phase velocity is thus wavelength-dependent. In fact, all deep waves are dispersive. Shorter deep gravity waves are thus slower. The group velocity is *half* the phase velocity. Therefore, as surfers know, waves on a shore that are due to a distant storm arrive separately: first the long period waves, then the short period waves. The general effects of dispersion on wave groups are shown in Figure 211.

The typical *wake* generated by a ship is made of waves that have the phase velocity of the ship. These waves form a wave group, and it travels with half that speed. Therefore, from a ship's point of view, the wake trails the ship. Wakes are behind the ship because

---

* The expression for the phase velocity can be derived by solving for the motion of the liquid in the linear regime, but this leads us too far from our walk.

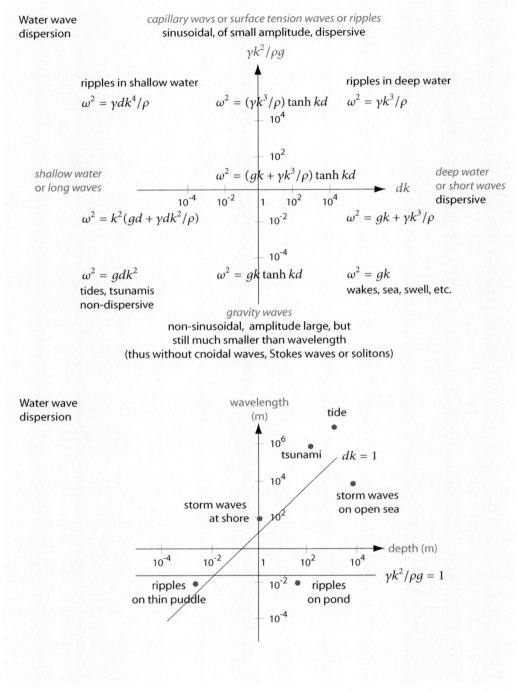

FIGURE 210 The different types of periodic water waves with low and moderate amplitude, visualized in two different diagrams using the depth d, the wave number $k = 2\pi/\lambda$ and the surface tension $\gamma = 72$ mPa.

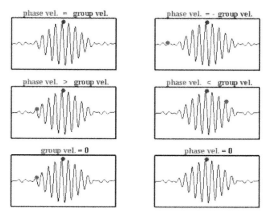

FIGURE 211 A visualisation of group velocity (blue) and phase velocity (red) for different types of wave dispersion. (QuickTime film © ISVR, University of Southampton)

their group velocity is lower than their phase velocity. We will explore wakes in more detail below.

The second limiting regime are *shallow water* or *long* gravity waves. They appear when the depth is less than 1/20th of the wavelength; in this case, the phase velocity is $c \approx \sqrt{gd}$, there is no dispersion, and the group velocity is the *same* as the phase velocity. In shallow water waves, water particles move on very flat elliptic paths.

The tide is an example of a shallow gravity wave. The most impressive shallow gravity waves are *tsunamis*, the large waves triggered by submarine earthquakes. (The Japanese term is composed of *tsu*, meaning harbour, and *nami*, meaning wave.) Since tsunamis are shallow waves, they show no, or little, dispersion and thus travel over long distances; they can go round the Earth several times. Typical oscillation times of tsunamis are between 6 and 60 minutes, giving wavelengths between 70 and 700 km and speeds in the open sea of 200 to 250 m/s, similar to that of a jet plane. Their amplitude on the open sea is often of the order of 10 cm; however, the amplitude scales with depth d as $1/d^4$ and heights up to 40 m have been measured at the shore. This was the order of magnitude of the large and disastrous tsunami observed in the Indian Ocean on 26 December 2004 and the one in Japan in 2011 that destroyed several nuclear power plants. Tsunamis can also be used to determine the ocean depth by measuring the speed of tsunamis. This allowed researchers to deduce, long before sonar and other high-tech systems were available, that the North Pacific has a depth of around 4 to 4.5 km.

The upper two limit regimes in Figure 210 are the surface tension waves, also called *capillary waves* or *ripples*. The value of the surface tension of water is 72 mPa. The first of these limit regimes are ripples on deep water. Their phase velocity is $c = \sqrt{\gamma k/\rho}$. As mentioned, all deep waves are dispersive. Indeed, the group velocity of ripples on deep water is 3/2 times the phase velocity. Therefore ripples steam ahead of a boat, whereas wakes trail behind.

Deep ripples have a minimum speed. . The minimum speed of short ripples is the reason for what we see when throw a *pebble* in a lake. A typical pebble creates ripples with a wavelength of about 1 cm. For water waves in this region, there is a minimum group velocity of 17.7 cm/s and a minimum phase velocity of around 23 cm/s. When a pebble falls into the water, it creates waves of various wavelengths; those with a wavelength around 1 cm are the slowest ones and are seen most clearly. The existence of a minimum phase velocity for ripples also means that insects walking on water generate no waves if they move more slowly than the minimum phase velocity; thus they feel little drag and can walk easily.

The final limit case of waves are ripples on shallow water. An example are the waves emitted by raindrops falling in a shallow puddle, say with a depth of 1 mm or less. The phase velocity is $c = \sqrt{\gamma d k^2/\rho}$; the group velocity has twice that value. Ripples on shallow water are thus dispersive, as are ripples in deep water.

Figure 210 shows the four types of water surface waves discussed so far. The general dispersion relation for all these waves is

$$\omega^2 = (gk + \gamma k^3/\rho) \tanh kd \,. \tag{96}$$

Several additional types of water waves also exist, such as *seiches*, i.e., standing waves in lakes, *internal waves* in stratified water systems, and *solitons*, i.e., non-periodic travelling waves. We will explore the last case later on.

Waves and their motion

Waves move. Therefore, any study of motion must include the study of wave motion. We know from experience that waves can hit or even damage targets; thus every wave carries energy and momentum, even though (on average) no matter moves along the wave propagation direction. The *energy E* of a wave is the sum of its kinetic and potential energy. The kinetic energy (density) depends on the temporal change of the displacement u at a given spot: rapidly changing waves carry a larger kinetic energy. The potential energy (density) depends on the gradient of the displacement, i.e., on its spatial change: steep waves carry a larger potential energy than shallow ones. (Can you explain why the potential energy does not depend on the displacement itself?) For harmonic, i.e., sinusoidal, waves propagating along the direction z, each type of energy is proportional to the square of its respective displacement change:

$$E \sim \left(\frac{\partial u}{\partial t}\right)^2 + v^2 \left(\frac{\partial u}{\partial z}\right)^2 \,. \tag{97}$$

How is the energy density related to the frequency?

The *momentum* of a wave is directed along the direction of wave propagation. The momentum value depends on both the temporal and the spatial change of displacement u. For harmonic waves, the momentum (density) P is proportional to the product of these two quantities:

$$P_z \sim \frac{\partial u}{\partial t} \frac{\partial u}{\partial z} \,. \tag{98}$$

When two linear wave trains collide or interfere, the total momentum is conserved throughout the collision. An important consequence of momentum conservation is that waves that are reflected by an obstacle do so with an outgoing angle equal to minus the

OSCILLATIONS AND WAVES

infalling angle. What happens to the phase?

In summary, waves, like moving bodies, carry energy and momentum. In simple terms, if you shout against a wall, the wall is hit by the sound waves. This hit, for example, can start avalanches on snowy mountain slopes. In the same way, waves, like bodies, can carry also angular momentum. (What type of wave is necessary for this to be possible?) However, the motion of waves differs from the motion of bodies. Six main properties distinguish the motion of waves from the motion of bodies.

1. Waves can add up or cancel each other out; thus they can interpenetrate each other. These effects, *superposition* and *interference*, are strongly tied to the linearity of many wave types.
2. Waves, such as sound, can go around corners. This is called *diffraction*. Diffraction is a consequence of interference, and it is thus not, strictly speaking, a separate effect.
3. Waves change direction when they change medium. This is called *refraction*. Refraction is central for producing images.
4. Waves can have a frequency-dependent propagation speed. This is called *dispersion*. It produces rainbows and the grumble of distant thunders.
5. Often, the wave amplitude decreases over time: waves show *damping*. Damping produces pauses between words.
6. Transverse waves in three dimensions can oscillate in different directions: they show *polarization*. Polarization is important for antennas and photographs.

In our everyday life, in contrast to moving waves, moving material bodies, such as stones, do not show any of these six effects. The six wave effects appear because wave motion is the motion of *extended* entities. The famous debate whether electrons are waves or particles thus requires us to check whether the six effects specific to waves can be observed for electrons or not. This exploration is part of quantum physics. Before we start it, can you give an example of an observation that automatically implies that the specific motion *cannot* be a wave?

As a result of having a frequency f and a propagation or phase velocity c, all sine waves are characterized by the distance λ between two neighbouring wave crests: this distance is called the wavelength λ. All waves obey the basic relation

$$\lambda f = c \, . \tag{99}$$

In many cases the phase velocity c depends on the wavelength of the wave. For example, this is the case for many water waves. This change of speed with wavelength is called *dispersion*. In contrast, the speed of sound in air does *not* depend on the wavelength (to a high degree of accuracy). Sound in air shows (almost) no dispersion. Indeed, if there were dispersion for sound, we could not understand each other's speech at larger distances.

Now comes a surprise. Waves can also exist in empty space. Light, radio waves and gravitational waves are examples. The exploration of electromagnetism and relativity will tell us more about their specific properties. Here is an appetizer. Light is a wave. Usually, we do not experience light as a wave, because its wavelength is only around one two-thousandth of a millimetre. But light shows all six effects typical of wave motion.

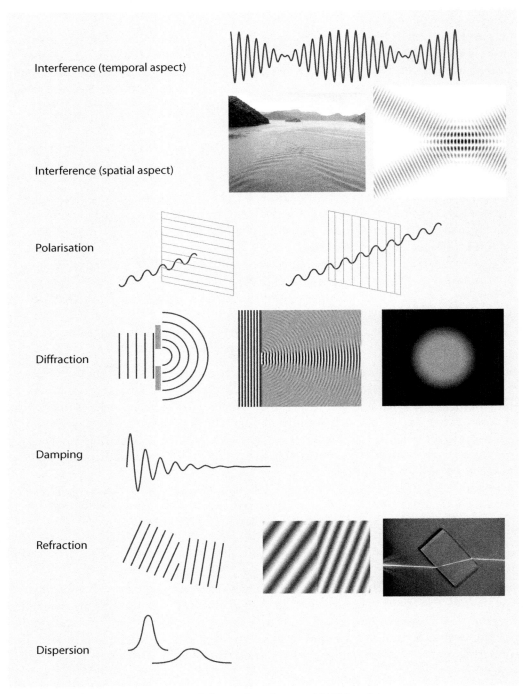

FIGURE 212 The six main properties of the motion of waves. (© Wikimedia)

A rainbow, for example, can only be understood fully when the last five wave effects are taken into account. Diffraction and interference of light can even be observed with your

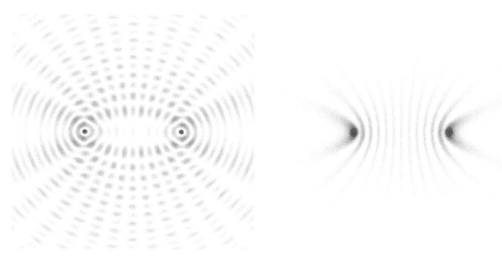

FIGURE 213 Interference of two circular or spherical waves emitted in phase: a snapshot of the amplitude (left), most useful to describe observations of water waves, and the distribution of the time-averaged intensity (right), most useful to describe interference of light waves (© Rüdiger Paschotta).

fingers only. Can you tell how?

For gravitational waves, superposition, damping and polarization have been observed. The other three wave effects are still unobserved.

Like every anharmonic oscillation, every anharmonic wave can be decomposed into sine waves. Figure 205 gives examples. If the various sine waves contained in a disturbance propagate differently, the original wave will change in shape while it travels. That is the reason why an echo does not sound exactly like the original sound; for the same reason, a nearby thunder and a far-away one sound different. These are effects of the – rather weak – dispersion of sound waves.

All systems which oscillate also *emit waves*. Any radio or TV receiver contains oscillators. As a result, any such receiver is also a (weak) transmitter; indeed, in some countries the authorities search for people who listen to radio without permission by listening to the radio waves emitted by these devices. Also, inside the human ear, numerous tiny structures, the hair cells, oscillate. As a result, the ear must also emit sound. This prediction, made in 1948 by Tommy Gold, was finally confirmed in 1979 by David Kemp. These so-called *otoacoustic emissions* can be detected with sensitive microphones; they are presently being studied in order to unravel the still unknown workings of the ear and in order to diagnose various ear illnesses without the need for surgery.

Since any travelling disturbance can be decomposed into sine waves, the term 'wave' is used by physicists for all travelling disturbances, whether they look like sine waves or not. In fact, the disturbances do not even have to be travelling. Take a standing wave: is it a wave or an oscillation? Standing waves do not travel; they are oscillations. But a standing wave can be seen as the superposition of two waves travelling in opposite directions. In fact, in nature, any object that we call 'oscillating' or 'vibrating' is extended, and its oscillation or vibration is always a standing wave (can you confirm this?); so we can say that in nature, *all oscillations are special forms of waves*.

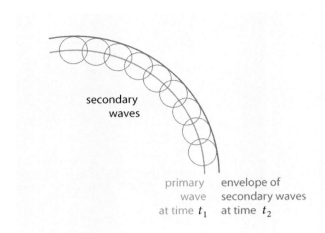

FIGURE 214 Wave propagation as a consequence of Huygens' principle.

The most important travelling disturbances are those that are localized. Figure 205 shows an example of a localized wave, also called a *wave group* or *pulse*, together with its decomposition into harmonic waves. Wave groups are used to talk and as signals for communication.

Why can we talk to each other? – Huygens' principle

The properties of our environment often disclose their full importance only when we ask simple questions. Why can we use the radio? Why can we talk on mobile phones? Why can we listen to each other? It turns out that a central part of the answer to these questions is that the space we live has an *odd* numbers of dimensions.

In spaces of *even* dimension, it is impossible to talk, because messages do not stop – or stop very slowly. This is an important result which is easily checked by throwing a stone into a lake: even after the stone has disappeared, waves are still emitted from the point at which it entered the water. Yet, when we stop talking, no waves are emitted any more from our mouth. In short, waves in two and in three dimensions behave differently.

In three dimensions, it is possible to describe the propagation of a wave in the following way: Every point on a wave front (of light or of sound) can be regarded as the source of *secondary waves*; the surface that is formed by the envelope of all the secondary waves determines the future position of the wave front. The idea is illustrated in Figure 214. This idea can be used to describe, without mathematics, the propagation of waves, their reflection, their refraction, and, with an extension due to Augustin Fresnel, their diffraction. (Try!)

This idea of secondary waves was first proposed by Christiaan Huygens in 1678 and is called *Huygens' principle*. Almost two hundred years later, Gustav Kirchhoff showed that the principle is a consequence of the wave equation in three dimensions, and thus, in the case of light, a consequence of Maxwell's field equations.

But the description of wave fronts as envelopes of secondary waves has an important limitation. It is *not* correct in two dimensions (even though Figure 214 is two-dimensional!). In particular, it does not apply to water waves. Water wave propagation

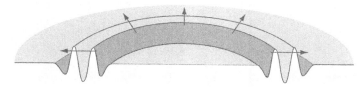

FIGURE 215 An impossible gravity water wave: the centre is never completely flat.

cannot be calculated in this way in an exact manner. (It is only possible in two dimensions if the situation is limited to a wave of a single frequency.) It turns out that for water waves, secondary waves do not only depend on the wave front of the primary waves, but depend also on their interior. The reason is that in two (and other even) dimensions, waves of different frequency necessarily have *different speeds*. (Tsunamis are not counter-examples; they non-dispersive only as a limit case.) And indeed, a stone falling into water is comparable to a banging sound: it generates waves of many frequencies. In contrast, in three (and larger odd) dimensions, waves of all frequencies have the same speed.

Mathematically, we can say that Huygens' principle holds if the wave equation is solved by a circular wave leaving no amplitude behind it. Mathematicians translate this by requiring that the evolving delta function $\delta(c^2 t^2 - r^2)$ satisfies the wave equation, i.e., that $\partial_t^2 \delta = c^2 \Delta \delta$. The delta function is that strange 'function' which is zero everywhere except at the origin, where it is infinite. A few more properties describe the precise way in which this happens.* In short, the delta function is an idealized, infinitely short and infinitely loud bang. It turns out that the delta function is a solution of the wave equation only if the space dimension is odd and at least three. In other words, while a *spherical* wave pulse is possible, a *circular* pulse is not: there is no way to keep the centre of an expanding wave quiet. (See Figure 215.) That is exactly what the stone experiment shows. You can try to produce a circular pulse (a wave that has only a few crests) the next time you are in the bathroom or near a lake: you will not succeed.

In summary, the reason a room gets dark when we switch off the light, is that we live in a space with a number of dimensions which is odd and larger than one.

Wave equations**

Waves are fascinating phenomena. Equally fascinating is their mathematical description.

The amplitude $A(\boldsymbol{x}, t)$ of a linear wave in one, two or three dimensions, the simplest of all wave types, results from

$$\frac{\partial^2 A(\boldsymbol{x}, t)}{\partial t^2} = v^2 \nabla^2 A(\boldsymbol{x}, t) \ . \tag{100}$$

The equation says that the acceleration of the amplitude (the left-hand term) is given by the square gradient, i.e., by the spatial variation, multiplied by the squared of the phase velocity v. In many cases, the amplitude is a vector, but the equation remains the same.

---

* The main property is $\int \delta(x) \, \mathrm{d}x = 1$. In mathematically precise terms, the delta 'function' is a *distribution*.
** This section can be skipped at first reading.

TABLE 40 Some signals.

System	Signal	Speed	Sensor
Matter signals			
Human	voltage pulses in nerves	up to 120 m/s	brain, muscles
	hormones in blood stream	up to 0.3 m/s	molecules on cell membranes
	immune system signals	up to 0.3 m/s	molecules on cell membranes
	singing	340 m/s	ear
Elephant, insects	soil trembling	c. 2 km/s	feet
Whale	singing, sonar	1500 m/s	ear
Dog	chemical tracks	1 m/s	nose
Butterfly	chemical mating signal carried by the wind	up to 10 m/s	antennae
Tree	chemical signal of attack carried by the air from one tree to the next	up to 10 m/s	leaves
Erratic block	carried by glacier	up to 0.1 μm/s	foot
Post	paper letters transported by trucks, ships and planes	up to 300 m/s	mail box
Electromagnetic fields			
Humans	yawning	300 Mm/s	eye
Electric eel	voltage pulse	up to 300 Mm/s	nerves
Insects, fish, molluscs	light pulse sequence	up to 300 Mm/s	eye
Flag signalling	orientation of flags	300 Mm/s	eye
Radio transmissions	electromagnetic field strength	up to 300 Mm/s	radio
Nuclear signals			
Supernovas	neutrino pulses	close to 300 Mm/s	specific chemical and radiation detectors
Nuclear reactions	glueballs, if they exist	close to 300 Mm/s	custom particle detectors

More correctly, the amplitude of a wave is what physicists call a *field*, because it is a number (or vector, or tensor) that depends on space and time.

Equation (100) is a *wave equation*. Mathematically, it is a *linear partial differential*

equation. It is called *linear* because it is linear in the amplitude A. Therefore, its solutions are sine and cosine waves of the type $A = \sin(x - vt + \varphi)$. Linear wave equations follow from elastic behaviour of some medium. The linearity also implies that the sum of two waves is also a possible wave; this so-called *superposition principle* is valid for all linear wave equations (and a few rare, but important non-linear wave equations). If linearity applies, any general wave can be *decomposed* into an infinite sum of sine and cosine waves. This discovery is due to Joseph Fourier (b. 1768, Auxerre, d. 1830, Paris).

The wave equation (100) is also *homogeneous*, meaning that there is no term independent of A, and thus, no energy source that drives the waves. Fourier decomposition also helps to understand and solve inhomogeneous wave equations, thus externally driven elastic media.

In several dimensions, the shape of the wave front is also of interest. In two dimensions, the simplest cases described by equation (100) are *linear* and *circular* waves. In three dimensions, the simplest cases described by equation (100) are *plane* and *spherical* waves.

In appropriate situations – thus when the elastic medium is finite and is excited in specific ways – equation (100) also leads to *standing* waves. Standing waves are superpositions of waves travelling in opposite directions.

Sound in gases, sound in liquids and solids, earthquakes, light in vacuum, certain water waves of small amplitude, and various other cases of waves with small amplitude are described by *linear* wave equations.

Mathematically, all wave equations, whether linear or not, are *hyperbolic* partial differential equations. This just means that the spatial second derivative has opposite sign to the temporal second derivative. By far the most interesting wave equations are *non-linear*. The most famous non-linear equation is the *Korteweg–de Vries equation*, which is the one-dimensional wave equation

$$\partial_t A + A\partial_x A + b\partial_{xxx} A = 0 \ . \tag{101}$$

It was discovered only very late that this evolution equation for the field $A(x,t)$ can be solved with paper and pencil.

Other non-linear wave equations describe specific situations. The Boussinesq equation, the sine–Gordon equation and many other wave equations have sparked a vast research field in mathematical and experimental physics. Non-linear partial differential equations are also essential in the study of self-organization.

Why are music and singing voices so beautiful?

Music works because it connects to emotions. And it does so, among others, by reminding us of the sounds (and emotions connected to them) that we experienced before birth. Percussion instruments remind us of the heart beat of our mother and ourselves, cord and wind instruments remind us of all the voices we heard back then. Musical instruments are especially beautiful if they are driven and modulated by the body and the art of the player. All classical instruments are optimized to allow this modulation and the expression of emotions. The connection between the musician and the instrument is most intense for the human voice; the next approximation are the wind instruments.

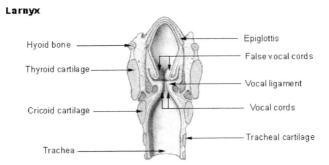
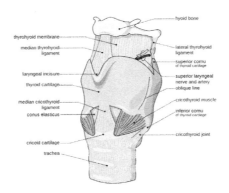

FIGURE 216 The human larynx is the part of the anatomy that contains the sound source of speech, the vocal folds (© Wikimedia).

Every musical instrument, the human voice included, consists of four elements: an energy source, an oscillating sound source, one or more resonators, and a radiating surface or orifice. In the human voice, the energy source is formed by the muscles of the thorax and belly, the sound source are vocal folds – also called vocal cords – the resonator is the vocal tract, and the mouth and nose form the orifice.

The breath of the singer or of the wind instrument player provides the energy for the sound generation and gives the input for the feedback loop that sets the pitch. While singing, the air passes the vocal folds. The rapid air flow reduces the air pressure, which attracts the cords to each other and thus reduces the cross section for the air flow. (This pressure reduction is described by the Bernoulli equation, as explained below.) As a result of the smaller cross section, the airflow is reduced, the pressure rises again, and the vocal cords open up again. This leads to larger airflow, and the circle starts again. The change between larger and smaller cord distance repeats so rapidly that sound is produced; the sound is then amplified in the mouth by the resonances that depend on the shape of the oral cavity. Using modern vocabulary, singing a steady note is a specific case of self-organization, namely an example of a limit cycle.

But how can a small instrument like the vocal tract achieve sounds more intense than that of a trombone, which is several metres long when unwound? How can the voice cover a range of 80 dB in intensity? How can the voice achieve up to five, even eight octaves in fundamental frequency with just two vocal folds? And how can the human voice produce its unmatched variation in timbre? Many details of these questions are still subject of research, though the general connections are now known.

The human vocal folds are on average the size of a thumb's nail; but they can vary in length and tension. The vocal folds have three components. Above all, they contain a ligament that can sustain large changes in tension or stress and forms the basic structure; such a ligament is needed to achieve a wide range of frequencies. Secondly, 90 % of the vocal folds is made of muscles, so that the stress and thus the frequency range can be increased even further. Finally, the cords are covered by a mucosa, a fluid-containing membrane that is optimized to enter in oscillation, through surface waves, when air passes by. This strongly non-linear system achieves, in exceptional singers, up to 5 octaves of fundamental pitch range.

Also the resonators of the human voice are exceptional. Despite the small size available, the non-linear properties of the resonators in the vocal tract – especially the effect called *inertive reactance* – allow to produce high intensity sound. This complex system, together with intense training, produces the frequencies, timbres and musical sequences that we enjoy in operas, in jazz, and in all other vocal performances. In fact, several results from research into the human voice – which were also deduced with the help of magnetic resonance imaging – are now regularly used to train and teach singers, in particular on when to use open mouth and when to use closed-mouth singing, or when to lower the larynx.

Singing is thus beautiful also because it is a non-linear effect. In fact, all instruments are non-linear oscillators. In *reed instruments*, such as the clarinet, the reed has the role of the vocal cords, and the pipe has the role of the resonator, and the mechanisms shift the opening that has the role of the mouth and lips. In *brass instruments*, such as the trombone, the lips play the role of the reed. In *airflow instruments*, such as the flute, the feedback loop is due to another effect: at the sound-producing edge, the airflow is deflected by the sound itself.

Ref. 221

The second reason that music is beautiful is due to the way the frequencies of the notes are selected. Certain frequencies sound agreeable to the ear when they are played together or closely after each other; other produce a sense of tension. Already the ancient Greek had discovered that these sensations depend exclusively on the *ratio* of the frequencies, or as musician say, on the *interval* between the pitches.

More specifically, a frequency ratio of 2 – musicians call the interval an *octave* – is the most agreeable consonance. A ratio of 3/2 (called perfect fifth) is the next most agreeable, followed by the ratio 4/3 (a perfect fourth), the ratio 5/4 (a major third) and the ratio 6/5 (a [third, minor]minor third). The choice of the first third in a scale has an important effect on the average emotions expressed by the music and is therefore also taken over in the name of the scale. Songs in C major generally have a more happy tune, whereas songs in A minor tend to sound sadder.

The least agreeable frequency ratios, the dissonances, are the *tritone* (7/5, also called *augmented fourth* or *diminished fifth* or *false quint*) and, to a lesser extent, the major and minor seventh (15/8 and 9/5). The tritone is used for the siren in German red cross vans. Long sequences of dissonances have the effect to induce trance; they are common in Balinese music and in jazz.

After centuries of experimenting, these results lead to a standardized arrangement of the notes and their frequencies that is shown in Figure 217. The arrangement, called the

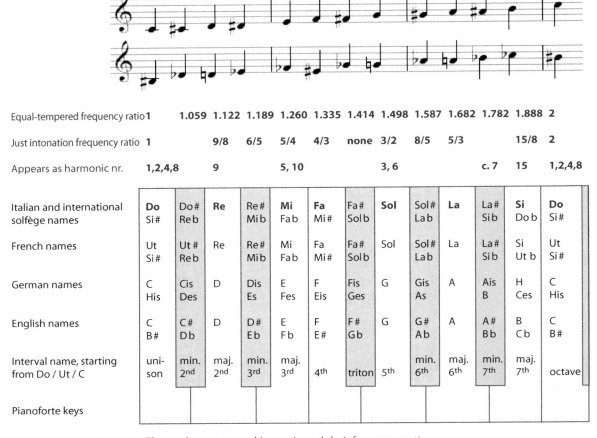

FIGURE 217 The twelve notes used in music and their frequency ratios.

equal intonation or *well-tempered intonation*, contains approximations to all the mentioned intervals; the approximations have the advantage that they allow the *transposition* of music to lower or higher notes. This is not possible with the ideal, so-called *just intonation*.

The next time you sing a song that you like, you might try to determine whether you use just or equal intonation – or a different intonation altogether. Different people have different tastes and habits.

Measuring sound

At every point in space, a sound wave in air produces two effects: a pressure change and a speed change. Figure 218 shows both of them: pressure changes induce changes in the density of the molecules, whereas velocity changes act on the average speed of the molecules.

The local sound pressure is measured with a *microphone* or an *ear*. The local molecular speed is measured with a *microanemometer* or *acoustic particle velocity sensor*. No such device existed until 1994, when Hans Elias de Bree invented a way to build one. As

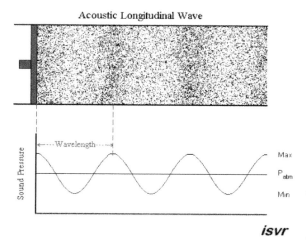

FIGURE 218 A schematic visualisation of the motion of molecules in sound wave in air (QuickTime film © ISVR, University of Southampton)

TABLE 41 Selected sound intensities.

Observation	Sound intensity
Sound threshold at 1 kHz	0 dB or 1 pW
Human speech	25 to 35 dB
Subway entering a subway station	100 dB
Ultrasound imaging of babies	over 100 dB
Conventional pain threshold	120 dB or 1 W
Rock concert with 400 000 W loudspeakers	135 to 145 dB
Fireworks	up to 150 dB
Gunfire	up to 155 dB
Missile launch	up to 170 dB
Blue whale singing	up to 175 dB
Volcanic eruptions, earthquakes, conventional bomb	up to 210 dB
Large meteoroid impact, large nuclear bomb	over 300 dB

shown in Figure 219, such a microanemometer is not easy to manufacture. Two tiny platinum wires are heated to 220°C; their temperature difference depends on the air speed and can be measured by comparing their electrical resistances. Due to the tiny dimensions, a frequency range up to about 20 kHz is possible. By putting three such devices at right angles to each other it is possible to localize the direction of a noise source. This is useful for the repair and development of cars, or to check trains and machinery. By arranging many devices on a square grid one can even build an 'acoustic camera'. It can even be used to pinpoint aircraft and drone positions, a kind of 'acoustic radar'. Because microanemometers act as extremely small and extremely directional microphones, and because they also work under water, the military and the spies are keen on them.

By the way: how does the speed of molecules due to sound compare to the speed of

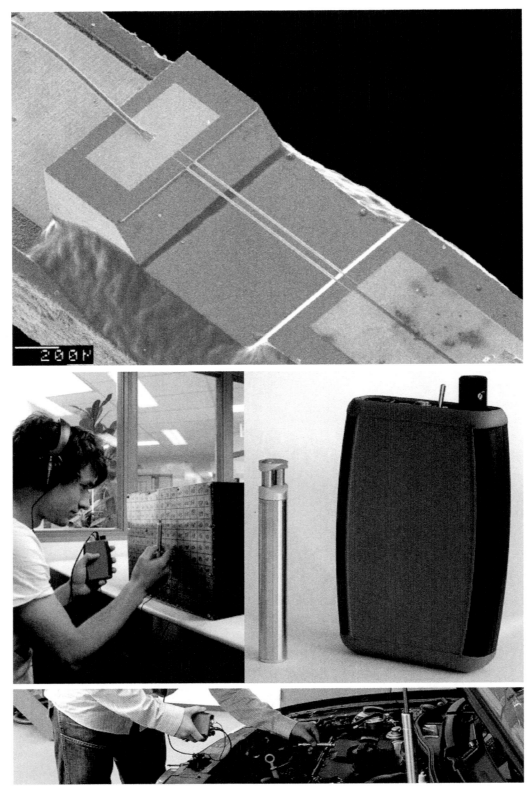

FIGURE 219 Top: two tiny heated platinum wires allow building a microanemometer. It can help to pinpoint exactly where a rattling noise comes from (© Microflown Technologies).

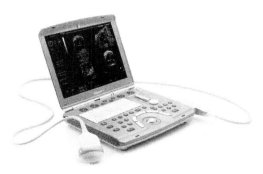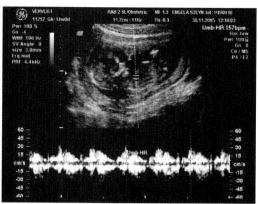

FIGURE 220 A modern ultrasound imaging system, and a common, but harmful ultrasound image of a foetus (© General Electric, Wikimedia).

molecules due to the air temperature?

Is ultrasound imaging safe for babies?

Ultrasound is used in medicine to explore the interior of human bodies. The technique, called *ultrasound imaging*, is helpful, convenient and widespread, as shown in Figure 220. However, it has a disadvantage. Studies at the Mayo Clinic in Minnesota have found that *pulsed* ultrasound, in contrast to *continuous* ultrasound, produces extremely high levels of *audible* sound inside the body. (Some sound intensities are listed in Table 41.)

Pulsed ultrasound is used in ultrasound imaging, and in some, but not all, foetal heartbeat monitors. Such machines thus produce high levels of sound in the *audible* range. This seems paradoxical; if you go to a gynaecologist and put the ultrasound head on your ear or head, you will only hear a very faint noise. In fact, it is this low intensity that tricks everybody to think that the noise level heard by babies is low. The noise level is only low because the human ear is full of air. In contrast, in a foetus, the ear is filled with liquid. This fact changes the propagation of sound completely: the sound generated by imaging machines is now fully focused and directly stimulates the inner ear. The total effect is similar to what happens if you put your finger in your ear: this can be very loud to yourself, but nobody else can hear what happens.

Recent research has shown that sound levels of over 100 dB, corresponding to a subway train entering the station, are generated by ultrasound imaging systems. Indeed, every gynaecologist will confirm that imaging disturbs the foetus. Questioned about this issue, several makers of ultrasound imaging devices confirmed that "a sound output of *only* 5 mW is used". That is 'only' the acoustic power of an oboe at full power! Since many ultrasound examinations take ten minutes and more, a damage to the ear of the foetus cannot be excluded. It is not sensible to expose a baby to this level of noise without good reason.

In short, ultrasound should be used for pregnant mothers only in case of necessity. Ultrasound is *not safe* for the ears of foetuses. (Statements by medical ultrasound societies saying the contrary are wrong.) In *all other* situations, ultrasound imaging is safe.

It should be noted however, that another potential problem of ultrasound imaging, the issue of tissue damage through cavitation, has not been explored in full detail yet.

Signals

A signal is the transport of information. Every signal, including those from Table 40, is motion of energy. Signals can be either objects or waves. A thrown stone can be a signal, as can a whistle. Waves are a more practical form of communication because they do not require transport of matter: it is easier to use electricity in a telephone wire to transport a statement than to send a messenger. Indeed, most modern technological advances can be traced to the separation between signal and matter transport. Instead of transporting an orchestra to transmit music, we can send radio signals. Instead of sending paper letters we write email messages. Instead of going to the library we browse the internet.

The greatest advances in communication have resulted from the use of signals for the transport of large amounts of energy. That is what electric cables do: they transport energy without transporting any (noticeable) matter. We do not need to attach our kitchen machines to the power station: we can get the energy via a copper wire.

For all these reasons, the term 'signal' is often meant to imply waves only. Voice, sound, electric signals, radio and light signals are the most common examples of wave signals.

Signals are characterized by their speed and their information content. Both quantities turn out to be limited. The limit on speed is the central topic of the theory of special relativity.

A simple limit on information content of wave signals can be expressed when noting that the information flow is given by the detailed shape of the signal. The shape is characterized by a frequency (or wavelength) and a position in time (or space). For every signal – and every wave – there is a relation between the time-of-arrival error Δt and the angular frequency error $\Delta \omega$:

$$\Delta t \, \Delta \omega \geqslant \tfrac{1}{2} \, . \tag{102}$$

This time–frequency indeterminacy relation expresses that, in a signal, it is impossible to specify both the time of arrival and the frequency with full precision. The two errors are (within a numerical factor) the inverse of each other. (We can also say that the time-bandwidth product is always larger than $1/4\pi$.) The limitation appears because on the one hand we need a wave as similar as possible to a long sine wave in order to precisely determine the frequency, but on the other hand we need a signal as short as possible to precisely determine its time of arrival. The contrast in the two requirements leads to the limit. The indeterminacy relation is thus a feature of every wave phenomenon. You might want to test this relation with any wave in your environment.

Similarly, there is a relation between the position error Δx and the wave vector error $\Delta k = 2\pi/\Delta\lambda$ of a signal:

$$\Delta x \, \Delta k \geqslant \tfrac{1}{2} \, . \tag{103}$$

Like the previous case, also this indeterminacy relation expresses that it is impossible to specify both the position of a signal and its wavelength with full precision. Also this position–wave-vector indeterminacy relation is a feature of any wave phenomenon.

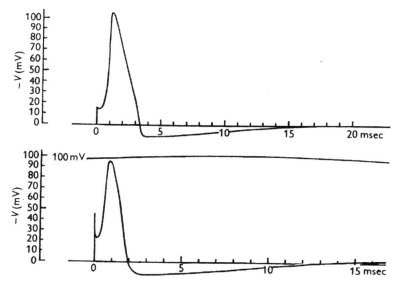

FIGURE 221 The electrical signals calculated (above) and measured (below) in a nerve, following Hodgkin and Huxley.

Every indeterminacy relation is the consequence of a smallest entity. In the case of waves, the smallest entity of the phenomenon is the period (or cycle, as it used to be called). Whenever there is a smallest unit in a natural phenomenon, an indeterminacy relation results. We will encounter other indeterminacy relations both in relativity and in quantum theory. As we will find out, they are due to smallest entities as well.

Whenever signals are sent, their content can be lost. Each of the six characteristics of waves listed on page 301 can lead to content degradation. Can you provide an example for each case? The energy, the momentum and all other conserved properties of signals are never lost, of course. The disappearance of signals is akin to the disappearance of motion. When motion disappears by friction, it only seems to disappear, and is in fact transformed into heat. Similarly, when a signal disappears, it only seems to disappear, and is in fact transformed into noise. *(Physical) noise* is a collection of numerous disordered signals, in the same way that heat is a collection of numerous disordered movements.

All signal propagation is described by a wave equation. A famous example is the set of equations found by Hodgkin and Huxley. It is a realistic approximation for the behaviour of electrical potential in nerves. Using facts about the behaviour of potassium and sodium ions, they found an elaborate wave equation that describes the voltage V in nerves, and thus the way the signals are propagated along them. The equation describes the characteristic voltage spikes measured in nerves, shown in Figure 221. The figure clearly shows that these waves differ from sine waves: they are not harmonic. Anharmonicity is one result of non-linearity. But non-linearity can lead to even stronger effects.

FIGURE 222 A solitary water wave followed by a motor boat, reconstructing the discovery by Scott Russel (© Dugald Duncan).

Solitary waves and solitons

In August 1834, the Scottish engineer John Scott Russell (b. 1808 Glasgow, d. 1882 London) recorded a strange observation in a water canal in the countryside near Edinburgh. When a boat pulled through the channel was suddenly stopped, a strange water wave departed from it. It consisted of a *single* crest, about 10 m long and 0.5 m high, moving at about 4 m/s. He followed that crest, shown in a reconstruction in Figure 222, with his horse for several kilometres: the wave died out only very slowly. Russell did not observe any dispersion, as is usual in deep water waves: the width of the crest remained constant. Russell then started producing such waves in his laboratory, and extensively studied their properties.

Russell showed that the speed depended on the amplitude, in contrast to linear, harmonic waves. He also found that the depth d of the water canal was an important parameter. In fact, the speed v, the amplitude A and the width L of these single-crested waves are related by

$$v = \sqrt{gd}\left(1 + \frac{A}{2d}\right) \quad \text{and} \quad L = \sqrt{\frac{4d^3}{3A}} \ . \tag{104}$$

As shown by these expressions, and noted by Russell, high waves are narrow and fast, whereas shallow waves are slow and wide. The shape of the waves is fixed during their motion. Today, these and all other stable waves with a single crest are called *solitary waves*. They appear only where the dispersion and the non-linearity of the system exactly compensate for each other. Russell also noted that the solitary waves in water channels can cross each other unchanged, even when travelling in opposite directions; solitary waves with this property are called *solitons*. In short, solitons are stable against encounters, as shown in Figure 223, whereas solitary waves in general are not.

Sixty years later, in 1895, Korteweg and de Vries found out that solitary waves in water

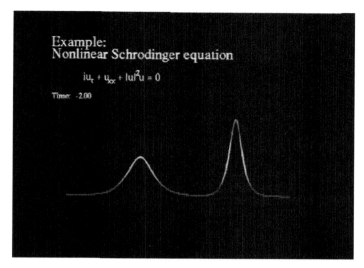

FIGURE 223 Solitons are stable against encounters. (QuickTime film © Jarmo Hietarinta)

channels have a shape described by

$$u(x,t) = A \operatorname{sech}^2 \frac{x-vt}{L} \quad \text{where} \quad \operatorname{sech} x = \frac{2}{e^x + e^{-x}}, \tag{105}$$

and that the relation found by Russell was due to the wave equation

$$\frac{1}{\sqrt{gd}} \frac{\partial u}{\partial t} + \left(1 + \frac{3}{2d} u\right) \frac{\partial u}{\partial x} + \frac{d^2}{6} \frac{\partial^3 u}{\partial x^3} = 0. \tag{106}$$

This equation for the elongation u is now called the *Korteweg–de Vries equation* in their honour.* The surprising stability of the solitary solutions is due to the opposite effect of the two terms that distinguish the equation from linear wave equations: for the solitary solutions, the non-linear term precisely compensates for the dispersion induced by the third-derivative term.

For many decades such solitary waves were seen as mathematical and physical curiosities. The reason was simple: nobody could solve the equations. All this changed almost a hundred years later, when it became clear that the Korteweg–de Vries equation is a universal model for weakly non-linear waves in the weak dispersion regime, and thus of basic importance. This conclusion was triggered by Kruskal and Zabusky, who in 1965 proved mathematically that the solutions (105) are *unchanged* in collisions. This discovery prompted them to introduce the term *soliton*. These solutions do indeed interpenetrate one another without changing velocity or shape: a collision only produces a small positional shift for each pulse.

Ref. 230

---

* The equation can be simplified by transforming the variable u; most concisely, it can be rewritten as $u_t + u_{xxx} = 6uu_x$. As long as the solutions are sech functions, this and other transformed versions of the equation are known by the same name.

Solitary waves play a role in many examples of fluid flows. They are found in ocean currents; and even the red spot on Jupiter, which was a steady feature of Jupiter photographs for many centuries, is an example.

Solitary waves also appear when extremely high-intensity sound is generated in solids. In these cases, they can lead to sound pulses of only a few nanometres in length. Solitary light pulses are also used inside certain optical communication fibres, where the lack of dispersion allows higher data transmission rates than are achievable with usual light pulses.

Towards the end of the twentieth century, mathematicians discovered that solitons obey a non-linear generalization of the superposition principle. (It is due to the Darboux–Backlund transformations and the structure of the Sato Grassmannian.) The mathematics of solitons is extremely interesting. The progress in mathematics triggered a second wave of interest in the mathematics of solitons arose, when quantum theorists became interested in them. The reason is simple: a soliton is a 'middle thing' between a particle and a wave; it has features of both concepts. For this reason, solitons were often seen – incorrectly though – as candidates for the description of elementary particles.

Curiosities and fun challenges about waves and oscillation

> Society is a wave. The wave moves onward, but the water of which it is composed does not.
> Ralph Waldo Emerson, *Self-Reliance*.

Sounds can be beautiful. If you enjoy the sound in the Pisa Baptistery, near the leaning tower, or like to feel the sound production in singing sand, or about The Whispering Gallery in St Paul's Cathedral, read the book by Trevor Cox, *Sonic Wonderland*. It opens the door to a new world.

**

When the frequency of a tone is doubled, one says that the tone is higher by an *octave*. Two tones that differ by an octave, when played together, sound pleasant to the ear. Two other agreeable frequency ratios – or 'intervals', as musicians say – are quarts and quints. What are the corresponding frequency ratios? The answer to this question was one of the oldest discoveries in physics and perception research; it is attributed to Pythagoras, around 500 BCE.

**

Many teachers, during their course, inhale helium in class. When they talk, they have a very *high* voice. The high pitch is due to the high value of the speed of sound in helium. A similar trick is also possible with SF_6; it leads to a very *deep* voice. What is less well known is that these experiments are *dangerous*! Helium is not a poison. But due to its small atomic radius it diffuses rapidly. Various people got an embolism from performing the experiment, and a few have died. Never do it yourself; watch a youtube video instead.

**

When a child on a swing is pushed by an adult, the build-up of amplitude is due to (direct) resonance. When, on the other hand, the child itself sets the swing into motion,

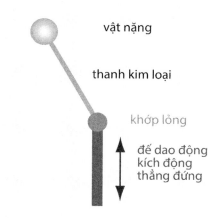

FIGURE 224 A vertically driven inverse pendulum is stable in the upright position at certain combinations of frequency and amplitude.

FIGURE 225 A particularly slow wave: a field of ski moguls (© Andreas Hallerbach).

it uses *twice* the natural frequency of the swing; this effect is called *parametric resonance*.

An astonishing effect of parametric resonance appears when an upside-down pendulum is attached to a vibrating base. Figure 224 shows the set-up; due to the joint, the mass is free to fall down on either side. Such a vertically driven inverse pendulum, sometimes also called a *Kapitza pendulum*, will remain firmly upright if the driving frequency of the joint is well chosen. For one of the many videos of the phenomenon, see www.youtube.com/watch?v=is_ejYsvAjY. Parametric resonance appears in many settings, including the sky. The Trojan asteroids are kept in orbit by parametric resonance.

* *

Also the bumps of skiing slopes, the so-called *ski moguls*, are waves: they move. Ski moguls are essential in many winter Olympic disciplines. Observation shows that ski moguls have a wavelength of typically 5 to 6 m and that they move with an average speed of 8 cm/day. Surprisingly, the speed is directed *upwards*, towards the top of the skiing slope. Can you explain why this is so? In fact, ski moguls are also an example of self-organization; this topic will be covered in more detail below.

* *

An orchestra is playing music in a large hall. At a distance of 30 m, somebody is listening to the music. At a distance of 3000 km, another person is listening to the music via the radio. Who hears the music first?

* *

What is the period of a simple pendulum, i.e., a mass m attached to a massless string of length l? What is the period if the string is much longer than the radius of the Earth?

* *

What path is followed by a body that moves without friction on a plane, but that is attached by a spring to a fixed point on the plane?

* *

The blue whale, *Balaenoptera musculus*, is the *loudest* animal found in nature: its voice can be heard at a distance of hundreds of kilometres.

* *

The exploration of sound in the sea, from the communication of whales to the sonar of dolphins, is a world of its own. As a start, explore the excellent www.dosits.org website.

* *

An interesting device that shows how rotation and oscillation are linked is the alarm *siren*. Find out how it works, and build one yourself.

* *

Jonathan Swift's Lilliputians are one twelfth of the size of humans. Show that the frequency of their voices must therefore be 144 times higher as that of humans, and thus be inaudible. Gulliver could not have heard what Lilliputians were saying. The same, most probably, would be true for Brobdingnagians, who were ten times taller than humans. Their sentences would also be a hundred times slower.

* *

Light is a wave, as we will discover later on. As a result, light reaching the Earth from space is refracted when it enters the atmosphere. Can you confirm that as a result, stars appear somewhat higher in the night sky than they really are?

* *

What are the highest sea waves? This question has been researched systematically only recently, using satellites. The surprising result is that sea waves with a height of 25 m and more are *common*: there are a few such waves on the oceans at any given time. This result confirms the rare stories of experienced ship captains and explains many otherwise unexplained ship sinkings.

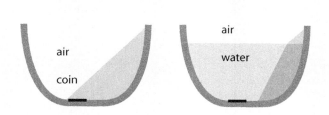

FIGURE 226 Shadows show the refraction of light.

Surfers may thus get many chances to ride 30 m waves. (The present record is just below this height.) But maybe the most impressive waves to surf are those of the Pororoca, a series of 4 m waves that move from the sea into the Amazon River every spring, against the flow of the river. These waves can be surfed for tens of kilometres.

* *

Interestingly, *every* water surface has waves, even if it seems completely flat. As a consequence of the finite temperature of water, its surface always has some roughness: there are *thermal* capillary waves. For water, with a surface tension of 72 mPa, the typical roughness at usual conditions is 0.2 nm. These thermal capillary waves, predicted since many centuries, have been observed only recently.

* *

All waves are damped, eventually. This effect is often frequency-dependent. Can you provide a confirmation of this dependence in the case of sound in air?

* *

When you make a hole with a needle in black paper, the hole can be used as a magnifying lens. (Try it.) Diffraction is responsible for the lens effect. By the way, the diffraction of light by holes was noted already by Francesco Grimaldi in the seventeenth century; he correctly deduced that *light is a wave*. His observations were later discussed by Newton, who wrongly dismissed them, because they did not fit his beliefs.

* *

Put an empty cup near a lamp, in such a way that the bottom of the cup remains in the shadow. When you fill the cup with water, some of the bottom will be lit, because of the refraction of the light from the lamp, as shown in Figure 226. The same effect allows us to build lenses. Refraction is thus at the basis of many optical instruments, such as the telescope or the microscope.

* *

Are water waves transverse or longitudinal?

* *

The speed of water waves limits the speed of ships. A surface ship cannot travel (much) faster than about $v_{crit} = \sqrt{0.16\,gl}$, where $g = 9.8\,\text{m/s}^2$, l is the boat's length, and 0.16 is a number determined experimentally, called the *critical Froude number*. This relation is valid for all vessels, from large tankers ($l = 100$ m gives $v_{crit} = 13$ m/s) down to ducks ($l = 0.3$ m gives $v_{crit} = 0.7$ m/s). The critical speed is that of a wave with the same wavelength as the ship. In fact, moving a ship at higher speeds than the critical value is possible, but requires much more energy. (A higher speed is also possible if the ship surfs on a wave.) How far away is the crawl *olympic swimming* record from the critical value?

Most water animals and ships are faster when they swim below the surface – where the limit due to surface waves does not exist – than when they swim on the surface. For example, ducks can swim three times as fast under water than on the surface.

* *

The group velocity of water waves (in deep water) is less than the velocity of the individual wave crests, the so-called *phase velocity*. As a result, when a group of wave crests travels, within the group the crests move from the back to the front: they appear at the back, travel forward and then die out at the front. The group velocity of water waves is *lower* than its phase velocity.

* *

We can hear the distant sea or a distant highway more clearly in the evening than in the morning. This is an effect of refraction. Sound speed increases with temperature. In the evening, the ground cools more quickly than the air above. As a result, sound leaving the ground and travelling upwards is refracted downwards, leading to the long hearing distance typical of evenings. In the morning, usually the air is cold above and warm below. Sound is refracted upwards, and distant sound does not reach a listener on the ground. Refraction thus implies that mornings are quiet, and that we can hear more distant sounds in the evenings. Elephants use the sound situation during evenings to communicate over distances of more than 10 km. (They also use sound waves in the ground to communicate, but that is another story.)

* *

Refraction also implies that there is a *sound channel* in the ocean, and one in the atmosphere. Sound speed increases with temperature, and increases with pressure. At an ocean depth of 1 km, or at an atmospheric height of 13 to 17 km (that is at the top of the tallest cumulonimbus clouds or equivalently, at the middle of the ozone layer) sound has *minimal* speed. As a result, sound that starts from that level and tries to leave is channelled back to it. Whales use this *sound!channel* to communicate with each other with beautiful songs; one can find recordings of these songs on the internet. The military successfully uses microphones placed at the sound channel in the ocean to locate submarines, and microphones on balloons in the atmospheric channel to listen for nuclear explosions.

In fact, sound experiments conducted by the military are the main reason why whales are deafened and lose their orientation, stranding on the shores. Similar experiments in the air with high-altitude balloons are often mistaken for flying saucers, as in the famous Roswell incident.

FIGURE 227 An artificial rogue wave – scaled down – created in a water tank (QuickTime film © Amin Chabchoub).

∗ ∗

Also small animals communicate by sound waves. In 2003, it was found that herring communicate using noises they produce when farting. When they pass wind, the gas creates a ticking sound whose frequency spectrum reaches up to 20 kHz. One can even listen to recordings of this sound on the internet. The details of the communication, such as the differences between males and females, are still being investigated. It is possible that the sounds may also be used by predators to detect herring, and they might even be used by future fishing vessels.

∗ ∗

Do plants produce sounds? Yes. Many plants, including pine trees and other trees, are known to produce low-power ultrasound when transporting sap; other plants, like corn, produce low-power audible sound in their roots.

Are plants sensitive to sound? Yes, they are. A large number of plants, including tomato plants, release pollen only when stimulated with a specific sound frequency emitted by their pollinators; in other plants, including corn again, roots seem to grow in the direction of certain sound sources.

But do plants communicate using sound? There is a remote possibility that root systems influence each other in this way; research is still ongoing. So far, no evidence was found. Maybe you can make a discovery, and jump-start the field of *phytoacoustics*?

∗ ∗

On windy seas, the white wave crests have several important effects. The noise stems from tiny exploding and imploding water bubbles. The noise of waves on the open sea is thus the superposition of many small explosions. At the same time, white crests are the events where the seas absorb carbon dioxide from the atmosphere, and thus reduce global warming.

∗ ∗

So-called *rogue waves* – also called *monster waves* or *freak waves* – are single waves on the open sea with a height of over 30 m that suddenly appear among much lower waves, have been a puzzling phenomenon for decades. For a long time it was not clear whether they really occurred. Only scattered reports by captains and mysteriously sunken ships pointed to their existence. Finally, from 1995 onwards, measurements started to confirm their existence. One reason for the scepticism was that the mechanism of their formation remained unclear. But experiments from 2010 onwards widened the understanding of non-linear water waves. These experiments first confirmed that under idealized conditions, water waves also show so-called *breather solutions*, or non-linear focussing. Finally, in 2014, Chabchoub and Fink managed to show, with a clever experimental technique based on time reversal, that non-linear focussing – including rogue waves – can appear in irregular water waves of much smaller amplitude. As Amin Chabchoub explains, the video proof looks like the one shown in Figure 227.

* *

Why are there many small holes in the ceilings of many office rooms?

* *

Which physical observable determines the wavelength of water waves emitted when a stone is thrown into a pond?

* *

Yakov Perelman lists the following four problems in his delightful physics problem book.
1. A stone falling into a lake produces circular waves. What is the shape of waves produced by a stone falling into a river, where the water flows?
2. It is possible to build a lens for sound, in the same way as it is possible to build lenses for light. What would such a lens look like?
3. What is the sound heard inside a shell?
4. Light takes about eight minutes to travel from the Sun to the Earth. What consequence does this have for the timing of sunrise?

* *

Typically, sound of a talking person produces a pressure variation of 20 mPa on the ear. How is this value determined?

The ear is indeed a sensitive device. As mentioned, most cases of sea mammals, like whales, swimming onto the shore are due to ear problems: usually some military device (either sonar signals or explosions) has destroyed their ear so that they became deaf and lose orientation.

* *

Why is the human ear, shown in Figure 228, so complex? The outer part, the *pinna* or *auricola*, concentrates the sound pressure at the tympanic membrane; it produces a gain of 3 dB. The tympanic membrane, or eardrum, is made in such a way as to always oscillate in fundamental mode, thus without any nodes. The tympanic membrane has a (very wide) resonance at 3 kHz, in the region where the ear is most sensitive. The eardrum transmits its motion, using the ossicles, into the inner ear. This mechanism thus transforms air waves into water waves in the inner ear, where they are detected. The efficiency

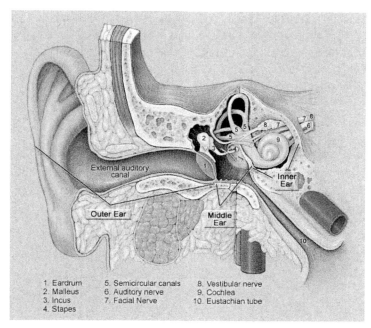

FIGURE 228 The human ear (© Northwestern University).

with which this transformation takes place is almost ideal; using the language of wave theory, ossicles are, above all, impedance transformers. Why does the ear transform air waves to water waves? Because water allows a smaller detector than air. Can you explain why?

* *

Infrasound, inaudible sound below 20 Hz, is a modern topic of research. In nature, infrasound is emitted by earthquakes, volcanic eruptions, wind, thunder, waterfalls, meteors and the surf. Glacier motion, seaquakes, avalanches and geomagnetic storms also emit infrasound. Human sources include missile launches, traffic, fuel engines and air compressors.

It is known that high intensities of infrasound lead to vomiting or disturbances of the sense of equilibrium (140 dB or more for 2 minutes), and even to death (170 dB for 10 minutes). The effects of lower intensities on human health are not yet known.

Infrasound can travel several times around the world before dying down, as the explosion of the Krakatoa volcano showed in 1883. With modern infrasound detectors, sea surf can be detected hundreds of kilometres away. Sea surf leads to a constant 'hum' of the Earth's crust at frequencies between 3 and 7 mHz. The *Global Infrasound Network* uses infrasound to detect nuclear weapon tests, earthquakes and volcanic eruptions, and can count meteors. Only very rarely can meteors be heard with the human ear. (See can-ndc.nrcan.gc.ca/is_infrasound-en.php.)

* *

The method used to deduce the sine waves contained in a signal, illustrated in Figure 205, is called the Fourier transformation. It is of importance throughout science and technology. In the 1980s, an interesting generalization became popular, called the *wavelet transformation*. In contrast to Fourier transformations, wavelet transformations allow us to localize signals in time. Wavelet transformations are used to compress digitally stored images in an efficient way, to diagnose aeroplane turbine problems, and in many other applications.

* *

If you like engineering challenges, here is one that is still open. How can one make a robust and efficient system that transforms the energy of sea waves into electricity?

* *

If you are interested in ocean waves, you might also enjoy the science of *oceanography*. For an introduction, see the open source textbooks at oceanworld.tamu.edu.

* *

What is the smallest structure in nature that has standing waves on its surface? The largest?

* *

In our description of extended bodies, we assumed that each spot of a body can be followed separately throughout its motion. Is this assumption justified? What would happen if it were not?

* *

A special type of waves appears in explosions and supersonic flight: *shock waves*. In a shock wave, the density or pressure of a gas changes abruptly, on distances of a few micrometers. Studying shock waves is a research field in itself; shock waves determine the flight of bullets, the snapping of whips and the effects of detonations.

Around a body moving with supersonic speed, the sound waves form a cone, as shown in Figure 229. When the cone passes an observer on the ground, the cone leads to a *sonic boom*. What is less well known is that the boom can be amplified. If an aeroplane *accelerates* through the sound barrier, certain observers at the ground will hear two booms or even a so-called superboom, because cones from various speeds can superpose at certain spots on the ground. A plane that performs certain manoeuvres, such as a curve at high speed, can even produce a superboom at a predefined spot on the ground. In contrast to normal sonic booms, superbooms can destroy windows, eardrums and lead to trauma, especially in children. Unfortunately, they are regularly produced on purpose by frustrated military pilots in various places of the world.

* *

What have swimming swans and ships have in common? The *wake* behind them. Despite the similarity, this phenomenon has no relation to the sonic boom. In fact, the angle of the wake is the *same* for ducks and ships: the angle is *independent* of the speed they travel or of the size of the moving body, provided the water is deep enough.

As explained above, water waves in *deep* water differ from sound waves: their group

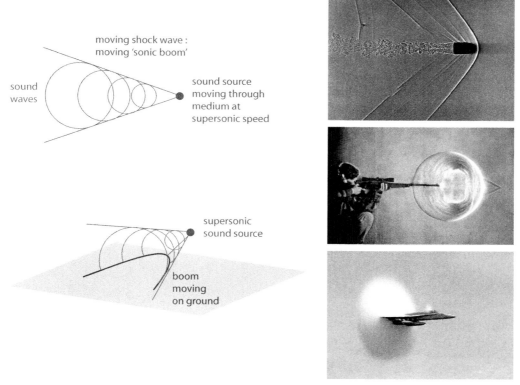

FIGURE 229 The shock wave created by a body in supersonic motion leads to a 'sonic boom' that moves through the air; it can be made visible by Schlieren photography or by water condensation (photo © Andrew Davidhazy, Gary Settles, NASA).

velocity is *one half* the phase velocity. (Can you deduce this from the dispersion relation $\omega = \sqrt{gk}$ between angular frequency and wave vector, valid for deep water waves?) Water waves will interfere where most of the energy is transported, thus around the group velocity. For this reason, in the graph shown in Figure 230, the diameter of each wave circle is always *half* the distance of their leftmost point O to the apex A. As a result, the half angle of the wake apex obeys

$$\sin \alpha = \tfrac{1}{3} \quad \text{giving a wake angle} \quad 2\alpha = 38.942° \, . \tag{107}$$

Figure 230 also allows deducing the curves that make up the wave pattern of the wake, using simple geometry.

It is essential to note that the fixed wake angle is valid only in *deep* water, i.e., only in water that is much deeper than the wavelength of the involved waves. In other words, for a given depth, the wake has the fixed shape only up to a maximum source speed. For high speeds, the wake angle narrows, and the pattern inside the wake changes.

* *

Bats fly at night using *echolocation*. Dolphins also use it. Sonar, used by fishing vessels

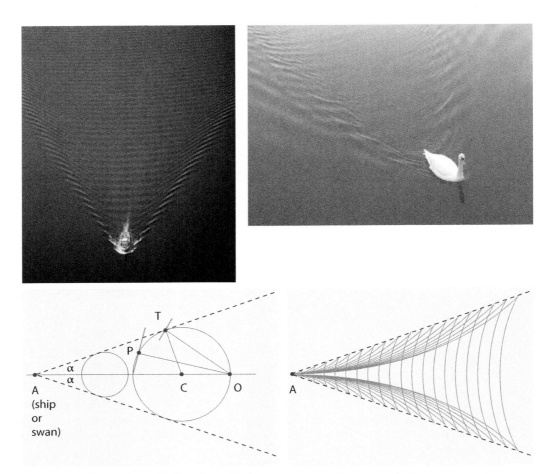

FIGURE 230 The wakes behind a ship and behind a swan, and the way to deduce the shape (photos © Wikimedia, Christopher Thorn).

to look for fish, copies the system of dolphins. Less well known is that humans have the same ability. Have you ever tried to echolocate a wall in a completely dark room? You will be surprised at how easily this is possible. Just make a loud hissing or whistling noise that stops abruptly, and listen to the echo. You will be able to locate walls reliably.

* *

Birds sing. If you want to explore how this happens, look at the impressive X-ray film of a singing bird found at the www.indiana.edu/~songbird/multi/cineradiography_index.html website.

* *

Every soliton is a one-dimensional structure. Do two-dimensional analogues exist? This issue was open for many years. Finally, in 1988, Boiti, Leon, Martina and Pempinelli found that a certain evolution equation, the so-called Davey–Stewartson equation, can have solutions that are localized in two dimensions. These results were generalized by

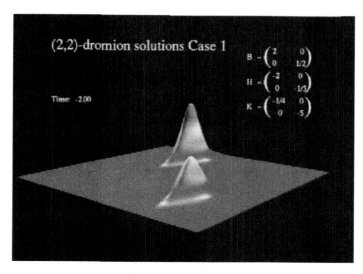

FIGURE 231 The calculated motion of a dromion across a two-dimensional substrate. (QuickTime film © Jarmo Hietarinta)

Fokas and Santini and further generalized by Hietarinta and Hirota. Such a solution is today called a *dromion*. Dromions are bumps that are localized in two dimensions and can move, without disappearing through diffusion, in non-linear systems. An example is shown in Figure 231. However, so far, no such solution has be observed in experiments; this is one of the most important experimental challenges left open in non-linear science.

* *

Water waves have not lost their interest up to this day. Most of all, two-dimensional solutions of solitonic water waves remain a topic of research. The experiments are simple, the mathematics is complicated and the issues are fascinating. In two dimensions, crests can even form hexagonal patterns! The relevant equation for shallow waves, the generalization of the Korteweg–de Vries equation to two dimensions, is called the *Kadomtsev–Petviashvili equation*. It leads to many unusual water waves, including cnoidal waves, solitons and dromions, some of which are shown in Figure 232. The issue of whether rectangular patterns exist is still open, and the exact equations and solutions for deep water waves are also unknown.

For moving water, waves are even more complex and show obvious phenomena, such as the Doppler effect, and less obvious ones, such as *bores*, i.e., the wave formed by the incoming tide in a river, and the subsequent *whelps*. Even a phenomenon as common as the water wave is still a field of research.

* *

How does the tone produced by blowing over a bottle depend on the dimension? For bottles that are bulky, the frequency f, the so-called *cavity resonance*, is found to depend

FIGURE 232 Unusual water waves in shallow water: (top) in an experimental water tank and in a storm in North Carolina, (bottom) an almost pure cnoidal wave near Panama and two such waves crossing at the Ile de Ré (photo © Diane Henderson, Anonymous, Wikimedia).

on the volume V of the bottle:

$$f = \frac{c}{2\pi}\sqrt{\frac{A}{VL}} \quad \text{or} \quad f \sim \frac{1}{\sqrt{V}} \tag{108}$$

where c is the speed of sound, A is the area of the opening, and L is the length of the neck of the bottle. Does the formula agree with your observations?

In fact, tone production is a complicated process, and specialized books exist on the topic. For example, when overblowing, a saxophone produces a second harmonic, an octave, whereas a clarinet produces a third harmonic, a quint (more precisely, a twelfth).

Why is this the case? The theory is complex, but the result simple: instruments whose cross-section increases along the tube, such as horns, trumpets, oboes or saxophones, overblow to octaves. For air instruments that have a (mostly) cylindrical tube, the effect of overblowing depends on the tone generation mechanism. Flutes overblow to the octave, but clarinets to the twelfth.

* *

Many acoustical systems do not only produce harmonics, but also subharmonics. There is a simple way to observe production of subharmonics: sing with your ears below water, in the bathtub. Depending on the air left in your ears, you can hear subharmonics of your own voice. The effect is quite special.

* *

The origin of the sound of cracking joints, for example the knuckles of the fingers, is a well-known puzzle. How would you test the conjecture that it is due to cavitation? What would you do to find out definitively?

* *

When a kilometre-deep hole is drilled into the Earth, and the sound at the bottom of the pit is studied, one finds that the deep Earth is filled with low-level, low-frequency sound. The oscillation periods can be longer than 300 s, i.e., with frequencies as low as 3 mHz. The majority of the sounds are due to the waves of the ocean. The sounds are modified in frequency by non-linear effects and can be detected even at the bottom of the sea and in the middle of continents. These sounds are now regularly used to search for oil and gas. The origins of specific sound bands are still subject of research; some might be related to the atmosphere or even to the motion of magma in the mantle and in volcanoes.

* *

All babies cry. Research showed that their sounds are composed of four basic patterns. It has been shown that the composition of the four patterns depends on the mother's language, even for newborn babies.

* *

Among the most impressive sound experiences are the singing performances of *countertenors* and of the even higher singing *male sopranos*. If you ever have the chance to hear one, do not miss the occasion.

Summary on waves and oscillations

In nature, apart from the motion of bodies, we observe also the motion of waves. Wave groups are often used as signals. Waves have energy, momentum and angular momentum. Waves exist in solids, liquids and gases, as well as along material interfaces. Waves can interfere, diffract, refract, disperse, dampen out and, if transverse, can be polarized. Solitary waves, i.e., waves with only one crest, are a special case of waves in media with specific non-linear dispersion.

Oscillations are a special case of waves; usually they are standing waves. Oscillation and waves only appear in extended systems.

Chapter 11
DO EXTENDED BODIES EXIST?
– LIMITS OF CONTINUITY

We have just discussed the motion of bodies that are extended. We have found that all extended bodies, be they solid of fluid, show wave motion. But are extended bodies actually found in nature? Strangely enough, this question has been one of the most intensely discussed questions in physics. Over the centuries, it has reappeared again and again, at each improvement of the description of motion; the answer has alternated between the affirmative and the negative. Many thinkers have been imprisoned, and many still are being persecuted, for giving answers that are not politically correct! In fact, the issue already arises in everyday life.

Mountains and fractals

Whenever we climb a mountain, we follow the outline of its shape. We usually describe this outline as a curved two-dimensional surface. But is this correct? There are alternative possibilities. A popular one is the idea that mountains might be fractal surfaces. A *fractal* was defined by Benoît Mandelbrot as a set that is self-similar under a countable but infinite number of magnification values. We have already encountered fractal lines. An example of an algorithm for building a (random) fractal *surface* is shown on the right side of Figure 233. It produces shapes which look remarkably similar to real mountains. The results are so realistic that they are used in Hollywood films. If this description were correct, mountains would be extended, but not continuous.

But mountains could also be fractals of a different sort, as shown in the left side of Figure 233. Mountain surfaces could have an infinity of small and smaller holes. In fact, we could also imagine that mountains are described as three-dimensional versions of the left side of the figure. Mountains would then be some sort of mathematical Swiss cheese. Can you devise an experiment to decide whether fractals provide the correct description for mountains? To settle the issue, we study chocolate bars, bones, trees and soap bubbles.

Can a chocolate bar last forever?

Any child knows how to make a chocolate bar last forever: eat half the remainder every day. However, this method only works if matter is scale-invariant. In other words, the method only works if matter is either *fractal*, as it then would be scale-invariant for a discrete set of zoom factors, or *continuous*, in which case it would be scale-invariant for any zoom factor. Which case, if either, applies to nature?

We have already encountered a fact making continuity a questionable assumption: continuity would allow us, as Banach and Tarski showed, to multiply food and any other

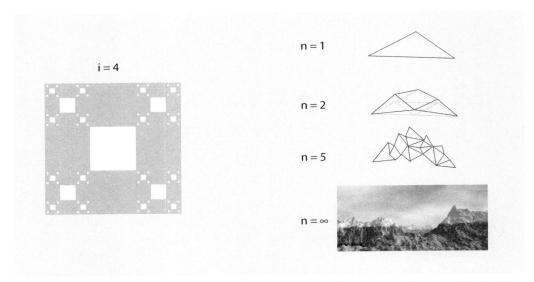

FIGURE 233 Floors (left) and mountains (right) could be fractals; for mountains this approximation is often used in computer graphics (image © Paul Martz).

matter by clever cutting and reassembling. Continuity would allow children to eat the *same* amount of chocolate every day, without ever buying a new bar.

▷ Matter is thus not continuous.

Now, fractal chocolate is not ruled out in this way; but other experiments settle the question. Indeed, we note that melted materials do not take up much smaller volumes than solid ones. We also find that even under the highest pressures, materials do not shrink. Thus we conclude again

▷ Matter is thus not a fractal.

What then is its structure?

To get an idea of the structure of matter we can take fluid chocolate, or even just some oil – which is the main ingredient of chocolate anyway – and spread it out over a large surface. For example, we can spread a drop of oil onto a pond on a day without rain or wind; it is not difficult to observe which parts of the water are covered by the oil and which are not: they change in reflection properties and the sky reflecte din the water has different colour on the water and on the oil. A small droplet of oil cannot cover a surface larger than – can you guess the value? The oil-covered water and the uncovered water have different colours. Trying to spread the oil film further inevitably *rips it apart*. The child's method of prolonging chocolate thus does not work for ever: it comes to a sudden end. The oil experiment, which can even be conducted at home, shows that there is a *minimum* thickness of oil films, with a value of about 2 nm. The experiment shows*

---

* The oil experiment was popularized by Thomson-Kelvin, a few decades after Loschmidt's determination

FIGURE 234 The spreading of a droplet of half a microlitre of oil, delivered with a micropipette, on a square metre of water covered with thin lycopodium powder (© Wolfgang Rueckner).

that there is a smallest size in oil. Oil is made of tiny components. Is this valid for *all* matter?

THE CASE OF GALILEO GALILEI

After the middle ages, Galileo (b. 1564 Pisa, d. 1642 Arcetri) was the first to state that *all* matter was made of smallest parts, which he called *piccolissimi quanti*, i.e., smallest quanta. Today, they are called *atoms*. However, Galileo paid dearly for this statement.

Indeed, during his life, Galileo was under attack for two reasons: because of his ideas on the motion of the Earth, and because of his ideas about atoms.* The discovery of the

---

of the size of molecules. It is often claimed that Benjamin Franklin was the first to conduct the oil experiment; that is wrong. Franklin did not measure the thickness, and did not even consider the question of the thickness. He did pour oil on water, but missed the most important conclusion that could be drawn from it. Even geniuses do not discover everything.

* To get a clear view of the matters of dispute in the case of Galileo, especially those of interest to physicists, the best text is the excellent book by PIETRO REDONDI, *Galileo eretico*, Einaudi, 1983, translated into

importance of both issues is the merit of the great historian Pietro Redondi, a collaborator of another great historian, Pierre Costabel. One of Redondi's research topics is the history of the dispute between the Jesuits, who at the time defended orthodox theology, and Galileo and the other scientists. In the 1980s, Redondi discovered a document of that time, an anonymous denunciation called G3, that allowed him to show that the condemnation of Galileo to life imprisonment for his views on the Earth's motion was organized by his friend the Pope to *protect* him from a sure condemnation to death over a different issue: atoms.

Galileo defended the view, explained in detail shortly, that since matter is not scale invariant, it must be made of 'atoms' or, as he called them, *piccolissimi quanti*. This was and still is a heresy, because atoms of matter contradict the central Catholic idea that in the Eucharist the *sensible qualities* of bread and wine exist independently of their *substance*. The distinction between substance and sensible qualities, introduced by Thomas Aquinas, is essential to make sense of *transubstantiation*, the change of bread and wine into human flesh and blood, which is a central tenet of the Catholic faith. Indeed, a true Catholic is still not allowed to believe in atoms to the present day, because the idea that matter is made of atoms contradicts transubstantiation. In contrast, protestant faith usually does not support transubstantiation and thus has no problems with atoms.

In Galileo's days, church tribunals punished heresy, i.e., personal opinions deviating from orthodox theology, by the death sentence. But Galileo was not sentenced to death. Galileo's life was saved by the Pope by making sure that the issue of transubstantiation would not be a topic of the trial, and by ensuring that the trial at the Inquisition be organized by a papal commission led by his nephew, Francesco Barberini. But the Pope also wanted Galileo to be punished, because he felt that his own ideas had been mocked in Galileo's book *Il Dialogo* and also because, under attack for his foreign policy, he was not able to ignore or suppress the issue.

As a result, in 1633 the seventy-year-old Galileo was condemned to a prison sentence, 'after invoking the name of Jesus Christ', for 'suspicion of heresy' (and thus not for heresy), because he did not comply with an earlier promise not to teach that the Earth moves. Indeed, the motion of the Earth contradicts what the Christian bible states. Galileo was convinced that *truth* was determined by observation, the Inquisition that it was determined by a book – and by itself. In many letters that Galileo wrote throughout his life he expressed his conviction that observational truth could never be a heresy. The trial showed him the opposite: he was forced to state that he erred in teaching that the Earth moves. After a while, the Pope reduced the prison sentence to house arrest.

Galileo's condemnation on the motion of the Earth was not the end of the story. In the years after Galileo's death, also atomism was condemned in several trials against Galileo's ideas and his followers. But the effects of these trials were not those planned by the Inquisition. Only twenty years after the famous trial, around 1650, every astronomer in the world was convinced of the motion of the Earth. And the result of the trials against atomism was that at the end of the 17th century, practically every scientist in the world was convinced that atoms exist. The trials accelerated an additional effect: after Galileo and Descartes, the centre of scientific research and innovation shifted from

---

English as *Galileo Heretic*, Princeton University Press, 1987. It is also available in many other languages; an updated edition that includes the newest discoveries appeared in 2004.

Catholic countries, like Italy or France, to protestant countries. In these, such as the Netherlands, England, Germany or the Scandinavian countries, the Inquisition had no power. This shift is still felt today.

It is a sad story that in 1992, the Catholic church did *not* revoke Galileo's condemnation. In that year, Pope John Paul II gave a speech on the Galileo case. Many years before, he had asked a study commission to re-evaluate the trial, because he wanted to express his regrets for what had happened and wanted to rehabilitate Galileo. The commission worked for twelve years. But the bishop that presented the final report was a crook: he avoided citing the results of the study commission, falsely stated the position of both parties on the subject of truth, falsely stated that Galileo's arguments on the motion of the Earth were weaker than those of the church, falsely summarized the past positions of the church on the motion of the Earth, avoided stating that prison sentences are not good arguments in issues of opinion or of heresy, made sure that rehabilitation was not even discussed, and of course, avoided any mention of transubstantiation. At the end of this power struggle, Galileo was thus *not* rehabilitated, in contrast to what the Pope wanted and in contrast to what most press releases of the time said; the Pope only stated that 'errors were made on both sides', and the crook behind all this was rewarded with a promotion.*

But that is not the end of the story. The documents of the trial, which were kept locked when Redondi made his discovery, were later made accessible to scholars by Pope John Paul II. In 1999, this led to the discovery of a new document, called EE 291, an internal expert opinion on the atom issue that was written for the trial in 1632, a few months before the start of the procedure. The author of the document comes to the conclusion that Galileo was indeed a heretic in the matter of atoms. The document thus proves that the cover-up of the transubstantiation issue during the trial of Galileo must have been systematic and thorough, as Redondi had deduced. Indeed, church officials and the Catholic catechism carefully avoid the subject of atoms even today; you can search the Vatican website www.vatican.va for any mention of them.

But Galileo did not want to attack transubstantiation; he wanted to advance the idea of atoms. And he did. Despite being condemned to prison in his trial, Galileo's last book, the *Discorsi*, written as a blind old man under house arrest, includes the arguments that imply the existence of atoms, or *piccolissimi quanti*. It is an irony of history that today, quantum theory, named by Max Born after the term used by Galileo for atoms, has become the most precise description of nature yet.

Galileo devoted his life to finding and telling the truth about motion and atoms. Let us explore how he concluded that all matter is made of atoms.

How high can animals jump?

Fleas can jump to heights a hundred times their size, humans only to heights about their own size. In fact, biological studies yield a simple observation: most animals, regardless of their size, achieve about the same jumping height, namely between 0.8 and 2.2 m, whether they are humans, cats, grasshoppers, apes, horses or leopards. We have explained this fact earlier on.

---

* We should not be too indignant: the same situation happens in many commercial companies every day; most industrial employees can tell similar stories.

At first sight, the observation of constant jumping height seems to be a simple example of scale invariance. But let us look more closely. There are some interesting exceptions at both ends of the mass range. At the small end, mites and other small insects do not achieve such heights because, like all small objects, they encounter the problem of air resistance. At the large end, elephants do not jump that high, because doing so would break their bones. But why do bones break at all?

Why are all humans of about the same size? Why are there no giant adults with a height of ten metres? Why aren't there any land animals larger than elephants? Galileo already gave the answer. The bones of which people and animals are made would not allow such changes of scale, as the bones of giants would collapse under the weight they have to sustain. A human scaled up by a factor 10 would weigh 1000 times as much, but its bones would only be 100 times as wide. But why do bones have a finite strength at all? There is only one explanation: because the constituents of bones stick to each other with a finite attraction. In contrast to bones, *continuous* matter – which exists only in cartoons – could not break at all, and *fractal* matter would be infinitely fragile. Galileo concluded:

> ▷ Matter breaks under finite loads because it is composed of small basic constituents.

Felling trees

Trees are fascinating structures. Take their size. Why do trees have limited size? Already in the sixteenth century, Galileo knew that it is not possible to increase tree height without limits: at some point a tree would not have the strength to support its own weight. He estimated the maximum height to be around 90 m; the actual record, unknown to him at the time, seems to be 150 m, for the Australian tree *Eucalyptus regnans*. But why does a limit exist at all? The answer is the same as for bones: wood has a finite strength because it is not scale invariant; and it is not scale invariant because it is made of small constituents, namely *atoms*.*

In fact, the derivation of the precise value of the height limit is more involved. Trees must not break under strong winds. Wind resistance limits the height-to-thickness ratio h/d to about 50 for normal-sized trees (for 0.2 m < d < 2 m). Can you say why? Thinner trees are limited in height to less than 10 m by the requirement that they return to the vertical after being bent by the wind.

Such studies of natural constraints also answer the question of why trees are made from wood and not, for example, from steel. You could check for yourself that the maximum height of a column of a given mass is determined by the ratio E/ρ^2 between the elastic module and the square of the mass density. For a long time, wood was actually the material for which this ratio was highest. Only recently have material scientists managed to engineer slightly better ratios: the fibre composites.

Why do materials break at all? All observations yield the same answer and confirm Galileo's reasoning: because there is a smallest size in materials. For example, bodies under stress are torn apart at the position at which their strength is minimal. If a body were completely homogeneous or continuous, it could not be torn apart; a crack could

---

* There is another important limiting factor: the water columns inside trees must not break. Both factors seem to yield similar limiting heights.

not start anywhere. If a body had a fractal, Swiss-cheese structure, cracks would have places to start, but they would need only an infinitesimal shock to do so.

Little hard balls

> I prefer knowing the cause of a single thing to being king of Persia.
>
> Democritus

Precise observations show that matter is neither continuous nor a fractal: all matter is made of smallest basic particles. Galileo, who deduced their existence by thinking about giants and trees, called them *smallest quanta*. Today they are called *atoms*, in honour of a famous argument of the ancient Greeks. Indeed, 2500 years ago, the Greeks asked the following question. If motion and matter are conserved, how can change and transformation exist? The philosophical school of Leucippus and Democritus of Abdera* studied two particular observations in special detail. They noted that *salt dissolves in water*. They also noted that *fish can swim in water*. In the first case, the volume of water does not increase when the salt is dissolved. In the second case, when fish advance, they must push water aside. Leucippus and Democritus deduced that there is only one possible explanation that satisfies these two observations and also reconciles conservation with transformation: nature is made of void and of small, indivisible and conserved particles.**

In short, since matter is hard, has a shape and is divisible, Leucippus and Democritus imagined it as being *made of atoms*. Atoms are particles which are hard, have a shape, but are indivisible. The Greek thus deduced that *every* example of motion, change and transformation is due to rearrangements of these particles; change and conservation are thus reconciled.

In other words, the Greeks imagined nature as a big Lego set. Lego pieces are first of all hard or *impenetrable*, i.e., repulsive at very small distances. Atoms thus explain why solids cannot be compressed much. Lego pieces are also *attractive* at small distances: they remain stuck together. Atoms this explain that solids exist. Finally, lego bricks have *no interaction* at large distances. Atoms thus explain the existence of gases. (Actually, what

---

* Leucippus of Elea (Λευκιππος) (*c.* 490 to *c.* 430 BCE), Greek philosopher; Elea was a small town south of Naples. It lies in Italy, but used to belong to the Magna Graecia. Democritus (Δεμοκριτος) of Abdera (*c.* 460 to *c.* 356 or 370 BCE), also a Greek philosopher, was arguably the greatest philosopher who ever lived. Together with his teacher Leucippus, he was the founder of the atomic theory; Democritus was a much admired thinker, and a contemporary of Socrates. The vain Plato never even mentions him, as Democritus was a danger to his own fame. Democritus wrote many books which all have been lost; they were not copied during the Middle Ages because of his scientific and rational world view, which was felt to be a danger by religious zealots who had the monopoly on the copying industry. Nowadays, it has become common to claim – incorrectly – that Democritus had no proof for the existence of atoms. That is a typical example of disinformation with the aim of making us feel superior to the ancients.

** The story is told by Lucretius, in full Titus Lucretius Carus, in his famous text *De rerum natura*, around 60 BCE. (An English translation can be found on www.perseus.tufts.edu/hopper/text?doc=Lucr.+1.1.) Lucretius relates many other proofs; in Book 1, he shows that there is vacuum in solids – as proven by porosity and by density differences – and in gases – as proven by wind. He shows that smells are due to particles, and that so is evaporation. (Can you find more proofs?) He also explains that the particles cannot be seen due to their small size, but that their effects can be felt and that they allow explaining all observations consistently.

Especially if we imagine particles as little balls, we cannot avoid calling this a typically male idea. (What would be the female approach?)

the Greeks called 'atoms' partly corresponds to what today we call 'molecules'. The latter term was introduced in 1811 by Amedeo Avogadro* in order to clarify the distinction. But we can forget this detail for the moment.)

Since atoms are invisible, it took many years before all scientists were convinced by the experiments showing their existence. In the nineteenth century, the idea of atoms was beautifully verified by a large number of experiments, such as the discovery of the 'laws' of chemistry and those of gas behaviour. We briefly explore the most interesting ones.

The sound of silence

Climbing the slopes of any mountain, we arrive in a region of the forest covered with deep snow. We stop for a minute and look around. It is already dark; all the animals are asleep; there is no wind and there are no sources of sound. We stand still, without breathing, and listen to the silence. (You can have the same experience also in a sound studio such as those used for musical recordings, or in a quiet bedroom at night.) In situations of complete silence, the ear automatically becomes more sensitive**; we then have a strange experience. We hear two noises, a lower- and a higher-pitched one, which are obviously generated inside the ear. Experiments show that the higher note is due to the activity of the nerve cells in the inner ear. The lower note is due to pulsating blood streaming through the head. But why do we hear a noise at all?

Many similar experiments confirm that whatever we do, we can never eliminate noise, i.e., random fluctuations, from measurements. This unavoidable type of random fluctuations is called *shot noise* in physics. The statistical properties of this type of noise actually correspond precisely to what would be expected if flows, instead of being motions of continuous matter, were transportation of a large number of equal, small and discrete entities. Therefore, the precise measurement of noise can be used to prove that air and liquids are made of molecules, that electric current is made of electrons, and even that light is made of photons. Shot noise is the sound of silence.

> ▷ Shot noise is the sound of atoms.

If matter were continuous or fractal, shot noise would not exist.

How to count what cannot be seen

In everyday life, atoms cannot be counted, mainly because they cannot be seen. Interestingly, the progress of physics allowed scholars to count atoms nevertheless. As mentioned, many of these methods use the measurement of noise.***

---

* Amedeo Avogadro (b. 1776 Turin, d. 1856 Turin) was an important physicist and chemist. Avogadro's number is named for him.
** The human ear can detect, in its most sensitive mode, pressure variations at least as small as 20 μPa and ear drum motions as small as 11 pm.
*** There are also various methods to count atoms by using electrolysis and determining the electron charge, by using radioactivity, X-ray scattering or by determining Planck's constant $\hbar$. We leave them aside here, because these methods actually *count* atoms. They are more precise, but also less interesting.

In physics, the term *noise* is not only used for the acoustical effect; it is used for *any process that is random*. The most famous kind of noise is Brownian motion, the motion of small particles, such as dust or pollen, floating in liquids. But small particles falling in air, such as mercury globules or selenium particles, and these fluctuations can be observed, for example with the help of air flows.

A mirror glued on a quartz fibre hanging in the air, especially at low pressure, changes orientation randomly, by small amounts, due to the collision of the air molecules. The random orientation changes, again a kind of noise, can be followed by reflecting a beam of light on the mirror and watching the light spot at a large distance.

Also density fluctuations, critical opalescence, and critical miscibility of liquids are forms of noise. It turns out that *every* kind of noise can be used to count atoms. The reason is that all noise in nature is related to the particle nature of matter or radiation. Indeed, all the mentioned methods have been used to count atoms and molecules, and to determine their sizes. Since the colour of the sky is a noise effect, we can indeed count air molecules by looking at the sky!

When Rayleigh explored scattering and the request of Maxwell, he deduced a relation between the scattering coefficient β, the index of refraction of air n and the particle density N/V:

$$\beta = 32\pi^3 \frac{(n-1)^2}{3\lambda^4 N/V} \tag{109}$$

The scattering coefficient is the inverse of the distance at which light is reduced in intensity by a factor $1/e$. Now, this distance can be estimated by determining the longest distance at which one can see a mountain in clear weather. Observation shows that this distance is at least 160 km. If we insert this observation in the equation, we can deduce a value for the number of molecules in air, and thus of Avogadro's constant!

The result of all precise measurements is that a *mol* of matter – for any gas, that is the amount of matter contained in 22.4 l of that gas at standard pressure – always contains the same number of atoms.

▷ One *mol* contains $6.0 \cdot 10^{23}$ particles.

The number is called *Avogadro's number*, after the first man who understood that volumes of gases have equal number of molecules, or *Loschmidt's number*, after the first man who measured it.* All the methods to determine Avogadro's number also allow us to deduce that most atoms have a size in the range between 0.1 and 0.3 nm. Molecules are composed of several or many atoms and are correspondingly larger.

How did Joseph Loschmidt** manage to be the first to determine his and Avogadro's number, and be the first to determine reliably the size of the components of matter?

Loschmidt knew that the dynamic viscosity μ of a gas was given by $\mu = \rho l v/3$, where ρ is the density of the gas, v the average speed of the components and l their mean free path. With Avogadro's prediction (made in 1811 without specifying any value) that a

---

* The term 'Loschmidt's number' is sometimes also used to designate the number of molecules in one cubic centimetre of gas.
** Joseph Loschmidt (b. 1821 Putschirn, d. 1895 Vienna) chemist and physicist.

FIGURE 235 Soap bubbles show visible effects of molecular size: before bursting, soap bubbles show small transparent spots; they appear black in this picture due to a black background. These spots are regions where the bubble has a thickness of only two molecules, with no liquid in between (© LordV).

volume V of any gas always contains the same number N of components, one also has $l = V/\sqrt{2\pi N\sigma^2}$, where σ is the cross-section of the components. (The cross-section is roughly the area of the shadow of an object.) Loschmidt then assumed that when the gas is liquefied, the volume of the liquid is the sum of the volumes of the particles. He then measured all the involved quantities, for mercury, and determined N. He thus determined the number of particles in one mole of matter, in one cubic centimetre of matter, and also the size of these particles.

Experiencing atoms

Matter is not continuous nor fractal. Matter contains smallest components with a characteristic size. Can we see effects of single atoms or molecules in everyday life? Yes, we can. We just need to watch *soap bubbles*. Soap bubbles have colours. But just before they burst, on the upper side of the bubble, the colours are interrupted by small transparent spots, as shown in Figure 235. Why? Inside a bubble, the liquid flows downwards, so that over time, the bubble gets thicker at the bottom and thinner at the top. After a while, in some regions all the liquid is gone, and in these regions, the bubble consists only of two molecular layers of soap molecules.

In fact, the arrangement of soap or oil molecules on water surfaces can be used to measure Avogadro's number, as mentioned. This has been done in various ingenious ways, and yields an extremely precise value with very simple means.

A simple experiment showing that solids have smallest components is shown in Fig-

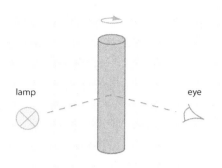
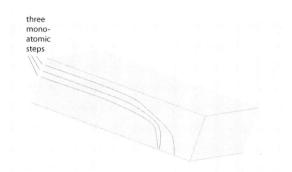

FIGURE 236 Atoms exist: rotating an illuminated, perfectly round single crystal aluminium rod leads to brightness oscillations because of the atoms that make it up

FIGURE 237 Atomic steps in broken gallium arsenide crystals (wafers) can be seen under a light microscope.

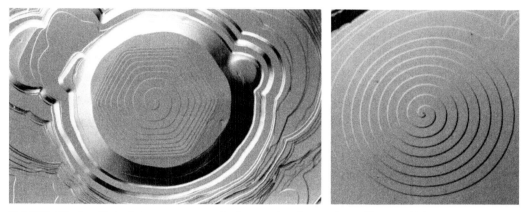

FIGURE 238 An effect of atoms: steps on single crystal surfaces – here silicon carbide grown on a carbon-terminated substrate (left) and on a silicon terminated substrate (right) observed in a simple light microscope (© Dietmar Siche).

ure 236. A cylindrical rod of pure, single crystal aluminium shows a surprising behaviour when it is illuminated from the side: its brightness depends on how the rod is oriented, even though it is completely round. This angular dependence is due to the atomic arrangement of the aluminium atoms in the rod.

It is not difficult to confirm experimentally the existence of smallest size in crystals. It is sufficient to break a single crystal, such as a gallium arsenide wafer, in two. The breaking surface is either completely flat or shows extremely small steps, as shown in Figure 237. These steps are visible under a normal light microscope. (Why?) Similarly, Figure 238 shows a defect that appeared in crystal growth. It turns out that all such step heights are multiples of a smallest height: its value is about 0.2 nm. The existence of a smallest height, corresponding to the height of an atom, contradicts all possibilities of scale invariance in matter.

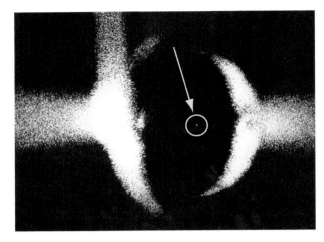

FIGURE 239 A single barium ion levitated in a Paul trap (image size around 2 mm) at the centre of the picture, visible also to the naked eye in the original experiment, performed in 1985 (© Werner Neuhauser).

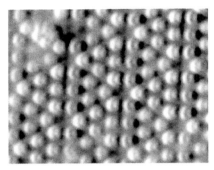

FIGURE 240 The atoms on the surface of a silicon crystal, mapped with an atomic force microscope (© Universität Augsburg)

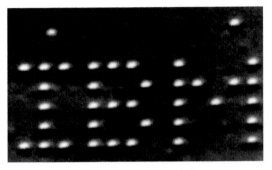

FIGURE 241 The result of moving helium atoms on a metallic surface. Both the moving and the imaging was performed with an atomic force microscope (© IBM).

Seeing atoms

Nowadays, with advances in technology, *single* atoms can be seen, photographed, hologrammed, counted, touched, moved, lifted, levitated and thrown around. And all these manipulations confirm that like everyday matter, atoms have mass, size, shape and colour. Single atoms have even been used as lamps and as lasers. Some experimental results are shown in Figure 239, Figure 240, Figure 240 and Figure 241.

The Greek imagined nature as a Lego set. And indeed, many modern researchers in several fields have fun playing with atoms in the same way that children play with Lego. A beautiful demonstration of these possibilities is provided by the many applications of the atomic force microscope. If you ever have the opportunity to use one, do not miss it! An atomic force microscope is a simple table-top device which follows the surface of an object with an atomically sharp needle;* such needles, usually of tungsten, are easily manufactured with a simple etching method. The changes in the height of the needle

---

* A cheap version costs only a few thousand euro, and will allow you to study the difference between a silicon wafer – crystalline – a flour wafer – granular-amorphous – and a consecrated wafer.

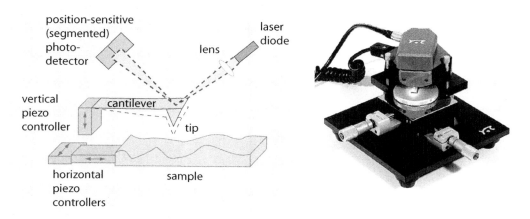

FIGURE 242 The principle and a realization of an atomic force microscope (photograph © Nanosurf).

along its path over the surface are recorded with the help of a deflected light ray, as shown in Figure 242. With a little care, the atoms of the object can be felt and made visible on a computer screen. With special types of such microscopes, the needle can be used to move atoms one by one to specified places on the surface. It is also possible to scan a surface, pick up a given atom and throw it towards a mass spectrometer to determine what sort of atom it is.

Incidentally, the construction of atomic force microscopes is only a small improvement on what nature is building already by the millions; when we use our ears to listen, we are actually detecting changes in eardrum position of down to 11 pm. In other words, we all have two 'atomic force microscopes' built into our heads.

Why is it useful to know that matter is made of atoms? Given only the *size* of atoms, we are able to understand and deduce many material properties. Atomic size determines the mass density, the elastic modulus, the surface tension, the thermal expansion coefficient, the heat of vaporization, the heat of fusion, the viscosity, the specific heat, the thermal diffusivity and the thermal conductivity. Just try.

Curiosities and fun challenges about solids and atoms

Glass is a solid. Nevertheless, many textbooks suggest that glass is a liquid. This error has been propagated for about a hundred years, probably originating from a mistranslation of a sentence in a German textbook published in 1933 by Gustav Tamman, *Der Glaszustand*. Can you give at least three reasons why glass is a solid and not a liquid?

* *

What is the maximum length of a vertically hanging wire? Could a wire be lowered from a suspended geostationary satellite down to the Earth? This would mean we could realize a space 'lift'. How long would the cable have to be? How heavy would it be? How would you build such a system? What dangers would it face?

* *

FIGURE 243 Rubik's cube: the complexity of simple three-dimensional motion (© Wikimedia).

2-Euro coin
cigarette
beer mat
glass

FIGURE 244 How can you move the coin into the glass without touching anything?

Every student probably knows *Rubik's cube*. Can you or did you deduce how Rubik built the cube without looking at its interior? Also for cubes with other numbers of segments? Is there a limit to the number of segments? These puzzles are even tougher than the search for the rearrangement of the cube.

* *

Arrange six matches in such a way that every match touches all the others.

* *

Physics is often good to win bets. See Figure 244 for a way to do so, due to Wolfgang Stalla.

* *

Matter is made of atoms. Over the centuries the stubborn resistance of many people to this idea has lead to the loss of many treasures. For over a thousand years, people thought that genuine pearls could be distinguished from false ones by hitting them with a hammer: only false pearls would break. However, *all* pearls break. (Also diamonds break in this situation.) Due to this belief, over the past centuries, all the most beautiful pearls in the world have been smashed to pieces.

* *

Comic books have difficulties with the concept of atoms. Could Asterix really throw Romans into the air using his fist? Are Lucky Luke's precise revolver shots possible? Can Spiderman's silk support him in his swings from building to building? Can the Roadrunner stop running in three steps? Can the Sun be made to stop in the sky by command? Can space-ships hover using fuel? Take any comic-book hero and ask yourself whether matter made of atoms would allow him the feats he seems capable of. You will find that most cartoons are comic precisely because they assume that matter is not made of atoms, but continuous! In a sense, atoms make life a serious adventure.

* *

Can humans start earthquakes? Yes. In fact, several strong earthquakes *have* been triggered by humans. This has happened when water dams have been filled, or when water has been injected into drilling holes. It has also been suggested that the extraction of deep underground water also causes earthquakes. If this is confirmed by future research, a sizeable proportion of all earthquakes could be human-triggered. Here is a simple question on the topic: What would happen if 1000 million Indians, triggered by a television programme, were to jump at the same time from the kitchen table to the floor?

* *

Many caves have *stalactites*. They form under two conditions: the water dripping from the ceiling of a cave must contain calcium carbonate, $CaCO_3$, and the difference between the carbon dioxide CO_2 concentrations in the water and in the air of the cave must be above a certain minimum value. If these conditions are fulfilled, calcareous sinter is deposited, and stalactites can form. How can the tip of a stalactite growing down from the ceiling be distinguished from the tip of a stalagmite rising from the floor? Does the difference exist also for icicles?

* *

Fractals do not exist. Which structures in nature approximate them most closely? One candidate is the lung, as shown in Figure 245. Its bronchi divide over and over, between 26 and 28 times. Each end then arrives at one of the 300 million *alveoli*, the 0.25 mm cavities in which oxygen is absorbed into the blood and carbon dioxide is expelled in to the air.

* *

How much more weight would your bathroom scales show if you stood on them in a vacuum?

* *

One of the most complex extended bodies is the human body. In modern simulations

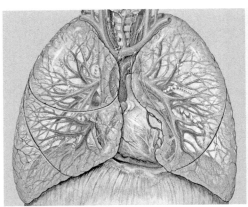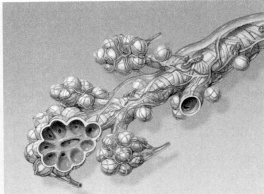

FIGURE 245 The human lung and its alveoli. (© Patrick J. Lynch, C. Carl Jaffe)

of the behaviour of humans in car accidents, the most advanced models include ribs, vertebrae, all other bones and the various organs. For each part, its specific deformation properties are taken into account. With such models and simulations, the protection of passengers and drivers in cars can be optimized.

* *

The human body is a remarkable structure. It is stiff and flexible, as the situation demands. Additionally, most stiff parts, the bones, are not attached to other stiff parts. Since a few years, artists and architects have started exploring such structures. An example of such a structure, a tower, is shown in Figure 246. It turns out that similar structures – sometimes called *tensegrity* structures – are good models for the human spine, for example. Just search the internet for more examples.

* *

The deepest hole ever drilled into the Earth is 12 km deep. In 2003, somebody proposed to enlarge such a hole and then to pour millions of tons of liquid iron into it. He claimed that the iron would sink towards the centre of the Earth. If a measurement device communication were dropped into the iron, it could send its observations to the surface using sound waves. Can you give some reasons why this would not work?

* *

The economic power of a nation has long been associated with its capacity to produce high-quality steel. Indeed, the Industrial Revolution started with the mass production of steel. Every scientist should know the basics facts about steel. *Steel* is a combination of iron and carbon to which other elements, mostly metals, may be added as well. One can distinguish three main types of steel, depending on the crystalline structure. *Ferritic steels* have a body-centred cubic structure, as shown in Figure 247, *austenitic steels* have a face-centred cubic structure, and *martensitic steels* have a body-centred tetragonal structure. Table 42 gives further details.

* *

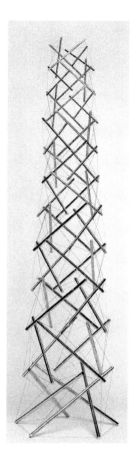

FIGURE 246 A stiff structure in which no rigid piece is attached to any other one (© Kenneth Snelson).

A simple phenomenon which requires a complex explanation is the cracking of a whip. Since the experimental work of Peter Krehl it has been known that the whip cracks when the tip reaches a velocity of *twice* the speed of sound. Can you imagine why?

* *

A bicycle chain is an extended object with no stiffness. However, if it is made to rotate rapidly, it acquires dynamical stiffness, and can roll down an inclined plane or along the floor. This surprising effect can be watched at the www.iwf.de/iwf/medien/infothek?Signatur=C+14825 website.

* *

Mechanical devices are not covered in this text. There is a lot of progress in the area even at present. For example, people have built robots that are able to ride a unicycle. But even the physics of human unicycling is not simple. Try it; it is an excellent exercise to stay young.

* *

TABLE 42 Steel types, properties and uses.

FERRITIC STEEL	AUSTENITIC STEEL	MARTENSITIC STEEL
Definition		
'usual' steel	'soft' steel	hardened steel, brittle
body centred cubic (bcc)	face centred cubic (fcc)	body centred tetragonal (bct)
iron and carbon	iron, chromium, nickel, manganese, carbon	carbon steel and alloys
Examples		
construction steel	most stainless (18/8 Cr/Ni) steels	knife edges
car sheet steel	kitchenware	drill surfaces
ship steel	food industry	spring steel, crankshafts
12 % Cr stainless ferrite	Cr/V steels for nuclear reactors	
Properties		
phases described by the iron-carbon phase diagram	phases described by the Schaeffler diagram	phases described by the iron-carbon diagram and the TTT (time–temperature transformation) diagram
in equilibrium at RT	some alloys in equilibrium at RT	not in equilibrium at RT, but stable
mechanical properties and grain size depend on heat treatment	mechanical properties and grain size depend on thermo-mechanical pre-treatment	mechanical properties and grain size strongly depend on heat treatment
hardened by reducing grain size, by forging, by increasing carbon content or by nitration	hardened by cold working only	hard anyway – made by laser irradiation, induction heating, etc.
grains of ferrite and paerlite, with cementite (Fe_3C)	grains of austenite	grains of martensite
ferromagnetic	not magnetic or weakly magnetic	ferromagnetic

There are many arguments against the existence of atoms as hard balls. Thomson-Kelvin put it in writing and spoke of "the monstrous assumption of infinitely strong and infinitely rigid pieces of matter". Even though Thomson was right in his comment, atoms do exist. Why?

* *

Sand has many surprising ways to move, and new discoveries are still made regularly. In 2001, Sigurdur Thoroddsen and Amy Shen discovered that a steel ball falling on a bed of sand produces, after the ball has sunk in, a *granular jet* that jumps out upwards from the sand. Figure 248 shows a sequence of photographs of the effect. The discovery has led to a stream of subsequent research.

FIGURE 247 Ferritic steels are bcc (body centred cubic), as shown by the famous *Atomium* in Brussels, a section of an iron crystal magnified to a height of over 100 m (photo and building are © Asbl Atomium Vzw – SABAM Belgium 2007).

* *

Engineering is not a part of this text. Nevertheless, it is an interesting topic. A few examples of what engineers do are shown in Figure 249 and Figure 250.

Summary on atoms

Matter is *not* scale invariant: in particular, it is neither smooth (continuous) nor fractal. There are no arbitrary small parts in matter. Everyday matter is made of countable components: *everyday matter is made of atoms*. This has been confirmed for all solids, liquids and gases. Pictures from atomic force microscopes show that the size and arrangement of atoms produce the *shape* and the *extension* of objects, confirming the Lego model of matter due to the ancient Greek. Different types of atoms, as well as their various combinations, produce different types of substances.

Studying matter in even more detail – as will be done later on – yields the now well-known idea that matter, at higher and higher magnifications, is made of molecules, atoms, nuclei, protons and neutrons, and finally, quarks. Atoms also contain electrons. A final type of matter, neutrino, is observed coming from the Sun and from certain types of radioactive materials. Even though the fundamental bricks have become somewhat smaller in the twentieth century, this will *not* happen in the future. The basic idea of the ancient Greek remains: matter is made of smallest entities, nowadays called *elementary particles*. In the parts of our adventure on quantum theory we will explore the consequences in detail. We will discover later on that the discreteness of matter is itself a consequence of the existence of a smallest change in nature.

Due to the existence of atoms, the description of everyday motion of extended objects can be reduced to the description of the motion of their atoms. Atomic motion will be

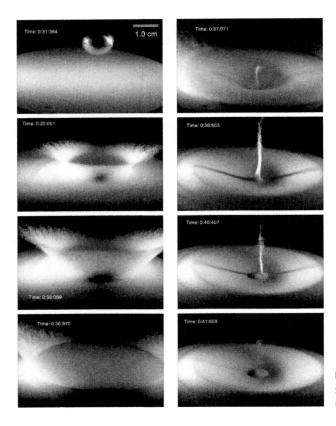

FIGURE 248 An example of a surprising motion of sand: granular jets (© Amy Shen).

FIGURE 249 Modern engineering highlights: a lithography machine for the production of integrated circuits and a paper machine (© ASML, Voith).

a major theme in the following pages. Two of its consequences are especially important: pressure and heat. We study them now.

LIMITS OF MATTER CONTINUITY

FIGURE 250 Sometimes unusual moving objects cross German roads (© RWE).

Chapter 12
FLUIDS AND THEIR MOTION

Fluids can be liquids or gases, including plasmas. And the motion of fluids can be exceedingly intricate, as Figure 251 shows. In fact, fluid motion is common and important – think about breathing, blood circulation or the weather. Exploring it is worthwhile.

What can move in nature? – Flows of all kinds

Before we continue with the exploration of fluids, we have a closer look at all the various types of motion in world around us. An overview about the types of motion in nature is given in Table 43. It lists motion in the various domains of natural sciences that belong to everyday life: motion of fluids, of matter, of matter types, of heat, of light and of charge.

All these types of motion are the domains of continuum physics. For each domain, the table gives the quantity whose motion is central to that domain. The moving quantity – be it mass, volume, entropy or charge – is described by an *extensive* observable, i.e., an observable that can accumulate. The motion of an extensive observable is called a *flow*. Every type of motion in nature is a flow. The extensive quantity is the 'stuff' that flows.

Within continuum physics, there are two domains we have not yet studied: the motion of charge and light, called *electrodynamics*, and the motion of heat, called *thermodynamics*. Once we have explored these domains, we will have completed the first step of our description of motion: continuum physics. In continuum physics, motion and moving entities are described with continuous quantities that can take any value, including arbitrarily small or arbitrarily large ones.

But nature is *not* continuous. We have already seen that matter cannot be indefinitely divided into ever-smaller entities. In fact, we will discover that there are precise experiments that provide limits to the observed values for *every* domain of continuum physics. We will discover that there are both lower and upper limits to mass, to speed, to angular momentum, to force, to entropy and to charge. The consequences of these discoveries lead to the next legs of our description of motion: relativity and quantum theory. Relativity is based on upper limits, quantum theory on lower limits. By the way, can you argue that all extensive quantities have a smallest value? The last leg of our description of motion will explore the unification of quantum theory and general relativity.

Every domain of physics, regardless of which one of the above legs it belongs to, describes change in terms an extensive quantity characteristic of the domain and energy. An observable quantity is called *extensive* if it increases with system size.

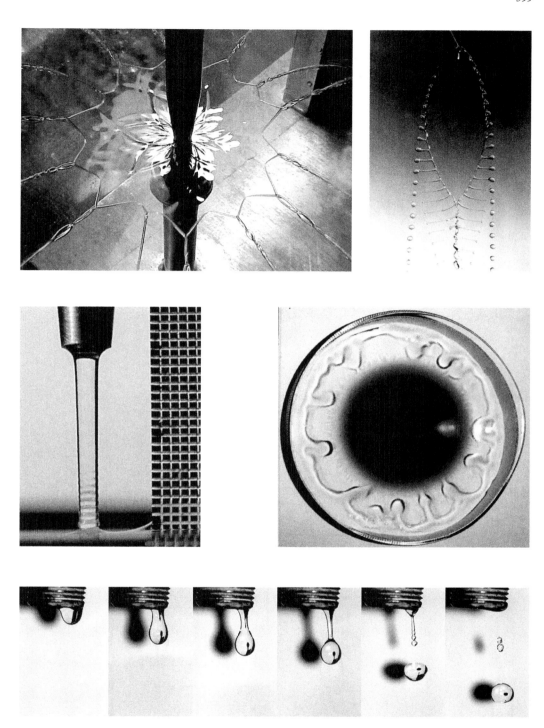

FIGURE 251 Examples of *laminar* fluid motion: a vertical water jet striking a horizontal impactor, two jets of a glycerol–water mixture colliding at an oblique angle, a water jet impinging on a reservoir in *stationary* motion, a glass of wine showing tears (all © John Bush, MIT) and a dripping water tap (© Andrew Davidhazy).

TABLE 43 Flows in nature: all extensive quantities, i.e., observable quantities that *accumulate* and thus move and *flow*.

Domain	Extensive Quantity (energy carrier)	Current (flow rate)	Intensive Quantity (driving strength)	Energy Flow (power)	Resistance to Transport (intensity of entropy generation)
Rivers	mass m	mass flow m/t	height (difference) gh	$P = gh\,m/t$	$R_m = ght/m$ [m²/s kg]
Gases	volume V	volume flow V/t	pressure p	$P = pV/t$	$R_V = pt/V$ [kg/s m⁵]
Mechanics	momentum $\mathbf{p}$	force $\mathbf{F} = d\mathbf{p}/dt$	velocity $\mathbf{v}$	$P = \mathbf{v}\mathbf{F}$	$R_p = \Delta V/F = t/m$ [s/kg]
	angular momentum $\mathbf{L}$	torque $\mathbf{M} = d\mathbf{L}/dt$	angular velocity $\boldsymbol{\omega}$	$P = \boldsymbol{\omega}\mathbf{M}$	$R_L = t/mr^2$ [s/kg m²]
Chemistry	amount of substance n	substance flow $I_n = dn/dt$	chemical potential μ	$P = \mu I_n$	$R_n = \mu t/n$ [Js/mol²]
Thermo-dynamics	entropy S	entropy flow $I_S = dS/dt$	temperature T	$P = T I_S$	$R_S = Tt/S$ [K²/W]
Electricity	charge q	electrical current $I = dq/dt$	electrical potential U	$P = UI$	$R = U/I$ [Ω]
Magnetism	no accumulable magnetic sources are found in nature				
Light	like all massless radiation, it can flow but cannot accumulate				
Nuclear physics	extensive quantities exist, but do not appear in everyday life				
Gravitation	empty space can move and flow, but the motion is not observed in everyday life				

▷ An *extensive* quantity describes what can flow in nature.

Examples are mass, volume, momentum, charge and entropy. Table 43 provides a full overview. In every flow there is also an physical observable that makes things move:

▷ An *intensive* quantity describes the driving strength of flow; it quantifies why it flows.

Examples are height difference, pressure, speed, potential, or temperature.

For every physical system, the *state* is described by a set of extensive and intensive ob-

12 FLUIDS AND THEIR MOTION

servables. Any complete description of motion and any complete observation of a system thus always requires both types of observables.

The extensive and intensive quantities – what flows and why it flows – for the case of *fluids* are volume and pressure. They are central to the description and the understanding of fluid flows. In the case of heat, the extensive and intensive quantities are *entropy* – the quantity that flows – and *temperature* – the quantity that makes heat flow. We will explore them shortly. We also note:

> ▷ The product of the extensive and intensive quantities is always energy per time, i.e., always the *power* of the corresponding flow.

The analogies of Table 43 can be carried even further. In all domains,

> ▷ The *capacity* of a system is defined by the extensive quantity divided by the intensive quantity.

The capacity measures how easy it is to let 'stuff' flow into a system. For electric charge, the capacity is what is usually called *electric capacity*. If the number is very large, it is very easy to add charge. For momentum, the capacity is called *mass*. Mass measures how easily momentum can be added to a system. Can you determine the quantities that measure capacity in the other cases?

In all fields it is possible to store energy by using the intensive quantity – such as $E = CU^2/2$ in a capacitor or $E = mv^2/2$ in a moving body – or by using the extensive quantity – such as $E = LI^2/2$ in a coil or $E = F^2/2k$ in a spring. Combining and alternating the two storage methods, we get oscillations.

> ▷ Every flow can produce oscillations.

Can you explain oscillations for the other cases of flow?

The state of a fluid

To describe motion means, first of all, to describe the state of the moving system. For most fluids, the state is described by specifying, at every point in space, the mass density, the velocity, the temperature and the pressure. We thus have two new observables: temperature, which we will explore in the next chapter, and pressure.

Here we concentrate on the other new observable:

> ▷ The ***pressure*** *p* at a point in a fluid is the force per area that a body of negligible size feels at that point: $p = F/A$.

Equivalently,

> ▷ ***Pressure*** is momentum change per area.

The unit of pressure is the pascal: 1 Pa is defined as $1\,\text{N/m}^2$. A selection of pressure values

FIGURE 252 Different pressure sensors: an antique barograph, an industrial water pressure sensor, a pressure sensitive resistor, the lateral line of a fish, and modern MEMS pressure sensors. (© Anonymous, ifm, Conrad, Piet Spaans, Proton Mikrotechnik)

TABLE 44 Some measured pressure values.

Observation	Pressure
Record negative pressure (tension) measured in water, after careful purification Ref. 270	-140 MPa $= -1400$ bar
Negative pressure measured in tree sap (xylem) Ref. 271, Ref. 254	up to -10 MPa $= -100$ bar
Negative pressure in gases	does not exist
Negative pressure in solids	is called tension
Record vacuum pressure achieved in laboratory	10 pPa (10^{-13} torr)
Pressure variation at hearing threshold	20 µPa
Pressure variation at hearing pain	100 Pa
Atmospheric pressure in La Paz, Bolivia	51 kPa
Atmospheric pressure in cruising passenger aircraft	75 kPa
Time-averaged pressure in pleural cavity in human thorax	0.5 kPa 5 mbar below atmospheric pressure
Standard sea-level atmospheric pressure	101.325 kPa or 1013.25 mbar or 760 torr
Healthy human arterial blood pressure at height of the heart: systolic, diastolic	17 kPa, 11 kPa above atmospheric pressure
Record pressure produced in laboratory, using a diamond anvil	$c.$ 200 GPa
Pressure at the centre of the Earth	$c.$ 370(20) GPa
Pressure at the centre of the Sun	$c.$ 24 PPa
Pressure at the centre of a neutron star	$c.\ 4 \cdot 10^{33}$ Pa
Planck pressure (maximum pressure possible in nature)	$4.6 \cdot 10^{113}$ Pa

found in nature is given in Table 44. Pressure is measured with the help of barometers or similar instruments, such as those shown in Figure 252. Also the human body is full of pressure sensors; the Merkel cells in the fingertips are examples. We will explore them later on.

Inside a fluid, pressure has no preferred direction. At the boundary of a fluid, pressure acts on the container wall. Pressure is not a simple property. Can you explain the observations of Figure 254? If the hydrostatic paradox – an effect of the so-called *communicating vases* – would not be valid, it would be easy to make perpetuum mobiles. Can

FIGURE 253 Daniel Bernoulli (1700–1782)

FIGURE 254 The hydrostatic and the hydrodynamic paradox (© IFE).

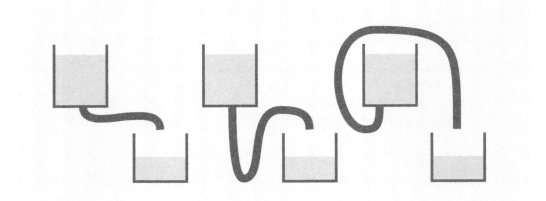

FIGURE 255 A puzzle: Which method of emptying a container is fastest? Does the method at the right hand side work at all?

you think about an example? Another puzzle about pressure is given in Figure 255.

The air around us has a considerable pressure, of the order of 100 kPa. As a result, it is not easy to make a vacuum; indeed, everyday forces are often too weak to overcome air pressure. This is known since several centuries, as Figure 256 shows. Your favorite

12 FLUIDS AND THEIR MOTION

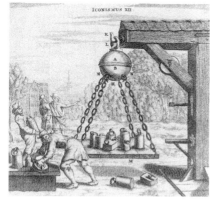

FIGURE 256 The pressure of air leads to surprisingly large forces, especially for large objects that enclose a vacuum. This was regularly demonstrated in the years from 1654 onwards by Otto von Guericke with the help of his so-called *Magdeburg hemispheres* and, above all, the various vacuum pumps that he invented (© Deutsche Post, Otto-von-Guericke-Gesellschaft, Deutsche Fotothek).

physics laboratory should possess a vacuum pump and a pair of (smaller) *Magdeburg hemispheres*; enjoy performing the experiment yourself.

Laminar and turbulent flow

Like all motion, fluid motion obeys energy conservation. For fluids in which no energy is transformed into heat, the conservation of energy is particularly simple. Motion that does not generate heat is motion without vortices; such fluid motion is called *laminar*. In the special case that, in addition, the speed of the fluid does not depend on time at all positions, the fluid motion is called *stationary*. Non-laminar flow is called *turbulent*. Figure 251 and Figure 257 show examples.

For motion that is both laminar and stationary, energy conservation can be expressed with the help of speed v and pressure p:

$$\tfrac{1}{2}\rho v^2 + p + \rho g h = \text{const} \tag{110}$$

where h is the height above ground, ρ is the mass density and $g = 9.8 \text{ m/s}^2$ is the gravitational acceleration. This is called *Bernoulli's equation*.* In this equation, the last term is

---

* Daniel Bernoulli (b. 1700 Bâle, d. 1782 Bâle), important mathematician and physicist. Also his father Jo-

FIGURE 257 Left: *non-stationary* and *stationary* laminar flows; right: an example of *turbulent* flow (© Martin Thum, Steve Butler).

only important if the fluid rises against ground. The first term is the kinetic energy (per volume) of the fluid, and the other two terms are potential energies (per volume). Indeed, the second term is the potential energy (per volume) resulting from the compression of the fluid. This is due to a further way to define pressure:

▷ *Pressure* is potential energy per volume.

Energy conservation implies that the lower the pressure is, the larger the speed of a fluid becomes. We can use this relation to measure the speed of a stationary water flow in a tube. We just have to narrow the tube somewhat at one location along the tube, and measure the pressure difference before and at the tube restriction. The speed v far from the constriction is then given as $v = k\sqrt{p_1 - p_2}$. (What is the constant k?) A device using this method is called a Venturi gauge.

Now think about flowing water. If the geometry is kept fixed and the water speed is increased – or the relative speed of a body in water is increased – at a certain speed we observe a transition: the water loses its clarity, the flow is not stationary and not laminar any more. We can observe the transition whenever we open a water tap: at a certain speed, the flow changes from laminar to turbulent. From this point onwards, Bernoulli's equation (110) is *not* valid any more.

The precise description of turbulence has not yet been achieved. This might be the toughest of all open problems in physics. When the young Werner Heisenberg was asked

---

hann and his uncle Jakob were famous mathematicians, as were his brothers and some of his nephews. Daniel Bernoulli published many mathematical and physical results. In physics, he studied the separation of compound motion into translation and rotation. In 1738 he published the *Hydrodynamique*, in which he deduced all results from a single principle, namely the conservation of energy. The so-called *Bernoulli equation* states how the pressure of a fluid decreases when its speed increases. He studied the tides and many complex mechanical problems, and explained the Boyle–Mariotte gas 'law'. For his publications he won the prestigious prize of the French Academy of Sciences – a forerunner of the Nobel Prize – ten times.

FIGURE 258 The *moth* sailing class: a 30 kg boat that sails above the water using *hydrofoils*, i.e., underwater wings (© Bladerider International).

to continue research on turbulence, he refused – rightly so – saying it was too difficult; he turned to something easier and he discovered and developed quantum theory instead. Turbulence is such a vast topic, with many of its concepts still not settled, that despite the number and importance of its applications, only now, at the beginning of the twenty-first century, are its secrets beginning to be unravelled.

It is thought that the equations of motion describing fluids in full generality, the so-called *Navier–Stokes equations*, are sufficient to understand turbulence.* But the mathematics behind these equations is mind-boggling. There is even a prize of one million dollars offered by the Clay Mathematics Institute for the completion of certain steps on the way to solving the equations.

Important systems which show laminar flow, vortices and turbulence at the same time are wings and sails. (See Figure 258.) All wings and sails work best in laminar mode. The essence of a wing is that it imparts air a downward velocity with as little turbulence as possible. (The aim to minimize turbulence is the reason that wings are curved. If the engine is very powerful, a flat wing at an angle also keeps an aeroplane in the air. However, the fuel consumption increases dramatically. On the other hand, strong turbulence is of advantage for landing safely.) Around a wing of a flying bird or aeroplane, the downward velocity of the trailing air leads to a centrifugal force acting on the air that passes above the wing. This leads to a lower pressure, and thus to lift. (Wings thus do *not* rely on the Bernoulli equation, where lower pressure *along* the flow leads to higher air speed, as unfortunately, many books used to say. Above a wing, the higher speed is related to lower pressure *across* the flow.)

The different speeds of the air above and below the wing lead to vortices at the end of every wing. These vortices are especially important for the take-off of any insect, bird and aeroplane. More aspects of wings are explored later on.

---

* They are named after Claude Navier (b. 1785 Dijon, d. 1836 Paris), important engineer and bridge builder, and Georges Gabriel Stokes (b. 1819 Skreen, d. 1903 Cambridge), important physicist and mathematician.

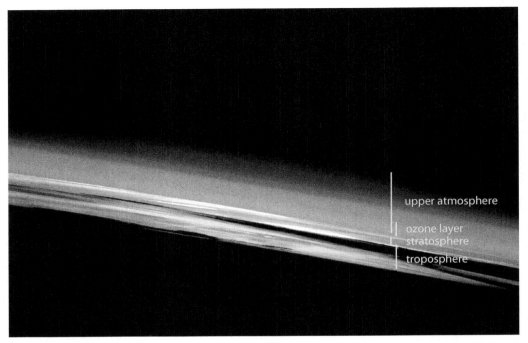

FIGURE 259 Several layers of the atmosphere are visble in this sunset photograph taken from the International Space Station, flying at several hundred km of altitude (courtesy NASA).

The atmosphere

The atmosphere, a thin veil around our planet that is shown in Figure 259, keeps us alive. The atmosphere is a fluid that surrounds the Earth and consists of $5 \cdot 10^{18}$ kg of gas. The density decreases with height: 50 % of the mass is below 5.6 km of height, 75 % within 11 km, 90 % within 16 km and 99,999 97 % within 100 km.*

At sea level, the atmospheric density is, on average, 1.29 kg/m^3 – about 1/800th of that of water – and the pressure is 101.3 kPa; both values decrease with altitude. The composition of the atmosphere at sea level is given on page 515. Also the composition varies with altitude; furthermore, the composition depends on the weather and on the pollution level.

The structure of the atmosphere is given in Table 45. The atmosphere ceases to behave as a gas above the thermopause, somewhere between 500 and 1000 km; above that altitude, there are no atomic collisions any more. In fact, we could argue that the atmosphere ceases to behave as an everyday gas above 150 km, when no audible sound is transmitted any more, not even at 20 Hz, due to the low atomic density.

---

* The last height is called the *Kármán line*; it is the conventional height at which a flying system cannot use lift to fly any more, so that it is often used as boundary between aeronautics and astronautics.

12 FLUIDS AND THEIR MOTION

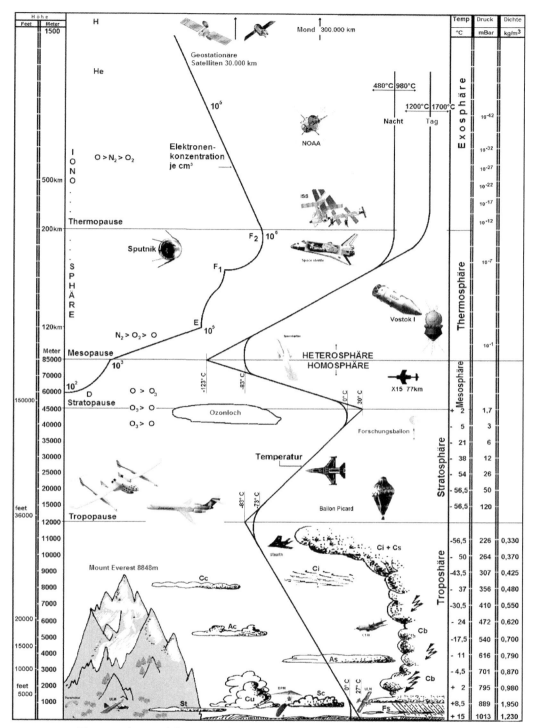

FIGURE 260 The main layers of the atmosphere (© Sebman81).

TABLE 45 The layers of the atmosphere.

Layer	Altitude	Details
Exosphere	> 500 to about 10 000 km	mainly composed of hydrogen and helium, includes the *magnetosphere*, temperature above 1000°C, contains many artificial satellites and sometimes aurora phenomena, includes, at its top, the luminous *geocorona*
Boundary: *thermopause* or *exobase*	between 500 and 1000 km	above: no 'gas' properties, no atomic collisions; below: gas properties, friction for satellites; altitude varies with solar activity
Thermosphere	from 85 km to thermopause	composed of oxygen, helium, hydrogen and ions, temperature of up to 2500°C, pressure 1 to 10 µPa; infrasound speed around 1000 m/s; no transmission of sound above 20 Hz at altitudes above 150 km; contains the International Space Station and many satellites; featured the Sputnik and the Space Shuttle
Heterosphere	all above turbopause	separate concept that includes all layers that show diffusive mixing, i.e., most of the thermosphere and the exosphere
Boundary: *turbopause* or *homopause*	100 km	boundary between diffusive mixing (above) and turbulent mixing (below)
Homosphere	everything below turbopause	separate concept that includes the lowest part of the thermosphere and all layers below it
Boundary: *mesopause*	85 km	temperature between −100°C and −85°C, lowest temperature 'on' Earth; temperature depends on season; contains ions, includes a sodium layer that is used to make guide stars for telescopes
Mesosphere	from stratopause to mesopause	temperature decreases with altitude, mostly hydrogen, contains noctilucent clouds, sprites, elves, ions; burns most meteors, shows atmospheric tides and a circulation from summer to winter pole
Ionosphere or *magnetosphere*	60 km to 1000 km	a separate concept that includes all layers that contain ions, thus the exosphere, the thermosphere and a large part of the mesosphere
Boundary: *stratopause* (or mesopeak)	50 to 55 km	maximum temperature between stratosphere and mesosphere; pressure around 100 Pa, temperature −15°C to −3°C
Stratosphere	up to the stratopause	stratified, no weather phenomena, temperature increases with altitude, dry, shows quasi-biennial oscillations, contains the *ozone layer* in its lowest 20 km, as well as aeroplanes and some balloons

TABLE 45 (Continued) The layers of the atmosphere.

LAYER	ALTITUDE	DETAILS
Boundary: *tropopause*	6 to 9 km at the poles, 17 to 20 km at the equator	temperature −50°C, temperature gradient vanishes, no water any more
Troposphere	up to the tropopause	contains water and shows weather phenomena; contains life, mountains and aeroplanes; makes stars flicker; temperature generally decreases with altitude; speed of sound is around 340 m/s
Boundary: *planetary boundary layer* or *peplosphere*	0.2 to 2 km	part of the troposphere that is influenced by friction with the Earth's surface; thickness depends on landscape and time of day

THE PHYSICS OF BLOOD AND BREATH

Fluid motion is of vital importance. There are at least four fluid circulation systems inside the human body. First, *blood* flows through the blood system by the heart. Second, *air* is circulated inside the lungs by the diaphragm and other chest muscles. Third, *lymph* flows through the lymphatic vessels, moved passively by body muscles. Fourth, the *cerebrospinal fluid* circulates around the brain and the spine, moved by motions of the head. For this reason, medical doctors like the simple statement: every illness is ultimately due to bad circulation.

Why do living beings have circulation systems? Circulation is necessary because diffusion is too slow. Can you detail the argument? We now explore the two main circulation systems in the human body.

Blood keeps us alive: it transports most chemicals required for our metabolism to and from the various parts of our body. The flow of blood is almost always laminar; turbulence only exists in the venae cavae, near the heart. The heart pumps around 80 ml of blood per heartbeat, about 5 l/min. At rest, a heartbeat consumes about 1.2 J. The consumption is sizeable, because the dynamic viscosity of blood ranges between $3.5 \cdot 10^{-3}$ Pa s (3.5 times higher than water) and 10^{-2} Pa s, depending on the diameter of the blood vessel; it is highest in the tiny capillaries. The speed of the blood is highest in the aorta, where it flows with 0.5 m/s, and lowest in the capillaries, where it is as low as 0.3 mm/s. As a result, a substance injected in the arm arrives in the feet between 20 and 60 s after the injection.

In fact, all animals have similar blood circulation speeds, usually between 0.2 m/s and 0.4 m/s. Why?

To achieve blood circulation, the heart produces a (systolic) pressure of about 16 kPa, corresponding to a height of about 1.6 m of blood. This value is needed by the heart to pump blood through the brain. When the heart relaxes, the elasticity of the arteries keeps the (diastolic) pressure at around 10 kPa. These values are measured at the height of the heart.* The values vary greatly with the position and body orientation at which they are

---

* The blood pressure values measured on the two upper arms also differ; for right handed people, the pressure in the right arm is higher.

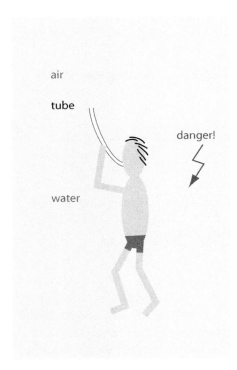

FIGURE 261 Attention, danger! Trying to do this will destroy your lung irreversibly and possibly kill you.

measured: the systolic pressure at the feet of a standing adult reaches 30 kPa, whereas it is 16 kPa in the feet of a lying person. For a standing human, the pressure in the veins in the foot is 18 kPa, larger than the systolic pressure in the heart. The high pressure values in the feet and legs is one of the reasons that leads to varicose veins. Nature uses many tricks to avoid problems with blood circulation in the legs. Humans leg veins have *valves* to avoid that the blood flows downwards; giraffes have extremely thin legs with strong and tight skin in the legs for the same reason. The same happens for other large animals.

At the end of the capillaries, the pressure is only around 2 kPa. The lowest blood pressure is found in veins that lead back from the head to the heart, where the pressure can even be slightly negative. Because of blood pressure, when a patient receives a (intravenous) *infusion*, the bag must have a minimum height above the infusion point where the needle enters the body; values of about 0.8 to 1 m cause no trouble. (Is the height difference also needed for person-to-person transfusions of blood?) Since arteries have higher blood pressure, for the more rare arterial infusions, hospitals usually use arterial pumps, to avoid the need for unpractical heights of 2 m or more.

Recent research has demonstrated what was suspected for a long time: in the capillaries, the red blood cells change shape and motion. The shape change depends on the capillary diameter and the flow speed. In larges vessels, red blood cells usually tumble in the blood stream. In smaller blood vessels, they roll and in still smaller ones they deform in various ways. These changes explain how blood flows more easily, i.e., with lower viscosity, in thinner vessels, It is even conjectured that disturbing this shape changes might be related to specific symptoms and illnesses.

12 FLUIDS AND THEIR MOTION

The physics of breathing is equally interesting. A human cannot breathe at any depth under water, even if he has a tube going to the surface, as shown in Figure 261. At a few metres of depth, trying to do so is inevitably *fatal*! Even at a depth of 50 cm only, the human body can only breathe in this way for a few minutes, and can get badly hurt for life. Why?

Inside the lungs, the gas exchange with the blood occurs in around 300 millions of little spheres, the *alveoli*, with a diameter between 0.2 and 0.6 mm. They are shown in Figure 245. To avoid that the large one grow and the small ones collapse – as in the experiment of Figure 276– the alveoli are covered with a phospholipid surfactant that reduces their surface tension. In newborns, the small radius of the alveoli and the low level of surfactant is the reason that the first breaths, and sometimes also the subsequent ones, require a large effort.

We need around 2 % of our energy for breathing alone. The speed of air in the throat is 3 m/s for normal breathing; when coughing, it can be as high as 50 m/s. The flow of air in the bronchi is turbulent; the noise can be heard in a quiet environment. In normal breathing, the breathing muscles, in the thorax and in the belly, exchange 0.5 l of air; in a deep breath, the volume can reach 4 l.

Breathing is especially tricky in unusual situations. After scuba diving* at larger depths than a few meters for more than a few minutes, it is important to rise slowly, to avoid a potentially fatal embolism. Why? The same can happen to participants in high altitude flights with balloons or aeroplanes, to high altitude parachutists and to cosmonauts.

Curiosities and fun challenges about fluids

What happens if people do not know the rules of nature? The answer is the same since 2000 years ago: taxpayer's money is wasted or health is in danger. One of the oldest examples, the aqueducts from Roman time, is shown in Figure 262. Aqueducts only exist because Romans did not know how fluids move. The figure tells why there are no aqueducts any more.

We note that using a 1 or 2 m water hose in the way shown in Figure 262 or in Figure 255 to transport gasoline can be dangerous. Why?

* *

Take an empty milk carton, and make a hole on one side, 1 cm above the bottom. Then make two holes above it, each 5 cm above the previous one. If you fill the carton with water and put it on a table, which of the three streams will reach further away? And if you put the carton on the edge on the table, so that the streams fall down on the floor?

* *

Your bathtub is full of water. You have an unmarked 3-litre container and an unmarked 5-litre container. How can you get 4 litres of water from the bathtub?

---

* Originally, 'scuba' is the abbreviation of 'self-contained underwater breathing apparatus'. The central device in it, the 'aqua lung', was invented by Emile Gagnan and Jacques Cousteau; it keeps the air pressure always at the same level as the water pressure.

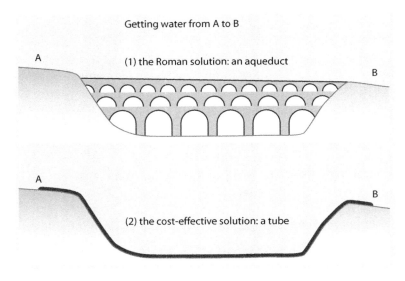

FIGURE 262 Wasting money because of lack of knowledge about fluids.

∗ ∗

What is the easiest way to create a supersonic jet of air? Simply drop a billiard ball into a bucket full of water. It took a long time to discover this simple method. Enjoy researching the topic.

∗ ∗

Fluids are important for motion. Spiders have muscles to flex their legs, but no muscles to extend them. How do they extend their legs?

In 1944, Ellis discovered that spiders extend their legs by hydraulic means: they increase the pressure of a fluid inside their legs; this pressure stretches the leg like the water pressure stiffens a garden hose. If you prefer, spider legs thus work a bit like the arm of an escavator. That is why spiders have bent legs when they are dead. The fluid mechanism works well: it is also used by jumping spiders.

∗ ∗

Where did the water in the oceans – whose amount is illustrated in Figure 263 – come from? Interestingly enough, this question is not fully settled! In the early age of the Earth, the high temperatures made all water evaporate and escape into space. So where did today's water come from? (For example, could the hydrogen come from the radioactivity of the Earth's core?) The most plausible proposal is that the water comes from comets. Comets are made, to a large degree, of ice. Comets hitting the Earth in the distant past seem have formed the oceans. In 2011, it was shown for the first time, by the Herschel infrared space telescope of the European Space Agency, that comets from the Kuiper belt – in contrast to comets from the inner Solar System – have ice of the same oxygen isotope composition as the Earth's oceans. The comet origin of oceans seems settled.

∗ ∗

The physics of under water diving, in particular the physics of apnoea diving, is full of

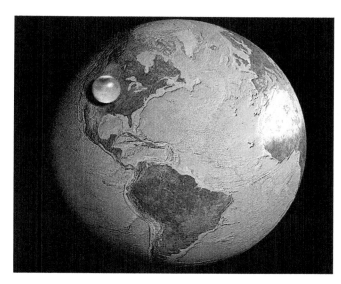

FIGURE 263 All the water on Earth would form a sphere with a radius of about 700 km, as illustrated in this computer-generated graphic. (© Jack Cook, Adam Nieman, Woods Hole Oceanographic Institution, Howard Perlman, USGS)

wonders and of effects that are not yet understood. For example, every apnoea champion knows that it is quite hard to hold the breath for five or six minutes while sitting in a chair. But if the same is done in a swimming pool, the feat becomes readily achievable for modern apnoea champions. It is still not fully clear why this is the case.

There are many apnoea diving disciplines. In 2009, the no-limit apnoea diving record is at the incredible depth of 214 m, achieved by Herbert Nitsch. The record static apnoea time is over eleven minutes, and, with hyperventilation with pure oxygen, over 22 minutes. The dynamic apnoea record, without fins, is 213 m.

When an apnoea diver reaches a depth of 100 m, the water pressure corresponds to a weight of over 11 kg on each square centimetre of his skin. To avoid the problems of ear pressure compensation at great depths, a diver has to flood the mouth and the trachea with water. His lungs have shrunk to one eleventh of their original size, to the size of apples. The water pressure shifts almost all blood from the legs and arms into the thorax and the brain. At 150 m, there is no light, and no sound – only the heart beat. And the heart beat is slow: there is only a beat every seven or eight seconds. He becomes relaxed and euphoric at the same time. None of these fascinating observations is fully understood.

Sperm whales, *Physeter macrocephalus*, can stay below water more than half an hour, and dive to a depth of more than 3000 m. Weddell seals, *Leptonychotes weddellii*, can stay below water for an hour and a half. The mechanisms are unclear, and but seem to involve haemoglobine and neuroglobine. The research into the involved mechanisms is interesting because it is observed that diving capability strengthens the brain. For example, bowhead whales, *Balaena mysticetus*, do not suffer strokes nor brain degeneration, even though they reach over 200 years in age.

* *

Apnoea records show the beneficial effects of oxygen on human health. An *oxygen bottle* is therefore a common item in professional first aid medical equipment.

*

What is the speed record for motion under water? Probably only few people know: it is a military secret. In fact, the answer needs to be split into two. The fastest published speed for a *projectile* under water, almost fully enclosed in a gas bubble, is 1550 m/s, faster than the speed of sound in water, achieved over a distance of a few metres in a military laboratory in the 1990s. The fastest system *with an engine* seems to be a torpedo, also moving mainly in a gas bubble, that reaches over 120 m/s, thus faster than any formula 1 racing car. The exact speed achieved is higher and secret, as the method of enclosing objects under water in gas bubbles, called *supercavitation*, is a research topic of military engineers all over the world.

The fastest fish, the sailfish *Istiophorus platypterus*, reaches 22 m/s, but speeds up to 30 m/s are suspected. Under water, the fastest *manned objects* are military submarines, whose speeds are secret, but believed to be around 21 m/s. (All military naval engineers in this world, with the enormous budgets they have, are not able to make submarines that are faster than fish. The reason that aeroplanes are faster than birds is evident: aeroplanes were not developed by military engineers, but by civilian engineers.) The fastest human-powered submarines reach around 4 m/s. We can estimate that if human-powered submarine developers had the same development budget as military engineers, their machines would probably be faster than nuclear submarines.

There are no record lists for swimming under water. Underwater swimming is known to be faster than above-water breast stroke, back stroke or dolphin stroke: that is the reason that swimming underwater over long distances is forbidden in competitions in these styles. However, it is not known whether crawl-style records are faster or slower than records for the fastest swimming style below water. Which one is faster in your own case?

*

How much water is necessary to moisten the air in a room in winter? At 0 °C, the vapour pressure of water is 6 mbar, 20 °C it is 23 mbar. As a result, heating air in the winter gives at most a humidity of 25 %. To increase the humidity by 50 %, about 1 litre of water per 100 m^3 is needed.

*

Surface tension can be dangerous. A man coming out of a swimming pool is wet. He carries about half a kilogram of water on his skin. In contrast, a wet insect, such as a house fly, carries many times its own weight. It is unable to fly and usually dies. Therefore, most insects stay away from water as much as they can – or at least use a long proboscis.

*

The human heart pumps blood at a rate of about 0.1 l/s. A typical *capillary* has the diameter of a red blood cell, around 7 μm, and in it the blood moves at a speed of half a millimetre per second. How many capillaries are there in a human?

*

You are in a boat on a pond with a stone, a bucket of water and a piece of wood. What happens to the water level of the pond after you throw the stone in it? After you throw

12 FLUIDS AND THEIR MOTION

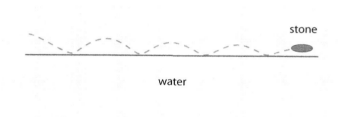

FIGURE 264 What is your personal stone-skipping record?

the water into the pond? After you throw the piece of wood?

* *

A ship leaves a river and enters the sea. What happens?

* *

Put a rubber air balloon over the end of a bottle and let it hang inside the bottle. How much can you blow up the balloon inside the bottle?

* *

Put a rubber helium balloon in your car. You accelerate and drive around bends. In which direction does the balloon move?

* *

Put a small paper ball into the neck of a horizontal bottle and try to blow it into the bottle. The paper will fly *towards* you. Why?

* *

It is possible to blow an egg from one egg-cup to a second one just behind it. Can you perform this trick?

* *

In the seventeenth century, engineers who needed to pump water faced a challenge. To pump water from mine shafts to the surface, no water pump managed more than 10 m of height difference. For twice that height, one always needed two pumps in series, connected by an intermediate reservoir. Why? How then do trees manage to pump water upwards for larger heights?

* *

When hydrogen and oxygen are combined to form water, the amount of hydrogen needed is exactly twice the amount of oxygen, if no gas is to be left over after the reaction. How does this observation confirm the existence of atoms?

* *

How are alcohol-filled chocolate pralines made? Note that the alcohol is not injected into them afterwards, because there would be no way to keep the result tight enough.

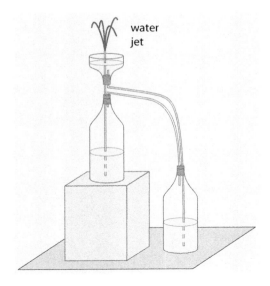

FIGURE 265 Heron's fountain in operation.

* *

How often can a stone jump when it is thrown over the surface of water? The present world record was achieved in 2002: 40 jumps. More information is known about the previous world record, achieved in 1992: a palm-sized, triangular and flat stone was thrown with a speed of 12 m/s (others say 20 m/s) and a rotation speed of about 14 revolutions per second along a river, covering about 100 m with 38 jumps. (The sequence was filmed with a video recorder from a bridge.)

What would be necessary to increase the number of jumps? Can you build a machine that is a better thrower than yourself?

* *

The most abundant component of air is nitrogen (about 78 %). The second component is oxygen (about 21 %). What is the third one?

* *

Which everyday system has a pressure lower than that of the atmosphere and usually kills a person if the pressure is raised to the usual atmospheric value?

* *

Water can flow uphill: Heron's fountain shows this most clearly. Heron of Alexandria (*c.* 10 to *c.* 70) described it 2000 years ago; it is easily built at home, using some plastic bottles and a little tubing. How does it work? How is it started?

* *

A light bulb is placed, underwater, in a stable steel cylinder with a diameter of 16 cm. An original Fiat Cinquecento car (500 kg) is placed on a piston pushing onto the water surface. Will the bulb resist?

FIGURE 266 Two wind measuring systems: a sodar system and a lidar system (© AQSystems, Leosphere).

∗ ∗

What is the most dense gas? The most dense vapour?

∗ ∗

Every year, the Institute of Maritime Systems of the Rostock, University of]University of Rostock organizes a contest. The challenge is to build a paper boat with the highest carrying capacity. The paper boat must weigh at most 10 g and fulfil a few additional conditions; the carrying capacity is measured by pouring small lead shot onto it, until the boat sinks. The 2008 record stands at 5.1 kg. Can you achieve this value? (For more information, see the www.paperboat.de website.)

∗ ∗

Is it possible to use the wind to move against the wind, head-on?

∗ ∗

Measuring wind speed is an important task. Two methods allow to measure the wind speed at an altitude of about 100 m above the ground: *sodar*, i.e., sound detection and ranging, and *lidar*, i.e., light detection and ranging. Two typical devices are shown in Figure 266. Sodar works also for clear air, whereas lidar needs aerosols.

∗ ∗

A modern version of an old question – originally posed by the physicist Daniel Colladon (b. 1802 Geneva, d. 1893 Geneva) – is the following. A ship of mass m in a river is pulled by horses walking along the river bank attached by ropes. If the river is of superfluid helium, meaning that there is no friction between ship and river, what energy is necessary to pull the ship upstream along the river until a height h has been gained?

∗ ∗

An urban legend pretends that at the bottom of large waterfalls there is not enough air to breathe. Why is this wrong?

∗ ∗

The Swiss physicist and inventor Auguste Piccard (b. 1884 Basel, d. 1962 Lausanne) was a famous explorer. Among others, he explored the stratosphere: he reached the record height of 16 km in his *aerostat*, a hydrogen gas balloon. Inside the airtight cabin hanging under his balloon, he had normal air pressure. However, he needed to introduce several ropes attached at the balloon into the cabin, in order to be able to pull and release them, as they controlled his balloon. How did he get the ropes into the cabin while at the same time preventing air from leaving?

∗ ∗

A human in air falls with a limiting speed of about 50 m/s (the precise value depends on clothing). How long does it take to fall from a plane at 3000 m down to a height of 200 m?

∗ ∗

To get an idea of the size of Avogadro's and Loschmidt's number, two questions are usually asked. First, on average, how many molecules or atoms that you breathe in with every breath have previously been exhaled by Caesar? Second, on average, how many atoms of Jesus do you eat every day? Even though the Earth is large, the resulting numbers are still telling.

∗ ∗

A few drops of tea usually flow along the underside of the spout of a teapot (or fall onto the table). This phenomenon has even been simulated using supercomputer simulations of the motion of liquids, by Kistler and Scriven, using the Navier–Stokes equations. Teapots are still shedding drops, though.

∗ ∗

The best giant soap bubbles can be made by mixing 1.5 l of water, 200 ml of corn syrup

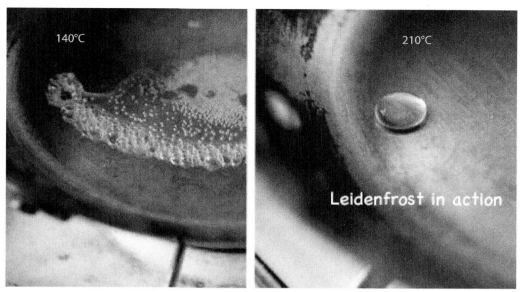

FIGURE 267 A water droplet on a pan: an example of the Leidenfrost effect (© Kenji Lopez-Alt).

and 450 ml of washing-up liquid. Mix everything together and then let it rest for four hours. You can then make the largest bubbles by dipping a metal ring of up to 100 mm diameter into the mixture. But why do soap bubbles burst?

* *

A drop of water that falls into a pan containing moderately hot oil evaporates immediately. However, if the oil is really hot, i.e., above 210°C, the water droplet dances on the oil surface for a considerable time. Cooks test the temperature of oil in this way. Why does this so-called *Leidenfrost effect* take place? The effect is named after the theologian and physician Johann Gottlob Leidenfrost (b. 1715 Rosperwenda, d. 1794 Duisburg). For an instructive and impressive demonstration of the Leidenfrost effect with water droplets, see the video featured at www.thisiscolossal.com/2014/03/water-maze/. The video also shows water droplets running uphill and running through a maze.

The Leidenfrost effect also allows one to plunge the bare hand into molten lead or liquid nitrogen, to keep liquid nitrogen in one's mouth, to check whether a pressing iron is hot, or to walk over hot coal – if one follows several safety rules, as explained by Jearl Walker. (Do *not* try this yourself! Many things can go wrong.) The main condition is that the hand, the mouth or the feet must be wet. Walker lost two teeth in a demonstration and badly burned his feet in a walk when the condition was not met. You can see some videos of the effect for a hand in liquid nitrogen on www.popsci.com/diy/article/2010-08/cool-hand-theo and for a finger in molten lead on www.popsci.com/science/article/2012-02/our-columnist-tests-his-trust-science-dipping-his-finger-molten-lead.

* *

Why don't air molecules fall towards the bottom of the container and stay there?

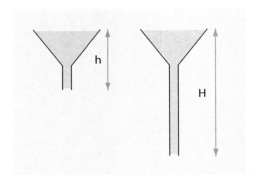

FIGURE 268 Which funnel empties more rapidly?

* *

Challenge 625 s
Ref. 282

Which of the two water funnels in Figure 268 is emptied more rapidly? Apply energy conservation to the fluid's motion (the Bernoulli equation) to find the answer.

* *

Challenge 626 s

As we have seen, fast flow generates an underpressure. How do fish prevent their eyes from popping when they swim rapidly?

* *

Challenge 627 ny

Golf balls have dimples for the same reasons that tennis balls are hairy and that shark and dolphin skin is not flat: deviations from flatness reduce the flow resistance because many small eddies produce less friction than a few large ones. Why?

* *

Challenge 628 s

The recognized record height reached by a helicopter is 12 442 m above sea level, though 12 954 m has also been claimed. (The first height was reached in 1972, the second in 2002, both by French pilots in French helicopters.) Why, then, do people still continue to use their legs in order to reach the top of Mount Sagarmatha, the highest mountain in the world?

* *

Challenge 629 e

A loosely knotted sewing thread lies on the surface of a bowl filled with water. Putting a bit of washing-up liquid into the area surrounded by the thread makes it immediately become circular. Why?

* *

Challenge 630 s

How can you put a handkerchief under water using a glass, while keeping it dry?

* *

Are you able to blow a ping-pong ball out of a funnel? What happens if you blow through a funnel towards a burning candle?

* *

FIGURE 269 A smoke ring, around 100 m in size, ejected from Mt. Etna's Bocca Nova in 2000 (© Daniela Szczepanski at www.vulkanarchiv.de and www.vulkane.net).

The fall of a leaf, with its complex path, is still a topic of investigation. We are far from being able to predict the time a leaf will take to reach the ground; the motion of the air around a leaf is not easy to describe. One of the simplest phenomena of hydrodynamics remains one of its most difficult problems.

∗ ∗

Fluids exhibit many interesting effects. Soap bubbles in air are made of a thin spherical film of liquid with air on both sides. In 1932, anti-bubbles, thin spherical films of air with liquid on both sides, were first observed. In 2004, the Belgian physicist Stéphane Dorbolo and his team showed that it is possible to produce them in simple experiments, and in particular, in Belgian beer.

∗ ∗

Have you ever dropped a Mentos candy into a Diet Coca Cola bottle? You will get an interesting effect. (Do it at your own risk...) Is it possible to build a rocket in this way?

∗ ∗

A needle can swim on water, if you put it there carefully. Just try, using a fork. Why does it float?

∗ ∗

The Rhine emits about 2 300 m³/s of water into the North Sea, the Amazon River about 120 000 m³/s into the Atlantic. How much is this less than $c^3/4G$?

∗ ∗

Fluids exhibit many complex motions. To see an overview, have a look at the beautiful

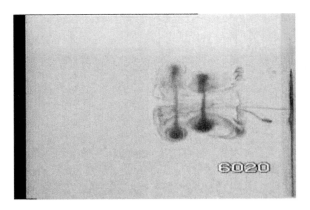

FIGURE 270 Two leapfrogging vortex rings. (QuickTime film © Lim Tee Tai)

collection on the website serve.me.nus.edu.sg/limtt. Among fluid motion, vortex rings, as emitted by smokers or volcanoes, have often triggered the imagination. (See Figure 269.) One of the most famous examples of fluid motion is the leapfrogging of vortex rings, shown in Figure 270. Lim Tee Tai explains that more than two leapfrogs are extremely hard to achieve, because the slightest vortex ring misalignment leads to the collapse of the system.

* *

A surprising effect can be observed when pouring *shampoo* on a plate: sometimes a thin stream is ejected from the region where the shampoo hits the plate. This so-called Kaye effect is best enjoyed in the beautiful movie produced by the University of Twente found on the youtube.com/watch?v=GX4_3cV_3Mw website.

* *

Most mammals take around 30 seconds to urinate. Can you find out why?

* *

Aeroplanes toilets are dangerous places. In the 1990s, a fat person sat on the toilet seat and pushed the 'flush' button while sitting. (Never try this yourself.) The underpressure exerted by the toilet was so strong that it pulled out the intestine and the person had to be brought into hospital. (Everything ended well, by the way.)

* *

If one surrounds water droplets with the correct type of dust, the droplets can roll along inclined planes. They can roll with a speed of up to 1 m/s, whereas on the same surface, water would flow hundred times more slowly. When the droplets get too fast, they become flat discs; at even higher speed, they get a doughnut shape. Such droplets can even jump and swim.

* *

It is well known that it is easier to ride a bicycle behind a truck, a car or a motorcycle,

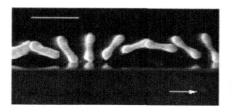 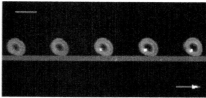

FIGURE 271 Water droplets covered with pollen rolling over inclined planes at 35 degrees (inclination not shown) have unexpected shapes. The grey lines have a length of 1 cm; the spherical droplets had initial radii of 1.3 mm and 2.5 mm, respectively, and the photographs were taken at 9 ms and 23 ms time intervals (© David Quéré).

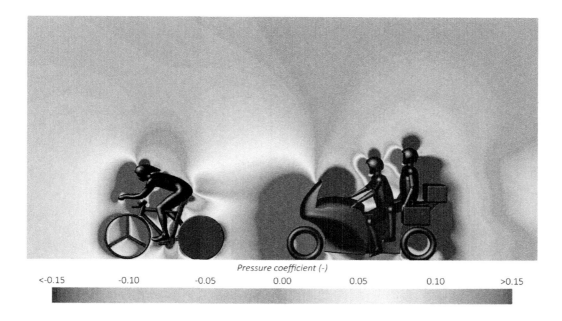

FIGURE 272 A computer visualization of the pressure regions surrounding a racing bicycle and a motorcycle following it (© Technical University Eindhoven).

because wind drag is lower in such situations. Indeed, the cycling speed record, achieved by Fred Rompelberg in 1990, is 74.7 m/s, or 268.8 km/h, and was achieved with a car driving in front of the bicycle. It is more surprising that a motorcycle driving *behind* a bicycle also reduces the drag, up to nine percent. In 2016, Bert Blocken and his team confirmed this effect in the wind tunnel and with numeric simulations. The effect is thus sufficient to determine the winner of racing prologues or time trials. In fact, three motorcycles riding in a row behind the bicycle can reduce the drag for the bicycle riding in front by up to 14 percent. The effect occurs because the following motorcycles cause a reduction of the low pressure region – shown in red in Figure 272 – behind the rider; this reduces the aerodynamic drag for the bicycle rider.

* *

All fluids can sustain vortices. Forexample, in air there are cyclones, tornados and sea hoses. But vortices also exist in the sea. They have a size between 100 m and 10 km; their short life last from a few hours up to a day. Sea vortices mix the water layers in the sea and they transport heat, nutrients and seaweed. Research on their properties and effects is just at its beginnings.

Summary on fluids

The motion of fluids is the motion of its constituent particles. The motion of fluids allows for swimming, flying, breathing, blood circulation, vortices and turbulence. Fluid motion can be *laminar* or *turbulent*. Laminar flow that lacks any internal friction is described by Bernoulli's equation, i.e., by energy conservation. Laminar flow with internal friction is beyond the scope of this text; so is turbulent flow. The exact description of turbulent fluid motion is the most complicated problem of physics and not yet fully solved.

Chapter 13
ON HEAT AND MOTION REVERSAL INVARIANCE

Spilled milk never returns into its container by itself. Any hot object, left alone, starts to cool down with time; it never heats up. These and many other observations how that numerous processes in nature are *irreversible*. On the other hand, for every a stone that flies along a path, it is possible to throw a stone that follows the path in the reverse direction. The motion of stones is *reversible*. Further observations show that irreversibility is only found in systems composed of a *many* particles, and that all irreversible systems involve *heat*.

Our everyday experience, including human warmth, ovens and stoves, shows that heat flows. *Heat moves.* But the lack of reversibility also shows that heat moves in a special way. Since heat appears in many-particle systems, we are led to explore the next global approach for the description of motion: *statistical physics.* Statistical physics, which includes *thermodynamics*, the study of heat and temperature, explains the origin of many material properties, and also the observed irreversibility of heat flow.

Does irreversibility mean that motion, at a fundamental level, is not invariant under reversal, as Nobel Prize winner Ilya Prigogine, one of the fathers of self-organization, thought? In this chapter we show that despite his other achievements, he was wrong.** To deduce this result, we first need to know the basic facts about temperature and heat; then we discuss irreversibility and motion reversal.

Temperature

Macroscopic bodies, i.e., bodies made of many atoms, have *temperature*. Only bodies made of few atoms do not have a temperature. Ovens have high temperature, refrigerators low temperature. Temperature changes have important effects: matter changes from solid to liquid to gaseous to plasma state. With a change in temperature, matter also changes size, colour, magnetic properties, stiffness and many other properties.

Temperature is an aspect of the *state* of a body. In other words, two otherwise identical bodies can be characterized and distinguished by their temperature. This is well-known to criminal organizations around the world that rig lotteries. When a blind-folded child is asked to draw a numbered ball from a set of such balls, such as in Figure 274, it is often told beforehand to draw only *hot* or *cold* balls. The blindfolding also helps to hide the

---

** Many even less serious thinkers often ask the question in the following term: is motion *time-invariant*? The cheap press goes even further, and asks whether motion has an 'arrow' or whether time has a preferred 'direction of flow'. We have already shown above that this is nonsense. We steer clear of such phrases in the following.

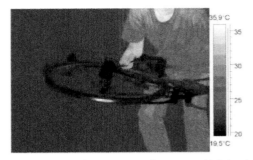

FIGURE 273 Braking generates heat on the floor and in the tire (© Klaus-Peter Möllmann and Michael Vollmer).

FIGURE 274 A rigged lottery shows that temperature is an aspect of the state of a body (© ISTA).

tears due to the pain.

The temperature of a macroscopic body is an aspect of its state. In particular, temperature is an *intensive* quantity or variable. In short, temperature describes the intensity of heat flow. An overview of measured temperature values is given in Table 46.

We observe that any two bodies in contact tend towards the *same* temperature: temperature is *contagious*. In other words, temperature describes a situation of *equilibrium*. Temperature thus behaves like pressure and any other intensive variable: it is the same for all parts of a system. We call *heating* the increase of temperature, and *cooling* its decrease.

How is temperature measured? The eighteenth century produced the clearest answer: temperature is best defined and measured by the *expansion of gases*. For the simplest, so-called *ideal* gases, the product of pressure p and volume V is proportional to temperature:

$$pV \sim T. \tag{111}$$

The proportionality constant is fixed by the *amount* of gas used. (More about it shortly.) The ideal gas relation allows us to determine temperature by measuring pressure and volume. This is the way (absolute) temperature has been defined and measured for about a century. To define the *unit* of temperature, we only have to fix the amount of gas used. It is customary to fix the amount of gas at 1 mol; for example, for oxygen this is 32 g. The proportionality constant for 1 mol, called the *ideal gas constant R*, is defined to be

$$R = 8.314\,4598(48)\,\text{J/mol K}\,. \tag{112}$$

This numerical value has been chosen in order to yield the best approximation to the independently defined Celsius temperature scale. Fixing the ideal gas constant in this way defines 1 K, or one Kelvin, as the unit of temperature.

In general, if we need to determine the temperature of an object, we thus need to take a mole of ideal gas, put it in contact with the object, wait a while, and then measure the pressure and the volume of the gas. The ideal gas relation (111) then gives the temperature of the object.

For every substance, a characteristic material property is the triple point. The *triple point* is the only temperature at which the solid, liquid, and gaseous phase coexists. For water, it lies at 0.01 °C (and at a partial vapour pressure of 611.657 Pa). Using the triple point of water, we can define the Kelvin in simple terms: a temperature increase of one Kelvin is the temperature increase that makes an ideal gas at the triple point of water increase in volume, keeping the pressure fixed, by a fraction of 1/273.16 or 0.366 09 %.

Most importantly, the ideal gas relation shows that there is a lowest temperature in nature, namely that temperature at which an ideal gas would have a vanishing volume. For an ideal gas, this would happen at $T = 0$ K. In reality, other effects, like the volume of the atoms themselves, prevent the volume of a real gas from ever reaching zero exactly.

▷ The unattainability of zero temperature is called the *third principle of thermodynamics*.

In fact, the temperature values achieved by a civilization can be used as a measure of its technological achievements. We can define in this way the *Bronze Age* (1.1 kK, 3500 BCE), the *Iron Age* (1.8 kK, 1000 BCE), the *Electric Age* (3 kK from *c.* 1880) and the *Atomic Age* (several MK, from 1944) in this way. Taking into account also the quest for lower temperatures, we can define the *Quantum Age* (4 K, starting 1908). All these thoughts lead to a simple question: what *exactly* are heating and cooling? What happens in these processes?

TABLE 46 Some temperature values.

OBSERVATION	TEMPERATURE
Lowest, but unattainable, temperature	0 K = −273.15 °C
In the context of lasers, it can make (almost) sense to talk about negative temperature.	
Temperature a perfect vacuum would have at Earth's surface	40 zK
Sodium gas in certain laboratory experiments – coldest matter system achieved by man and possibly in the universe	0.45 nK
Temperature of neutrino background in the universe	*c.* 2 K

TABLE 46 (Continued) Some temperature values.

Observation	Temperature
Temperature of (photon) cosmic background radiation in the universe	2.7 K
Liquid helium	4.2 K
Oxygen triple point	54.3584 K
Liquid nitrogen	77 K
Coldest weather ever measured (Antarctic)	185 K = −88 °C
Freezing point of water at standard pressure	273.15 K = 0.00 °C
Triple point of water	273.16 K = 0.01 °C
Average temperature of the Earth's surface	287.2 K
Smallest uncomfortable skin temperature	316 K (10 K above normal)
Interior of human body	310.0 ± 0.5 K = 36.8 ± 0.5 °C
Temperature of most land mammals	310 ± 3 K = 36.8 ± 2 °C
Hottest weather ever measured	343.8 K = 70.7 °C
Boiling point of water at standard pressure	373.13 K or 99.975 °C
Temperature of hottest living things: thermophile bacteria	395 K = 122 °C
Large wood fire, liquid bronze	c. 1100 K
Freezing point of gold	1337.33 K
Liquid, pure iron	1810 K
Bunsen burner flame	up to 1870 K
Light bulb filament	2.9 kK
Melting point of hafnium carbide	4.16 kK
Earth's centre	5(1) kK
Sun's surface	5.8 kK
Air in lightning bolt	30 kK
Hottest star's surface (centre of NGC 2240)	250 kK
Space between Earth and Moon (no typo)	up to 1 MK
Centre of white dwarf	5 to 20 MK
Sun's centre	20 MK
Centre of the accretion disc in X-ray binary stars	10 to 100 MK
Inside the JET fusion tokamak	100 MK
Centre of hottest stars	1 GK
Maximum temperature of systems without electron–positron pair generation	ca. 6 GK
Universe when it was 1 s old	100 GK
Hagedorn temperature	1.9 TK
Heavy ion collisions – highest man-made value	up to 3.6 TK
Planck temperature – nature's upper temperature limit	10^{32} K

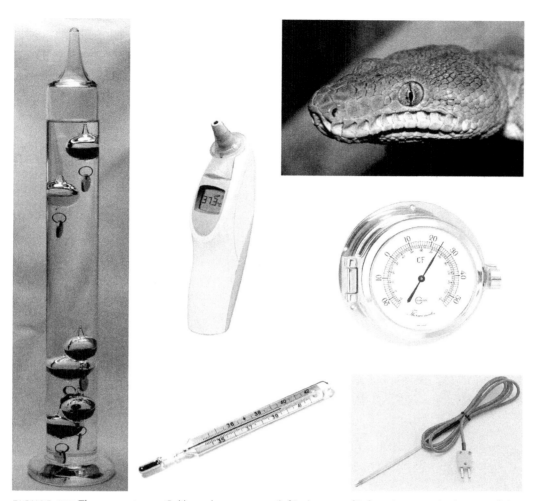

FIGURE 275 Thermometers: a Galilean thermometer (left), the row of infrared sensors in the jaw of the emerald tree boa *Corallus caninus*, an infrared thermometer to measure body temperature in the ear, a nautical thermometer using a bimetal, a mercury thermometer, and a thermocouple that is attached to a voltmeter for read-out (© Wikimedia, Ron Marcus, Braun GmbH, Universum, Wikimedia, Thermodevices).

Thermal energy

Around us, friction slows down moving bodies, and, while doing so, heats them up. The 'creation' of heat by friction can be observed in many experiments. An example is shown in Figure 273. Such experiments show that heat can be generated from friction, just by continuous rubbing, *without any limit*. This endless 'creation' of heat implies that heat is neither a material fluid nor a substance extracted from the body – which in this case would be consumed after a certain time – but something else. Indeed, today we know that heat, even though it behaves in some ways like a fluid, is due to disordered motion of particles. The conclusion from all these explorations is:

▷ Friction is the transformation of mechanical (i.e., ordered) energy into (disordered) *thermal energy*, i.e., into disordered motion of the particles making up a material.

Heating and cooling is thus the flow of disordered energy. In order to increase the temperature of 1 kg of water by 1 K using friction, 4.2 kJ of mechanical energy must be supplied. The first to measure this quantity with precision was, in 1842, the physician Julius Robert Mayer (b. 1814 Heilbronn, d. 1878 Heilbronn). He described his experiments as proofs of the conservation of energy; indeed, he was the first person to state energy conservation! It is something of an embarrassment to modern physics that a medical doctor was the first to show the conservation of energy, and furthermore, that he was ridiculed by most physicists of his time. Worse, conservation of energy was accepted by scientists only when it was publicized many years later by two authorities: Hermann von Helmholtz – himself also a physician turned physicist – and William Thomson, who also cited similar, but later experiments by James Joule.* All of them acknowledged Mayer's priority. Marketing by William Thomson eventually led to the naming of the unit of energy after Joule. In summary, two medical doctors proved to all experts on motion:

▷ In a closed system, the sum of mechanical energy and thermal energy is constant. This is called the *first principle of thermodynamics*.

Equivalently, it is impossible to produce mechanical energy without paying for it with some other form of energy. This is an important statement, because among others it means that humanity will stop living one day. Indeed, we live mostly on energy from the Sun; since the Sun is of finite size, its energy content will eventually be consumed. Can you estimate when this will happen?

The first principle of thermodynamics, the conservation of energy, implies:

▷ There is *no* perpetuum mobile 'of the first kind'.

In particular, no machine can run without energy input. For this very reason, we need food to eat: the energy in the food keeps us alive. If we stop eating, we die. The conservation of energy also makes most so-called 'wonders' impossible: in nature, energy cannot be created, but is conserved.

Thermal energy is a form of energy. Thermal energy can be stored, accumulated, transferred, transformed into mechanical energy, electrical energy or light. In short, thermal energy can be transformed into motion, into work, and thus into money.

The first principle of thermodynamics also allows us to formulate what a car engine achieves. Car engines are devices that transform hot matter – the hot exploding fuel inside the cylinders – into motion of the car wheels. Car engines, like steam engines, are thus examples of *heat engines*.

---

* Hermann von Helmholtz (b. 1821 Potsdam, d. 1894 Berlin), important scientist. William Thomson-Kelvin (b. 1824 Belfast, d. 1907 Netherhall), important physicist. James Prescott Joule (b. 1818 Salford, d. 1889 Sale), physicist. Joule is pronounced so that it rhymes with 'cool', as his descendants like to stress. (The pronunciation of the name 'Joule' varies from family to family.)

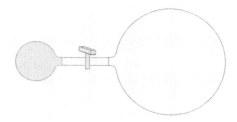

FIGURE 276 Which balloon wins when the tap is opened? Note: this is also how aneurisms grow in arteries.

The study of heat and temperature is called *thermostatics* if the systems concerned are at equilibrium, and *thermodynamics* if they are not. In the latter case, we distinguish situations *near* equilibrium, when equilibrium concepts such as temperature can still be used, from situations *far* from equilibrium, where such concepts often cannot be applied.

Does it make sense to distinguish between thermal energy and heat? It does. Many older texts use the term 'heat' to mean the *same* as thermal energy. However, this is confusing to students and experts:

> ▷ In this text, 'heat' is used, in accordance with modern approaches, as the everyday term for *entropy*.

Both thermal energy and heat flow from one body to another, and both accumulate. Both have no measurable mass.* Both the amount of thermal energy and the amount of heat inside a body increase with increasing temperature. The precise relation will be given shortly. But heat has many other interesting properties and stories to tell. Of these, two are particularly important: first, heat is due to particles; and secondly, heat is at the heart of the difference between past and future. These two stories are intertwined.

Why do balloons take up space? – The end of continuity

Heat properties are material-dependent. Studying thermal properties therefore should enable us to understand something about the constituents of matter. Now, the simplest materials of all are gases.** Gases need space: any amount of gas has pressure and volume. It did not take a long time to show that gases *could not* be continuous. One of the first scientists to think about gases as made up of atoms or molecules was Daniel Bernoulli. Bernoulli reasoned that if gases are made up of small particles, with mass and momentum, he should be able to make quantitative predictions about the behaviour of gases, and check them with experiment. If the particles fly around in a gas, then the *pressure* of a gas in a container is produced by the steady flow of particles hitting the wall. Bernoulli understood that if he reduced the volume to one half, the particles in the gas would need only to travel half as long to hit a wall: thus the pressure of the gas

---

* This might change in future, when mass measurements improve in precision, thus allowing the detection of relativistic effects. In this case, temperature increase may be detected through its related mass increase. However, such changes are noticeable only with twelve or more digits of precision in mass measurements.
** By the way, the word *gas* is a modern construct. It was coined by the alchemist and physician Johan Baptista van Helmont (b. 1579 Brussels, d. 1644 Vilvoorde), to sound similar to 'chaos'. It is one of the few words which have been invented by one person and then adopted all over the world.

FIGURE 277 What happened here? (© Johan de Jong)

would double. He also understood that if the temperature of a gas is increased while its volume is kept constant, the speed of the particles would increase. Combining these results, Bernoulli concluded that if the particles are assumed to behave as tiny, hard and perfectly elastic balls, the pressure p, the volume V and the temperature T must be related by

$$pV = nRT = kNT \ . \tag{113}$$

In this so-called *ideal gas relation*, R is the *ideal gas constant*; it has the value $R = 8.314\,4598(48)$ J/(mol K) and n is the number of moles. In the alternative writing of the gas relation, N is the number of particles contained in the gas and k is the Boltzmann constant, one of the fundamental constants of nature, is given by $k = 1.380\,648\,52(79) \cdot 10^{-23}$ J/K. (More about this constant is told below.) A gas made of particles with such a textbook behaviour is called an *ideal gas*. Relation (113), often also called the *ideal gas 'law'*, was known before Bernoulli; the relation has been confirmed by experiments at room and higher temperatures, for all known gases.

Bernoulli derived the ideal gas relation, with a specific prediction for the proportionality constant R, from the single assumption that gases are made of small particles with mass. This derivation provides a clear argument for the existence of atoms and for their behaviour as normal, though small objects. And indeed, we have already seen above how N can be determined experimentally.

The ideal gas model helps us to answer questions such as the one illustrated in Figure 276. Two *identical* rubber balloons, one filled up to a larger size than the other, are connected via a pipe and a valve. The valve is opened. Which one deflates?

The ideal gas relation states that hotter gases, at given pressure, need more volume. The relation thus explains why winds and storms exist, why hot air balloons rise – even those of Figure 277 – why car engines work, why the ozone layer is destroyed by certain gases, or why during the extremely hot summer of 2001 in the south of Turkey, oxygen masks were necessary to walk outside during the day.

The ideal gas relation also explains why on the 21st of August 1986, over a thousand people and three-thousand livestock where found dead in their homes in Cameroon. They were living below a volcano whose crater contains a lake, *Lake Nyos*. It turns out that the volcano continuously emits carbon dioxide, or CO_2, into the lake. The carbon dioxide is usually dissolved in the water. But in August 1986, an unknown event triggered the release of a bubble of around one million tons of CO_2, about a cubic kilometre, into the atmosphere. Because carbon dioxide ($2.0\,\text{kg/m}^3$) is denser than air ($1.2\,\text{kg/m}^3$), the gas flowed down into the valleys and villages below the volcano. The gas has no colour and smell, and it leads to asphyxiation. It is unclear whether the outgassing system installed in the lake after the event is sufficiently powerful to avoid a recurrence of the event.

Using the ideal gas relation you are now able to explain why balloons increase in size as they rise high up in the atmosphere, even though the air is colder there. The largest balloon built so far had a diameter, at high altitude, of 170 m, but only a fraction of that value at take-off. How much?

Now you can also take up the following challenge: how can you measure the weight of a car or a bicycle with a ruler only?

The picture of gases as being made of hard constituents without any long-distance interactions breaks down at very low temperatures. However, the ideal gas relation (113) can be improved to overcome these limitations, by taking into account the deviations due to interactions between atoms or molecules. This approach is now standard practice and allows us to measure temperatures even at extremely low values. The effects observed below 80 K, such as the solidification of air, frictionless transport of electrical current, or frictionless flow of liquids, form a fascinating world of their own, the beautiful domain of low-temperature physics. The field will be explored later on.

Not long after Bernoulli, chemists found strong arguments confirming the existence of atoms. They discovered that chemical reactions occur under 'fixed proportions': only specific ratios of amounts of chemicals react. Many researchers, including John Dalton, deduced that this property occurs because in chemistry, all reactions occur atom by atom. For example, two hydrogen atoms and one oxygen atoms form one water molecule in this way – even though these terms did not exist at the time. The relation is expressed by the chemical formula H_2O. These arguments are strong, but did not convince everybody. Finally, the existence of atoms was confirmed by observing the effects of their motion even more directly.

Brownian motion

If fluids are made of particles moving randomly, this random motion should have observable effects. An example of the observed motion is shown in Figure 280. The particles seem to follow a random zig zag movement. The first description is by Lucretius, in the year 60 BCE, in his poem *De rerum natura*. In it, Lucretius tells about a common observation: in air that is illuminated by the Sun, dust partciles seem to dance.

In 1785, Jan Ingenhousz saw that coal dust particles never come to rest. Indeed, under a microscope it is easy to observe that coal dust or other small particles in or on a liquid never come to rest. Ingenhousz discovered what is called *Brownian motion* today. 40 years after him, the botanist Robert Brown was the first Englishman to repeat the observation, this time for small particles floating in vacuoles *inside* pollen. Further ex-

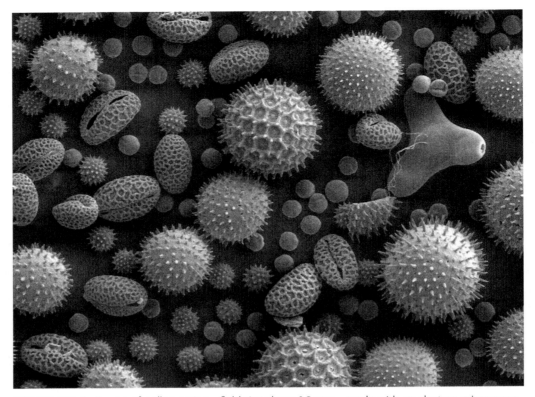

FIGURE 278 An image of pollen grains – field size about 0.3 mm – made with an electron microscope (Dartmouth College Electron Microscope Facility).

periments by many other researchers showed that the observation of a random motion is independent of the type of particle and of the type of liquid. In other words, Ingenhousz had discovered a fundamental form of *noise* in nature.

Around 1860, the random motion of particles in liquids was attributed by various researchers to the molecules of the liquid that were colliding with the particles. In 1905 and 1906, Marian von Smoluchowski and, independently, Albert Einstein argued that this attribution could be tested experimentally, even though at that time nobody was able to observe molecules directly. The test makes use of the specific properties of thermal noise.

It had already been clear for a long time that if molecules, i.e., indivisible matter particles, really existed, then thermal energy had to be disordered motion of these constituents and temperature had to be the average energy per degree of freedom of the constituents. Bernoulli's model of Figure 279 implies that for monatomic gases the kinetic energy T_{kin} per particle is given by

$$T_{\text{kin}} = \tfrac{3}{2} kT \qquad (114)$$

where T is temperature. The so-called *Boltzmann constant* $k = 1.4 \cdot 10^{-23}$ J/K is the standard conversion factor between temperature and energy.* At a room temperature

---

* The Boltzmann constant k was discovered and named by Max Planck, in the same work in which he also

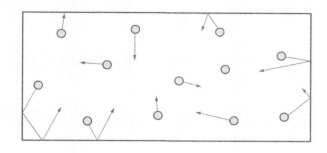

FIGURE 279 The basic idea of statistical mechanics about gases: gases are systems of moving particles, and pressure is due to their collisions with the container.

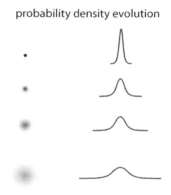

FIGURE 280 Example paths for particles in Brownian motion and their displacement distribution.

of 293 K, the kinetic energy of a particle is thus 6 zJ.

If you use relation (114) to calculate the speed of air molecules at room temperature, you get values of several hundred metres per second, about the speed of sound! Given this large speed, why does smoke from a candle take so long to diffuse through a room that has no air currents? Rudolph Clausius (b. 1822 Köslin, d. 1888 Bonn) answered this question in the mid-nineteenth century: smoke diffusion is slowed by the collisions with air molecules, in the same way as pollen particles collide with molecules in liquids. Since flows are usually more effective than diffusion, the materials that show no flows at all are those where the importance of diffusion is most evident: solids. Metal hardening and semiconductor production are examples.

The description of Brownian motion can be tested by following the displacement of pollen particles under the microscope. At first sight, we might guess that the average

---

discovered what is now called Planck's constant $\hbar$, the quantum of action. For more details on Max Planck, see later on.

Planck named the Boltzmann constant after the important physicist Ludwig Boltzmann (b. 1844 Vienna, d. 1906 Duino), who is most famous for his work on thermodynamics. Boltzmann explained all thermodynamic phenomena and observables, above all entropy itself, as results of the behaviour of molecules. It seems that Boltzmann committed suicide partly because of the animosities of his fellow physicists towards his ideas and himself. Nowadays, his work is standard textbook material.

distance the pollen particle has moved after n collisions should be zero, because the molecule velocities are random. However, this is wrong, as experiment shows.

An increasing average *square* displacement, written $\langle d^2 \rangle$, is observed for the pollen particle. It cannot be predicted in which direction the particle will move, but it does move. If the distance the particle moves after one collision is l, the average square displacement after n collisions is given, as you should be able to show yourself, by

$$\langle d^2 \rangle = nl^2 \ . \tag{115}$$

For molecules with an average velocity v over time t this gives

$$\langle d^2 \rangle = nl^2 = vlt \ . \tag{116}$$

In other words, the average square displacement increases proportionally with time. Of course, this is only valid because the liquid is made of separate molecules. Repeatedly measuring the position of a particle should give the distribution shown in Figure 280 for the probability that the particle is found at a given distance from the starting point. This is called the *(Gaussian) normal distribution*. In 1908, Jean Perrin* performed extensive experiments in order to test this prediction. He found that equation (116) corresponded completely with observations, thus convincing everybody that Brownian motion is indeed due to collisions with the molecules of the surrounding liquid, as had been expected.** Perrin received the 1926 Nobel Prize in Physics for these experiments.

Einstein also showed that the same experiment could be used to determine the number of molecules in a litre of water (or equivalently, the Boltzmann constant k). Can you work out how he did this?

Why stones can be neither smooth nor fractal, nor made of little hard balls

The exploration of temperature yields another interesting result. Researchers first studied gases, and measured how much energy was needed to heat them by 1 K. The result is simple: all gases share only a few values, when the number of molecules N is taken into account. Monatomic gases (in a container with constant volume and at sufficient temperature) require $3Nk/2$, diatomic gases (and those with a linear molecule) $5Nk/2$, and almost all other gases $3Nk$, where $k = 1.4 \cdot 10^{-23}$ J/K is the Boltzmann constant.

The explanation of this result was soon forthcoming: each thermodynamic degree of freedom*** contributes the energy $kT/2$ to the total energy, where T is the temperature.

---

* Jean Perrin (b. 1870 Lille, d. 1942 New York), important physicist, devoted most of his career to the experimental proof of the atomic hypothesis and the determination of Avogadro's number; in pursuit of this aim he perfected the use of emulsions, Brownian motion and oil films. His Nobel Prize speech (nobelprize.org/physics/laureates/1926/perrin-lecture.html) tells the interesting story of his research. He wrote the influential book *Les atomes* and founded the Centre National de la Recherche Scientifique. He was the first to speculate, in 1901, that an atom might be similar to a small solar system.
** In a delightful piece of research, Pierre Gaspard and his team showed in 1998 that Brownian motion is also chaotic, in the strict physical sense given later on.
*** A *thermodynamic degree of freedom* is, for each particle in a system, the number of dimensions in which

FIGURE 281 The fire pump.

So the number of degrees of freedom in physical bodies is finite. Bodies are not continuous, nor are they fractals: if they were, their specific thermal energy would be infinite. Matter is indeed made of small basic entities.

All degrees of freedom contribute to the specific thermal energy. At least, this is what classical physics predicts. Solids, like stones, have 6 thermodynamic degrees of freedom and should show a specific thermal energy of $3Nk$. At high temperatures, this is indeed observed. But measurements of solids at room temperature yield lower values, and the lower the temperature, the lower the values become. Even gases show values lower than those just mentioned, when the temperature is sufficiently low. In other words, molecules and atoms behave differently at low energies: atoms are not immutable little hard balls. The deviation of these values is one of the first hints of quantum theory.

ENTROPY

> – It's irreversible.
> – Like my raincoat!
> Mel Brooks, *Spaceballs*, 1987

Every domain of physics describes change in terms of three quantities: energy, as well as an intensive and an extensive quantity characteristic of the domain. In the domain of thermal physics, the intensive quantity is temperature. What is the corresponding extensive quantity?

The obvious guess would be 'heat'. Unfortunately, the quantity that physicists usually call 'heat' is not the same as what we call 'heat' in our everyday speech. For this historical reason, we need to introduce a new term. The extensive quantity corresponding to what we call 'heat' in everyday speech is called *entropy* in physics.*

---

it can move plus the number of dimensions in which it is kept in a potential. Atoms in a solid have six, particles in monatomic gases have only three; particles in diatomic gases or rigid linear molecules have five. The number of degrees of freedom of larger molecules depends on their shape.

* The term 'entropy' was invented by the physicist Rudolph Clausius (b. 1822 Köslin, d. 1888 Bonn) in 1865. He formed it from the Greek ἐν 'in' and τρόπος 'direction', to make it sound similar to 'energy'. The term

Entropy describes the *amount* of everyday heat. Entropy is measured in joule per kelvin or J/K; some example values (per amount of matter) are listed in Table 47 and Table 48. Entropy describes everyday heat in the same way as momentum describes everyday motion. Correspondingly, temperature describes the *intensity* of everyday heat or entropy, in the same way that speed describes the intensity of motion.

When two objects of different speeds collide, a flow of momentum takes place between them. Similarly, when two objects differing in temperature are brought into contact, an entropy flow takes place between them. We now define the concept of entropy – 'everyday heat' – more precisely and explore its properties in some more detail.

Entropy measures the degree to which energy is *mixed up* inside a system, that is, the degree to which energy is spread or shared among the components of a system. When all components of a system – usually the molecules or atoms – move in the same way, *in concert*, the entropy of the system is low. When the components of the system move completely independently, *randomly*, the entropy is large. In short, entropy measures the amount of disordered energy content per temperature in a system. That is the reason that it is measured in J/K.

Entropy is an extensive quantity, like charge and momentum. It is measured by transferring it to a measurement apparatus. The simplest measurement apparatus is a mixture of water and ice. When an amount S of entropy is transferred to the mixture, the amount of melted ice is a measure for the transferred entropy.

More precisely, the entropy ΔS flowing into a system is measured by measuring the energy E flowing into the system, and recording the temperature T that occurs during the process:

$$\Delta S = \int_{T_{\text{start}}}^{T_{\text{end}}} \frac{dE}{T} \ . \tag{117}$$

Often, this can be approximated as $\Delta S = P \Delta t / \overline{T}$, where P is the power of the heating device, Δt is the heating time, and $\overline{T}$ is the average temperature.

Since entropy measures an amount, an extensive quantity, and not an intensity, entropy *adds up* when identical systems are composed into one. When two one-litre bottles of water at the same temperature are poured together, the entropy of the water adds up. Again, this corresponds to the behaviour of momentum: it also adds up when systems are composed.

Like any other extensive quantity, entropy can be accumulated in a body, and entropy can flow into or out of bodies. When we transform water into steam by heating it, we say that we add entropy to the water. We also add entropy when we transform ice into liquid water. After either transformation, the added entropy is contained in the warmer phase. Indeed, we can *measure* the entropy we add by measuring how much ice melts or how much water evaporates. In short, entropy is the exact term for what we call 'heat' in everyday speech.

Whenever we dissolve a block of salt in water, the entropy of the total system must increase, because the disorder increases. We now explore this process.

---

entropy has always had the meaning given here.

In contrast, what physicists traditionally called 'heat' is a form of energy and not an extensive quantity in general.

TABLE 47 Some measured specific entropy values.

Process/System	Entropy value
Carbon, solid, in diamond form	2.43 J/K mol
Carbon, solid, in graphite form	5.69 J/K mol
Melting of ice	1.21 kJ/K kg = 21.99 J/K mol
Iron, solid, under standard conditions	27.2 J/K mol
Magnesium, solid, under standard conditions	32.7 J/K mol
Water, liquid, under standard conditions	70.1(2) J/K mol
Boiling of 1 kg of liquid water at 101.3 kPa	6.03 kJ/K = 110 J/K mol
Helium gas under standard conditions	126.15 J/K mol
Hydrogen gas under standard conditions	130.58 J/K mol
Carbon gas under standard conditions	158 J/K mol
Water vapour under standard conditions	188.83 J/K mol
Oxygen O_2 under standard conditions	205.1 J/K mol
C_2H_6 gas under standard conditions	230 J/K mol
C_3H_8 gas under standard conditions	270 J/K mol
C_4H_{10} gas under standard conditions	310 J/K mol
C_5H_{12} gas under standard conditions	348.9 J/K mol
$TiCl_4$ gas under standard conditions	354.8 J/K mol

TABLE 48 Some typical entropy values per particle at *standard* temperature and pressure as multiples of the Boltzmann constant.

Material	Entropy per particle
Monatomic solids	$0.3\,k$ to $10\,k$
Diamond	$0.29\,k$
Graphite	$0.68\,k$
Lead	$7.79\,k$
Monatomic gases	15-$25\,k$
Helium	$15.2\,k$
Radon	$21.2\,k$
Diatomic gases	$15\,k$ to $30\,k$
Polyatomic solids	$10\,k$ to $60\,k$
Polyatomic liquids	$10\,k$ to $80\,k$
Polyatomic gases	$20\,k$ to $60\,k$
Icosane	$112\,k$

ENTROPY FROM PARTICLES

Once it had become clear that heat and temperature are due to the motion of microscopic particles, people asked what entropy *was* microscopically. The answer can be formulated in various ways. The two most extreme answers are:

▷ Entropy measures the (logarithm of the) number W of possible microscopic states. A given macroscopic state can have many microscopic realizations. The logarithm of this number, multiplied by the Boltzmann constant k, gives the entropy.*
▷ Entropy is the expected number of yes-or-no questions, multiplied by $k \ln 2$, the answers of which would tell us everything about the system, i.e., about its microscopic state.

In short, the higher the entropy, the more microstates are possible. Through either of these definitions, entropy measures the quantity of randomness in a system. In other words, entropy measures the transformability of energy: higher entropy means lower transformability. Alternatively, entropy measures the *freedom* in the choice of microstate that a system has. High entropy means high freedom of choice for the microstate. For example, when a molecule of glucose (a type of sugar) is produced by photosynthesis, about 40 bits of entropy are released. This means that after the glucose is formed, 40 additional yes-or-no questions must be answered in order to determine the full microscopic state of the system. Physicists often use a macroscopic unit; most systems of interest are large, and thus an entropy of 10^{23} bits is written as 1 J/K. (This is only approximate. Can you find the precise value?)

To sum up, entropy is thus a specific measure for the characterization of disorder of thermal systems. Three points are worth making here. First of all, entropy is not *the* measure of disorder, but *one* measure of disorder. It is therefore *not* correct to use entropy as a *synonym* for the concept of disorder, as is often done in the popular literature. Entropy is only defined for systems that have a temperature, in other words, only for systems that are in or near equilibrium. (For systems far from equilibrium, no measure of disorder has been found yet; probably none is possible.) In fact, the use of the term entropy has degenerated so much that sometimes one has to call it *thermodynamic* entropy for clarity.

Secondly, entropy is related to information *only if* information is defined also as $-k \ln W$. To make this point clear, take a book with a mass of one kilogram. At room temperature, its entropy content is about 4 kJ/K. The printed information inside a book, say 500 pages of 40 lines with each containing 80 characters out of 64 possibilities, corresponds to an entropy of $4 \cdot 10^{-17}$ J/K. In short, what is usually called 'information' in everyday life is a negligible fraction of what a physicist calls information. Entropy is defined using the *physical* concept of information.

Finally, entropy is *not* a measure for what in normal life is called the *complexity* of a situation. In fact, nobody has yet found a quantity describing this everyday notion. The task is surprisingly difficult. Have a try!

---

* When Max Planck went to Austria to search for the anonymous tomb of Boltzmann in order to get him buried in a proper grave, he inscribed the formula $S = k \ln W$ on the tombstone. (Which physicist would finance the tomb of another, nowadays?)

In summary, if you hear the term entropy used with a different meaning than the expression $S = k \ln W$, beware. Somebody is trying to get you, probably with some ideology.

THE MINIMUM ENTROPY OF NATURE – THE QUANTUM OF INFORMATION

Before we complete our discussion of thermal physics we must point out in another way the importance of the Boltzmann constant k. We have seen that this constant appears whenever the granularity of matter plays a role; it expresses the fact that matter is made of small basic entities. The most striking way to put this statement is the following:

▷ There is a smallest entropy in nature: $S \geq k$.

This result is almost 100 years old; it was stated most clearly (with a different numerical factor) by Leo Szilard. The same point was made by Léon Brillouin (again with a different numerical factor). The statement can also be taken as the *definition* of the Boltzmann constant k.

The existence of a smallest entropy in nature is a strong idea. It eliminates the possibility of the continuity of matter and also that of its fractality. A smallest entropy implies that matter is made of a finite number of small components. The lower limit to entropy expresses the fact that matter is made of particles.* The limit to entropy also shows that Galilean physics cannot be correct: Galilean physics assumes that arbitrarily small quantities do exist. The entropy limit is the first of several limits to motion that we will encounter in our adventure. After we have found all limits, we can start the final leg that leads to the unified description of motion.

The existence of a smallest quantity implies a limit on the precision of measurements. Measurements cannot have infinite precision. This limitation is usually stated in the form of an indeterminacy relation. Indeed, the existence of a smallest entropy can be rephrased as an indeterminacy relation between the temperature T and the inner energy U of a system:

$$\Delta \frac{1}{T} \Delta U \geq \frac{k}{2} . \quad (118)$$

This relation** was given by Niels Bohr; it was discussed by Werner Heisenberg, who called it one of the basic indeterminacy relations of nature. The Boltzmann constant (divided by 2) thus fixes the smallest possible entropy value in nature. For this reason, Gilles Cohen-Tannoudji calls it the *quantum of information* and Herbert Zimmermann calls it the *quantum of entropy*.

The relation (118) points towards a more general pattern. For every minimum value for an observable, there is a corresponding indeterminacy relation. We will come across this several times in the rest of our adventure, most importantly in the case of the quantum

---

* The minimum entropy implies that matter is made of tiny spheres; the minimum *action*, which we will encounter in quantum theory, implies that these spheres are actually small clouds.
** It seems that the historical value for the right hand side, k, has to be corrected to $k/2$, for the same reason that the quantum of action $\hbar$ appears with a factor 1/2 in Heisenberg's indeterminacy relations.

of action and Heisenberg's indeterminacy relation.

The existence of a smallest entropy has numerous consequences. First of all, it sheds light on the third principle of thermodynamics. A smallest entropy implies that absolute zero temperature is not achievable. Secondly, a smallest entropy explains why entropy values are finite instead of infinite. Thirdly, it fixes the absolute value of entropy for every system; in continuum physics, entropy, like energy, is only defined up to an additive constant. The quantum of entropy settles all these issues.

The existence of a minimum value for an observable implies that an indeterminacy relation appears for any two quantities whose product yields that observable. For example, entropy production rate and time are such a pair. Indeed, an indeterminacy relation connects the entropy production rate $P = dS/dt$ and the time t:

$$\Delta P \, \Delta t \geqslant \frac{k}{2} \,. \tag{119}$$

From this and the previous relation (118) it is possible to deduce all of statistical physics, i.e., the precise theory of thermostatics and thermodynamics. We will not explore this further here. (Can you show that the third principle follows from the existence of a smallest entropy?) We will limit ourselves to one of the cornerstones of thermodynamics: the second principle.

Is everything made of particles?

> A physicist is the atom's way of knowing about atoms.
> George Wald

Historically, the study of statistical mechanics has been of fundamental importance for physics. It provided the first demonstration that physical objects are made of interacting particles. The story of this topic is in fact a long chain of arguments showing that all the properties we ascribe to objects, such as size, stiffness, colour, mass density, magnetism, thermal or electrical conductivity, result from the interaction of the many particles they consist of.

▷ All objects are made of interacting particles.

This discovery has often been called the main result of modern science.

How was this discovery made? Table 43 listed the main extensive quantities used in physics. Extensive quantities are able to flow. It turns out that all flows in nature are *composed* of elementary processes, as shown in Table 49. We have already seen that flows of mass, volume, charge, entropy and substance are composed. Later, quantum theory will show the same for flows of angular momentum and of the nuclear quantum numbers.

▷ All flows are made of particles.

The success of this idea has led many people to generalize it to the statement: 'Everything we observe is made of parts.' This approach has been applied with success to chem-

TABLE 49 Some minimum flow values found in nature.

OBSERVATION	MINIMUM FLOW
Matter flow	one molecule or one atom or one particle
Volume flow	one molecule or one atom or one particle
Angular momentum flow	Planck's quantum of action
Chemical amount of substance	one molecule, one atom or one particle
Entropy flow	the minimum entropy
Charge flow	one elementary charge
Light flow	one single photon, Planck's quantum of action

FIGURE 282 A 111 crystal surface of a gold single crystal, every bright dot being an atom, with a surface dislocation (© CNRS).

istry with molecules, materials science and geology with crystals, electricity with electrons, atoms with elementary particles, space with points, time with instants, light with photons, biology with cells, genetics with genes, neurology with neurons, mathematics with sets and relations, logic with elementary propositions, and even to linguistics with morphemes and phonemes. All these sciences have flourished on the idea that everything is made of *related parts*. The basic idea seems so self-evident that we find it difficult even to formulate an alternative; try!

However, in the case of the *whole* of nature, the idea that nature is a sum of related parts is incorrect and only approximate. It turns out to be a prejudice, and a prejudice so entrenched that it retarded further developments in physics in the latter decades of the twentieth century. In particular, it does *not* apply to elementary particles or to space:

▷ Elementary particles and space are not made of parts.

Finding the correct description for the whole of nature, for elementary particles and for space is the biggest challenge of our adventure. It requires a complete change in thinking habits. There is a lot of fun ahead.

> Jede Aussage über Komplexe läßt sich in eine Aussage über deren Bestandteile und in diejenigen Sätze zerlegen, welche die Komplexe vollständig beschreiben.*
> Ludwig Wittgenstein, *Tractatus*, 2.0201

The second principle of thermodynamics

In contrast to several other important extensive quantities, entropy is *not* conserved. On the one hand, in closed systems, entropy accumulates and never decreases; the sharing or mixing of energy among the components of a system cannot be undone. On the other hand, the sharing or mixing can increase spontaneously over time. Entropy is thus only 'half conserved'. What we call thermal equilibrium is simply the result of the highest possible mixing. Entropy allows us to define the concept of *equilibrium* more precisely as the state of maximum entropy, or maximum energy sharing among the components of a system. In short, the entropy of a closed system increases until it reaches the maximum possible value, the equilibrium value.

The non-conservation of entropy has far-reaching consequences. When a piece of rock is detached from a mountain, it falls, tumbles into the valley, heating up a bit, and eventually stops. The opposite process, whereby a rock cools and tumbles upwards, is never observed. Why? We could argue that the opposite motion does not contradict any rule or pattern about motion that we have deduced so far.

Rocks never fall upwards because mountains, valleys and rocks are made of *many* particles. Motions of many-particle systems, especially in the domain of thermodynamics, are called *processes*. Central to thermodynamics is the distinction between *reversible* processes, such as the flight of a thrown stone, and *irreversible* processes, such as the afore-mentioned tumbling rock. Irreversible processes are all those processes in which friction and its generalizations play a role. Irreversible processes are those processes that increase the sharing or mixing of energy. They are important: if there were no friction, shirt buttons and shoelaces would not stay fastened, we could not walk or run, coffee machines would not make coffee, and maybe most importantly of all, we would have no memory.

Irreversible processes, in the sense in which the term is used in thermodynamics, transform macroscopic motion into the disorganized motion of all the small microscopic components involved: they increase the sharing and mixing of energy. Irreversible processes are therefore not *strictly* irreversible – but their reversal is *extremely* improbable. We can say that entropy measures the 'amount of irreversibility': it measures the degree of mixing or decay that a collective motion has undergone.

---

* 'Every statement about complexes can be resolved into a statement about their constituents and into the propositions that describe the complexes completely.'

Entropy is not conserved. Indeed, entropy – 'heat' – can appear out of nowhere, spontaneously, because energy sharing or mixing can happen by itself. For example, when two different liquids of the same temperature are mixed – such as water and sulphuric acid – the final temperature of the mix can differ. Similarly, when electrical current flows through material at room temperature, the system can heat up or cool down, depending on the material.

All experiments on heat agree with the so-called *second principle of thermodynamics*, which states:

▷ The entropy in a closed system tends towards its maximum.

In sloppy terms, 'entropy ain't what it used to be.' In this statement, a *closed system* is a system that does not exchange energy or matter with its environment. Can you think of an example?

In a closed system, entropy never decreases. Even everyday life shows us that in a closed system, such as a room, the disorder increases with time, until it reaches some maximum. To reduce disorder, we need effort, i.e., work and energy. In other words, in order to reduce the disorder in a system, we need to connect the system to an energy source in some clever way. For this reason, refrigerators need electrical current or some other energy source.

In 1866, Ludwig Boltzmann showed that the second principle of thermodynamics results from the principle of least action. Can you imagine and sketch the general ideas?

Because entropy never decreases in closed systems, *white colour does not last*. Whenever disorder increases, the colour white becomes 'dirty', usually grey or brown. Perhaps for this reason white objects, such as white clothes, white houses and white underwear, are valued in our society. White objects defy decay.

The second principle implies that heat cannot be transformed to work completely. In other words, every heat engine needs cooling: that is the reason for the holes in the front of cars. The first principle of thermodynamics then states that the mechanical power of a heat engine is the difference between the inflow of thermal energy at high temperature and the outflow of thermal energy at low temperature. If the cooling is insufficient – for example, because the weather is too hot or the car speed too low – the power of the engine is reduced. Every driver knows this from experience.

In summary, the concept of entropy, corresponding to what is called 'heat' in everyday life – but *not* to what is traditionally called 'heat' in physics! – describes the randomness of the internal motion in matter. Entropy is not conserved: in a closed system, entropy never decreases, but it can increase until it reaches a maximum value. The non-conservation of entropy is due to the many components inside everyday systems. The large number of components lead to the non-conservation of entropy and therefore explain, among many other things, that many processes in nature never occur backwards, even though they could do so in principle.

Why can't we remember the future?

> It's a poor sort of memory which only works backwards.
> Lewis Carroll, *Alice in Wonderland*

When we first discussed time, we ignored the difference between past and future. But obviously, a difference exists, as we do not have the ability to remember the future. This is not a limitation of our brain alone. All the devices we have invented, such as tape recorders, photographic cameras, newspapers and books, only tell us about the past. Is there a way to build a video recorder with a 'future' button? Such a device would have to solve a deep problem: how would it distinguish between the near and the far future? It does not take much thought to see that any way to do this would conflict with the second principle of thermodynamics. That is unfortunate, as we would need precisely the same device to show that there is faster-than-light motion. Can you find the connection?

In summary, the future cannot be remembered because entropy in closed systems tends towards a maximum. Put even more simply, memory exists because the brain is made of many particles, and so the brain is limited to the past. However, for the most simple types of motion, when only a few particles are involved, the difference between past and future disappears. For few-particle systems, there is no difference between times gone by and times approaching. We could say that the future differs from the past only in our brain, or equivalently, only because of friction. Therefore the difference between the past and the future is not mentioned frequently in this walk, even though it is an essential part of our human experience. But the fun of the present adventure is precisely to overcome our limitations.

Flow of entropy

We know from daily experience that transport of an extensive quantity always involves friction. Friction implies generation of entropy. In particular, the flow of entropy itself produces additional entropy. For example, when a house is heated, entropy is produced in the wall. Heating means to keep a temperature difference ΔT between the interior and the exterior of the house. The heat flow J traversing a square metre of wall is given by

$$J = \kappa \Delta T = \kappa (T_i - T_e) \qquad (120)$$

where κ is a constant characterizing the ability of the wall to conduct heat. While conducting heat, the wall also *produces* entropy. The entropy production σ is proportional to the difference between the interior and the exterior entropy flows. In other words, one has

$$\sigma = \frac{J}{T_e} - \frac{J}{T_i} = \kappa \frac{(T_i - T_e)^2}{T_i T_e} \ . \qquad (121)$$

Note that we have assumed in this calculation that everything is near equilibrium in each slice parallel to the wall, a reasonable assumption in everyday life. A typical case of a good wall has $\kappa = 1 \text{ W/m}^2\text{K}$ in the temperature range between 273 K and 293 K. With this value, one gets an entropy production of

$$\sigma = 5 \cdot 10^{-3} \text{ W/m}^2\text{K} \,. \tag{122}$$

Can you compare the amount of entropy that is produced in the flow with the amount that is transported? In comparison, a good goose-feather duvet has $\kappa = 1.5 \text{ W/m}^2\text{K}$, which in shops is also called 15 tog.*

The insulation power of materials is usually measured by the constant $\lambda = \kappa d$ which is independent of the thickness d of the insulating layer. Values in nature range from about 2000 W/K m for diamond, which is the best conductor of all, down to between 0.1 W/K m and 0.2 W/K m for wood, between 0.015 W/K m and 0.05 W/K m for wools, cork and foams, and the small value of $5 \cdot 10^{-3}$ W/K m for krypton gas.

Entropy can be transported in three ways: through *heat conduction*, as just mentioned, via *convection*, used for heating houses, and through *radiation*, which is possible also through empty space. For example, the Earth radiates about 1.2 W/m²K into space, in total thus about 0.51 PW/K. The entropy is (almost) the same that the Earth receives from the Sun. If more entropy had to be radiated away than received, the temperature of the surface of the Earth would have to increase. This is called the *greenhouse effect* or *global warming*. Let's hope that it remains small in the near future.

Do isolated systems exist?

In all our discussions so far, we have assumed that we can distinguish the system under investigation from its environment. But do such *isolated* or *closed* systems, i.e., systems not interacting with their environment, actually exist? Probably our own human condition was the original model for the concept: we do experience having the possibility to act independently of our environment. An isolated system may be simply defined as a system not exchanging any energy or matter with its environment. For many centuries, scientists saw no reason to question this definition.

The concept of an isolated system had to be refined somewhat with the advent of quantum mechanics. Nevertheless, the concept of isolated system provides useful and precise descriptions of nature also in the quantum domain, if some care is used. Only in the final part of our walk will the situation change drastically. There, the investigation of whether the universe is an isolated system will lead to surprising results. (What do you think? A strange hint: your answer is almost surely wrong.) We'll take the first steps towards the answer shortly.

Curiosities and fun challenges about reversibility and heat

Running backwards is an interesting sport. The 2006 world records for running backwards can be found on www.recordholders.org/en/list/backwards-running.html. You will be astonished how much these records are faster than your best personal *forward*-running time.

---

* The unit tog is not as bad as the official unit (not a joke) BthU · h/sqft/cm/°F used in some remote provinces of our galaxy.

∗ ∗

In 1912, Emile Borel noted that if a gram of matter on Sirius was displaced by one centimetre, it would change the gravitational field on Earth by a tiny amount only. But this tiny change would be sufficient to make it impossible to calculate the path of molecules in a gas after a fraction of a second.

∗ ∗

If heat really is disordered motion of atoms, a big problem appears. When two atoms collide head-on, in the instant of smallest distance, neither atom has velocity. Where does the kinetic energy go? Obviously, it is transformed into potential energy. But that implies that atoms can be deformed, that they have internal structure, that they have parts, and thus that they can in principle be split. In short, if heat is disordered atomic motion, *atoms are not indivisible*! In the nineteenth century this argument was put forward in order to show that heat cannot be atomic motion, but must be some sort of fluid. But since we know that heat really is kinetic energy, *atoms must be divisible*, even though their name means 'indivisible'. We do not need an expensive experiment to show this! We will discover more about them later on in our exploration.

∗ ∗

Compression of air increases its temperature. This is shown directly by the fire pump, a variation of a bicycle pump, shown in Figure 281. (For a working example, see the web page www.de-monstrare.nl). A match head at the bottom of an air pump made of transparent material is easily ignited by the compression of the air above it. The temperature of the air after compression is so high that the match head ignites spontaneously.

∗ ∗

In the summer, temperature of the air can easily be measured with a clock. Indeed, the rate of chirping of most crickets depends on temperature. For example, for a cricket species most common in the United States, by counting the number of chirps during 8 seconds and adding 4 yields the air temperature in degrees Celsius.

∗ ∗

How long does it take to cook an egg? This issue has been researched in extensive detail; of course, the time depends on what type of cooked egg you want, how large it is, and whether it comes from the fridge or not. There is even a formula for calculating the cooking time! Egg white starts hardening at 62 °C, the yolk starts hardening at 65 °C. The best-tasting hard eggs are formed at 69 °C, half-hard eggs at 65 °C, and soft eggs at 63 °C. If you cook eggs at 100 °C (for a long time) , the white gets the consistency of rubber and the yolk gets a green surface that smells badly, because the high temperature leads to the formation of the smelly H_2S, which then bonds to iron and forms the green FeS. Note that when temperature is controlled, the time plays no role; 'cooking' an egg at 65 °C for 10 minutes or 10 hours gives the *same* result.

∗ ∗

It is possible to cook an egg in such a way that the white is hard but the yolk remains liquid. Can you achieve the opposite? Research has even shown how you can cook an

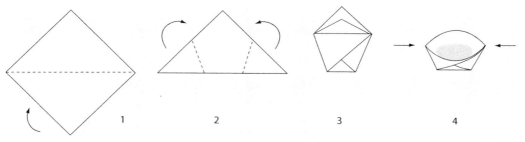

FIGURE 283 Can you boil water in this paper cup?

egg so that the yolk remains at the centre. Can you imagine the method?

* *

Not only gases, but also most other materials expand when the temperature rises. As a result, the electrical wires supported by pylons hang much lower in summer than in winter. True?

* *

The following is a famous problem asked by Fermi. Given that a human corpse cools down in four hours after death, what is the minimum number of calories needed per day in our food?

* *

The energy contained in thermal motion is not negligible. For example, a 1 g bullet travelling at the speed of sound has a kinetic energy of only 0.04 kJ= 0.01 kcal. What is its thermal energy content?

* *

How does a typical, 1500 m^3 hot-air balloon work?

* *

If you do not like this text, here is a proposal. You can use the paper to make a cup, as shown in Figure 283, and boil water in it over an open flame. However, to succeed, you have to be a little careful. Can you find out in what way?

* *

Mixing 1 kg of water at 0 °C and 1 kg of water at 100 °C gives 2 kg of water at 50 °C. What is the result of mixing 1 kg of *ice* at 0 °C and 1 kg of water at 100 °C?

* *

Temperature has many effects. In the past years, the World Health Organization found that drinking liquids that are hotter than 65 °C – including coffee, chocolate or tea – causes cancer in the oesophagus, independently of the type of liquid.

* *

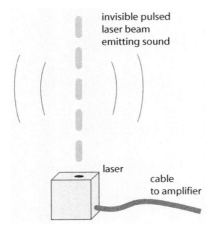

FIGURE 284 The invisible loudspeaker.

Ref. 314

The highest recorded air temperature in which a man has survived is 127°C. This was tested in 1775 in London, by the secretary of the Royal Society, Charles Blagden, together with a few friends, who remained in a room at that temperature for 45 minutes. Interestingly, the raw steak which he had taken in with him was cooked ('well done') when he and his friends left the room. What condition had to be strictly met in order to avoid cooking the people in the same way as the steak?

Challenge 667 s

* *

Is the *Boltzmann constant k* really the smallest possible value for entropy in nature? How then can the entropy per particle of Krypton be as low as $0.3k$ per particle? The answer to this paradox is that a single free particle has more entropy, in fact more than k, than a bound one. The limit entropy k is thus valid for a physical system, such as the crystal as a whole, but not separately for the entropy value for each bound particle that is part of a system.

* *

The influential astronomer Anders Celsius (b. 1701 Uppsala, d. 1744 Uppsala) originally set the freezing point of water at 100 degrees and the boiling point at 0 degrees. Shortly afterwards, the scale was reversed to the one in use now. However, this is not the whole story. With the official definition of the kelvin and the degree Celsius, at the standard pressure of 101 325 Pa, water boils at 99.974°C. Can you explain why it is not 100°C any more?

Ref. 315

Challenge 668 s

* *

Challenge 669 s Can you fill a bottle precisely with 1 ± 10^{-30} kg of water?

* *

One gram of fat, either butter or human fat, contains 38 kJ of chemical energy (or, in ancient units more familiar to nutritionists, 9 kcal). That is the same value as that of petrol.

Challenge 670 s Why are people and butter less dangerous than petrol?

**

In 1992, the Dutch physicist Martin van der Mark invented a loudspeaker which works by heating air with a laser beam. He demonstrated that with the right wavelength and with a suitable modulation of the intensity, a laser beam in air can generate sound. The effect at the basis of this device, called the *photoacoustic effect*, appears in many materials. The best laser wavelength for air is in the infrared domain, on one of the few absorption lines of water vapour. In other words, a properly modulated infrared laser beam that shines through the air generates sound. Such light can be emitted from a small matchbox-sized semiconductor laser hidden in the ceiling and shining downwards. The sound is emitted in all directions perpendicular to the beam. Since infrared laser light is not (usually) visible, Martin van der Mark thus invented an invisible loudspeaker! Unfortunately, the efficiency of present versions is still low, so that the power of the speaker is not yet sufficient for practical applications. Progress in laser technology should change this, so that in the future we should be able to hear sound that is emitted from the centre of an otherwise empty room.

**

A famous exam question: How can you measure the height of a building with a barometer, a rope and a ruler? Find at least six different ways.

**

What is the approximate probability that out of one million throws of a coin you get exactly 500 000 heads and as many tails? You may want to use Stirling's formula $n! \approx \sqrt{2\pi n}(n/e)^n$ to calculate the result.*

**

Does it make sense to talk about the entropy of the universe?

**

Can a helium balloon lift the tank which filled it?

**

All friction processes, such as osmosis, diffusion, evaporation, or decay, are *slow*. They take a characteristic time. It turns out that any (macroscopic) process with a time-scale is irreversible. This is no real surprise: we know intuitively that undoing things always takes more time than doing them. That is again the second principle of thermodynamics.

**

It turns out that *storing* information is possible with negligible entropy generation. However, *erasing* information requires entropy. This is the main reason why computers, as well as brains, require energy sources and cooling systems, even if their mechanisms would otherwise need no energy at all.

---

* There are many improvements to Stirling's formula. A simple one is Gosper's formula $n! \approx \sqrt{(2n+1/3)\pi}(n/e)^n$. Another is $\sqrt{2\pi n}(n/e)^n e^{1/(12n+1)} < n! < \sqrt{2\pi n}(n/e)^n e^{1/(12n)}$.

When mixing hot rum and cold water, how does the increase in entropy due to the mixing compare with the entropy increase due to the temperature difference?

* *

Why aren't there any small humans, say 10 mm in size, as in many fairy tales? In fact, there are no warm-blooded animals of that size at all. Why not?

* *

Shining a light onto a body and repeatedly switching it on and off produces sound. This is called the *photoacoustic effect*, and is due to the thermal expansion of the material. By changing the frequency of the light, and measuring the intensity of the noise, one reveals a characteristic photoacoustic spectrum for the material. This method allows us to detect gas concentrations in air of one part in 10^9. It is used, among other methods, to study the gases emitted by plants. Plants emit methane, alcohol and acetaldehyde in small quantities; the photoacoustic effect can detect these gases and help us to understand the processes behind their emission.

* *

What is the rough probability that all oxygen molecules in the air would move away from a given city for a few minutes, killing all inhabitants?

* *

If you pour a litre of water into the sea, stir thoroughly through all the oceans and then take out a litre of the mixture, how many of the original atoms will you find?

* *

How long would you go on breathing in the room you are in if it were airtight?

* *

Heat loss is a larger problem for smaller animals, because the surface to volume ratio increases when size decreases. As a result, small animals are found in hot climate, large animals are found in cold climates. This is true for bears, birds, rabbits, insects and many other animal families. For the same reason, small living beings need high amounts of food per day, when calculated in body weight, whereas large animals need far less food.

* *

What happens if you put some ash onto a piece of sugar and set fire to the whole? (Warning: this is dangerous and not for kids.)

* *

Entropy calculations are often surprising. For a system of N particles with two states each, there are $W_{\text{all}} = 2^N$ states. For its most probable configuration, with exactly half the particles in one state, and the other half in the other state, we have $W_{\max} = N!/((N/2)!)^2$. Now, for a macroscopic system of particles, we might typically have $N = 10^{24}$. That gives $W_{\text{all}} \gg W_{\max}$; indeed, the former is 10^{12} times larger than the latter. On the other hand, we find that ln W_{all} and ln W_{max} agree for the first 20 digits! Even though the configuration with exactly half the particles in each state is much more rare than the general case, where the ratio is allowed to vary, the entropy turns out to be the same. Why?

* *

If heat is due to motion of atoms, our built-in senses of heat and cold are simply detectors of motion. How could they work?

By the way, the senses of smell and taste can also be seen as motion detectors, as they signal the presence of molecules flying around in air or in liquids. Do you agree?

* *

The Moon has an atmosphere, although an extremely thin one, consisting of sodium (Na) and potassium (K). This atmosphere has been detected up to nine Moon radii from its surface. The atmosphere of the Moon is generated at the surface by the ultraviolet radiation from the Sun. Can you estimate the Moon's atmospheric density?

* *

Does it make sense to add a line in Table 43 for the quantity of physical action? A column? Why?

* *

Diffusion provides a length scale. For example, insects take in oxygen through their skin. As a result, the interiors of their bodies cannot be much more distant from the surface than about a centimetre. Can you list some other length scales in nature implied by diffusion processes?

* *

Rising warm air is the reason why many insects are found in tall clouds in the evening. Many insects, especially that seek out blood in animals, are attracted to warm and humid air.

* *

Thermometers based on mercury can reach 750°C. How is this possible, given that mercury boils at 357°C?

* *

What does a burning candle look like in weightless conditions?

* *

It is possible to build a power station by building a large chimney, so that air heated by the Sun flows upwards in it, driving a turbine as it does so. It is also possible to make a power station by building a long vertical tube, and letting a gas such as ammonia rise into it which is then liquefied at the top by the low temperatures in the upper atmosphere; as it falls back down a second tube as a liquid – just like rain – it drives a turbine. Why are such schemes, which are almost completely non-polluting, not used yet?

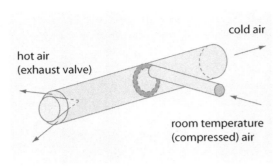
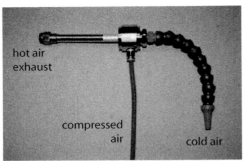

FIGURE 285 The design of the Wirbelrohr or Ranque–Hilsch vortex tube, and a commercial version, about 40 cm in size, used to cool manufacturing processes (© Coolquip).

* *

One of the most surprising devices ever invented is the *Wirbelrohr* or Ranque–Hilsch vortex tube. By blowing compressed air at room temperature into it at its midpoint, two flows of air are formed at its ends. One is extremely cold, easily as low as −50°C, and one extremely hot, up to 200°C. No moving parts and no heating devices are found inside. How does it work?

* *

Thermoacoustic engines, pumps and refrigerators provide many strange and fascinating applications of heat. For example, it is possible to use loud sound in closed metal chambers to move heat from a cold place to a hot one. Such devices have few moving parts and are being studied in the hope of finding practical applications in the future.

* *

Does a closed few-particle system contradict the second principle of thermodynamics?

* *

What happens to entropy when gravitation is taken into account? We carefully left gravitation out of our discussion. In fact, gravitation leads to many new problems – just try to think about the issue. For example, Jacob Bekenstein has discovered that matter reaches its highest possible entropy when it forms a black hole. Can you confirm this?

* *

The numerical values – but not the units! – of the Boltzmann constant $k = 1.38 \cdot 10^{-23}$ J/K and the combination h/ce – where h is Planck's constant, c the speed of light and e the electron charge – agree in their exponent and in their first three digits. How can you dismiss this as mere coincidence?

* *

Mixing is not always easy to perform. The experiment of Figure 286 gives completely different results with water and glycerine. Can you guess them?

* *

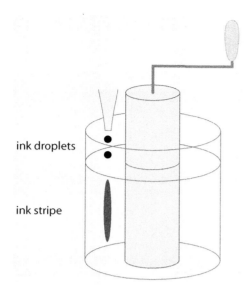

FIGURE 286 What happens to the ink stripe if the inner cylinder is turned a few times in one direction, and then turned back by the same amount?

How can you get rid of chewing gum in clothes?

* *

With the knowledge that air consist of molecules, Maxwell calculated that "each particle makes 8 077 200 000 collisions per second". How did he do this?

* *

A perpetuum mobile 'of the second kind' is a machine converts heat into motion without the use of a second, cooler bath. Entropy implies that such a device does not exist. Can you show this?

* *

There are less-well known arguments proving the existence atoms. In fact, two everyday observations prove the existence of atoms: reproduction and memory. Why?

* *

In the context of lasers and of spin systems, it is fun to talk about negative temperature. Why is this not really sensible?

Summary on heat and time-invariance

Microscopic motion due to gravity and electric interactions, thus all microscopic motion in everyday life, is *reversible*: such motion can occur backwards in time. In other words, motion due to gravity and electromagnetism is *symmetric under motion reversal* or, as is often incorrectly stated, under 'time reversal'.

Nevertheless, everyday motion is *irreversible*, because there are no completely closed systems in everyday life. Lack of closure leads to fluctuations; fluctuations lead to friction.

Equivalently, irreversibility results from the extremely low probability of motion reversal in many-particle systems. Macroscopic irreversibility does not contradict microscopic reversibility.

For these reasons, in everyday life, entropy in closed systems never decreases. This leads to a famous issue: how can biological evolution be reconciled with entropy increase? Let us have a look.

Chapter 14

SELF-ORGANIZATION AND CHAOS – THE SIMPLICITY OF COMPLEXITY

> To speak of non-linear physics is like calling zoology the study of non-elephant animals.
> Stanislaw Ulam

In our list of global descriptions of motion, the study of self-organization is the high point. Self-organization is the appearance of order. In physics, *order* is a term that includes *shapes*, such as the complex symmetry of snowflakes; *patterns*, such as the stripes of zebras and the ripples on sand; and *cycles*, such as the creation of sound when singing. When we look around us, we note that every example of what we call *beauty* is a combination of shapes, patterns and cycles. (Do you agree?) Self-organization can thus be called the study of the origin of beauty. Table 50 shows how frequently the appearance of order shapes our environment.

TABLE 50 Some rhythms, patterns and shapes observed in nature.

Observation	Driving 'force'	Restoring 'force'	Typ. Scale
Fingerprint	chemical reactions	diffusion	0.1 mm
Clock ticking	falling weight	friction	1 s
Chalk squeaking due to stick-slip instability	motion	friction	600 Hz
Musical note generation in violin	bow motion	friction	600 Hz
Musical note generation in flute	air flow	turbulence	400 Hz
Train oscillations transversally to the track	motion	friction	0.3 Hz
Flow structures in waterfalls and fountains	water flow	turbulence	10 cm
Jerky detachment of scotch tape	pulling speed	sticking friction	0.1 Hz
Radius oscillations in spaghetti and polymer fibre production	extrusion speed	friction	10 cm
Patterns on buckled metal plates and foils	deformation	stiffness	depend on thickness

TABLE 50 (Continued) Some rhythms, patterns and shapes observed in nature.

Observation	Driving 'force'	Restoring 'force'	Typ. Scale
Flapping of flags in steady wind	air flow	stiffness	20 cm
Dripping of water tap	water flow	surface tension	1 Hz
Bubble stream from a beer glass irregularity	dissolved gas pressure	surface tension	0.1 Hz, 1 mm
Raleigh–Bénard instability	temperature gradient	diffusion	0.1 Hz, 1 mm
Couette–Taylor flow	speed gradient	friction	0.1 Hz, 1 mm
Bénard–Marangoni flow, sea wave generation	surface tension	viscosity	0.1 Hz, 1 mm
Karman wakes, Emmon spots, Osborne Reynolds flow	momentum	viscosity	from mm to km
Regular bangs in a car exhaustion pipe	flow	pressure resonances	0.3 Hz
Regular cloud arrangements	flow	diffusion	0.5 km
El Niño	flow	diffusion	5 to 7 years
Wine arcs on glass walls	surface tension	binary mixture	0.1 Hz, 1 mm
Ferrofluids surfaces in magnetic fields	magnetic energy	gravity	3 mm
Patterns in liquid crystals	electric energy	stress	1 mm, 3 s
Flickering of aging fluorescence lights	electron flow	diffusion	1 Hz
Surface instabilities of welding	electron flow	diffusion	1 cm
Tokamak plasma instabilities	electron flow	diffusion	10 s
Snowflake formation and other dendritic growth processes	concentration gradient	surface diffusion	10 µm
Solidification interface patterns, e.g. in CBr_4	entropy flow	surface tension	1 mm
Periodic layers in metal corrosion	concentration gradients	diffusion	10 µm
Hardening of steel by cold working	strain	dislocation motion	5 µm
Labyrinth structures in proton irradiated metals	particle flow	dislocation motion	5 µm
Patterns in laser irradiated Cd-Se alloys	laser irradiation	diffusion	50 µm

TABLE 50 (Continued) Some rhythms, patterns and shapes observed in nature.

Observation	Driving 'force'	Restoring 'force'	Typ. Scale
Dislocation patterns and density oscillations in fatigued Cu single crystals	strain	dislocation motion	10 μm 100 s
Laser light emission, its cycles and chaotic regimes	pumping energy	light losses	10 ps to 1 ms
Rotating patterns from shining laser light on the surface of certain electrolytes	light energy	diffusion	1 mm
Belousov-Zhabotinski reaction patterns and cycles	concentration gradients	diffusion	1 mm, 10 s
Flickering of a burning candle	heat and concentration gradients	thermal and substance diffusion	0.1 s
Regular sequence of hot and cold flames in carbohydrate combustion	heat and concentration gradients	thermal and substance diffusion	1 cm
Feedback whistle from microphone to loudspeaker	amplifiers	electric losses	1 kHz
Any electronic oscillator in radio sets, television sets, computers, mobile phones, etc.	power supply	resistive losses	1 kHz to 30 GHz
Periodic geyser eruptions	underground heating	evaporation	10 min
Periodic earthquakes at certain faults	tectonic motion	ruptures	1 Ms
Hexagonal patterns in basalt rocks	heating	heat diffusion	1 m
Hexagonal patterns on dry soil	regular temperature changes	water diffusion	0.5 m
Periodic intensity changes of the Cepheids and other stars	nuclear fusion	energy emission	3 Ms
Convection cells on the surface of the Sun	nuclear fusion	energy emission	1000 km
Formation and oscillations of the magnetic field of the Earth and other celestial bodies	charge separation due to convection and friction	resistive losses	100 ka
Wrinkling/crumpling transition	strain	stiffness	1 mm

TABLE 50 (Continued) Some rhythms, patterns and shapes observed in nature.

Observation	Driving 'force'	Restoring 'force'	Typ. scale
Patterns of animal furs	chemical concentration	diffusion	1 cm
Growth of fingers and limbs	chemical concentration	diffusion	1 cm
Symmetry breaking in embryogenesis, such as the heart on the left	probably molecular chirality plus chemical concentration	diffusion	1 m
Cell differentiation and appearance of organs during growth	chemical concentration	diffusion	10 μm to 30 m
Prey–predator oscillations	reproduction	hunger	3 to 17 a
Thinking	neuron firing	heat dissipation	1 ms, 100 μm

Appearance of order

The appearance of order is a general observation across nature. *Fluids* in particular exhibit many phenomena where order appears and disappears. Examples include the more or less regular flickering of a burning candle, the flapping of a flag in the wind, the regular stream of bubbles emerging from small irregularities in the surface of a beer or champagne glass, and the regular or irregular dripping of a water tap. Figure 251 shows some additional examples, and so do the figures in this chapter. Other examples include the appearance of clouds and of regular cloud arrangements in the sky. It can be fascinating to ponder, during an otherwise boring flight, the mechanisms behind the formation of the cloud shapes and patterns you see from the aeroplane. A typical cloud has a mass density of 0.3 to 5 g/m^3, so that a large cloud can contain several thousand tons of water.

Other cases of self-organization are mechanical, such as the formation of mountain ranges when continents move, the creation of earthquakes, or the formation of laughing folds at the corners of human eyes.

All *growth* processes are self-organization phenomena. The appearance of order is found from the cell differentiation in an embryo inside a woman's body; the formation of colour patterns on tigers, tropical fish and butterflies; the symmetrical arrangements of flower petals; the formation of biological rhythms; and so on.

Have you ever pondered the incredible way in which teeth grow? A practically inorganic material forms shapes in the upper and the lower rows fitting exactly into each other. How this process is controlled is still a topic of research. Also the formation, before and after birth, of neural networks in the brain is another process of self-organization. Even the physical processes at the basis of thinking, involving changing electrical signals, is to be described in terms of self-organization.

Biological evolution is a special case of growth. Take the evolution of animal shapes. It turns out that snake tongues are forked because that is the most efficient shape for following chemical trails left by prey and other snakes of the same species. (Snakes smell

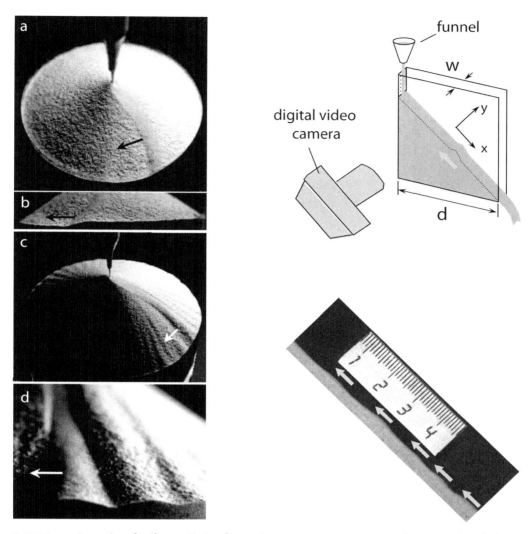

FIGURE 287 Examples of self-organization for sand: spontaneous appearance of a temporal cycle (a and b), spontaneous appearance of a periodic pattern (b and c), spontaneous appearance of a spatiotemporal pattern, namely solitary waves (right) (© Ernesto Altshuler et al.).

with the help of their tongue.) How many tips would the tongues of flying reptiles need, such as flying dragons?

The fixed number of fingers in human hands are also consequence of self-organization. The number of petals of flowers may or may not be due to self-organization.

Studies into the conditions required for the appearance or disappearance of order have shown that their description requires only a few common concepts, independently of the details of the physical system. This is best seen looking at a few simple examples.

TABLE 51 Patterns and a cycle on horizontal sand and on sand-like surfaces in the sea and on land.

Pattern/cycle	Period	Amplitude	Origin
Under water			
Ripples	5 cm	5 mm	water waves
Megaripples	1 m	0.1 m	tides
Sand waves	100 to 800 m	5 m	tides
Sand banks	2 to 10 km	2 to 20 m	tides
In air			
Ripples	0.1 m	0.05 m	wind
Singing sand	65 to 110 Hz	up to 105 dB	wind on sand dunes, avalanches making the dune vibrate
Road corrugations	0.3 to 0.9 m	0.05 m	wheels
Ski moguls	5 to 6 m	up to 1 m	skiers
Elsewhere			
On Mars	a few km	few tens of m	wind

FIGURE 288 Road corrugations (courtesy David Mays).

Self-organization in sand

All the richness of self-organization reveals itself in the study of plain sand. Why do sand dunes have ripples, as does the sand floor at the bottom of the sea? How do avalanches occur on steep heaps of sand? How does sand behave in hourglasses, in mixers, or in vibrating containers? The results are often surprising.

An overview of self-organization phenomena in sand is given in Table 51. For example, as recently as 2006, the Cuban research group of Ernesto Altshuler and his colleagues discovered solitary waves on sand flows (shown in Figure 287). They had already discovered the revolving river effect on sand piles, shown in the same figure, in 2002. Even

THE SIMPLICITY OF COMPLEXITY

FIGURE 289 Oscillons formed by shaken bronze balls; horizontal size is about 2 cm (© Paul Umbanhowar)

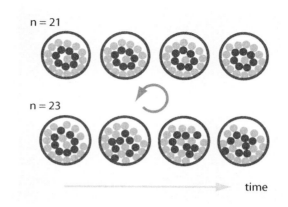

FIGURE 290 Magic numbers: 21 spheres, when swirled in a dish, behave differently from non-magic numbers, like 23, of spheres (redrawn from photographs © Karsten Kötter).

more surprisingly, these effects occur only for Cuban sand, and a few rare other types of sand. The reasons are still unclear.

Similarly, in 1996 Paul Umbanhowar and his colleagues found that when a flat container holding tiny bronze balls (around 0.165 mm in diameter) is shaken up and down in vacuum at certain frequencies, the surface of this bronze 'sand' forms stable heaps. They are shown in Figure 289. These heaps, so-called *oscillons*, also bob up and down. The oscillons can move and interact with one another.

Oscillons in bronze sand are a simple example for a general effect in nature:

> ▷ *Discrete* systems with non-linear interactions can exhibit *localized excitations*.

This fascinating topic is just beginning to be researched. It might well be that one day it will yield results relevant to our understanding of the growth of organisms.

Sand shows many other pattern-forming processes.

— A mixture of sand and sugar, when poured onto a heap, forms regular layered structures that in cross section look like zebra stripes.
— Horizontally rotating cylinders with binary mixtures inside them separate the mixture out over time.
— Take a container with two compartments separated by a 1 cm wall. Fill both halves with sand and rapidly shake the whole container with a machine. Over time, all the sand will spontaneously accumulate in one half of the container.
— In sand, people have studied the various types of sand dunes that 'sing' when the wind blows over them.

FIGURE 291 Self-organization: a growing snowflake. (QuickTime film © Kenneth Libbrecht)

— Also the corrugations formed by traffic on roads without tarmac, the washboard roads shown in Figure 288, are an example of self-organization. These corrugation patterns often move, over time, *against* the traffic direction. Can you explain why? The moving ski moguls mentioned above also belong here.

In fact, the behaviour of sand and dust is proving to be such a beautiful and fascinating topic that the prospect of each human returning to dust does not look so grim after all.

Self-organization of spheres

A stunningly simple and beautiful example of self-organization is the effect discovered in 1999 by Karsten Kötter and his group. They found that the behaviour of a set of spheres swirled in a dish depends on the number of spheres used. Usually, all the spheres get continuously mixed up. But for certain 'magic' numbers, such as 21, stable ring patterns emerge, for which the outside spheres remain outside and the inside ones remain inside. The rings, best seen by colouring the spheres, are shown in Figure 290.

Conditions for the appearance of order

The many studies of self-organizing systems have changed our understanding of nature in a number of ways. First of all, they have shown that patterns and shapes are similar to cycles: all are *due to motion*. Without motion, and thus without history, there is no order, neither patterns nor shapes nor rhythms. Every pattern has a history; every pattern is a result of *motion*. As an example, Figure 291 shows how a snowflake grows.

Secondly, patterns, shapes and rhythms are due to the organized motion of *large numbers of small constituents*. Systems which self-organize are always composite: they are *cooperative structures*.

Thirdly, all these systems obey evolution equations which are *non-linear* in the macroscopic configuration variables. Linear systems do not self-organize.

Fourthly, the appearance and disappearance of order depends on the strength of a driving force or driving process, the so-called *order parameter*.

Finally, all order and all structure appears when two general types of motion compete with each other, namely a 'driving', energy-adding process, and a *'dissipating', braking mechanism*. Thermodynamics thus plays a role in all self-organization. Self-organizing systems are always *dissipative systems*, and are always far from equilibrium. When the driving and the dissipation are of the same order of magnitude, and when the key behaviour of the system is not a linear function of the driving action, order may appear.*

THE MATHEMATICS OF ORDER APPEARANCE

Every pattern, every shape and every rhythm or cycle can be described by some observable A that describes the amplitude of the pattern, shape or rhythm. For example, the amplitude A can be a length for sand patterns, or a chemical concentration for biological systems, or a sound pressure for sound appearance.

Order appears when the amplitude A differs from zero. To understand the appearance of order, we have to understand the evolution of the amplitude A. The study of order has shown that this amplitude always follows similar evolution equations, *independently* of the physical mechanism of system. This surprising result unifies the whole field of self-organization.

All self-organizing systems at the onset of order appearance can be described by equations for the pattern amplitude A of the general form

$$\frac{\partial A(t, x)}{\partial t} = \lambda A - \mu |A|^2 A + \kappa \Delta A + \text{higher orders} \:. \tag{123}$$

Here, the observable A – which can be a real or a complex number, in order to describe phase effects – is the observable that appears when order appears, such as the oscillation amplitude or the pattern amplitude. The first term λA is the *driving term*, in which λ is a parameter describing the strength of the driving. The next term is a typical *non-linearity* in A, with μ a parameter that describes its strength, and the third term $\kappa \Delta A = \kappa(\partial^2 A/\partial x^2 + \partial^2 A/\partial y^2 + \partial^2 A/\partial z^2)$ is a typical diffusive and thus *dissipative* term.

We can distinguish two main situations. In cases where the dissipative term plays no role ($\kappa = 0$), we find that when the driving parameter λ increases above zero, a *temporal* oscillation appears, i.e., a stable *limit cycle* with non-vanishing amplitude. In cases where the diffusive term does play a role, equation (123) describes how an amplitude for a *spatial* oscillation appears when the driving parameter λ becomes positive, as the solution $A = 0$ then becomes spatially unstable.

In both cases, the onset of order is called a *bifurcation*, because at this critical value of the driving parameter λ the situation with amplitude zero, i.e., the homogeneous (or unordered) state, becomes unstable, and the ordered state becomes stable. *In non-linear systems, order is stable.* This is the main conceptual result of the field. Equation (123) and its numerous variations allow us to describe many phenomena, ranging from spirals,

---

* To describe the 'mystery' of human life, terms like 'fire', 'river' or 'tree' are often used as analogies. These are all examples of self-organized systems: they have many degrees of freedom, have competing driving and braking forces, depend critically on their initial conditions, show chaos and irregular behaviour, and sometimes show cycles and regular behaviour. Humans and human life resemble them in all these respects; thus there is a solid basis to their use as metaphors. We could even go further and speculate that pure beauty *is* pure self-organization. The lack of beauty indeed often results from a disturbed equilibrium between external braking and external driving.

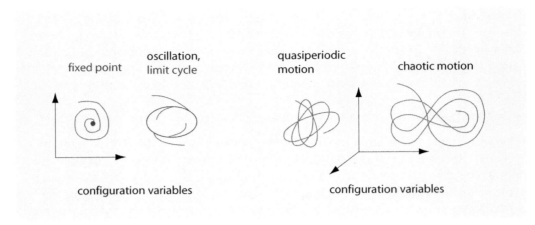

FIGURE 292 Examples of different types of motion in configuration space.

Ref. 329   waves, hexagonal patterns, and topological defects, to some forms of turbulence. For every physical system under study, the main task is to distil the observable A and the parameters λ, μ and κ from the underlying physical processes.

In summary, the appearance of order can be described to generally valid equations. Self-organization is, fundamentally, a simple process. In short: beauty is simple.

Self-organization is a vast field which is yielding new results almost by the week. To discover new topics of study, it is often sufficient to keep your eye open; most effects are

Challenge 707 e   comprehensible without advanced mathematics. Enjoy the hunting!

Chaos

Most systems that show self-organization also show another type of motion. When the driving parameter of a self-organizing system is increased to higher and higher values, order becomes more and more irregular, and in the end one usually finds *chaos*.

For physicists, chaotic motion is the most irregular type of motion.* Chaos can be defined independently of self-organization, namely as that motion of systems for which small changes in initial conditions evolve into large changes of the motion (exponentially with time). This is illustrated in Figure 293. More precisely,

> ▷ (Physical) *chaos* is irregular motion characterized by a positive *Lyapounov exponent* in the presence of a strictly valid evolution.

A simple chaotic system is the damped pendulum above three magnets. Figure 294 shows how regions of predictability (around the three magnet positions) gradually change into a chaotic region, i.e., a region of effective unpredictability, for higher initial amplitudes. The weather is also a chaotic system, as are dripping water-taps, the fall of dice, and many

---

* On the topic of chaos, see the beautiful book by Heinz-Otto Peitgen, Hartmut Jürgens & Dietmar Saupe, *Chaos and Fractals*, Springer Verlag, 1992. It includes stunning pictures, the necessary mathematical background, and some computer programs allowing personal exploration of the topic. 'Chaos' is an old word: according to Greek mythology, the first goddess, Gaia, i.e., the Earth, emerged from the chaos existing at the beginning. She then gave birth to the other gods, the animals and the first humans.

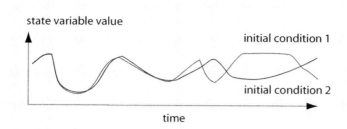

FIGURE 293 Chaos as sensitivity to initial conditions.

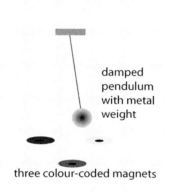
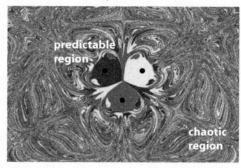

FIGURE 294 A simple chaotic system: a metal pendulum over three magnets (fractal © Paul Nylander).

other everyday systems. For example, research on the mechanisms by which the heart beat is generated has shown that the heart is not an oscillator, but a chaotic system with irregular cycles. This allows the heart to be continuously ready for demands for changes in beat rate which arise once the body needs to increase or decrease its efforts.

There is chaotic motion also in machines: chaos appears in the motion of trains on the rails, in gear mechanisms, and in fire-fighter's hoses. The precise study of the motion in a zippo cigarette lighter will probably also yield an example of chaos. The mathematical description of chaos – simple for some textbook examples, but extremely involved for others – remains an important topic of research.

Incidentally, can you give a simple argument to show that the so-called *butterfly effect* does not exist? This 'effect' is often cited in newspapers. The claim is that non-linearities imply that a small change in initial conditions can lead to large effects; thus a butterfly wing beat is alleged to be able to induce a tornado. Even though non-linearities do indeed lead to growth of disturbances, the butterfly 'effect' has *never* been observed. Thus it does *not* exist. This 'effect' exists only to sell books and to get funding.

All the steps from disorder to order, quasiperiodicity and finally to chaos, are examples of self-organization. These types of motion, illustrated in Figure 292, are observed in many fluid systems. Their study should lead, one day, to a deeper understanding of the mysteries of turbulence. Despite the fascination of this topic, we will not explore it further, because it does not help solving the mystery of motion.

Emergence

> From a drop of water a logician could predict
> an Atlantic or a Niagara.
> Arthur Conan Doyle, *A Study in Scarlet*

Self-organization is of interest also for a more general reason. It is sometimes said that our ability to formulate the patterns or rules of nature from observation does not imply the ability to predict *all* observations from these rules. According to this view, so-called 'emergent' properties exist, i.e., properties appearing in complex systems as something *new* that cannot be deduced from the properties of their parts and their interactions. (The ideological backdrop to this view is obvious; it is the latest attempt to fight the idea of determinism.) The study of self-organization has definitely settled this debate. The properties of water molecules do allow us to predict Niagara Falls.* Similarly, the diffusion of signal molecules do determine the development of a single cell into a full human being: in particular, cooperative phenomena determine the places where arms and legs are formed; they ensure the (approximate) right–left symmetry of human bodies, prevent mix-ups of connections when the cells in the retina are wired to the brain, and explain the fur patterns on zebras and leopards, to cite only a few examples. Similarly, the mechanisms at the origin of the heart beat and many other cycles have been deciphered. Several cooperative fluid phenomena have been simulated even down to the molecular level.

Self-organization provides general principles which allow us in principle to predict the behaviour of complex systems of any kind. They are presently being applied to the most complex system in the known universe: the human brain. The details of how it learns to coordinate the motion of the body, and how it extracts information from the images in the eye, are being studied intensely. The ongoing work in this domain is fascinating. (A neglected case of self-organization is *laughter*, but also *humour* itself.) If you plan to become a scientist, consider taking this path.

Self-organization research provided the final arguments that confirmed what J. Offrey de la Mettrie stated and explored in his famous book *L'homme machine* in 1748: humans are complex machines. Indeed, the lack of understanding of complex systems in the past was due mainly to the restrictive teaching of the subject of motion, which usually concentrated – as we do in this walk – on examples of motion in *simple* systems. The concepts of self-organization allow us to understand and to describe what happens during the functioning and the growth of organisms.

Even though the subject of self-organization provides fascinating insights, and will do so for many years to come, we now leave it. We continue with our own adventure,

---

* Already small versions of Niagara Falls, namely dripping water taps, show a large range of cooperative phenomena, including the chaotic, i.e., non-periodic, fall of water drops. This happens when the water flow rate has the correct value, as you can verify in your own kitchen.

THE SIMPLICITY OF COMPLEXITY

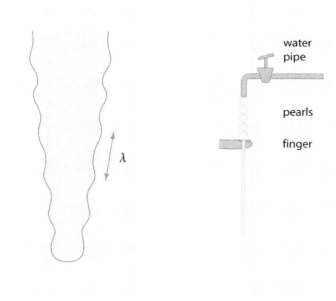
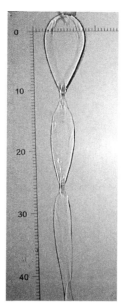

FIGURE 295 The wavy surface of icicles.

FIGURE 296 Water pearls.

FIGURE 297 A braiding water stream (© Vakhtang Putkaradze).

namely to explore the *basics* of motion.

> Ich sage euch: man muss noch Chaos in sich haben, um einen tanzenden Stern gebären zu können. Ich sage euch: ihr habt noch Chaos in euch.*
> Friedrich Nietzsche, *Also sprach Zarathustra*.

Curiosities and fun challenges about self-organization

All icicles have a wavy surface, with a crest-to-crest distance of about 1 cm, as shown in Figure 295. The distance is determined by the interplay between water flow and surface cooling. How? (Indeed, stalactites do not show the effect.)

* *

When a fine stream of water leaves a water tap, putting a finger in the stream leads to a wavy shape, as shown in Figure 296. Why?

* *

The research on sand has shown that it is often useful to introduce the concept of *granular temperature*, which quantifies how fast a region of sand moves. Research into this field is still in full swing.

---

* 'I tell you: one must have chaos inside oneself, in order to give birth to a dancing star. I tell you: you still have chaos inside you.'

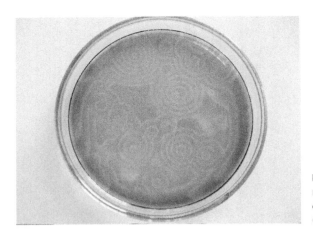

FIGURE 298 The Belousov-Zhabotinski reaction: the liquid periodically changes colour, both in space and time (© Yamaguchi University).

∗ ∗

When water emerges from a oblong opening, the stream forms a braid pattern, as shown in Figure 297. This effect results from the interplay and competition between inertia and surface tension: inertia tends to widen the stream, while surface tension tends to narrow it. Predicting the distance from one narrow region to the next is still a topic of research.

If the experiment is done in free air, without a plate, one usually observes an additional effect: there is a *chiral* braiding at the narrow regions, induced by the asymmetries of the water flow. You can observe this effect in the toilet! Scientific curiosity knows no limits: are you a right-turner or a left-turner, or both? On every day?

∗ ∗

When wine is made to swirl in a wine glass, after the motion has calmed down, the wine flowing down the glass walls forms little arcs. Can you explain in a few words what forms them?

∗ ∗

How does the average distance between cars parked along a street change over time, assuming a constant rate of cars leaving and arriving?

∗ ∗

A famous case of order appearance is the *Belousov-Zhabotinski reaction*. This mixture of chemicals spontaneously produces spatial and temporal patterns. Thin layers produce slowly rotating spiral patterns, as shown in Figure 298; Large, stirred volumes oscillate back and forth between two colours. A beautiful movie of the oscillations can be found on www.uni-r.de/Fakultaeten/nat_Fak_IV/Organische_Chemie/Didaktik/Keusch/D-oscill-d.htm. The exploration of this reaction led to the Nobel Prize in Chemistry for Ilya Prigogine in 1997.

∗ ∗

Gerhard Müller has discovered a simple but beautiful way to observe self-organization in solids. His system also provides a model for a famous geological process, the formation of

FIGURE 299 A famous correspondence: on the left, hexagonal columns in starch, grown in a kitchen pan (the red lines are 1 cm in length), and on the right, hexagonal columns in basalt, grown from lava in Northern Ireland (top right, view of around 300 m, and middle right, view of around 40 m) and in Iceland (view of about 30 m, bottom right) (© Gerhard Müller, Raphael Kessler - www.raphaelk.co.uk, Bob Pohlad, and Cédric Hüsler).

hexagonal columns in basalt, such as the Giant's Causeway in Northern Ireland. Similar formations are found in many other places of the Earth. Just take some rice flour or corn starch, mix it with about half the same amount of water, put the mixture into a pan and dry it with a lamp: hexagonal columns form. The analogy with basalt structures is possible because the drying of starch and the cooling of lava are diffusive processes governed by the same equations, because the boundary conditions are the same, and because both materials respond to cooling with a small reduction in volume.

* *

Water flow in pipes can be laminar (smooth) or turbulent (irregular and disordered).

The transition depends on the diameter d of the pipe and the speed v of the water. The transition usually happens when the so-called *Reynolds number* – defined as Re = vd/η becomes greater than about 2000. (The Reyonolds number is one of the few physical observables with a conventional abbreviation made of two letters.) Here, η is the *kinematic viscosity* of the water, around 1 mm²/s; in contrast, the *dynamic viscosity* is defined as $\mu = \eta\rho$, where ρ is the density of the fluid. A high Reynolds number means a high ratio between inertial and dissipative effects and specifies a turbulent flow; a low Reynolds number is typical of *viscous* flow.

Modern, careful experiments show that with proper handling, laminar flows can be produced up to Re = 100 000. A linear analysis of the equations of motion of the fluid, the Navier–Stokes equations, even predicts stability of laminar flow for *all* Reynolds numbers. This riddle was solved only in the years 2003 and 2004. First, a complex mathematical analysis showed that the laminar flow is not always stable, and that the transition to turbulence in a long pipe occurs with travelling waves. Then, in 2004, careful experiments showed that these travelling waves indeed appear when water is flowing through a pipe at large Reynolds numbers.

* *

For more beautiful pictures on self-organization in fluids, see the mentioned serve.me.nus.edu.sg/limtt website.

* *

Chaos can also be observed in simple (and complicated) electronic circuits. If the electronic circuit that you have designed behaves erratically, check this option!

* *

Also *dance* is an example of self-organization. This type of self-organization takes place in the brain. Like for all complex movements, learning them is often a challenge. Nowadays there are beautiful books that tell how physics can help you improve your dancing skills and the grace of your movements.

* *

Do you want to enjoy working on your PhD? Go into a scientific toy shop, and look for any toy that moves in a complex way. There are high chances that the motion is chaotic; explore the motion and present a thesis about it. For example, go to the extreme: explore the motion of a hanging rope whose upper end is externally driven. This simple system is fascinating in its range of complex motion behaviours.

* *

Self-organization is also observed in liquid corn starch–water mixtures. Enjoy the film at www.youtube.com/watch?v=f2XQ97XHjVw and watch even more bizarre effects, for humans walking over a pool filled with the liquid, on www.youtube.com/watch?v=nq3ZjY0Uf-g.

* *

Snowflakes and snow crystals have already been mentioned as examples of self-

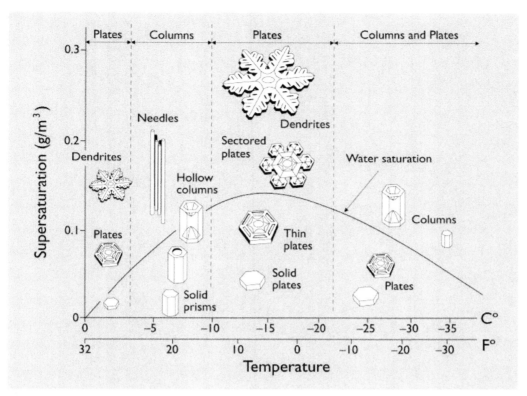

FIGURE 300 How the shape of snow crystals depend on temperature and saturation (© Kenneth Libbrecht).

organization. Figure 300 shows the general connection. To learn more about this fascinating topic, explore the wonderful website snowcrystals.com by Kenneth Libbrecht. A complete classification of snow crystals has also been developed.

* *

A famous example of self-organization whose mechanisms are not well-known so far, is the *hiccup*. It is known that the vagus nerve plays a role in it. Like for many other examples of self-organization, it takes quite some energy to get rid of a hiccup. Modern experimental research has shown that orgasms, which strongly stimulate the vagus nerve, are excellent ways to overcome hiccups. One of these researchers has won the 2006 IgNobel Prize for medicine for his work.

* *

Another important example of self-organization is the *weather*. If you want to know more about the known connections between the weather and the quality of human life on Earth, free of any ideology, read the wonderful book by Reichholf. It explains how the weather between the continents is connected and describes how and why the weather changed in the last one thousand years.

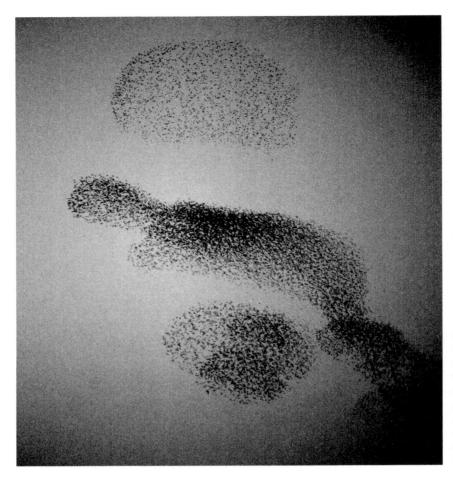

FIGURE 301 A typical swarm of starlings that visitors in Rome can observe every autumn (© Andrea Cavagna, Physics Today).

* *

Does self-organization or biological evolution contradict the second principle of thermodynamics? Of course not.

Self-organizing systems are, by definition, open systems, and often far from equilibrium. The concept of entropy cannot be defined, in general, for such systems, and the second principle of thermodynamics does not apply to them. However, entropy can be defined for systems *near* equilibrium. In such systems the second principle of thermodynamics must be modified; it is then possible to explore and confirm that self-organization does not contradict, but in fact *follows* from the modified second principle.

In particular, evolution does not contradict thermodynamics, as the Earth is not a closed thermodynamic system in equilibrium. Statements of the opposite are only made by crooks.

* *

In 2015, three physicists predicted that there is a largest Lyapunov exponent in nature, so

that one always has

$$\lambda \leqslant e\pi \frac{kT}{\hbar} \qquad (124)$$

where T is temperature, k is Boltzmann's constant, and $\hbar$ is the quantum of action. The growth of disorder is thus limited.

* *

Are systems that show self-organization the most complex ones that can be studied with evolution equations? No. The most complex systems are those that consist of many interacting self-organizing systems. The obvious example are *swarms*. Swarms of birds, as shown in Figure 301, of fish, of insects and of people – for example in a stadium or in cars on a highway – have been studied extensively and are still a subject of research. Their beauty is fascinating.

The other example of many interconnected self-organized systems is the brain; the exploration of how the interconnected neurons work to produce our *thoughts* will occupy researchers for many years. We will explore some aspects in the next volumes.

Summary on self-organization and chaos

Appearance of *order*, in form of patterns, shapes and cycles, is not due to a decrease in entropy, but to a competition between driving causes and dissipative effects in open systems. Such appearance of order is a common process, is often automatic, and is predictable with (quite) simple equations. Also the growth of living systems and biological evolution are examples of appearance of order.

Chaos, the sensitivity of evolution to initial conditions, is common in strongly driven open systems. Chaos is at the basis of everyday chance, and often is described by simple equations as well. In nature, complexity is apparent. Motion is simple.

Chapter 15
FROM THE LIMITATIONS OF PHYSICS TO THE LIMITS OF MOTION

> I only know that I know nothing.
> Socrates, as cited by Plato

We have explored, in our environment, the concept of motion. We called this exploration of moving objects and fluids *Galilean physics*. We found that in everyday life, motion is *predictable*: nature shows no surprises and no miracles. In particular, we have found six important aspects of this predictability:

1. Everyday motion is *continuous*. Motion allows us to define space and time.
2. Everyday motion *conserves* mass, momentum, energy and angular momentum. Nothing appears out of nothing.
3. Everyday motion is *relative*: motion depends on the observer.
4. Everyday motion is *reversible*: everyday motion can occur backwards.
5. Everyday motion is *mirror-invariant*: everyday motion can occur in a mirror-reversed way.

The final property is the most important and, in addition, contains the five previous ones:

6. **Everyday motion is *lazy*:** motion happens in a way that minimizes change, i.e., physical action.

This Galilean description of nature made engineering possible: textile machines, steam engines, combustion motors, kitchen appliances, watches, many children toys, fitness machines, medical devices and all the progress in the quality of life that came with these devices are due to the results of Galilean physics. But despite these successes, Socrates' saying, cited above, still applies to Galilean physics: we still know almost nothing. Let us see why.

RESEARCH TOPICS IN CLASSICAL DYNAMICS

Even though mechanics and thermodynamics are now several hundred years old, research into its details is still ongoing. For example, we have already mentioned above that it is unclear whether the Solar System is stable. The long-term future of the planets is unknown! In general, the behaviour of few-body systems interacting through gravitation is still a research topic of mathematical physics. Answering the simple question of how long a given set of bodies gravitating around each other will stay together is a formidable challenge. The history of this so-called *many-body problem* is long and involved.

TABLE 52 Examples of errors in state-of-the art measurements (numbers in brackets give one standard deviation in the last digits), partly taken from physics.nist.gov/constants.

Observation	Measurement	Precision / accuracy
Highest precision achieved: ratio between the electron magnetic moment and the Bohr magneton μ_e/μ_B	$-1.001\,159\,652\,180\,76(24)$	$2.6 \cdot 10^{-13}$
High precision: Rydberg constant	$10\,973\,731.568\,539(55)\,\text{m}^{-1}$	$5.0 \cdot 10^{-12}$
High precision: astronomical unit	$149\,597\,870.691(30)\,\text{km}$	$2.0 \cdot 10^{-10}$
Industrial precision: typical part dimension tolerance in car engine	$2\,\mu\text{m}$ of $20\,\text{cm}$	$5 \cdot 10^{-6}$
Low precision: gravitational constant G	$6.674\,28(67) \cdot 10^{-11}\,\text{Nm}^2/\text{kg}^2$	$1.0 \cdot 10^{-4}$
Everyday precision: human circadian clock governing sleep	$15\,\text{h}$ to $75\,\text{h}$	2

Interesting progress has been achieved, but the final answer still eludes us.

Many challenges remain in the fields of self-organization, of non-linear evolution equations and of chaotic motion. In these fields, *turbulence* is a famous example: a precise description of turbulence has not yet been achieved, despite intense efforts. Large prizes are offered for its solution.

Many other challenges motivate numerous researchers in mathematics, physics, chemistry, biology, medicine and the other natural sciences. But apart from these research topics, classical physics leaves unanswered several basic questions.

What is contact?

> " Democritus declared that there is a unique sort of motion: that ensuing from collision.
> Simplicius, *Commentary on the Physics of Aristotle*, 42, 10 "

Of the questions unanswered by classical physics, the details of *contact* and *collisions* are among the most pressing. Indeed, we defined mass in terms of velocity changes during collisions. But why do objects change their motion in such instances? Why are collisions between two balls made of chewing gum different from those between two stainless-steel balls? What happens during those moments of contact?

Contact is related to material properties, which in turn influence motion in a complex way. The complexity is such that the sciences of material properties developed independently from the rest of physics for a long time; for example, the techniques of metallurgy (often called the oldest science of all), of chemistry and of cooking were related to the properties of motion only in the twentieth century, after having been independently pursued for thousands of years. Since material properties determine the essence of contact, we *need* knowledge about matter and about materials to understand the notion of mass, of contact and thus of motion. The parts of our adventure that deal with quantum theory will reveal these connections.

What determines precision and accuracy?

Precision has its own fascination. How many digits of π, the ratio between circumference and diameter of a circle, do you know by heart? What is the largest number of digits of π you have calculated yourself?

Is it possible to draw or cut a rectangle for which the ratio of lengths is a number, e.g. of the form 0.131520091514001315211420010914..., whose digits encode a full book? (A simple method would code a space as 00, the letter 'a' as 01, 'b' as 02, 'c' as 03, etc. Even more interestingly, could the number be printed inside its own book?)

Why are so many measurement results, such as those of Table 52, of *limited* precision, even if the available financial budget for the measurement apparatus is almost unlimited? These are all questions about precision.

When we started climbing Motion Mountain, we explained that gaining height means increasing the *precision* of our description of nature. To make even this statement itself more precise, we distinguish between two terms: *precision* is the degree of reproducibility; *accuracy* is the degree of correspondence to the actual situation. Both concepts apply to measurements,* to statement and to physical concepts.

Statements with false accuracy and false precision abound. What should we think of a car company – Ford – who claim that the drag coefficient c_w of a certain model is 0.375? Or of the official claim that the world record in fuel consumption for cars is 2315.473 km/l? Or of the statement that 70.3 % of all citizens share a certain opinion? One lesson we learn from investigations into measurement errors is that we should never provide more digits for a result than we can put our hand into fire for.

In short, precision and accuracy are *limited*. At present, the record number of reliable digits ever measured for a physical quantity is 13. Why so few? Galilean physics doesn't provide an answer at all. What is the maximum number of digits we can expect in measurements; what determines it; and how can we achieve it? These questions are still open at this point in our adventure. They will be covered in the parts on quantum theory.

In our walk we aim for highest possible precision and accuracy, while avoiding false accuracy. Therefore, concepts have mainly to be *precise*, and descriptions have to be *accurate*. Any inaccuracy is a proof of lack of understanding. To put it bluntly, in our adventure, 'inaccurate' means *wrong*. Increasing the accuracy and precision of our description of nature implies leaving behind us all the mistakes we have made so far. This quest raises several issues.

Can all of nature be described in a book?

> Darum kann es in der Logik auch *nie* Überraschungen geben.**
> Ludwig Wittgenstein, *Tractatus*, 6.1251

Could the perfect physics publication, one that describes *all* of nature, exist? If it does, it must also describe itself, its own production – including its readers and its author – and most important of all, its own contents. Is such a book possible? Using the concept

---

* For measurements, both precision and accuracy are best described by their *standard deviation*, as explained on page 458.
** 'Hence there can *never* be surprises in logic.'

of information, we can state that such a book must contain all information contained in the universe. Is this possible? Let us check the options.

If nature requires an *infinitely* long book to be fully described, such a publication obviously cannot exist. In this case, only approximate descriptions of nature are possible and a perfect physics book is impossible.

If nature requires a *finite* amount of information for its description, there are two options. One is that the information of the universe is so large that it cannot be summarized in a book; then a perfect physics book is again impossible. The other option is that the universe does contain a finite amount of information and that it can be summarized in a few short statements. This would imply that the rest of the universe would not add to the information already contained in the perfect physics book.

We note that the answer to this puzzle also implies the answer to another puzzle: whether a brain can contain a full description of nature. In other words, the real question is: can we understand nature? Can we reach our aim to understand motion? We usually believe this. But the arguments just given imply that we effectively believe that the universe does not contain more information than what our brain could contain or even contains already. What do you think? We will solve this puzzle later in our adventure. Until then, do make up your own mind.

Something is wrong about our description of motion

> Je dis seulement qu'il n'y a point d'espace, où il
> n'y a point de matière; et que l'espace lui-même
> n'est point une réalité absolue. *
> Leibniz

We described nature in a rather simple way. *Objects* are permanent and massive entities localized in space-time. *States* are changing properties of objects, described by position in space and instant in time, by energy and momentum, and by their rotational equivalents. *Time* is the relation between events measured by a clock. *Clocks* are devices in undisturbed motion whose position can be observed. *Space* and position is the relation between objects measured by a metre stick. *Metre sticks* are devices whose shape is subdivided by some marks, fixed in an invariant and observable manner. *Motion* is change of position with time (times mass); it is determined, does not show surprises, is conserved (even in death), and is due to gravitation and other interactions.

Even though this description works rather well in practice, it contains a circular definition. Can you spot it? Each of the two central concepts of motion is defined with the help of the other. Physicists worked for about 200 years on classical mechanics without noticing or wanting to notice the situation. Even thinkers with an interest in discrediting science did not point it out. Can an exact science be based on a circular definition? Obviously yes, and physics has done quite well so far. Is the situation unavoidable in principle?

---

* 'I only say that there is no space where there is no matter; and that space itself is not an absolute reality.' Gottfried Wilhelm Leibniz writes this already in 1716, in section 61 of his famous fifth letter to Clarke, the assistant and spokesman of Newton. Newton, and thus Clarke, held the opposite view; and as usual, Leibniz was right.

Undoing the circular definition of Galilean physics is one of the aims of the rest of our walk. The search for a solution is part of the last leg of our adventure. To achieve the solution, we need to increase substantially the level of precision in our description of motion.

Whenever precision is increased, imagination is restricted. We will discover that many types of motion that seem possible are not. Motion is *limited*. Nature limits speed, size, acceleration, mass, force, power and many other quantities. Continue reading the other parts of this adventure only if you are prepared to exchange fantasy for precision. It will be no loss, because exploring the precise working of nature will turn out to be more fascinating than any fantasy.

Why is measurement possible?

In the description of gravity given so far, the one that everybody learns – or should learn – at school, acceleration is connected to mass and distance via $a = GM/r^2$. That's all. But this simplicity is deceiving. In order to check whether this description is correct, we have to measure lengths and times. However, it is *impossible* to measure lengths and time intervals with any clock or any ruler based on the gravitational interaction alone! Try to conceive such an apparatus and you will be inevitably be disappointed. You always need a non-gravitational method to start and stop the stopwatch. Similarly, when you measure length, e.g. of a table, you have to hold a ruler or some other device near it. The interaction necessary to line up the ruler and the table cannot be gravitational.

A similar limitation applies even to mass measurements. Try to measure mass using gravitation alone. Any scale or balance needs other – usually mechanical, electromagnetic or optical – interactions to achieve its function. Can you confirm that the same applies to speed and to angle measurements? In summary, whatever method we use,

> ▷ In order to measure velocity, length, time, and mass, interactions other than gravity are needed.

Our ability to measure shows that gravity is not all there is. And indeed, we still need to understand charge and colours.

In short, Galilean physics does not explain our ability to measure. In fact, it does not even explain the existence of measurement standards. Why do objects have fixed lengths? Why do clocks work with regularity? Galilean physics cannot explain these observations; we will need relativity and quantum physics to find out.

Is motion unlimited?

Galilean physics suggests that linear motion could go on forever. In fact, Galilean physics tacitly assumes that the universe is infinite in space and time. Indeed, finitude of any kind contradicts the Galilean description of motion. On the other hand, we know from observation that the universe is *not* infinite: if it were infinite, the night would not be dark.

Galilean physics also suggests that speeds can have any value. But the existence of infinite speeds in nature would not allow us to define time sequences. Clocks would be impossible. In other words, a description of nature that allows unlimited speeds is not precise. Precision and measurements require limits.

Because Galilean physics disregards limits to motion, Galilean physics is inaccurate, and thus wrong.

To achieve the highest possible precision, and thus to find the correct description of motion, we need to discover all of *motion's limits*. So far, we have discovered one: there is a smallest entropy in nature. We now turn to another, more striking limit: the speed limit for energy, objects and signals. To observe and understand the speed limit, the next volume explores the most rapid motion of energy, objects and signals that is known: the motion of light.

Appendix A
NOTATION AND CONVENTIONS

Newly introduced concepts are indicated, throughout this text, by *italic typeface*. New definitions are also referred to in the index. In this text, naturally we use the international SI units; they are defined in Appendix B. Experimental results are cited with limited precision, usually only two digits, as this is almost always sufficient for our purposes. High-precision reference values for important quantities can also be found in Appendix B. Additional precision values on composite physical systems are given in volume V.

But the information that is provided in this volume uses some additional conventions that are worth a second look.

The Latin alphabet

> " What is written without effort is in general read without pleasure. "
> Samuel Johnson

Books are collections of symbols. *Writing* was probably invented between 3400 and 3300 BCE by the Sumerians in Mesopotamia (though other possibilities are also discussed). It then took over a thousand years before people started using symbols to represent sounds instead of concepts: this is the way in which the first alphabet was created. This happened between 2000 and 1600 BCE (possibly in Egypt) and led to the Semitic alphabet. The use of an alphabet had so many advantages that it was quickly adopted in all neighbouring cultures, though in different forms. As a result, the Semitic alphabet is the forefather of all alphabets used in the world.

This text is written using the Latin alphabet. At first sight, this seems to imply that its pronunciation *cannot* be explained in print, in contrast to the pronunciation of other alphabets or of the International Phonetic Alphabet (IPA). (They can be explained using the alphabet of the main text.) However, it *is* in principle possible to write a text that describes exactly how to move lips, mouth and tongue for each letter, using physical concepts where necessary. The descriptions of pronunciations found in dictionaries make indirect use of this method: they refer to the memory of pronounced words or sounds found in nature.

Historically, the Latin alphabet was derived from the Etruscan, which itself was a derivation of the Greek alphabet. An overview is given in Figure 302. There are two main forms of the Latin alphabet.

A NOTATION AND CONVENTIONS

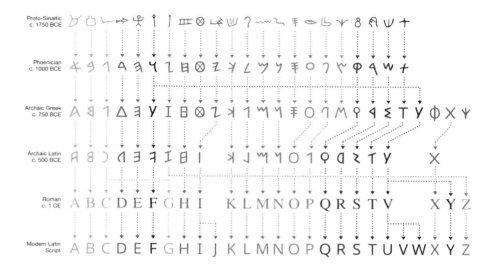

FIGURE 302 A summary of the history of the Latin alphabet (© Matt Baker at usefulcharts.com).

The *ancient* Latin alphabet,
used from the sixth century BCE onwards:

A B C D E F Z H I K L M N O P Q R S T V X

The *classical* Latin alphabet,
used from the second century BCE until the eleventh century:

A B C D E F G H I K L M N O P Q R S T V X Y Z

The letter G was added in the third century BCE by the first Roman to run a fee-paying school, Spurius Carvilius Ruga. He added a horizontal bar to the letter C and substituted the letter Z, which was not used in Latin any more, for this new letter. In the second century BCE, after the conquest of Greece, the Romans included the letters Y and Z from the Greek alphabet at the end of their own (therefore effectively reintroducing the Z) in order to be able to write Greek words. This classical Latin alphabet was stable for the next thousand years.*

The classical Latin alphabet was spread around Europe, Africa and Asia by the Romans during their conquests; due to its simplicity it began to be used for writing in nu-

---

* To meet Latin speakers and writers, go to www.alcuinus.net.

merous other languages. Most modern 'Latin' alphabets include a few other letters. The letter W was introduced in the eleventh century in French and was then adopted in most European languages.* The letter U was introduced in the mid fifteenth century in Italy, the letter J at the end of that century in Spain, to distinguish certain sounds which had previously been represented by V and I. The distinction proved a success and was already common in most European languages in the sixteenth century. The contractions æ and œ date from the Middle Ages. The German alphabet includes the *sharp s*, written ß, a contraction of 'ss' or 'sz', and the Nordic alphabets added *thorn*, written Þ or þ, and *eth*, written Ð or ð, both taken from the futhorc,** and other signs.

Lower-case letters were not used in classical Latin; they date only from the Middle Ages, from the time of Charlemagne. Like most accents, such as ê, ç or ä, which were also first used in the Middle Ages, lower-case letters were introduced to save the then expensive paper surface by shortening written words.

> " Outside a dog, a book is a man's best friend.
> Inside a dog, it's too dark to read. "
> Groucho Marx

The Greek alphabet

The Greek alphabet is central to modern culture and civilization. It is at the origin of the Etruscan alphabet, from which the Latin alphabet was derived. The Greek alphabet was itself derived from the Phoenician or a similar northern Semitic alphabet in the tenth century BCE. The Greek alphabet, for the first time, included letters also for vowels, which the Semitic alphabets lacked (and often still lack).

In the Phoenician alphabet and in many of its derivatives, such as the Greek alphabet, each letter has a proper name. This is in contrast to the Etruscan and Latin alphabets. The first two Greek letter names are, of course, the origin of the term *alphabet* itself.

In the tenth century BCE, the Ionian or *ancient* (eastern) Greek alphabet consisted of the upper-case letters only. In the sixth century BCE several letters were dropped, while a few new ones and the lower-case versions were added, giving the *classical* Greek alphabet. Still later, accents, subscripts and breathings were introduced. Table 53 also gives the values signified by the letters when they were used as numbers. For this special use, the obsolete ancient letters were kept during the classical period; thus they also acquired lower-case forms.

The Latin correspondence in the table is the standard classical one, used for writing Greek words. The question of the correct *pronunciation* of Greek has been hotly debated

---

* In Turkey, still in 2013, you can be convoked in front of a judge if you use the letters w, q or x in an official letter; these letters only exist in the Kurdish language, not in Turkish. Using them is 'unturkish' behaviour, support of terrorism and punishable by law. It is not generally known how physics and mathematics teachers cope with this situation.

** The Runic script, also called *Futhark* or *Futhorc*, a type of alphabet used in the Middle Ages in Germanic, Anglo–Saxon and Nordic countries, probably also derives from the Etruscan alphabet. The name derives from the first six letters: f, u, th, a (or o), r, k (or c). The third letter is the letter thorn mentioned above; it is often written 'Y' in Old English, as in 'Ye Olde Shoppe.' From the runic alphabet Old English also took the letter *wyn* to represent the 'w' sound, and the already mentioned eth. (The other letters used in Old English – not from futhorc – were the *yogh*, an ancient variant of g, and the ligatures æ or Æ, called *ash*, and œ or Œ, called *ethel*.)

A NOTATION AND CONVENTIONS

TABLE 53 The ancient and classical Greek alphabets, and the correspondence with Latin letters and Indian digits.

Anc.	Class.		Name	Corresp.		Anc.	Class.		Name	Corresp.	
A	A	α	alpha	a	1	N	N	ν	nu	n	50
B	B	β	beta	b	2	Ξ	Ξ	ξ	xi	x	60
Γ	Γ	γ	gamma	g, n[1]	3	O	O	o	omicron	o	70
Δ	Δ	δ	delta	d	4	Π	Π	π	pi	p	80
E	E	ε	epsilon	e	5	Ϙ	ϙ, ϟ		qoppa[3]	q	90
F	Ϝ, ϛ		digamma, stigma[2]	w	6	P	P	ρ	rho	r, rh	100
						Σ	Σ	σ, ς	sigma[4]	s	200
Z	Z	ζ	zeta	z	7	T	T	τ	tau	t	300
H	H	η	eta	e	8	Υ		υ	upsilon	y, u[5]	400
Θ	Θ	θ	theta	th	9	Φ		φ	phi	ph, f	500
I	I	ι	iota	i, j	10	X		χ	chi	ch	600
K	K	κ	kappa	k	20	Ψ		ψ	psi	ps	700
Λ	Λ	λ	lambda	l	30	Ω		ω	omega	o	800
M	M	μ	mu	m	40	Ϡ	ϡ		sampi[6]	s	900

The regional archaic letters yot, sha and san are not included in the table. The letter san was the ancestor of sampi.

1. Only if before velars, i.e., before kappa, gamma, xi and chi.
2. 'Digamma' is the name used for the F-shaped form. It was mainly used as a letter (but also sometimes, in its lower-case form, as a number), whereas the shape and name 'stigma' is used only for the number. Both names were derived from the respective shapes; in fact, the stigma is a medieval, uncial version of the digamma. The name 'stigma' is derived from the fact that the letter looks like a sigma with a tau attached under it – though unfortunately not in all modern fonts. The original letter name, also giving its pronunciation, was 'waw'.
3. The version of qoppa that looks like a reversed and rotated z is still in occasional use in modern Greek. Unicode calls this version 'koppa'.
4. The second variant of sigma is used only at the end of words.
5. Uspilon corresponds to 'u' only as the second letter in diphthongs.
6. In older times, the letter sampi was positioned between pi and qoppa.

in specialist circles; the traditional *Erasmian* pronunciation does not correspond either to the results of linguistic research, or to modern Greek. In classical Greek, the sound that sheep make was βη–βη. (Erasmian pronunciation wrongly insists on a narrow η; modern Greek pronunciation is different for β, which is now pronounced 'v', and for η, which is now pronounced as 'i:' – a long 'i'.) Obviously, the pronunciation of Greek varied from region to region and over time. For Attic Greek, the main dialect spoken in the classical period, the question is now settled. Linguistic research has shown that chi, phi and theta were less aspirated than usually pronounced in English and sounded more like the initial sounds of 'cat', 'perfect' and 'tin'; moreover, the zeta seems to have been pronounced more like 'zd' as in 'buzzed'. As for the vowels, contrary to tradition, epsilon is closed and short whereas eta is open and long; omicron is closed and short whereas omega is wide and long, and upsilon is really a sound like a French 'u' or German 'ü.'

The Greek vowels can have rough or smooth *breathings, subscripts,* and acute, grave, circumflex or diaeresis *accents*. Breathings – used also on ρ – determine whether the letter is aspirated. Accents, which were interpreted as stresses in the Erasmian pronunciation, actually represented pitches. Classical Greek could have up to three of these added signs per letter; modern Greek never has more than one.

Another descendant of the Greek alphabet* is the *Cyrillic alphabet*, which is used with slight variations in many Slavic languages, such as Russian and Bulgarian. There is no standard transcription from Cyrillic to Latin, so that often the same Russian name is spelled differently in different countries or even in the same country on different occasions.

TABLE 54 The beginning of the Hebrew abjad.

Letter	Names	Correspondence	
א	aleph	a	1
ב	beth	b	2
ג	gimel	g	3
ד	daleth	d	4
etc.			

The Hebrew alphabet and other scripts

The Phoenician alphabet is also the origin of the Hebrew consonant alphabet or *abjad*. Its first letters are given in Table 54. Only the letter aleph is commonly used in mathematics, though others have been proposed.

Around one hundred writing systems are in use throughout the world. Experts classify them into five groups. *Phonemic alphabets*, such as Latin or Greek, have a sign for each consonant and vowel. *Abjads* or consonant alphabets, such as Hebrew or Arabic, have a sign for each consonant (sometimes including some vowels, such as aleph), and do not write (most) vowels; most abjads are written from right to left. *Abugidas*, also called *syllabic alphabets* or *alphasyllabaries*, such as Balinese, Burmese, Devanagari, Tagalog, Thai, Tibetan or Lao, write consonants and vowels; each consonant has an inherent vowel which can be changed into the others by diacritics. *Syllabaries*, such as Hiragana or Ethiopic, have a sign for each syllable of the language. Finally, *complex scripts*, such as Chinese, Mayan or the Egyptian hieroglyphs, use signs which have both sound and meaning. Writing systems can have text flowing from right to left, from bottom to top, and can count book pages in the opposite sense to this book.

---

* The Greek alphabet is also the origin of the *Gothic alphabet*, which was defined in the fourth century by Wulfila for the Gothic language, using also a few signs from the Latin and futhorc scripts.

The Gothic alphabet is not to be confused with the so-called *Gothic letters*, a style of the *Latin* alphabet used all over Europe from the eleventh century onwards. In Latin countries, Gothic letters were replaced in the sixteenth century by the *Antiqua*, the ancestor of the type in which this text is set. In other countries, Gothic letters remained in use for much longer. They were used in type and handwriting in Germany until 1941, when the National Socialist government suddenly abolished them, in order to comply with popular demand. They remain in sporadic use across Europe. In many physics and mathematics books, Gothic letters are used to denote vector quantities.

A NOTATION AND CONVENTIONS

Even though there are about 6000 languages on Earth, there are only about one hundred writing systems in use today. About fifty other writing systems have fallen out of use.* For physical and mathematical formulae, though, the sign system used in this text, based on Latin and Greek letters, written from left to right and from top to bottom, is a standard the world over. It is used independently of the writing system of the text containing it.

Numbers and the Indian digits

Both the digits and the method used in this text to write numbers originated in India. The Indian digits date from around 600 BCE and were brought to the Mediterranean by Arabic mathematicians in the Middle Ages. The number system used in this text is thus much younger than the alphabet. The Indian numbers were made popular in Europe by Leonardo of Pisa, called Fibonacci,** in his book *Liber Abaci* or 'Book of Calculation', which he published in 1202. That book revolutionized mathematics. Anybody with paper and a pen (the pencil had not yet been invented) was now able to calculate and write down numbers as large as reason allowed, or even larger, and to perform calculations with them. Fibonacci's book started:

> Novem figure indorum he sunt 9 8 7 6 5 4 3 2 1. Cum his itaque novem figuris, et cum hoc signo 0, quod arabice zephirum appellatur, scribitur quilibet numerus, ut inferius demonstratur.***

The Indian method of writing numbers, the *Indian number system*, introduced two innovations: a large one, the *positional system*, and a small one, the digit zero.**** The positional system, as described by Fibonacci, was so much more efficient that it completely replaced the previous *Roman number system*, which writes 1996 as IVMM or MCMIVC or MCMXCVI, as well as the *Greek number system*, in which the Greek letters were used for numbers in the way shown in Table 53, thus writing 1996 as ͵αϡϟϛ′. Compared to these systems, the Indian numbers are a much better technology.

The Indian number system proved so practical that calculations done on paper completely eliminated the need for the *abacus*, which therefore fell into disuse. The abacus is still in use in countries, for example in Asia, America or Africa, and by people who do not use a positional system to write numbers. It is also useful for the blind.

The Indian number system also eliminated the need for systems to represent numbers with fingers. Such ancient systems, which could show numbers up to 10 000 and more,

---

* A well-designed website on the topic is www.omniglot.com. The main present and past writing systems are encoded in the Unicode standard, which at present contains 52 writing systems. See www.unicode.org. An eccentric example of a writing system is the alphabet proposed by George Bernard Shaw, presented at en.wikipedia.org/wiki/Shavian_alphabet.
** Leonardo di Pisa, called Fibonacci (b. *c.* 1175 Pisa, d. 1250 Pisa), was the most important mathematician of his time.
*** 'The nine figures of the Indians are: 9 8 7 6 5 4 3 2 1. With these nine figures, and with this sign 0 which in Arabic is called zephirum, any number can be written, as will be demonstrated below.'
**** It is thus *not* correct to call the digits 0 to 9 *Arabic*. Both the digits used in Arabic texts and the digits used in Latin texts such as this one derive from the *Indian digits*. You can check this by yourself: only the digits 0, 2, 3 and 7 resemble those used in Arabic writing, and then only if they are turned clockwise by 90°.

have left only one trace: the term 'digit' itself, which derives from the Latin word for finger.

The power of the positional number system is often forgotten. For example, only a positional number system allows mental calculations and makes calculating prodigies possible.*

The symbols used in the text

> To avoide the tediouse repetition of these woordes: is equalle to: I will sette as I doe often in woorke use, a paire of paralleles, or Gemowe lines of one lengthe, thus: = , bicause noe .2. thynges, can be moare equalle
>
> Robert Recorde**

Besides text and numbers, physics books contain other symbols. Most symbols have been developed over hundreds of years, so that only the clearest and simplest are now in use. In this adventure, the symbols used as abbreviations for *physical* quantities are all taken from the Latin or Greek alphabets and are always defined in the context where they are used. The symbols designating units, constants and particles are defined in Appendix B and in Appendix B of volume V. The symbols used in this text are those in common use in the practice and the teaching of physics.

There is even an international standard for the symbols in physical formulae – ISO EN 80000, formerly ISO 31 – but it is shamefully expensive, virtually inaccessible and incredibly useless: the symbols listed in it are those in common use anyway, and their use is not binding anywhere, not even in the standard itself! ISO 80000 is a prime example of bureaucracy gone wrong.

The *mathematical* symbols used in this text, in particular those for mathematical operations and relations, are given in the following list, together with their historical origin. The details of their history have been extensively studied in by scholars.

TABLE 55 The history of mathematical notation and symbols.

Symbol	Meaning	Origin
+, −	plus, minus	Johannes Widmann 1489; the plus sign is derived from Latin 'et'.
√	read as 'square root'	used by Christoff Rudolff in 1525; the sign evolved from a point.
=	equal to	Robert Recorde 1557

* Currently, the shortest time for finding the thirteenth (integer) root of a hundred-digit (integer) number, a result with 8 digits, is below 4 seconds, and 70.2 seconds for a 200 digit number, for which the result has sixteen digits. Both records are held by Alexis Lemaire. For more about the stories and the methods of calculating prodigies, see the bibliography.

** Robert Recorde (b. *c.* 1510 Tenby, d. 1558 London), mathematician and physician; he died in prison because of debts. The quotation is from his *The Whetstone of Witte*, 1557. An image showing the quote can be found at en.wikipedia.org/wiki/Equals_sign. It is usually suggested that the quote is the first introduction of the equal sign; claims that Italian mathematicians used the equal sign before Recorde are not backed up by convincing examples.

TABLE 55 (Continued) The history of mathematical notation and symbols.

Symbol	Meaning	Origin
$\{\}, [\,], (\,)$	grouping symbols	use starts in the sixteenth century
$>, <$	larger than, smaller than	Thomas Harriot 1631
$\times$	multiplied with, times	England c. 1600, made popular by William Oughtred 1631
a^n	a to the power n, $a \cdot \ldots \cdot a$ (n factors)	René Descartes 1637
x, y, z	coordinates, unknowns	René Descartes 1637
$ax + by + c = 0$	constants and equations for unknowns	René Descartes 1637
∞	infinity	John Wallis 1655
d/dx, dx, $\int y\, dx$	derivative, differential, integral	Gottfried Wilhelm Leibniz 1675
$:$	divided by	Gottfried Wilhelm Leibniz 1684
$\cdot$	multiplied with, times	Gottfried Wilhelm Leibniz c. 1690
a_1, a_n	indices	Gottfried Wilhelm Leibniz c. 1690
$\sim$	similar to	Gottfried Wilhelm Leibniz c. 1690
π	circle number, $4 \arctan 1$	William Jones 1706
φx	function of x	Johann Bernoulli 1718
$fx, f(x)$	function of x	Leonhard Euler 1734
e	$\sum_{n=0}^{\infty} \frac{1}{n!} = \lim_{n \to \infty}(1 + 1/n)^n$	Leonhard Euler 1736
$f'(x)$	derivative of function at x	Giuseppe Lagrangia 1770
$\Delta x, \sum$	difference, sum	Leonhard Euler 1755
$\prod$	product	Carl Friedrich Gauss 1812
i	imaginary unit, $+\sqrt{-1}$	Leonhard Euler 1777
$\neq$	is different from	Leonhard Euler eighteenth century
$\partial/\partial x$	partial derivative, read like 'd/dx'	it was derived from a cursive form of 'd' or of the letter 'dey' of the Cyrillic alphabet by Adrien-Marie Legendre in 1786 and made popular by Carl Gustav Jacobi in 1841
$n!$	factorial, $1 \cdot 2 \cdot \ldots \cdot$	Christian Kramp 1808
Δ	Laplace operator	Robert Murphy 1833
$|x|$	absolute value	Karl Weierstrass 1841
∇	read as 'nabla' (or 'del')	introduced by William Hamilton in 1853 and Peter Tait in 1867, named after the shape of an old Egyptian musical instrument
$\subset, \supset$	set inclusion	Ernst Schröder in 1890
$\cup, \cap$	set union and intersection	Giuseppe Peano 1888
$\in$	element of	Giuseppe Peano 1888
$\otimes$	dyadic product or tensor product or outer product	unknown
$\langle\psi|, |\psi\rangle$	bra and ket state vectors	Paul Dirac 1930

TABLE 55 (Continued) The history of mathematical notation and symbols.

Symbol	Meaning	Origin
∅	empty set	André Weil as member of the Nicolas Bourbaki group in the early twentieth century
[x]	the measurement unit of a quantity x	twentieth century

Other signs used here have more complicated origins. The & sign is a contraction of Latin *et* meaning 'and', as is often more clearly visible in its variations, such as *&*, the common italic form.

Each of the punctuation signs used in sentences with modern Latin alphabets, such as , . ; : ! ? ' ' » « – ( ) ... has its own history. Many are from ancient Greece, but the question mark is from the court of Charlemagne, and exclamation marks appear first in the sixteenth century.* The @ or *at-sign* probably stems from a medieval abbreviation of Latin *ad*, meaning 'at', similarly to how the & sign evolved from Latin *et*. In recent years, the *smiley* :-) and its variations have become popular. The smiley is in fact a new version of the 'point of irony' which had been proposed in 1899, without success, by the poet Alcanter de Brahm (b. 1868 Mulhouse, d. 1942 Paris).

The section sign § dates from the thirteenth century in northern Italy, as was shown by the palaeographer Paul Lehmann. It was derived from ornamental versions of the capital letter C for *capitulum*, i.e., 'little head' or 'chapter.' The sign appeared first in legal texts, where it is still used today, and then spread into other domains.

The paragraph sign ¶ was derived from a simpler ancient form looking like the Greek letter Γ, a sign which was used in manuscripts from ancient Greece until well into the Middle Ages to mark the start of a new text paragraph. In the Middle Ages it took the modern form, probably because a letter c for *caput* was added in front of it.

One of the most important signs of all, the *white space* separating words, was due to Celtic and Germanic influences when these people started using the Latin alphabet. It became commonplace between the ninth and the thirteenth century, depending on the language in question.

Calendars

The many ways to keep track of time differ greatly from civilization to civilization. The most common calendar, and the one used in this text, is also one of the most absurd, as it is a compromise between various political forces who tried to shape it.

In ancient times, independent localized entities, such as tribes or cities, preferred *lunar* calendars, because lunar timekeeping is easily organized locally. This led to the use of the month as a calendar unit. Centralized states imposed *solar* calendars, based on the year. Solar calendars require astronomers, and thus a central authority to finance them. For various reasons, farmers, politicians, tax collectors, astronomers, and some, but not all, religious groups wanted the calendar to follow the solar year as precisely as possible. The compromises necessary between days and years are the origin of leap days.

---

* On the parenthesis see the beautiful book by J. Lennard, *But I Digress*, Oxford University Press, 1991.

The compromises necessary between months and year led to the varying lengths of the months; they are different in different calendars. The most commonly used year–month structure was organized over 2000 years ago by Caesar, and is thus called the *Julian calendar*.

The system was destroyed only a few years later: August was lengthened to 31 days when it was named after Augustus. Originally, the month was only 30 days long; but in order to show that Augustus was as important as Caesar, after whom July is named, all month lengths in the second half of the year were changed, and February was shortened by one additional day.

The *week* is an invention of the Babylonians. One day in the Babylonian week was 'evil' or 'unlucky', so it was better to do nothing on that day. The modern week cycle with its resting day descends from that superstition. (The way astrological superstition and astronomy cooperated to determine the order of the weekdays is explained in the section on gravitation.) Although about three thousand years old, the week was fully included into the Julian calendar only around the year 300, towards the end of the Western Roman Empire. The final change in the Julian calendar took place between 1582 and 1917 (depending on the country), when more precise measurements of the solar year were used to set a new method to determine leap days, a method still in use today. Together with a reset of the date and the fixation of the week rhythm, this standard is called the *Gregorian calendar* or simply the *modern calendar*. It is used by a majority of the world's population.

Despite its complexity, the modern calendar does allow you to determine the day of the week of a given date in your head. Just execute the following six steps:
 1. take the last two digits of the year, and divide by 4, discarding any fraction;
 2. add the last two digits of the year;
 3. subtract 1 for January or February of a leap year;
 4. add 6 for 2000s or 1600s, 4 for 1700s or 2100s,
      2 for 1800s and 2200s, and 0 for 1900s or 1500s;
 5. add the day of the month;
 6. add the month key value, namely 144 025 036 146 for JFM AMJ JAS OND.

The remainder after division by 7 gives the day of the week, with the correspondence 1-2-3-4-5-6-0 meaning Sunday-Monday-Tuesday-Wednesday-Thursday-Friday-Saturday.*

When to start counting the years is a matter of choice. The oldest method not attached to political power structures was that used in ancient Greece, when years were counted from the first Olympic games. People used to say, for example, that they were born in the first year of the twenty-third Olympiad. Later, political powers always imposed the counting of years from some important event onwards.** Maybe reintroducing the

---

* Remembering the intermediate result for the current year can simplify things even more, especially since the dates 4.4, 6.6, 8.8, 10.10, 12.12, 9.5, 5.9, 7.11, 11.7 and the last day of February all fall on the same day of the week, namely on the year's intermediate result plus 4.
** The present counting of years was defined in the Middle Ages by setting the date for the foundation of Rome to the year 753 BCE, or 753 *before the Common Era*, and then counting backwards, so that the BCE years behave almost like negative numbers. However, the year 1 follows directly after the year 1 BCE: there was no year 0.
  Some other standards set by the Roman Empire explain several abbreviations used in the text:
   - c. is a Latin abbreviation for *circa* and means 'roughly';

Olympic counting is worth considering?

People Names

In the Far East, such as *Corea**, *Japan* or *China*, family names are put in front of the given name. For example, the first Japanese winner of the Nobel Prize in Physics was Yukawa Hideki. In *India*, often, but not always, there is no family name; in those cases, the father's first name is used. In *Russia*, the family name is rarely used in conversation; instead, the first name of the father is. For example, Lev Landau was addressed as Lev Davidovich ('son of David'). In addition, Russian transliteration is not standardized; it varies from country to country and from tradition to tradition. For example, one finds the spellings Dostojewski, Dostoevskij, Dostoïevski and Dostoyevsky for the same person. In the *Netherlands*, the official given names are never used; every person has a semi-official first name by which he is called. For example, Gerard 't Hooft's official given name is Gerardus. In *Germany*, some family names have special pronunciations. For example, Voigt is pronounced 'Fohgt'. In *Italy*, during the Middle Age and the Renaissance, people were called by their first name only, such as Michelangelo or Galileo, or often by first name plus a personal surname that was not their family name, but was used like one, such as Niccolò Tartaglia or Leonardo Fibonacci. In *ancient Rome*, the name by which people are known is usually their surname. The family name was the middle name. For example, Cicero's family name was Tullius. The law introduced by Cicero was therefore known as 'lex Tullia'. In *ancient Greece*, there were no family names. People had only one name. In the English language, the Latin version of the Greek name is used, such as Democritus.

Abbreviations and eponyms or concepts?

Sentences like the following are the scourge of modern physics:

> The EPR paradox in the Bohm formulation can perhaps be resolved using the GRW approach, with the help of the WKB approximation of the Schrödinger equation.

- i.e. is a Latin abbreviation for *id est* and means 'that is';
- e.g. is a Latin abbreviation for *exempli gratia* and means 'for the sake of example';
- ibid. is a Latin abbreviation for *ibidem* and means 'at that same place';
- inf. is a Latin abbreviation for *infra* and means '(see) below';
- op. cit. is a Latin abbreviation for *opus citatum* and means 'the cited work';
- et al. is a Latin abbreviation for *et alii* and means 'and others'.

By the way, *idem* means 'the same' and *passim* means 'here and there' or 'throughout'. Many terms used in physics, like frequency, acceleration, velocity, mass, force, momentum, inertia, gravitation and temperature, are derived from Latin. In fact, it is arguable that the language of science has been Latin for over two thousand years. In Roman times it was Latin vocabulary with Latin grammar, in modern times it switched to Latin vocabulary with French grammar, then for a short time to Latin vocabulary with German grammar, after which it changed to Latin vocabulary with British/American grammar.

Many units of measurement also date from Roman times, as explained in the next appendix. Even the infatuation with Greek technical terms, as shown in coinages such as 'gyroscope', 'entropy' or 'proton', dates from Roman times.

* Corea was temporarily forced to change its spelling to 'Korea' by the Japanese Army because the generals could not bear the fact that Corea preceded Japan in the alphabet. This is not a joke.

Using such vocabulary is the best way to make language unintelligible to outsiders. (In fact, the sentence is nonsense anyway, because the 'GRW approach' is false.) First of all, the sentence uses abbreviations, which is a shame. On top of this, the sentence uses people's names to characterize concepts, i.e., it uses *eponyms*. Originally, eponyms were intended as tributes to outstanding achievements. Today, when formulating radical new laws or variables has become nearly impossible, the spread of eponyms intelligible to a steadily decreasing number of people simply reflects an increasingly ineffective drive to fame.

Eponyms are a proof of scientist's lack of imagination. We avoid them as much as possible in our walk and give *common* names to mathematical equations or entities wherever possible. People's names are then used as appositions to these names. For example, 'Newton's equation of motion' is never called 'Newton's equation'; 'Einstein's field equations' is used instead of 'Einstein's equations'; and 'Heisenberg's equation of motion' is used instead of 'Heisenberg's equation'.

However, some exceptions are inevitable: certain terms used in modern physics have no real alternatives. The Boltzmann constant, the Planck scale, the Compton wavelength, the Casimir effect and Lie groups are examples. In compensation, the text makes sure that you can look up the definitions of these concepts using the index. In addition, the text tries to provide pleasurable reading.

Ref. 367

Appendix B
UNITS, MEASUREMENTS AND CONSTANTS

Measurements are comparisons with standards. Standards are based on *units*. Many different systems of units have been used throughout the world. Most of these standards confer power to the organization in charge of them. Such power can be misused; this is the case today, for example in the computer industry, and was so in the distant past. The solution is the same in both cases: organize an independent and global standard. For measurement units, this happened in the eighteenth century: in order to avoid misuse by authoritarian institutions, to eliminate problems with differing, changing and irreproducible standards, and – this is not a joke – to simplify tax collection and to make it more just, a group of scientists, politicians and economists agreed on a set of units. It is called the *Système International d'Unités*, abbreviated *SI*, and is defined by an international treaty, the 'Convention du Mètre'. The units are maintained by an international organization, the 'Conférence Générale des Poids et Mesures', and its daughter organizations, the 'Commission Internationale des Poids et Mesures' and the 'Bureau International des Poids et Mesures' (BIPM). All originated in the times just before the French revolution.

SI units

All SI units are built from seven *base units*. Their simplest definitions, translated from French into English, are the following ones, together with the dates of their formulation and a few comments:

- 'The *second* is the duration of 9 192 631 770 periods of the radiation corresponding to the transition between the two hyperfine levels of the ground state of the caesium 133 atom.' (1967) The 2019 definition is equivalent, but much less clear.*
- 'The *metre* is the length of the path travelled by light in vacuum during a time interval of 1/299 792 458 of a second.' (1983) The 2019 definition is equivalent, but much less clear.*
- 'The *kilogram*, symbol kg, is the SI unit of mass. It is defined by taking the fixed numerical value of the Planck constant h to be $6.626\,070\,15 \cdot 10^{-34}$ when expressed in the unit J · s, which is equal to kg · m^2 · s^{-1}.' (2019)*
- 'The *ampere*, symbol A, is the SI unit of electric current. It is defined by taking the fixed numerical value of the elementary charge e to be $1.602\,176\,634 \cdot 10^{-19}$ when expressed in the unit C, which is equal to A · s.' (2019)*
- 'The *kelvin*, symbol K, is the SI unit of thermodynamic temperature. It is defined by taking the fixed numerical value of the Boltzmann constant k to be $1.380\,649 \cdot 10^{-23}$ when

expressed in the unit J · K^{-1}.' (2019)*

- 'The mole, symbol mol, is the SI unit of amount of substance. One mole contains exactly 6.02214076 · 10^{23} elementary entities.' (2019)*

- 'The *candela* is the luminous intensity, in a given direction, of a source that emits monochromatic radiation of frequency 540 · 10^{12} hertz and has a radiant intensity in that direction of (1/683) watt per steradian.' (1979) The 2019 definition is equivalent, but much less clear.*

We note that both time and length units are defined as certain properties of a standard example of motion, namely light. In other words, also the Conférence Générale des Poids et Mesures makes the point that the observation of motion is a *prerequisite* for the definition and construction of time and space. *Motion is the fundament of every observation and of all measurement.* By the way, the use of light in the definitions had been proposed already in 1827 by Jacques Babinet.**

From these basic units, all other units are defined by multiplication and division. Thus, all SI units have the following properties:

- SI units form a system with *state-of-the-art precision*: all units are defined with a precision that is higher than the precision of commonly used measurements. Moreover, the precision of the definitions is regularly being improved. The present relative uncertainty of the definition of the second is around 10^{-14}, for the metre about 10^{-10}, for the kilogram about 10^{-9}, for the ampere 10^{-7}, for the mole less than 10^{-6}, for the kelvin 10^{-6} and for the candela 10^{-3}.

- SI units form an *absolute* system: all units are defined in such a way that they can be reproduced in every suitably equipped laboratory, independently, and with high precision. This avoids as much as possible any misuse by the standard-setting organization. In fact, the SI units are as now as near as possible to Planck's natural units, which are presented below. In practice, the SI is now an international standard defining the numerical values of the seven constants c, $\hbar$, e, k, N_A and K_{cd}. After over 200 years of discussions, the CGPM has little left to do.

- SI units form a *practical* system: the base units are quantities of everyday magnitude. Frequently used units have standard names and abbreviations. The complete list includes the seven base units just given, the supplementary units, the derived units and the admitted units.

The *supplementary* SI units are two: the unit for (plane) angle, defined as the ratio of arc length to radius, is the *radian* (rad). For solid angle, defined as the ratio of the subtended area to the square of the radius, the unit is the *steradian* (sr).

The *derived* units with special names, in their official English spelling, i.e., without capital letters and accents, are:

---

* The symbols of the seven units are s, m, kg, A, K, mol and cd. The full offical definitions are found at www.bipm.org. For more details about the levels of the caesium atom, consult a book on atomic physics. The Celsius scale of temperature θ is defined as: $\theta/°C = T/K - 273.15$; note the small difference with the number appearing in the definition of the kelvin. In the definition of the candela, the frequency of the light corresponds to 555.5 nm, i.e., green colour, around the wavelength to which the eye is most sensitive.

** Jacques Babinet (1794–1874), French physicist who published important work in optics.

Name	Abbreviation	Name	Abbreviation
hertz	Hz = 1/s	newton	N = kg m/s²
pascal	Pa = N/m² = kg/m s²	joule	J = Nm = kg m²/s²
watt	W = kg m²/s³	coulomb	C = As
volt	V = kg m²/As³	farad	F = As/V = A²s⁴/kg m²
ohm	Ω = V/A = kg m²/A²s³	siemens	S = 1/Ω
weber	Wb = Vs = kg m²/As²	tesla	T = Wb/m² = kg/As² = kg/Cs
henry	H = Vs/A = kg m²/A²s²	degree Celsius	°C (see definition of kelvin)
lumen	lm = cd sr	lux	lx = lm/m² = cd sr/m²
becquerel	Bq = 1/s	gray	Gy = J/kg = m²/s²
sievert	Sv = J/kg = m²/s²	katal	kat = mol/s

We note that in all definitions of units, the kilogram only appears to the powers of 1, 0 and −1. Can you try to formulate the reason?

The *admitted* non-SI units are *minute*, *hour*, *day* (for time), *degree* 1° = π/180 rad, *minute* 1′ = π/10 800 rad, *second* 1″ = π/648 000 rad (for angles), *litre*, and *tonne*. All other units are to be avoided.

All SI units are made more practical by the introduction of standard names and abbreviations for the powers of ten, the so-called *prefixes*:*

Power	Name		Power	Name		Power	Name		Power	Name	
10^{1}	deca	da	10^{-1}	deci	d	10^{18}	Exa	E	10^{-18}	atto	a
10^{2}	hecto	h	10^{-2}	centi	c	10^{21}	Zetta	Z	10^{-21}	zepto	z
10^{3}	kilo	k	10^{-3}	milli	m	10^{24}	Yotta	Y	10^{-24}	yocto	y
10^{6}	Mega	M	10^{-6}	micro	μ	unofficial:			Ref. 370		
10^{9}	Giga	G	10^{-9}	nano	n	10^{27}	Xenta	X	10^{-27}	xenno	x
10^{12}	Tera	T	10^{-12}	pico	p	10^{30}	Wekta	W	10^{-30}	weko	w
10^{15}	Peta	P	10^{-15}	femto	f	10^{33}	Vendekta	V	10^{-33}	vendeko	v
						10^{36}	Udekta	U	10^{-36}	udeko	u

- SI units form a *complete* system: they cover in a systematic way the full set of observables of physics. Moreover, they fix the units of measurement for all other sciences as well.

---

* Some of these names are invented (yocto to sound similar to Latin *octo* 'eight', zepto to sound similar to Latin *septem*, yotta and zetta to resemble them, exa and peta to sound like the Greek words ἑξάκις and πεντάκις for 'six times' and 'five times', the unofficial ones to sound similar to the Greek words for nine, ten, eleven and twelve); some are from Danish/Norwegian (atto from *atten* 'eighteen', femto from *femten* 'fifteen'); some are from Latin (from *mille* 'thousand', from *centum* 'hundred', from *decem* 'ten', from *nanus* 'dwarf'); some are from Italian (from *piccolo* 'small'); some are Greek (micro is from μικρός 'small', deca/deka from δέκα 'ten', hecto from ἑκατόν 'hundred', kilo from χίλιοι 'thousand', mega from μέγας 'large', giga from γίγας 'giant', tera from τέρας 'monster').

Translate: I was caught in such a traffic jam that I needed a microcentury for a picoparsec and that my car's average fuel consumption was huge: it was two tenths of a square millimetre.

- SI units form a *universal* system: they can be used in trade, in industry, in commerce, at home, in education and in research. They could even be used by extraterrestrial civilizations, if they existed.
- SI units form a *self-consistent* system: the product or quotient of two SI units is also an SI unit. This means that in principle, the same abbreviation, e.g. 'SI', could be used for every unit.

The SI units are not the only possible set that could fulfil all these requirements, but they are the only existing system that does so.*

The meaning of measurement

Every measurement is a comparison with a standard. Therefore, any measurement requires *matter* to realize the standard (even for a speed standard), and *radiation* to achieve the comparison. The concept of measurement thus assumes that matter and radiation exist and can be clearly separated from each other.

Every measurement is a comparison. Measuring thus implies that space and time exist, and that they differ from each other.

Every measurement produces a measurement result. Therefore, every measurement implies the *storage* of the result. The process of measurement thus implies that the situation before and after the measurement can be distinguished. In other terms, every measurement is an *irreversible* process.

Every measurement is a process. Thus every measurement takes a certain amount of time and a certain amount of space.

All these properties of measurements are simple but important. Beware of anybody who denies them.

Curiosities and fun challenges about units

Not using SI units can be expensive. In 1999, NASA lost a satellite on Mars because some software programmers had used provincial units instead of SI units in part of the code. As a result of using feet instead of meters, the Mars Climate Orbiter crashed into the planet, instead of orbiting it; the loss was around 100 million euro.**

* *

The second does not correspond to 1/86 400th of the day any more, though it did in the year 1900; the Earth now takes about 86 400.002 s for a rotation, so that the *International Earth Rotation Service* must regularly introduce a leap second to ensure that the Sun is at

---

* Apart from international units, there are also *provincial* units. Most provincial units still in use are of Roman origin. The mile comes from *milia passum*, which used to be one thousand (double) strides of about 1480 mm each; today a nautical mile, once defined as minute of arc on the Earth's surface, is defined as exactly 1852 m. The inch comes from *uncia/onzia* (a twelfth – now of a foot). The pound (from *pondere* 'to weigh') is used as a translation of *libra* – balance – which is the origin of its abbreviation lb. Even the habit of counting in dozens instead of tens is Roman in origin. These and all other similarly funny units – like the system in which all units start with 'f', and which uses furlong/fortnight as its unit of velocity – are now officially defined as multiples of SI units.

** This story revived an old but false urban legend claiming that only three countries in the world do not use SI units: Liberia, the USA and Myanmar.

the highest point in the sky at 12 o'clock sharp.* The time so defined is called *Universal Time Coordinate*. The speed of rotation of the Earth also changes irregularly from day to day due to the weather; the average rotation speed even changes from winter to summer because of the changes in the polar ice caps; and in addition that average decreases over time, because of the friction produced by the tides. The rate of insertion of leap seconds is therefore higher than once every 500 days, and not constant in time.

* *

The most precise clock ever built, using microwaves, had a stability of 10^{-16} during a running time of 500 s. For longer time periods, the record in 1997 was about 10^{-15}; but values around 10^{-17} seem within technological reach. The precision of clocks is limited for short measuring times by *noise*, and for long measuring times by *drifts*, i.e., by systematic effects. The region of highest stability depends on the clock type; it usually lies between 1 ms for optical clocks and 5000 s for masers. Pulsars are the only type of clock for which this region is not known yet; it certainly lies at more than 20 years, the time elapsed at the time of writing since their discovery.

* *

The least precisely measured of the fundamental constants of physics are the gravitational constant G and the strong coupling constant α_s. Even less precisely known are the age of the universe and its density (see Table 60).

* *

The precision of mass measurements of solids is limited by such simple effects as the adsorption of water. Can you estimate the mass of a monolayer of water – a layer with thickness of one molecule – on a metal weight of 1 kg?

* *

In the previous millennium, thermal energy used to be measured using the unit *calorie*, written as cal. 1 cal is the energy needed to heat 1 g of water by 1 K. To confuse matters, 1 kcal was often written 1 Cal. (One also spoke of a large and a small calorie.) The value of 1 kcal is 4.1868 kJ.

* *

SI units are adapted to humans: the values of heartbeat, human size, human weight, human temperature and human substance are no more than a couple of orders of magnitude near the unit value. SI units thus (roughly) confirm what Protagoras said 25 centuries ago: 'Man is the measure of all things.'

* *

Some units systems are particularly badly adapted to humans. The most infamous is shoe

---

* Their website at hpiers.obspm.fr gives more information on the details of these insertions, as does maia.usno.navy.mil, one of the few useful military websites. See also www.bipm.fr, the site of the BIPM.

size S. It is a pure number calculated as

$$S_{\text{France}} = 1.5 \text{ cm}^{-1}(l + (1 \pm 1) \text{ cm})$$
$$S_{\text{central Europe}} = 1.5748 \text{ cm}^{-1}(l + (1 \pm 1) \text{ cm})$$
$$S_{\text{Anglo-saxon men}} = 1.181 \text{ cm}^{-1}(l + (1 \pm 1) \text{ cm}) - 22 \quad (125)$$

where l is the length of a foot and the correction length depends on the manufacturing company. In addition, the Anglo-saxon formula is not valid for women and children, where the first factor depends, for marketing reasons, both on manufacturer and size itself. The ISO standard for shoe size requires, unsurprisingly, to use foot length in millimetres.

∗ ∗

The table of SI prefixes covers 72 orders of magnitude. How many additional prefixes will be needed? Even an extended list will include only a small part of the infinite range of possibilities. Will the Conférence Générale des Poids et Mesures have to go on forever, defining an infinite number of SI prefixes? Why?

∗ ∗

In the 21st century, the textile industry uses three measurement systems to express how fine a fibre is. All three use the linear mass density m/l. The international system uses the unit 1 tex = 1 g/km. Another system is particular to silk and used the *Denier* as unit: 1 den = 1/9 g/km. The third system uses the unit *Number English*, defined as the number of hanks of cotton that weigh one pound. Here, a *hank* is 7 leas. A *lea* is 120 yards and a *yard* is three feet. In addition, the defining number of hanks differs for linnen and differs again for wool, and in addition it depends on the treatment method of the wool. Reading about textile units makes every comedy show feel like a boring lullaby.

∗ ∗

The French philosopher Voltaire, after meeting Newton, publicized the now famous story that the connection between the fall of objects and the motion of the Moon was discovered by Newton when he saw an apple falling from a tree. More than a century later, just before the French Revolution, a committee of scientists decided to take as the unit of force precisely the force exerted by gravity on a *standard apple*, and to name it after the English scientist. After extensive study, it was found that the mass of the standard apple was 101.9716 g; its weight was called 1 newton. Since then, visitors to the museum in Sèvres near Paris have been able to admire the standard metre, the standard kilogram and the standard apple.*

---

* To be clear, this is a joke; no standard apple exists. It is *not* a joke however, that owners of several apple trees in Britain and in the US claim descent, by rerooting, from the original tree under which Newton had his insight. DNA tests have even been performed to decide if all these derive from the same tree. The result was, unsurprisingly, that the tree at MIT, in contrast to the British ones, is a fake.

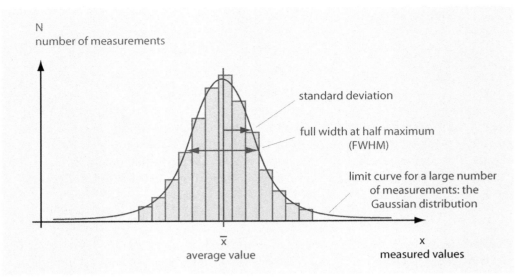

FIGURE 303 A precision experiment and its measurement distribution. The precision is high if the width of the distribution is narrow; the accuracy is high if the centre of the distribution agrees with the actual value.

Precision and accuracy of measurements

Measurements are the basis of physics. Every measurement has an *error*. Errors are due to lack of precision or to lack of accuracy. *Precision* means how well a result is reproduced when the measurement is repeated; *accuracy* is the degree to which a measurement corresponds to the actual value.

Lack of precision is due to accidental or *random errors*; they are best measured by the *standard deviation*, usually abbreviated σ; it is defined through

$$\sigma^2 = \frac{1}{n-1} \sum_{i=1}^{n} (x_i - \bar{x})^2 \;, \tag{126}$$

where $\bar{x}$ is the average of the measurements x_i. (Can you imagine why $n-1$ is used in the formula instead of n?)

For most experiments, the distribution of measurement values tends towards a normal distribution, also called *Gaussian distribution*, whenever the number of measurements is increased. The distribution, shown in Figure 303, is described by the expression

$$N(x) \approx e^{-\frac{(x-\bar{x})^2}{2\sigma^2}} \;. \tag{127}$$

The square σ^2 of the standard deviation is also called the *variance*. For a Gaussian distribution of measurement values, 2.35σ is the full width at half maximum.

Lack of accuracy is due to *systematic errors*; usually these can only be estimated. This estimate is often added to the random errors to produce a *total experimental error*, sometimes also called *total uncertainty*. The *relative* error or uncertainty is the ratio between the error and the measured value.

For example, a professional measurement will give a result such as 0.312(6) m. The number between the parentheses is the standard deviation σ, in units of the last digits. As above, a Gaussian distribution for the measurement results is assumed. Therefore, a value of 0.312(6) m implies that the actual value is expected to lie

— within 1σ with 68.3 % probability, thus in this example within 0.312 ± 0.006 m;
— within 2σ with 95.4 % probability, thus in this example within 0.312 ± 0.012 m;
— within 3σ with 99.73 % probability, thus in this example within 0.312 ± 0.018 m;
— within 4σ with 99.9937 % probability, thus in this example within 0.312 ± 0.024 m;
— within 5σ with 99.999 943 % probability, thus in this example within 0.312 ± 0.030 m;
— within 6σ with 99.999 999 80 % probability, thus in this example within 0.312 ± 0.036 m;
— within 7σ with 99.999 999 999 74 % probability, thus in this example within 0.312 ± 0.041 m.

(Do the latter numbers make sense?)

Note that standard deviations have one digit; you must be a world expert to use two, and a fool to use more. If no standard deviation is given, a (1) is assumed. As a result, among professionals, 1 km and 1000 m are *not* the same length!

What happens to the errors when two measured values A and B are added or subtracted? If all measurements are independent – or uncorrelated – the standard deviation of the sum *and* that of difference is given by $\sigma = \sqrt{\sigma_A^2 + \sigma_B^2}$. For both the product or ratio of two measured and uncorrelated values C and D, the result is $\rho = \sqrt{\rho_C^2 + \rho_D^2}$, where the ρ terms are the *relative* standard deviations.

Assume you measure that an object moves 1 m in 3 s: what is the measured speed value?

Limits to precision

What are the limits to accuracy and precision? There is no way, even in principle, to measure a length x to a *precision* higher than about 61 digits, because in nature, the ratio between the largest and the smallest measurable length is $\Delta x/x > l_{\rm Pl}/d_{\rm horizon} = 10^{-61}$. (Is this ratio valid also for force or for volume?) In the final volume of our text, studies of clocks and metre bars strengthen this theoretical limit.

But it is not difficult to deduce more stringent practical limits. No imaginable machine can measure quantities with a higher precision than measuring the diameter of the Earth within the smallest length ever measured, about 10^{-19} m; that is about 26 digits of precision. Using a more realistic limit of a 1000 m sized machine implies a limit of 22 digits. If, as predicted above, time measurements really achieve 17 digits of precision, then they are nearing the practical limit, because apart from size, there is an additional practical restriction: cost. Indeed, an additional digit in measurement precision often means an additional digit in equipment cost.

Physical constants

In physics, general observations are deduced from more fundamental ones. As a consequence, many measurements can be deduced from more fundamental ones. The most fundamental measurements are those of the physical constants.

The following tables give the world's best values of the most important physical constants and particle properties – in SI units and in a few other common units – as published in the standard references. The values are the world averages of the best measurements made up to the present. As usual, experimental errors, including both random and estimated systematic errors, are expressed by giving the standard deviation in the last digits. In fact, behind each of the numbers in the following tables there is a long story which is worth telling, but for which there is not enough room here.

In principle, all quantitative properties of matter can be calculated with quantum theory and the values of certain physical constants. For example, colour, density and elastic properties can be predicted using the equations of the standard model of particle physics and the values of the following basic constants.

TABLE 57 Basic physical constants.

Quantity	Symbol	Value in SI units	Uncert.[a]
Constants that define the SI measurement units			
Vacuum speed of light[c]	c	299 792 458 m/s	0
Vacuum permeability[c]	μ_0	$4\pi \cdot 10^{-7}$ H/m	0
		= 1.256 637 061 435 ... μH/m0	
Vacuum permittivity[c]	$\varepsilon_0 = 1/\mu_0 c^2$	8.854 187 817 620... pF/m	0
Original Planck constant	h	$6.626\,069\,57(52) \cdot 10^{-34}$ Js	$4.4 \cdot 10^{-8}$
Reduced Planck constant, quantum of action	$\hbar$	$1.054\,571\,726(47) \cdot 10^{-34}$ Js	$4.4 \cdot 10^{-8}$
Positron charge	e	0.160 217 656 5(35) aC	$2.2 \cdot 10^{-8}$
Boltzmann constant	k	$1.380\,6488(13) \cdot 10^{-23}$ J/K	$9.1 \cdot 10^{-7}$
Gravitational constant	G	$6.673\,84(80) \cdot 10^{-11}$ Nm^2/kg^2	$1.2 \cdot 10^{-4}$
Gravitational coupling constant	$\kappa = 8\pi G/c^4$	$2.076\,50(25) \cdot 10^{-43}$ s^2/kg m	$1.2 \cdot 10^{-4}$
Fundamental constants (of unknown origin)			
Number of space-time dimensions		3 + 1	0[b]
Fine-structure constant[d] or e.m. coupling constant	$\alpha = \frac{e^2}{4\pi\varepsilon_0 \hbar c}$ $= g_{em}(m_e^2 c^2)$	1/137.035 999 074(44) = 0.007 297 352 5698(24)	$3.2 \cdot 10^{-10}$ $3.2 \cdot 10^{-10}$
Fermi coupling constant[d] or weak coupling constant	$G_F/(\hbar c)^3$ $\alpha_w(M_Z) = g_w^2/4\pi$	$1.166\,364(5) \cdot 10^{-5}$ GeV^{-2} 1/30.1(3)	$4.3 \cdot 10^{-6}$ $1 \cdot 10^{-2}$
Weak mixing angle	$\sin^2 \theta_W(\overline{MS})$ $\sin^2 \theta_W$ (on shell) $= 1 - (m_W/m_Z)^2$	0.231 24(24) 0.2224(19)	$1.0 \cdot 10^{-3}$ $8.7 \cdot 10^{-3}$
Strong coupling constant[d]	$\alpha_s(M_Z) = g_s^2/4\pi$	0.118(3)	$25 \cdot 10^{-3}$
CKM quark mixing matrix	$\|V\|$	$\begin{pmatrix} 0.97428(15) & 0.2253(7) & 0.00347(16) \\ 0.2252(7) & 0.97345(16) & 0.0410(11) \\ 0.00862(26) & 0.0403(11) & 0.999152(45) \end{pmatrix}$	
Jarlskog invariant	J	$2.96(20) \cdot 10^{-5}$	

TABLE 57 (Continued) Basic physical constants.

Quantity	Symbol	Value in SI units	Uncert.[a]
PMNS neutrino mixing m.	P	$\begin{pmatrix} 0.82 & 0.55 & -0.15+0.038i \\ -0.36+0.020i & 0.70+0.013i & 0.61 \\ 0.44+0.026i & -0.45+0.017i & 0.77 \end{pmatrix}$	

Elementary particle masses (of unknown origin)

Quantity	Symbol	Value in SI units	Uncert.[a]
Electron mass	m_e	$9.10938291(40) \cdot 10^{-31}$ kg	$4.4 \cdot 10^{-8}$
		$5.4857990946(22) \cdot 10^{-4}$ u	$4.0 \cdot 10^{-10}$
		$0.510998928(11)$ MeV	$2.2 \cdot 10^{-8}$
Muon mass	m_μ	$1.883531475(96) \cdot 10^{-28}$ kg	$5.1 \cdot 10^{-8}$
		$0.1134289267(29)$ u	$2.5 \cdot 10^{-8}$
		$105.6583715(35)$ MeV	$3.4 \cdot 10^{-8}$
Tau mass	m_τ	$1.77682(16)$ GeV$/c^2$	
El. neutrino mass	m_{ν_e}	< 2 eV$/c^2$	
Muon neutrino mass	m_{ν_μ}	< 2 eV$/c^2$	
Tau neutrino mass	m_{ν_τ}	< 2 eV$/c^2$	
Up quark mass	u	1.8 to 3.0 MeV$/c^2$	
Down quark mass	d	4.5 to 5.5 MeV$/c^2$	
Strange quark mass	s	$95(5)$ MeV$/c^2$	
Charm quark mass	c	$1.275(25)$ GeV$/c^2$	
Bottom quark mass	b	$4.18(17)$ GeV$/c^2$	
Top quark mass	t	$173.5(1.4)$ GeV$/c^2$	
Photon mass	γ	$< 2 \cdot 10^{-54}$ kg	
W boson mass	$W^\pm$	$80.385(15)$ GeV$/c^2$	
Z boson mass	Z^0	$91.1876(21)$ GeV$/c^2$	
Higgs mass	H	$126(1)$ GeV$/c^2$	
Gluon mass	$g_{1...8}$	c. 0 MeV$/c^2$	

Composite particle masses

Quantity	Symbol	Value in SI units	Uncert.[a]
Proton mass	m_p	$1.672621777(74) \cdot 10^{-27}$ kg	$4.4 \cdot 10^{-8}$
		$1.007276466812(90)$ u	$8.9 \cdot 10^{-11}$
		$938.272046(21)$ MeV	$2.2 \cdot 10^{-8}$
Neutron mass	m_n	$1.674927351(74) \cdot 10^{-27}$ kg	$4.4 \cdot 10^{-8}$
		$1.00866491600(43)$ u	$4.2 \cdot 10^{-10}$
		$939.565379(21)$ MeV	$2.2 \cdot 10^{-8}$
Atomic mass unit	$m_u = m_{^{12}C}/12 = 1$ u	$1.660538921(73)$ yg	$4.4 \cdot 10^{-8}$

a. Uncertainty: standard deviation of measurement errors.
b. Only measured from to 10^{-19} m to 10^{26} m.
c. Defining constant.
d. All coupling constants depend on the 4-momentum transfer, as explained in the section on renormalization. *Fine-structure constant* is the traditional name for the electromagnetic coup-

ling constant g_{em} in the case of a 4-momentum transfer of $Q^2 = c^2 m_e^2$, which is the smallest one possible. At higher momentum transfers it has larger values, e.g., $g_{em}(Q^2 = c^2 M_W^2) \approx 1/128$. In contrast, the strong coupling constant has lower values at higher momentum transfers; e.g., $\alpha_s(34\,\text{GeV}) = 0.14(2)$.

Why do all these basic constants have the values they have? For any basic constant *with a dimension*, such as the quantum of action $\hbar$, the numerical value has only historical meaning. It is $1.054 \cdot 10^{-34}$ Js because of the SI definition of the joule and the second. The question why the value of a dimensional constant is not larger or smaller therefore always requires one to understand the origin of some dimensionless number giving the ratio between the constant and the corresponding *natural unit* that is defined with c, G, $\hbar$ and α. More details and the values of the natural units are given later. Understanding the sizes of atoms, people, trees and stars, the duration of molecular and atomic processes, or the mass of nuclei and mountains, implies understanding the ratios between these values and the corresponding natural units. The key to understanding nature is thus the understanding of all ratios, and thus of all dimensionless constants. The quest of understanding all ratios, including the fine structure constant α itself, is completed only in the final volume of our adventure.

The basic constants yield the following useful high-precision observations.

TABLE 58 Derived physical constants.

Quantity	Symbol	Value in SI units	Uncert.
Vacuum wave resistance	$Z_0 = \sqrt{\mu_0/\varepsilon_0}$	$376.730\,313\,461\,77...\,\Omega$	0
Avogadro's number	N_A	$6.022\,141\,29(27) \cdot 10^{23}$	$4.4 \cdot 10^{-8}$
Loschmidt's number at 273.15 K and 101 325 Pa	N_L	$2.686\,7805(24) \cdot 10^{23}$	$9.1 \cdot 10^{-7}$
Faraday's constant	$F = N_A e$	$96\,485.3365(21)$ C/mol	$2.2 \cdot 10^{-8}$
Universal gas constant	$R = N_A k$	$8.314\,4621(75)$ J/mol K	$9.1 \cdot 10^{-7}$
Molar volume of an ideal gas at 273.15 K and 101 325 Pa	$V = RT/p$	$22.413\,968(20)$ l/mol	$9.1 \cdot 10^{-7}$
Rydberg constant [a]	$R_\infty = m_e c \alpha^2/2h$	$10\,973\,731.568\,539(55)$ m^{-1}	$5 \cdot 10^{-12}$
Conductance quantum	$G_0 = 2e^2/h$	$77.480\,917\,346(25)$ μS	$3.2 \cdot 10^{-10}$
Magnetic flux quantum	$\varphi_0 = h/2e$	$2.067\,833\,758(46)$ pWb	$2.2 \cdot 10^{-8}$
Josephson frequency ratio	$2e/h$	$483.597\,870(11)$ THz/V	$2.2 \cdot 10^{-8}$
Von Klitzing constant	$h/e^2 = \mu_0 c/2\alpha$	$25\,812.807\,4434(84)\,\Omega$	$3.2 \cdot 10^{-10}$
Bohr magneton	$\mu_B = e\hbar/2m_e$	$9.274\,009\,68(20)$ yJ/T	$2.2 \cdot 10^{-8}$
Classical electron radius	$r_e = e^2/4\pi\varepsilon_0 c^2 m_e$	$2.817\,940\,3267(27)$ fm	$9.7 \cdot 10^{-10}$
Compton wavelength of the electron	$\lambda_C = h/m_e c$	$2.426\,310\,2389(16)$ pm	$6.5 \cdot 10^{-10}$
	$\lambdabar_c = \hbar/m_e c = r_e/\alpha$	$0.386\,159\,268\,00(25)$ pm	$6.5 \cdot 10^{-10}$
Bohr radius [a]	$a_\infty = r_e/\alpha^2$	$52.917\,721\,092(17)$ pm	$3.2 \cdot 10^{-10}$
Quantum of circulation	$h/2m_e$	$3.636\,947\,5520(24) \cdot 10^{-4}$ m^2/s	$6.5 \cdot 10^{-10}$
Specific positron charge	e/m_e	$1.758\,820\,088(39) \cdot 10^{11}$ C/kg	$2.2 \cdot 10^{-8}$
Cyclotron frequency of the electron	$f_c/B = e/2\pi m_e$	$27.992\,491\,10(62)$ GHz/T	$2.2 \cdot 10^{-8}$

B UNITS, MEASUREMENTS AND CONSTANTS 463

TABLE 58 (Continued) Derived physical constants.

Quantity	Symbol	Value in SI units	Uncert.
Electron magnetic moment	μ_e	$-9.28476430(21) \cdot 10^{-24}$ J/T	$2.2 \cdot 10^{-8}$
	μ_e/μ_B	$-1.00115965218076(27)$	$2.6 \cdot 10^{-13}$
	μ_e/μ_N	$-1.83828197090(75) \cdot 10^3$	$4.1 \cdot 10^{-10}$
Electron g-factor	g_e	$-2.00231930436153(53)$	$2.6 \cdot 10^{-13}$
Muon–electron mass ratio	m_μ/m_e	$206.7682843(52)$	$2.5 \cdot 10^{-8}$
Muon magnetic moment	μ_μ	$-4.49044807(15) \cdot 10^{-26}$ J/T	$3.4 \cdot 10^{-8}$
muon g-factor	g_μ	$-2.0023318418(13)$	$6.3 \cdot 10^{-10}$
Proton–electron mass ratio	m_p/m_e	$1836.15267245(75)$	$4.1 \cdot 10^{-10}$
Specific proton charge	e/m_p	$9.57883358(21) \cdot 10^7$ C/kg	$2.2 \cdot 10^{-8}$
Proton Compton wavelength	$\lambda_{C,p} = h/m_p c$	$1.32140985623(94)$ fm	$7.1 \cdot 10^{-10}$
Nuclear magneton	$\mu_N = e\hbar/2m_p$	$5.05078353(11) \cdot 10^{-27}$ J/T	$2.2 \cdot 10^{-8}$
Proton magnetic moment	μ_p	$1.410606743(33) \cdot 10^{-26}$ J/T	$2.4 \cdot 10^{-8}$
	μ_p/μ_B	$1.521032210(12) \cdot 10^{-3}$	$8.1 \cdot 10^{-9}$
	μ_p/μ_N	$2.792847356(23)$	$8.2 \cdot 10^{-9}$
Proton gyromagnetic ratio	$\gamma_p = 2\mu_p/\hbar$	$2.675222005(63) \cdot 10^8$ Hz/T	$2.4 \cdot 10^{-8}$
Proton g factor	g_p	$5.585694713(46)$	$8.2 \cdot 10^{-9}$
Neutron–electron mass ratio	m_n/m_e	$1838.6836605(11)$	$5.8 \cdot 10^{-10}$
Neutron–proton mass ratio	m_n/m_p	$1.00137841917(45)$	$4.5 \cdot 10^{-10}$
Neutron Compton wavelength	$\lambda_{C,n} = h/m_n c$	$1.3195909068(11)$ fm	$8.2 \cdot 10^{-10}$
Neutron magnetic moment	μ_n	$-0.96623647(23) \cdot 10^{-26}$ J/T	$2.4 \cdot 10^{-7}$
	μ_n/μ_B	$-1.04187563(25) \cdot 10^{-3}$	$2.4 \cdot 10^{-7}$
	μ_n/μ_N	$-1.91304272(45)$	$2.4 \cdot 10^{-7}$
Stefan–Boltzmann constant	$\sigma = \pi^2 k^4/60\hbar^3 c^2$	$56.70373(21)$ nW/m²K⁴	$3.6 \cdot 10^{-6}$
Wien's displacement constant	$b = \lambda_{max} T$	$2.8977721(26)$ mmK	$9.1 \cdot 10^{-7}$
		$58.789254(53)$ GHz/K	$9.1 \cdot 10^{-7}$
Electron volt	eV	$1.602176565(35) \cdot 10^{-19}$ J	$2.2 \cdot 10^{-8}$
Bits to entropy conversion const.	$k \ln 2$	10^{23} bit $= 0.9569945(9)$ J/K	$9.1 \cdot 10^{-7}$
TNT energy content		3.7 to 4.0 MJ/kg	$4 \cdot 10^{-2}$

a. For infinite mass of the nucleus.

Some useful properties of our local environment are given in the following table.

TABLE 59 Astronomical constants.

Quantity	Symbol	Value
Tropical year 1900 [a]	a	31556925.9747 s
Tropical year 1994	a	31556925.2 s
Mean sidereal day	d	$23^h 56' 4.09053''$
Average distance Earth–Sun [b]		$149597870.691(30)$ km

TABLE 59 (Continued) Astronomical constants.

Quantity	Symbol	Value
Astronomical unit [b]	AU	149 597 870 691 m
Light year, based on Julian year [b]	al	9.460 730 472 5808 Pm
Parsec	pc	30.856 775 806 Pm = 3.261 634 al
Earth's mass	$M_⊕$	$5.973(1) \cdot 10^{24}$ kg
Geocentric gravitational constant	GM	$3.986\,004\,418(8) \cdot 10^{14}$ m^3/s^2
Earth's gravitational length	$l_⊕ = 2GM/c^2$	8.870 056 078(16) mm
Earth's equatorial radius [c]	$R_{⊕\mathrm{eq}}$	6378.1366(1) km
Earth's polar radius [c]	$R_{⊕\mathrm{p}}$	6356.752(1) km
Equator–pole distance [c]		10 001.966 km (average)
Earth's flattening [c]	$e_⊕$	1/298.25642(1)
Earth's av. density	$ρ_⊕$	5.5 Mg/m^3
Earth's age	$T_⊕$	4.50(4) Ga = 142(2) Ps
Earth's normal gravity	g	9.806 65 m/s^2
Earth's standard atmospher. pressure	p_0	101 325 Pa
Moon's radius	$R_{☾\mathrm{v}}$	1738 km in direction of Earth
Moon's radius	$R_{☾\mathrm{h}}$	1737.4 km in other two directions
Moon's mass	$M_☾$	$7.35 \cdot 10^{22}$ kg
Moon's mean distance [d]	$d_☾$	384 401 km
Moon's distance at perigee [d]		typically 363 Mm, historical minimum 359 861 km
Moon's distance at apogee [d]		typically 404 Mm, historical maximum 406 720 km
Moon's angular size [e]		average 0.5181° = 31.08′, minimum 0.49°, maximum 0.55°
Moon's average density	$ρ_☾$	3.3 Mg/m^3
Moon's surface gravity	$g_☾$	1.62 m/s^2
Moon's atmospheric pressure	$p_☾$	from 10^{-10} Pa (night) to 10^{-7} Pa (day)
Jupiter's mass	$M_♃$	$1.90 \cdot 10^{27}$ kg
Jupiter's radius, equatorial	$R_♃$	71.398 Mm
Jupiter's radius, polar	$R_♃$	67.1(1) Mm
Jupiter's average distance from Sun	$D_♃$	778 412 020 km
Jupiter's surface gravity	$g_♃$	24.9 m/s^2
Jupiter's atmospheric pressure	$p_♃$	from 20 kPa to 200 kPa
Sun's mass	$M_⊙$	$1.988\,43(3) \cdot 10^{30}$ kg
Sun's gravitational length	$2GM_⊙/c^2$	2.953 250 08(5) km
Heliocentric gravitational constant	$GM_⊙$	$132.712\,440\,018(8) \cdot 10^{18}$ m^3/s^2
Sun's luminosity	$L_⊙$	384.6 YW
Solar equatorial radius	$R_⊙$	695.98(7) Mm

B UNITS, MEASUREMENTS AND CONSTANTS

TABLE 59 (Continued) Astronomical constants.

QUANTITY	SYMBOL	VALUE
Sun's angular size		0.53° average; minimum on fourth of July (aphelion) 1888 ″, maximum on fourth of January (perihelion) 1952 ″
Sun's average density	$\rho_\odot$	1.4 Mg/m^3
Sun's average distance	AU	149 597 870.691(30) km
Sun's age	$T_\odot$	4.6 Ga
Solar velocity around centre of galaxy	$v_{\odot g}$	220(20) km/s
Solar velocity against cosmic background	$v_{\odot b}$	370.6(5) km/s
Sun's surface gravity	$g_\odot$	274 m/s^2
Sun's lower photospheric pressure	$p_\odot$	15 kPa
Distance to Milky Way's centre		8.0(5) kpc = 26.1(1.6) kal
Milky Way's age		13.6 Ga
Milky Way's size		c. 10^{21} m or 100 kal
Milky Way's mass		10^{12} solar masses, c. 2 · 10^{42} kg
Most distant galaxy cluster known	SXDF-XCLJ 0218-0510	9.6 · 10^9 al

a. Defining constant, from vernal equinox to vernal equinox; it was once used to define the second. (Remember: π seconds is about a nanocentury.) The value for 1990 is about 0.7 s less, corresponding to a slowdown of roughly 0.2 ms/a. (Watch out: why?) There is even an empirical formula for the change of the length of the year over time.

b. The truly amazing precision in the average distance Earth–Sun of only 30 m results from time averages of signals sent from Viking orbiters and Mars landers taken over a period of over twenty years. Note that the International Astronomical Union distinguishes the average distance Earth–Sun from the *astronomical unit* itself; the latter is defined as a fixed and exact length. Also the *light year* is a unit defined as an exact number by the IAU. For more details, see www.iau.org/public/measuring.

c. The shape of the Earth is described most precisely with the World Geodetic System. The last edition dates from 1984. For an extensive presentation of its background and its details, see the www.wgs84.com website. The International Geodesic Union refined the data in 2000. The radii and the flattening given here are those for the 'mean tide system'. They differ from those of the 'zero tide system' and other systems by about 0.7 m. The details constitute a science in itself.

d. Measured centre to centre. To find the precise position of the Moon in the sky at a given date, see the www.fourmilab.ch/earthview/moon_ap_per.html page. For the planets, see the page www.fourmilab.ch/solar/solar.html and the other pages on the same site.

e. Angles are defined as follows: 1 degree = 1° = π/180 rad, 1 (first) minute = 1 ′ = 1°/60, 1 second (minute) = 1 ″ = 1 ′/60. The ancient units 'third minute' and 'fourth minute', each 1/60th of the preceding, are not in use any more. ('Minute' originally means 'very small', as it still does in modern English.)

Some properties of nature at large are listed in the following table. (If you want a challenge, can you determine whether any property of the universe itself is listed?)

TABLE 60 Cosmological constants.

QUANTITY	SYMBOL	VALUE
Cosmological constant	Λ	$c.\ 1 \cdot 10^{-52}$ m^{-2}
Age of the universe[a]	t_0	$4.333(53) \cdot 10^{17}$ s $= 13.8(0.1) \cdot 10^9$ a
(determined from space-time, via expansion, using general relativity)		
Age of the universe[a]	t_0	over $3.5(4) \cdot 10^{17}$ s $= 11.5(1.5) \cdot 10^9$ a
(determined from matter, via galaxies and stars, using quantum theory)		
Hubble parameter[a]	H_0	$2.3(2) \cdot 10^{-18}$ s^{-1} $= 0.73(4) \cdot 10^{-10}$ a^{-1}
		$= h_0 \cdot 100$ km/s Mpc $= h_0 \cdot 1.0227 \cdot 10^{-10}$ a^{-1}
Reduced Hubble parameter[a]	h_0	$0.71(4)$
Deceleration parameter[a]	$q_0 = -(\ddot{a}/a)_0/H_0^2$	$-0.66(10)$
Universe's horizon distance[a]	$d_0 = 3ct_0$	$40.0(6) \cdot 10^{26}$ m $= 13.0(2)$ Gpc
Universe's topology		trivial up to 10^{26} m
Number of space dimensions		3, for distances up to 10^{26} m
Critical density	$\rho_c = 3H_0^2/8\pi G$	$h_0^2 \cdot 1.878\,82(24) \cdot 10^{-26}$ kg/m^3
of the universe		$= 0.95(12) \cdot 10^{-26}$ kg/m^3
(Total) density parameter[a]	$\Omega_0 = \rho_0/\rho_c$	$1.02(2)$
Baryon density parameter[a]	$\Omega_{B0} = \rho_{B0}/\rho_c$	$0.044(4)$
Cold dark matter density parameter[a]	$\Omega_{CDM0} = \rho_{CDM0}/\rho_c$	$0.23(4)$
Neutrino density parameter[a]	$\Omega_{\nu 0} = \rho_{\nu 0}/\rho_c$	0.001 to 0.05
Dark energy density parameter[a]	$\Omega_{X0} = \rho_{X0}/\rho_c$	$0.73(4)$
Dark energy state parameter	$w = p_X/\rho_X$	$-1.0(2)$
Baryon mass	m_b	$1.67 \cdot 10^{-27}$ kg
Baryon number density		$0.25(1)$ /m^3
Luminous matter density		$3.8(2) \cdot 10^{-28}$ kg/m^3
Stars in the universe	n_s	$10^{22\pm 1}$
Baryons in the universe	n_b	$10^{81\pm 1}$
Microwave background temperature[b]	T_0	$2.725(1)$ K
Photons in the universe	n_γ	10^{89}
Photon energy density	$\rho_\gamma = \pi^2 k^4/15T_0^4$	$4.6 \cdot 10^{-31}$ kg/m^3
Photon number density		410.89 /cm^3 or 400 /cm$^3 (T_0/2.7$ K$)^3$
Density perturbation amplitude	$\sqrt{S}$	$5.6(1.5) \cdot 10^{-6}$
Gravity wave amplitude	$\sqrt{T}$	$< 0.71 \sqrt{S}$
Mass fluctuations on 8 Mpc	σ_8	$0.84(4)$
Scalar index	n	$0.93(3)$
Running of scalar index	$dn/d\ln k$	$-0.03(2)$
Planck length	$l_{Pl} = \sqrt{\hbar G/c^3}$	$1.62 \cdot 10^{-35}$ m
Planck time	$t_{Pl} = \sqrt{\hbar G/c^5}$	$5.39 \cdot 10^{-44}$ s
Planck mass	$m_{Pl} = \sqrt{\hbar c/G}$	21.8 µg

TABLE 60 (Continued) Cosmological constants.

Quantity	Symbol	Value
Instants in history[a]	t_0/t_{Pl}	$8.7(2.8) \cdot 10^{60}$
Space-time points inside the horizon[a]	$N_0 = (R_0/l_{Pl})^3 \cdot (t_0/t_{Pl})$	$10^{244\pm1}$
Mass inside horizon	M	$10^{54\pm1}$ kg

a. The index 0 indicates present-day values.
b. The radiation originated when the universe was 380 000 years old and had a temperature of about 3000 K; the fluctuations ΔT_0 which led to galaxy formation are today about $16 \pm 4\,\mu K = 6(2) \cdot 10^{-6}\, T_0$.

Useful numbers

e	2.71828 18284 59045 23536 02874 71352 66249 77572 47093 69995$_9$
π	3.14159 26535 89793 23846 26433 83279 50288 41971 69399 37510$_5$
π²	9.86960 44010 89358 61883 44909 99876 15113 53136 99407 24079$_0$
γ	0.57721 56649 01532 86060 65120 90082 40243 10421 59335 93992$_3$
ln 2	0.69314 71805 59945 30941 72321 21458 17656 80755 00134 36025$_5$
ln 10	2.30258 50929 94045 68401 79914 54684 36420 76011 01488 62877$_2$
√10	3.16227 76601 68379 33199 88935 44432 71853 37195 55139 32521$_6$

Appendix C
SOURCES OF INFORMATION ON MOTION

> "No place affords a more striking conviction of the vanity of human hopes than a public library."
> Samuel Johnson

> "In a consumer society there are inevitably two kinds of slaves: the prisoners of addiction and the prisoners of envy."
> Ivan Illich**

In the text, good books that introduce neighbouring domains are presented in the bibliography. The bibliography also points to journals and websites, in order to satisfy more intense curiosity about what is encountered in this adventure. All citations can also be found by looking up the author in the name index. To find additional information, either libraries or the internet can help.

In a library, review articles of recent research appear in journals such as Reviews of Modern Physics, Reports on Progress in Physics, Contemporary Physics and Advances in Physics. Good pedagogical introductions are found in the American Journal of Physics, the European Journal of Physics and Physik in unserer Zeit.

Overviews on research trends occasionally appear in magazines such as Physics World, Physics Today, Europhysics Journal, Physik Journal and Nederlands tijdschrift voor natuurkunde. For coverage of all the sciences together, the best sources are the magazines Nature, New Scientist, Naturwissenschaften, La Recherche and Science News.

Research papers on the foundations of motion appear mainly in Physics Letters B, Nuclear Physics B, Physical Review D, Physical Review Letters, Classical and Quantum Gravity, General Relativity and Gravitation, International Journal of Modern Physics and Modern Physics Letters. The newest results and speculative ideas are found in conference proceedings, such as the Nuclear Physics B Supplements. Research articles also appear in Fortschritte der Physik, European Physical Journal, La Rivista del Nuovo Cimento, Europhysics Letters, Communications in Mathematical Physics, Journal of Mathematical Physics, Foundations of Physics, International Journal of Theoretical Physics and Journal of Physics G.

There are only a few internet physics journals of quality: one is Living Reviews in Relativity, found at www.livingreviews.org, the other is the New Journal of Physics, which can be found at the www.njp.org website. There are, unfortunately, also many internet physics journals that publish incorrect research. They are easy to spot: they ask for money

---

** Ivan Illich (b. 1926 Vienna, d. 2002 Bremen), theologian and social and political thinker.

to publish a paper.

By far the simplest way to keep in touch with ongoing research on motion and modern physics is to use the *internet*, the international computer network. To start using it, ask a friend who knows.*

In the last decade of the twentieth century, the internet expanded into a combination of library, business tool, discussion platform, media collection, garbage collection and, above all, addiction provider. Do not use it too much. Commerce, advertising and – unfortunately – addictive material for children, youth and adults, as well as crime of all kind are also an integral part of the web. With a personal computer, a modem and free browser software, you can look for information in millions of pages of documents or destroy your professional career through addiction. The various parts of the documents are located in various computers around the world, but the user does not need to be aware of this.**

Most theoretical physics papers are available free of charge, as *preprints*, i.e., before official publication and checking by referees, at the arxiv.org website. A service for finding subsequent preprints that cite a given one is also available.

Research papers on the description of motion appear *after* this text is published can also be found via www.webofknowledge.com a site accessible only from libraries. It allows one to search for all publications which *cite* a given paper.

Searching the web for authors, organizations, books, publications, companies or simple keywords using search engines can be a rewarding experience or an episode of addiction, depending entirely on yourself. A selection of interesting servers about motion is given below.

---

* It is also possible to use the internet and to download files through FTP with the help of email only. But the tools change too often to give a stable guide here. Ask your friend.

** Several decades ago, the provocative book by IVAN ILLICH, *Deschooling Society*, Harper & Row, 1971, listed four basic ingredients for any educational system:

1. access to *resources* for learning, e.g. books, equipment, games, etc. at an affordable price, for everybody, at any time in their life;

2. for all who want to learn, access to *peers* in the same learning situation, for discussion, comparison, cooperation and competition;

3. access to *elders*, e.g. teachers, for their care and criticism towards those who are learning;

4. exchanges between students and *performers* in the field of interest, so that the latter can be models for the former. For example, there should be the possibility to listen to professional musicians and reading the works of specialist writers. This also gives performers the possibility to share, advertise and use their skills.

Illich develops the idea that if such a system were informal – he then calls it a 'learning web' or 'opportunity web' – it would be superior to formal, state-financed institutions, such as conventional schools, for the development of mature human beings. These ideas are deepened in his following works, *Deschooling Our Lives*, Penguin, 1976, and *Tools for Conviviality*, Penguin, 1973.

Today, any networked computer offers *email* (electronic mail), FTP (file transfers to and from another computer), access to discussion groups on specific topics, such as particle physics, and the *world-wide web*. In a rather unexpected way, all these facilities of the internet have transformed it into the backbone of the 'opportunity web' discussed by Illich. However, as in any school, it strongly depends on the user's discipline whether the internet actually does provide a learning web or an entry into addiction.

TABLE 61 Some interesting sites on the world-wide web.

Topic	Website address
General knowledge	
Innovation in science and technology	www.innovations-report.de
Book collections	www.ulib.org
	books.google.com
Entertaining science education by Theodore Gray	www.popsci.com/category/popsci-authors/theodore-gray
Entertaining and professional science education by Robert Krampf	thehappyscientist.com
Science Frontiers	www.science-frontiers.com
Science Daily News	www.sciencedaily.com
Science News	www.sciencenews.org
Encyclopedia of Science	www.daviddarling.info
Interesting science research	www.max-wissen.de
Quality science videos	www.vega.org.uk
ASAP Science videos	plus.google.com/101786231119207015313/posts
Physics	
Learning physics with toys from rubbish	www.arvindguptatoys.com
Official SI unit website	www.bipm.fr
Unit conversion	www.chemie.fu-berlin.de/chemistry/general/units.html
Particle data	pdg.web.cern.ch
Engineering data and formulae	www.efunda.com
Information on relativity	math.ucr.edu/home/baez/relativity.html
Research preprints	arxiv.org
	www.slac.stanford.edu/spires
Abstracts of papers in physics journals	www.osti.gov
Many physics research papers	sci-hub.tv, sci-hub.la
	libgen.pw, libgen.io
Physics news, weekly	www.aip.org/physnews/update
Physics news, daily	phys.org
Physics problems by Yacov KantorKantor, Yacov	www.tau.ac.il/~kantor/QUIZ/
Physics problems by Henry Greenside	www.phy.duke.edu/~hsg/physics-challenges/challenges.html
Physics 'question of the week'	www.physics.umd.edu/lecdem/outreach/QOTW/active
Physics 'miniproblem'	www.nyteknik.se/miniproblemet
Physikhexe	physik-verstehen-mit-herz-und-hand.de/html/de-6.html

C SOURCES OF INFORMATION ON MOTION

Topic	Website address
Magic science tricks	www.sciencetrix.com
Physics stack exchange	physics.stackexchange.com
'Ask the experts'	www.sciam.com/askexpert_directory.cfm
Nobel Prize winners	www.nobel.se/physics/laureates
Videos of Nobel Prize winner talks	www.mediatheque.lindau-nobel.org
Pictures of physicists	www.if.ufrj.br/famous/physlist.html
Physics organizations	www.cern.ch
	www.hep.net
	www.nikhef.nl
	www.het.brown.edu/physics/review/index.html
Physics textbooks on the web	www.physics.irfu.se/CED/Book
	www.biophysics.org/education/resources.htm
	www.lightandmatter.com
	www.physikdidaktik.uni-karlsruhe.de/index_en.html
	www.feynmanlectures.info
	hyperphysics.phy-astr.gsu.edu/hbase/hph.html
	www.motionmountain.net
Three beautiful French sets of notes on classical mechanics and particle theory	feynman.phy.ulaval.ca/marleau/notesdecours.htm
The excellent *Radical Freshman Physics* by David Raymond	www.physics.nmt.edu/~raymond/teaching.html
Physics course scripts from MIT	ocw.mit.edu/courses/physics/
Physics lecture scripts in German and English	www.akleon.de
'World lecture hall'	wlh.webhost.utexas.edu
Optics picture of the day	www.atoptics.co.uk/opod.htm
Living Reviews in Relativity	www.livingreviews.org
Wissenschaft in die Schulen	www.wissenschaft-schulen.de
Videos of Walter Lewin'sIndexLewin, Walter physics lectures	ocw.mit.edu/courses/physics/ 8-01-physics-i-classical-mechanics-fall-1999/
Physics videos of Matt Carlson	www.youtube.com/sciencetheater
Physics videos by the University of Nottingham	www.sixtysymbols.com
Physics lecture videos	www.coursera.org/courses?search=physics
	www.edx.org/course-list/allschools/physics/allcourses

Mathematics

Topic	Website address
'Math forum' internet resource collection	mathforum.org/library
Biographies of mathematicians	www-history.mcs.st-andrews.ac.uk/BiogIndex.html
Purdue math problem of the week	www.math.purdue.edu/academics/pow
Macalester College maths problem of the week	mathforum.org/wagon
Mathematical formulae	dlmf.nist.gov
Weisstein's World of Mathematics	mathworld.wolfram.com
Functions	functions.wolfram.com
Symbolic integration	www.integrals.com
Algebraic surfaces	www.mathematik.uni-kl.de/~hunt/drawings.html
Math lecture videos, in German	www.j3l7h.de/videos.html
Gazeta Matematica, in Romanian	www.gazetamatematica.net

Astronomy

ESA	sci.esa.int
NASA	www.nasa.gov
Hubble space telescope	hubble.nasa.gov
Sloan Digital Sky Survey	skyserver.sdss.org
The 'cosmic mirror'	www.astro.uni-bonn.de/~dfischer/mirror
Solar System simulator	space.jpl.nasa.gov
Observable satellites	liftoff.msfc.nasa.gov/RealTime/JPass/20
Astronomy picture of the day	antwrp.gsfc.nasa.gov/apod/astropix.html
The Earth from space	www.visibleearth.nasa.gov
From Stargazers to Starships	www.phy6.org/stargaze/Sintro.htm
Current solar data	www.n3kl.org/sun

Specific topics

Sonic wonders to visit in the world	www.sonicwonders.org
Encyclopedia of photonics	www.rp-photonics.com
Chemistry textbook, online	chemed.chem.wisc.edu/chempaths/GenChem-Textbook
Minerals	webmineral.com
	www.mindat.org
Geological Maps	onegeology.org
Optical illusions	www.sandlotscience.com
Rock geology	sandatlas.org
Petit's science comics	www.jp-petit.org
Physical toys	www.e20.physik.tu-muenchen.de/~cucke/toylinke.htm

C SOURCES OF INFORMATION ON MOTION

Topic	Website address
Physics humour	www.dctech.com/physics/humor/biglist.php
Literature on magic	www.faqs.org/faqs/magic-faq/part2
music library, searchable by tune	imslp.org
Making paper aeroplanes	www.pchelp.net/paper_ac.htm
	www.ivic.qc.ca/~aleexpert/aluniversite/klinevogelmann.html
Small flying helicopters	pixelito.reference.be
Science curiosities	www.wundersamessammelsurium.info
Ten thousand year clock	www.longnow.org
Gesellschaft Deutscher Naturforscher und Ärzte	www.gdnae.de
Pseudoscience	suhep.phy.syr.edu/courses/modules/PSEUDO/pseudo_main.html
Crackpots	www.crank.net
Periodic table with videos for each element	www.periodicvideos.com
Mathematical quotations	math.furman.edu/mwoodard/~mquot.html
The 'World Question Center'	www.edge.org/questioncenter.html
Plagiarism	www.plagiarized.com
Hoaxes	www.museumofhoaxes.com
Encyclopedia of Earth	www.eoearth.org
This is colossal	thisiscolossal.com

Do you want to study physics without actually going to university? Nowadays it is possible to do so via email and internet, in German, at the University of Kaiserslautern.* In the near future, a nationwide project in Britain should allow the same for English-speaking students. As an introduction, use the latest update of this physics text!

> Das Internet ist die offenste Form der geschlossenen Anstalt.**
> Matthias Deutschmann

> Si tacuisses, philosophus mansisses.***
> After Boethius.

* See the www.fernstudium-physik.de website.
** 'The internet is the most open form of a closed institution.'
*** 'If you had kept quiet, you would have remained a philosopher.' After the story Boethius (c. 480–c. 525) tells in De consolatione philosophiae, 2.7, 67 ff.

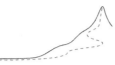 CHALLENGE HINTS AND SOLUTIONS

> " Never make a calculation before you know the answer. "
> John Wheeler's motto

John Wheeler wanted people to estimate, to try and to guess; but not saying the guess out loud. A correct guess reinforces the physics instinct, whereas a wrong one leads to the pleasure of surprise. Guessing is thus an important first step in solving every problem.

Teachers have other criteria to keep in mind. Good problems can be solved on different levels of difficulty, can be solved with words or with images or with formulae, activate knowledge, concern real world applications, and are open.

Challenge 1, page 10: Do not hesitate to be demanding and strict. The next edition of the text will benefit from it.

Challenge 2, page 16: There are many ways to distinguish real motion from an illusion of motion: for example, only real motion can be used to set something else into motion. In addition, the motion illusions of the figures show an important failure; nothing moves if the head and the paper remain fixed with respect to each other. In other words, the illusion only *amplifies* existing motion, it does not *create* motion from nothing.

Challenge 3, page 17: Without detailed and precise experiments, both sides can find examples to prove their point. Creation is supported by the appearance of mould or bacteria in a glass of water; creation is also supported by its opposite, namely traceless disappearance, such as the disappearance of motion. However, conservation is supported and creation falsified by all those investigations that explore assumed cases of appearance or disappearance in full detail.

Challenge 4, page 19: The amount of water depends on the shape of the bucket. The system chooses the option (tilt or straight) for which the centre of gravity is lowest.

Challenge 5, page 20: To simplify things, assume a cylindrical bucket. If you need help, do the experiment at home. For the reel, the image is misleading: the rim on which the reel advances has a *larger* diameter than the section on which the string is wound up. The wound up string does not touch the floor, like for the reel shown in Figure 304.

Challenge 6, page 19: Political parties, sects, helping organizations and therapists of all kinds are typical for this behaviour.

Challenge 7, page 24: The issue is not yet completely settled for the motion of empty space, such as in the case of gravitational waves. Thus, the motion of empty space might be an exception. In any case, empty space is not made of small particles of finite size, as this would contradict the transversality of gravity waves.

Challenge 8, page 26: Holes are not physical systems, because in general they cannot be tracked.

Challenge 9, page 26: The circular definition is: objects are defined as what moves with respect

FIGURE 304 The assumed shape for the reel puzzle.

FIGURE 305 A soap bubble while bursting (© Peter Wienerroither).

to the background, and the background is defined as what stays when objects change. We shall return to this important issue several times in our adventure. It will require a certain amount of patience to solve it, though.

Challenge 10, page 28: No, the universe does not have a state. It is not measurable, not even in principle. See the discussion on the issue in volume IV, on quantum theory.

Challenge 11, page 28: The final list of intrinsic properties for physical systems found in nature is given in volume V, in the section of particle physics. And of course, the universe has no intrinsic, permanent properties. None of them are measurable for the universe as a whole, not even in principle.

Challenge 12, page 31: Hint: yes, there is such a point.

Challenge 13, page 31: See Figure 305 for an intermediate step. A bubble bursts at a point, and then the rim of the hole increases rapidly, until it disappears on the antipodes. During that process the remaining of the bubble keeps its spherical shape, as shown in the figure. For a film of the process, see www.youtube.com/watch?v=dIZwQ24_OU0 (or search for 'bursting soap bubble'). In other words, the final droplets that are ejected stem from the point of the bubble which is opposite to the point of puncture; they are never ejected from the centre of the bubble.

Challenge 14, page 31: A ghost can be a moving image; it cannot be a moving object, as objects cannot interpenetrate.

Challenge 15, page 31: If something could stop moving, motion could disappear into nothing. For a precise proof, one would have to show that no atom moves any more. So far, this has never been observed: motion is conserved. (Nothing in nature can disappear into nothing.)

Challenge 16, page 31: This would indeed mean that space is infinite; however, it is impossible to observe that something moves 'forever': nobody lives that long. In short, there is no way to prove that space is infinite in this way. In fact, there is no way to prove that space if infinite in any other way either.

Challenge 17, page 31: The necessary rope length is nh, where n is the number of wheels/pulleys. And yes, the farmer is indeed doing something sensible.

Challenge 19, page 31: How would you measure this?

Challenge 20, page 31: The number of reliable digits of a measurement result is a simple quantification of precision. More details can be found by looking up 'standard deviation' in the index.

Challenge 21, page 31: No; memory is needed for observation and measurements. This is the case for humans and measurement apparatus. Quantum theory will make this particularly clear.

Challenge 22, page 31: Note that you never have observed zero speed. There is always some measurement error which prevents one to say that something is zero. No exceptions!

Challenge 23, page 32: $(2^{64} - 1)$ = 18 446 744 073 700 551 615 grains of wheat, with a grain weight of 40 mg, are 738 thousand million tons. Given a world harvest in 2006 of 606 million tons, the grains amount to about 1200 years of the world's wheat harvests.

The grain number calculation is simplified by using the formula $1 + m + m^2 + m^3 + ...m^n = (m^{n+1} - 1)/(m - 1)$, that gives the sum of the so-called *geometric sequence*. The name is historical and is used as a contrast to the *arithmetic sequence* $1 + 2 + 3 + 4 + 5 + ...n = n(n + 1)/2$. Can you prove the two expressions?

The chess legend is mentioned first by Ibn Khallikan (b. 1211 Arbil, d. 1282 Damascus). King Shiram and king Balhait, also mentioned in the legend, are historical figures that lived between the second and fourth century CE. The legend appears to have combined two different stories. Indeed, the calculation of grains appears already in the year 947, in the famous text *Meadows of Gold and Mines of Precious Stones* by Al-Masudi (b. c. 896 Baghdad, d. 956 Cairo).

Challenge 24, page 32: In clean experiments, the flame leans forward. But such experiments are not easy, and sometimes the flame leans backward. Just try it. Can you explain both observations?

Challenge 25, page 32: Accelerometers are the simplest motion detectors. They exist in form of piezoelectric devices that produce a signal whenever the box is accelerated and can cost as little as one euro. Another accelerometer that might have a future is an interference accelerometer that makes use of the motion of an interference grating; this device might be integrated in silicon. Other, more precise accelerometers use gyroscopes or laser beams running in circles.

Velocimeters and position detectors can also detect motion; they need a wheel or at least an optical way to look out of the box. Tachographs in cars are examples of velocimeters, computer mice are examples of position detectors.

CHALLENGE HINTS AND SOLUTIONS

A cheap enough device would be perfect to measure the speed of skiers or skaters. No such device exists yet.

Challenge 26, page 32: The ball rolls (or slides) towards the centre of the table, as the table centre is somewhat nearer to the centre of the Earth than the border; then the ball shoots over, performing an oscillation around the table centre. The period is 84 min, as shown in challenge 405. (This has never been observed, so far. Why?)

Challenge 27, page 32: Only if the acceleration never vanishes. Accelerations can be felt. Accelerometers are devices that measure accelerations and then deduce the position. They are used in aeroplanes when flying over the atlantic. If the box does not accelerate, it is impossible to say whether it moves or sits still. It is even impossible to say in which direction one moves. (Close your eyes in a train at night to confirm this.)

Challenge 28, page 32: The block moves twice as fast as the cylinders, independently of their radius.

Challenge 29, page 32: This methods is known to work with other fears as well.

Challenge 30, page 33: Three couples require 11 passages. Two couples require 5. For four or more couples there is no solution. What is the solution if there are n couples and $n-1$ places on the boat?

Challenge 31, page 33: Hint: there is an infinite number of such shapes. These curves are called also *Reuleaux curves*. Another hint: The 20 p and 50 p coins in the UK have such shapes. And yes, other shapes than cylinders are also possible: take a twisted square bar, for example.

Challenge 32, page 33: If you do not know, ask your favourite restorer of old furniture.

Challenge 33, page 33: For this beautiful puzzle, see arxiv.org/abs/1203.3602.

Challenge 34, page 33: Conservation, relativity and minimization are valid generally. In some rare processes in nuclear physics, motion invariance (reversibility) is broken, as is mirror invariance. Continuity is known not to be valid at smallest length and time intervals, but no experiments has yet probed those domains, so that it is still valid in practice.

Challenge 35, page 34: In everyday life, this is correct; what happens when quantum effects are taken into account?

Challenge 36, page 36: Take the average distance change of two neighbouring atoms in a piece of quartz over the last million years. Do you know something still slower?

Challenge 37, page 37: There is only one way: compare the velocity to be measured with the speed of light – using cleverly placed mirrors. In fact, almost all physics textbooks, both for schools and for university, start with the definition of space and time. Otherwise excellent relativity textbooks have difficulties avoiding this habit, even those that introduce the now standard k-calculus (which is in fact the approach mentioned here). Starting with speed is the most logical and elegant approach. But it is possible to compare speeds without metre sticks and clocks. Can you devise a method?

Challenge 38, page 37: There is no way to sense your own motion if you are in a vacuum. No way in principle. This result is often called the *principle of relativity*.

In fact, there is a way to measure your motion in space (though not in vacuum): measure your speed with respect to the cosmic background radiation. So we have to be careful about what is implied by the question.

Challenge 39, page 37: The wing load W/A, the ratio between weight W and wing area A, is obviously proportional to the third root of the weight. (Indeed, $W \sim l^3$, $A \sim l^2$, l being the dimension of the flying object.) This relation gives the green trend line.

The wing load W/A, the ratio between weight W and wing area A, is, like all forces in fluids, proportional to the square of the cruise speed v: we have $W/A = v^2 0.38 \text{ kg/m}^3$. The unexplained

FIGURE 306 Sunbeams in a forest (© Fritz Bieri and Heinz Rieder).

factor contains the density of air and a general numerical coefficient that is difficult to calculate. This relation connects the upper and lower horizontal scales in the graph.

As a result, the cruise speed scales as the *sixth root* of weight: $v \sim W^{1/6}$. In other words, an Airbus A380 is 750 000 million times heavier than a fruit fly, but only a hundred times as fast.

Challenge 41, page 41: Equivalently: do points in space exist? The final part of our adventure explores this issue in detail.

Vol. VI, page 65

Challenge 42, page 42: All electricity sources must use the same phase when they feed electric power into the net. Clocks of computers on the internet must be synchronized.

Challenge 43, page 42: Note that the shift increases quadratically with time, not linearly.

Challenge 44, page 43: Galileo measured time with a scale (and with other methods). His stopwatch was a water tube that he kept closed with his thumb, pointing into a bucket. To start the stopwatch, he removed his thumb, to stop it, he put it back on. The volume of water in the bucket then gave him a measure of the time interval. This is told in his famous book GA-LILEO GALILEI, *Discorsi e dimostrazioni matematiche intorno a due nuove scienze attenenti alla mecanica e i movimenti locali*, usually simply called the 'Discorsi', which he published in 1638 with Louis Elsevier in Leiden, in the Netherlands.

Challenge 45, page 44: Natural time is measured with natural motion. Natural motion is the motion of light. Natural time is thus defined with the motion of light.

Challenge 46, page 48: There is no way to define a local time at the poles that is consistent with all neighbouring points. (For curious people, check the website www.arctic.noaa.gov/gallery_np.html.)

Challenge 48, page 50: The forest is full of light and thus of light rays: they are straight, as shown by the sunbeams in Figure 306.

Challenge 49, page 50: One pair of muscles moves the lens along the third axis by deforming the eye from prolate to spherical to oblate.

Challenge 50, page 50: You can solve this problem by trying to think in four dimensions. (Train using the well-known three-dimensional projections of four-dimensional cubes.) Try to imagine how to switch the sequence when two pieces cross. Note: it is usually *not* correct, in this domain, to use time instead of a fourth *spatial* dimension!

Challenge 51, page 52: Measure distances using light.

Challenge 54, page 56: It is easier to work with the unit torus. Take the unit interval [0, 1] and equate the end points. Define a set B in which the elements are a given real number b from the interval plus all those numbers who differ from that real by a rational number. The unit circle can be thought as the union of all the sets B. (In fact, every set B is a shifted copy of the rational numbers $\mathbb{Q}$.) Now build a set A by taking one element from each set B. Then build the set family consisting of the set A and its copies A_q shifted by a rational q. The union of all these sets is the unit torus. The set family is countably infinite. Then divide it into *two* countably infinite set families. It is easy to see that each of the two families can be renumbered and its elements shifted in such a way that each of the two families forms a unit torus.

Mathematicians say that there is no countably infinitely additive measure of $\mathbb{R}^n$ or that sets such as A are non-measurable. As a result of their existence, the 'multiplication' of lengths is possible. Later on we shall explore whether bread or *gold* can be multiplied in this way.

Challenge 55, page 56: Hint: start with triangles.

Challenge 56, page 56: An example is the region between the x-axis and the function which assigns 1 to every transcendental and 0 to every non-transcendental number.

Challenge 57, page 57: We use the definition of the function of the text. The dihedral angle of a regular tetrahedron is an irrational multiple of π, so the tetrahedron has a non-vanishing Dehn invariant. The cube has a dihedral angle of π/2, so the Dehn invariant of the cube is 0. Therefore, the cube is not equidecomposable with the regular tetrahedron.

Challenge 58, page 58: If you think you can show that empty space is continuous, you are wrong. Check your arguments. If you think you can prove the opposite, you *might* be right – but only if you already know what is explained in the final part of the text. If that is not the case, check your arguments. In fact, time is neither discrete nor continuous.

Challenge 60, page 59: Obviously, we use light to check that the plumb line is straight, so the two definitions must be the same. This is the case because the field lines of gravity are also possible paths for the motion of light. However, this is not always the case; can you spot the exceptions?

Another way to check straightness is along the surface of calm water.

A third, less precise way, way is to make use of the straightness sensors on the brain. The human brain has a built-in faculty to determine whether an objects seen with the eyes is straight. There are special cells in the brain that fire when this is the case. Any book on vision perception tells more about this topic.

Challenge 61, page 60: The hollow Earth theory is correct if the distance formula is used consistently. In particular, one has to make the assumption that objects get smaller as they approach the centre of the hollow sphere. Good explanations of all events are found on www.geocities.com/inversedearth. Quite some material can be found on the internet, also under the names of celestrocentric system, inner world theory or concave Earth theory. There is no way to prefer one description over the other, except possibly for reasons of simplicity or intellectual laziness.

Challenge 63, page 61: A hint is given in Figure 307. For the measurement of the speed of light with almost the same method, see volume II, on page 20.

Challenge 64, page 61: A fast motorbike is faster: a motorbike driver can catch an arrow, a stunt that was shown on the German television show 'Wetten dass' in the year 2001.

Challenge 65, page 61: The 'only' shape that prevents a cover to fall into the hole beneath is a circular shape. Actually, slight deviations from the circular shape are also allowed.

Challenge 68, page 61: The walking speed of older men depends on their health. If people walk faster than 1.4 m/s, they are healthy. The study concluded that the grim reaper walks with a preferred speed of 0.82 m/s, and with a maximum speed of 1.36 m/s.

Challenge 69, page 61: 72 stairs.

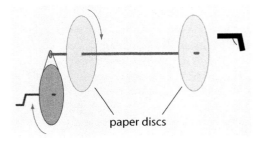

FIGURE 307 A simple way to measure bullet speeds.

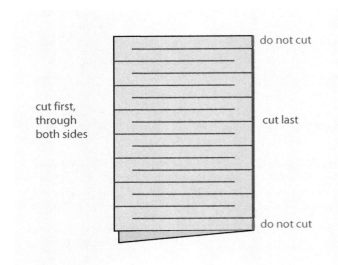

FIGURE 308 How to make a hole in a postcard that allows stepping through it.

Challenge 73, page 62: See Figure 308 for a way to realize the feat.

Challenge 74, page 62: Within 1 per cent, one *fifth* of the height must be empty, and four fifths must be filled; the exact value follows from $\sqrt[3]{2} = 1.25992...$

Challenge 75, page 62: One pencil draws a line of between 20 and 80 km, if no lead is lost when sharpening. Numbers for the newly invented plastic, flexible pencils are not available.

Challenge 79, page 63: The bear is white, because the obvious spot of the house is at the North pole. But there are infinitely many additional spots (without bears) near the South pole: can you find them?

Challenge 80, page 63: We call L the initial length of the rubber band, v the speed of the snail relative to the band and V the speed of the horse relative to the floor. The speed of the snail relative to the floor is given as

$$\frac{ds}{dt} = v + V\frac{s}{L + Vt} \,. \tag{128}$$

This is a so-called *differential equation* for the unknown snail position $s(t)$. You can check – by simple insertion – that its solution is given by

$$s(t) = \frac{v}{V}(L + Vt)\ln(1 + Vt/L) \,. \tag{129}$$

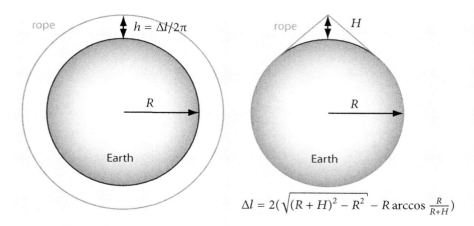

FIGURE 309 Two ways to lengthen a rope around the Earth.

Therefore, the snail reaches the horse at a time

$$t_{\text{reaching}} = \frac{L}{V}(e^{V/v} - 1) \tag{130}$$

which is finite for all values of L, V and v. You can check however, that the time is very large indeed, if realistic speed values are used.

Challenge 81, page 63: Colour is a property that applies only to objects, not to boundaries. In the mentioned case, only spots and backgrounds have colours. The question shows that it is easy to ask questions that make no sense also in physics.

Challenge 82, page 63: You can do this easily yourself. You can even find websites on the topic.

Challenge 84, page 63: Clocks with two hands: 22 times. Clocks with three hands: 2 times.

Challenge 85, page 64: 44 times.

Challenge 86, page 64: For two hands, the answer is 143 times.

Challenge 87, page 64: The Earth rotates with 15 minutes per minute.

Challenge 88, page 64: You might be astonished, but no reliable data exist on this question. The highest speed of a throw measured so far seems to be a 45 m/s cricket bowl. By the way, much more data are available for speeds achieved with the help of rackets. The c. 70 m/s of fast badminton smashes seem to be a good candidate for record racket speed; similar speeds are achieved by golf balls.

Challenge 89, page 64: A *spread out* lengthening by 1 m allows even many cats to slip through, as shown on the left side of Figure 309. But the right side of the figure shows a better way to use the extra rope length, as Dimitri Yatsenko points out: a *localized* lengthening by 1 mm then already yields a height of 1.25 m, allowing a child to walk through. In fact, a lengthening by 1 m performed in this way yields a peak height of 121 m!

Challenge 90, page 64: 1.8 km/h or 0.5 m/s.

Challenge 92, page 64: The question makes sense, especially if we put our situation in relation to the outside world, such as our own family history or the history of the universe. The different usage reflects the idea that we are able to determine our position by ourselves, but not the time in which we are. The section on determinism will show how wrong this distinction is.

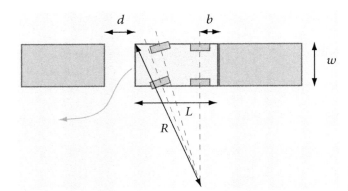

FIGURE 310 Leaving a parking space – the outer turning radius.

Challenge 93, page 64: Yes, there is. However, this is not obvious, as it implies that space and time are not continuous, in contrast to what we learn in primary school. The answer will be found in the final part of this text.

Challenge 94, page 64: For a curve, use, at each point, the curvature radius of the circle approximating the curve in that point; for a surface, define two directions in each point and use two such circles along these directions.

Challenge 95, page 64: It moves about 1 cm in 50 ms.

Challenge 96, page 65: The surface area of the lung is between 100 and 200 m^2, depending on the literature source, and that of the intestines is between 200 and 400 m^2.

Challenge 97, page 65: A limit does not exist in classical physics; however, there is one in nature which appears as soon as quantum effects are taken into account.

Challenge 98, page 65: The final shape is a full cube without any hole.

Challenge 99, page 65: The required gap d is

$$d = \sqrt{(L-b)^2 - w^2 + 2w\sqrt{R^2 - (L-b)^2}} - L + b \, , \qquad (131)$$

as deduced from Figure 310. See also R. HOYLE, *Requirements for a perfect s-shaped parallel parking maneuvre in a simple mathematical model*, 2003. In fact, the mathematics of parallel parking is beautiful and interesting. See, for example, the web page rigtriv.wordpress.com/2007/10/01/parallel-parking/ or the explanation in EDWARD NELSON, *Tensor Analysis*, Princeton University Press, 1967, pp. 33–36. Nelson explains how to define vector fields that change the four-dimensional configuration of a car, and how to use their algebra to show that a car can leave parking spaces with arbitrarily short distances to the cars in front and in the back.

Challenge 100, page 65: A smallest gap does not exist: any value will do! Can you show this?

Challenge 101, page 66: The following solution was proposed by Daniel Hawkins.

Assume you are sitting in car A, parked behind car B, as shown in Figure 311. There are two basic methods for exiting a parking space that requires the reverse gear: rotating the car to move the centre of rotation away from (to the right of) car B, and shifting the car downward to move the centre of rotation away from (farther below) car B. The first method requires car A to be partially diagonal, which means that the method will not work for d less than a certain value, essentially the value given above, when no reverse gear is needed. We will concern ourselves with the second method (pictured), which will work for an infinitesimal d.

In the case where the distance d is less than the minimum required distance to turn out of the parking space without using the reverse gear for a given geometry L, w, b, R, an attempt to

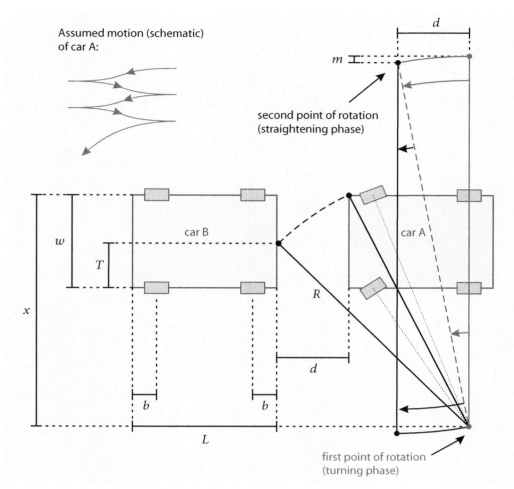

FIGURE 311 Solving the car parking puzzle (© Daniel Hawkins).

turn out of the parking space will result in the corner of car A touching car B at a distance T away from the edge of car B, as shown in Figure 311. This distance T is the amount by which car A must be translated downward in order to successfully turn out of the parking space.

The method to leave the parking space, shown in the top left corner of Figure 311, requires two phases to be successful: the initial **turning phase**, and the **straightening phase**. By turning and straightening out, we achieve a vertical shift downward and a horizontal shift left, while preserving the original orientation. That last part is key because if we attempted to turn until the corner of car A touched car B, car A would be rotated, and any attempt to straighten out would just follow the same arc backward to the initial position, while turning the wheel the other direction would rotate the car even more, as in the first method described above.

Our goal is to turn as far as we can and still be able to completely straighten out by time car A touches car B. To analyse just how much this turn should be, we must first look at the properties of a turning car.

Ackermann steering is the principle that in order for a car to turn smoothly, all four wheels must rotate about the same point. This was patented by Rudolph Ackermann in 1817. Some properties of Ackermann steering in relation to this problem are as follows:

- The back wheels stay in alignment, but the front wheels (which we control), must turn different amounts to rotate about the same centre.

- The centres of rotation for left and right turns are on opposite sides of the car

- For equal magnitudes of left and right turns, the centres of rotation are equidistant from the nearest edge of the car. Figure 311 makes this much clearer.

- All possible centres of rotation are on the same line, which also always passes through the back wheels.

- When the back wheels are 'straight' (straight will always mean in the same orientation as the initial position), they will be vertically aligned with the centres of rotation.

- When the car is turning about one centre, say the one associated with the maximum left turn, then the potential centre associated with the maximum right turn will rotate along with the car. Similarly, when the cars turns about the right centre, the left centre rotates.

Now that we know the properties of Ackermann steering, we can say that in order to maximize the shift downward while preserving the orientation, we must turn left about the 1st centre such that the 2nd centre rotates a *horizontal* distance d, as shown in Figure 311. When this is achieved, we brake, and turn the steering wheel the complete opposite direction so that we are now turning right about the 2nd centre. Because we shifted leftward d, we will straighten out at the exact moment car A comes in contact with car B. This results in our goal, a downward shift m and leftward shift d while preserving the orientation of car A. A similar process can be performed in reverse to achieve another downward shift m and a *rightward* shift d, effectively moving car A from its initial position (before any movement) downward $2m$ while preserving its orientation. This can be done indefinitely, which is why it is possible to get out of a parking space with an infinitesimal d between car A and car B. To determine how many times this procedure (both sets of turning and straightening) must be performed, we must only divide T (remember T is the amount by which car A must be shifted downward in order to turn out of the parking spot normally) by $2m$, the total downward shift for one iteration of the procedure. Symbolically,

$$n = \frac{T}{2m}. \tag{132}$$

In order to get an expression for n in terms of the geometry of the car, we must solve for T and

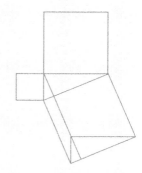

FIGURE 312 A simple drawing – one of the many possible one – that allows proving Pythagoras' theorem.

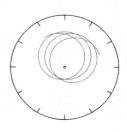

FIGURE 313 The trajectory of the middle point between the two ends of the hands of a clock.

$2m$. To simplify the derivations we define a new length x, also shown in Figure 311.

$$x = \sqrt{R^2 - (L-b)^2}$$

$$\begin{aligned}T &= \sqrt{R^2 - (L-b+d)^2} - x + w \\ &= \sqrt{R^2 - (L-b+d)^2} - \sqrt{R^2 - (L-b)^2} + w\end{aligned}$$

$$\begin{aligned}m &= 2x - w - \sqrt{(2x-w)^2 - d^2} \\ &= 2\sqrt{R^2 - (L-b)^2} - w - \sqrt{(2\sqrt{R^2-(L-b)^2} - w)^2 - d^2} \\ &= 2\sqrt{R^2 - (L-b)^2} - w - \sqrt{4(R^2-(L-b)^2) - 4w\sqrt{R^2-(L-b)^2} + w^2 - d^2} \\ &= 2\sqrt{R^2 - (L-b)^2} - w - \sqrt{4R^2 - 4(L-b)^2 - 4w\sqrt{R^2-(L-b)^2} + w^2 - d^2}\end{aligned}$$

We then get

$$n = \frac{T}{2m} = \frac{\sqrt{R^2 - (L-b+d)^2} - \sqrt{R^2 - (L-b)^2} + w}{4\sqrt{R^2-(L-b)^2} - 2w - 2\sqrt{4R^2 - 4(L-b)^2 - 4w\sqrt{R^2-(L-b)^2} + w^2 - d^2}}\ .$$

The value of n must always be rounded *up* to the next integer to determine how many times one must go backward and forward to leave the parking spot.

Challenge 102, page 66: Nothing, neither a proof nor a disproof.

Challenge 103, page 66: See volume II, on page 20. On extreme shutters, see also the discussion in Volume VI, on page 119.

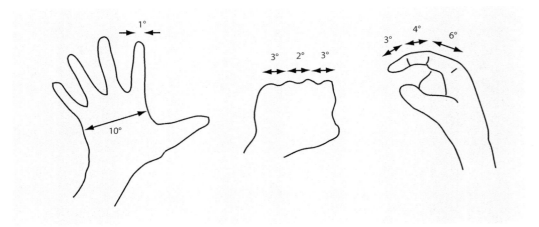

FIGURE 314 The angles defined by the hands against the sky, when the arms are extended.

Challenge 104, page 67: A hint for the solution is given in Figure 312.

Challenge 105, page 67: Because they are or were liquid.

Challenge 106, page 67: The shape is shown in Figure 313; it has eleven lobes.

Challenge 107, page 67: The cone angle φ, the angle between the cone axis and the cone border (or equivalently, *half* the apex angle of the cone) is related to the solid angle Ω through the relation $\Omega = 2\pi(1 - \cos\varphi)$. Use the surface area of a spherical cap to confirm this result.

Challenge 109, page 67: See Figure 314.

Challenge 113, page 69: Hint: draw all objects involved.

Challenge 114, page 69: The curve is obviously called a *catenary*, from Latin 'catena' for chain. The formula for a catenary is $y = a\cosh(x/a)$. If you approximate the chain by short straight segments, you can make wooden blocks that can form an arch without any need for glue. The St. Louis arch is in shape of a catenary. A suspension bridge has the shape of a catenary before it is loaded, i.e., before the track is attached to it. When the bridge is finished, the shape is in between a catenary and a parabola.

Challenge 115, page 69: The inverse radii, or curvatures, obey $a^2+b^2+c^2+d^2 = (1/2)(a+b+c+d)^2$. This formula was discovered by René Descartes. If one continues putting circles in the remaining spaces, one gets so-called circle packings, a pretty domain of recreational mathematics. They have many strange properties, such as intriguing relations between the coordinates of the circle centres and their curvatures.

Challenge 116, page 70: One option: use the three-dimensional analogue of Pythagoras's theorem. The answer is 9.

Challenge 117, page 70: There are two solutions. (Why?) They are the two positive solutions of $l^2 = (b+x)^2 + (b+b^2/x)^2$; the height is then given as $h = b + x$. The two solutions are 4.84 m and 1.26 m. There are closed formulas for the solutions; can you find them?

Challenge 118, page 70: The best way is to calculate first the height B at which the blue ladder touches the wall. It is given as a solution of $B^4 - 2hB^3 - (r^2-b^2)B^2 + 2h(r^2-b^2)B - h^2(r^2-b^2) = 0$. Integer-valued solutions are discussed in MARTIN GARDNER, *Mathematical Circus*, Spectrum, 1996.

Challenge 119, page 70: Draw a logarithmic scale, i.e., put every number at a distance corresponding to its natural logarithm. Such a device, called a *slide rule*, is shown in Figure 315. Slide

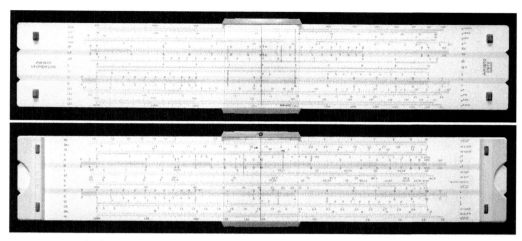

FIGURE 315 A high-end slide rule, around 1970 (© Jörn Lütjens).

rules were the precursors of electronic calculators; they were used all over the world in prehistoric times, i.e., until around 1970. See also the web page www.oughtred.org.

Challenge 120, page 71: Two more days. Build yourself a model of the Sun and the Earth to verify this. In fact, there is a small correction to the value 2, for the same reason that the makes the solar day shorter than 24 hours.

Challenge 121, page 71: The Sun is exactly behind the back of the observer; it is setting, and the rays are coming from behind and reach deep into the sky in the direction opposite to that of the Sun.

Challenge 123, page 71: The volume is given by $V = \int A dx = \int_{-1}^{1} 4(1 - x^2) dx = 16/3$.

Challenge 124, page 71: Yes. Try it with a paper model.

Challenge 125, page 71: Problems appear when quantum effects are added. A two-dimensional universe would have no matter, since matter is made of spin 1/2 particles. But spin 1/2 particles do not exist in two dimensions. Can you find additional reasons?

Challenge 126, page 72: Two dimensions of time do not allow ordering of events and observations. To say 'before' and 'afterwards' becomes impossible. In everyday life and all domains accessible to measurement, time is definitely one-dimensional.

Challenge 127, page 72: No experiment has ever found any hint. Can this be nevertheless? Probably not, as argued in the last volume of this book series.

Challenge 130, page 73: The best solution seems to be 23 extra lines. Can you deduce it? To avoid spoiling the fun of searching, no solution is given here. You can find solutions on blog.vixra.org/2010/12/26/a-christmas-puzzle.

Challenge 131, page 73: If you solve this so-called ropelength problem, you will become a famous mathematician. The length is known only with about 6 decimals of precision. No exact formula is known, and the exact shape of such ideal knots is unknown for all non-trivial knots. The problem is also unsolved for all non-trivial ideal *closed* knots, for which the two ends are glued together.

Challenge 132, page 76: From $x = gt^2/2$ you get the following rule: square the number of seconds, multiply by five and you get the depth in metres.

Challenge 133, page 76: Just experiment.

Challenge 134, page 77: The Academicians suspended one cannon ball with a thin wire just in front of the mouth of the cannon. When the shot was released, the second, flying cannon ball flew through the wire, thus ensuring that both balls started at the same time. An observer from far away then tried to determine whether both balls touched the Earth at the same time. The experiment is not easy, as small errors in the angle and air resistance confuse the results.

Challenge 135, page 77: A parabola has a so-called focus or focal point. All light emitted from that point and reflected exits in the same direction: all light rays are emitted in parallel. The name 'focus' – Latin for fireplace – expresses that it is the hottest spot when a parabolic mirror is illuminated. Where is the focus of the parabola $y = x^2$? (Ellipses have two foci, with a slightly different definition. Can you find it?)

Challenge 136, page 78: The long jump record could surely be increased by getting rid of the sand stripe and by measuring the true jumping distance with a photographic camera; that would allow jumpers to run more closely to their top speed. The record could also be increased by a small inclined step or by a spring-suspended board at the take-off location, to increase the take-off angle.

Challenge 137, page 78: It may be held by Roald Bradstock, who threw a golf ball over 155 m. Records for throwing mobile phones, javelins, people and washing machines are shorter.

Challenge 138, page 79: Walk or run in the rain, measure your own speed v and the angle from the vertical α with which the rain appears to fall. Then the speed of the rain is $v_{\text{rain}} = v/\tan\alpha$.

Challenge 139, page 79: In ice skating, quadruple jumps are now state of the art. In dance, no such drive for records exists.

Challenge 140, page 79: Neglecting air resistance and approximating the angle by 45°, we get $v = \sqrt{dg}$, or about 3.8 m/s. This speed is created by a steady pressure build-up, using blood pressure, which is suddenly released with a mechanical system at the end of the digestive canal. The cited reference tells more about the details.

Challenge 141, page 79: On horizontal ground, for a speed v and an angle from the horizontal α, neglecting air resistance and the height of the thrower, the distance d is $d = v^2 \sin 2\alpha/g$.

Challenge 142, page 79: Astonishingly, the answer is not clear. In 2012, the human record is eleven balls. For robots, the present record is three balls, as performed by the Sarcoman robot. The internet is full of material and videos on the topic. It is a challenge for people and robots to reach the maximum possible number of balls.

Challenge 143, page 79: It is said so, as rain drops would then be ice spheres and fall with high speed.

Challenge 144, page 80: Yes! People have gone to hospital and even died because a falling bullet went straight through their head. See S. MIRSKY, *It is high, it is far*, Scientific American p. 86, February 2004, or C. TUIJN, *Vallende kogels*, Nederlands tijdschrift voor natuurkunde 71, pp. 224–225, 2005. Firing a weapon into the air is a crime.

Challenge 145, page 80: This is a true story. The answer can only be given if it is known whether the person had the chance to jump while running or not. In the case described by R. CROSS, *Forensic physics 101: falls from a height*, American Journal of Physics 76, pp. 833–837, 2008, there was no way to run, so that the answer was: murder.

Challenge 146, page 80: For jumps of an animal of mass m the necessary energy E is given as $E = mgh$, and the work available to a muscle is roughly speaking proportional to its mass $W \sim m$. Thus one gets that the height h is independent of the mass of the animal. In other words, the specific mechanical energy of animals is around 1.5 ± 0.7 J/kg.

Challenge 147, page 80: Stones *never* follow parabolas: when studied in detail, i.e., when the change of g with height is taken into account, their precise path turns out to be an ellipse. This

CHALLENGE HINTS AND SOLUTIONS   489

shape appears most clearly for long throws, such as throws around a sizeable part of the Earth, or for orbiting objects. In short, stones follow parabolas only if the Earth is assumed to be flat. If its curvature is taken into account, they follow ellipses.

Challenge 148, page 81: The set of all rotations around a point in a plane is indeed a vector space. What about the set of all rotations around *all* points in a plane? And what about the three-dimensional cases?

Challenge 151, page 81: The scalar product between two vectors a and b is given by

$$ab = ab \cos \sphericalangle(a, b) \,. \tag{133}$$

How does this differ from the vector product?

Challenge 154, page 84: One candidate for the lowest practical acceleration of a physical system are the accelerations measured by gravitational wave detectors. They are below 10^{-13} m/s^2. But these low values are beaten by the acceleration of the continental drift after the continents 'snap' apart: they accelerate from 7 mm/a to 40 mm/a in a 'mere' 3 million years. This corresponds to a value of 10^{-23} m/s^2. Is there a theoretical lowest limit to acceleration?

Challenge 155, page 85: In free fall (when no air is present) or inside a space station orbiting the Earth, you are accelerated but do not feel anything. However, the issue is not so simple. On the one hand, constant and homogeneous accelerations are indeed not felt if there is no non-accelerated reference. This indistinguishability or equivalence between acceleration and 'feeling nothing' was an essential step for Albert Einstein in his development of general relativity. On the other hand, if our senses were sensitive enough, we would feel something: both in the free fall and in the space station, the acceleration is neither constant nor homogeneous. So we can indeed say that accelerations found in nature can always be felt.

Challenge 156, page 85: Professor to student: What is the derivative of velocity? Acceleration! What is the derivative of acceleration? I don't know. *Jerk*! The fourth, fifth and sixth derivatives of position are sometimes called *snap*, *crackle* and *pop*.

Challenge 158, page 87: One can argue that any source of light must have finite size.

Challenge 160, page 89: What the unaided human eye perceives as a tiny black point is usually about 50 μm in diameter.

Challenge 161, page 89: See volume III, page 170.

Challenge 162, page 89: One has to check carefully whether the conceptual steps that lead us to extract the concept of point from observations are correct. It will be shown in the final part of the adventure that this is not the case.

Challenge 163, page 89: One can rotate the hand in a way that the arm makes the motion described. See also volume IV, page 132.

Challenge 164, page 89: The number of cables has no limit. A visualization of tethered rotation with 96 connections is found in volume VI, on page 181.

Challenge 165, page 90: The blood and nerve supply is not possible if the wheel has an axle. The method shown to avoid tangling up connections only works when the rotating part has *no* axle: the 'wheel' must float or be kept in place by other means. It thus becomes impossible to make a wheel *axle* using a single piece of skin. And if a wheel without an axle could be built (which might be possible), then the wheel would periodically run over the connection. Could such a axle-free connection realize a propeller?

By the way, it is still thinkable that animals have wheels on axles, if the wheel is a 'dead' object. Even if blood supply technologies like continuous flow reactors were used, animals could not make such a detached wheel grow in a way tuned to the rest of the body and they would have difficulties repairing a damaged wheel. Detached wheels cannot be grown on animals; they must be dead.

Challenge 166, page 92: The brain in the skull, the blood factories inside bones or the growth of the eye are examples.

Challenge 167, page 92: In 2007, the largest big wheels for passengers are around 150 m in diameter. The largest wind turbines are around 125 m in diameter. Cement kilns are the longest wheels: they can be over 300 m along their axis.

Challenge 168, page 92: Air resistance reduces the maximum distance achievable with the soccer ball – which is realized for an angle of about $\pi/4 = 45°$ – from around $v^2/g = 91.7$ m down to around 50 m.

Challenge 173, page 98: One can also add the Sun, the sky and the landscape to the list.

Challenge 174, page 99: There is no third option. Ghosts, hallucinations, Elvis sightings, or extraterrestrials must all be objects or images. Also shadows are only special types of images.

Challenge 175, page 99: The issue was hotly discussed in the seventeenth century; even Galileo argued for them being images. However, they are objects, as they can collide with other objects, as the spectacular collision between Jupiter and the comet Shoemaker-Levy 9 in 1994 showed. In the meantime, satellites have been made to collide with comets and even to shoot at them (and hitting).

Challenge 176, page 100: The minimum speed is roughly the one at which it is possible to ride without hands. If you do so, and then *gently* push on the steering wheel, you can make the experience described above. Watch out: too strong a push will make you fall badly.

The *bicycle* is one of the most complex mechanical systems of everyday life, and it is still a subject of research. And obviously, the world experts are Dutch. An overview of the behaviour of a bicycle is given in Figure 316. The main result is that the bicycle is stable in the upright position at a range of medium speeds. Only at low and at large speeds must the rider actively steer to ensure upright position of the bicycle.

For more details, see the paper J. P. MEIJAARD, J. M. PAPADOPOULOS, A. RUINA & A. L. SCHWAB, *Linearized dynamics equations for the balance and steer of a bicycle: a benchmark and review*, Proceedings of the Royal Society A **463**, pp. 1955–1982, 2007, and J. D. G. KOOIJMAN, A. L. SCHWAB & J. P. MEIJAARD, *Experimental validation of a model of an uncontrolled bicycle*, Multibody System Dynamics **19**, pp. 115–132, 2008. See also the audiophile.tam.cornell.edu/~als93/Bicycle/index.htm website.

Challenge 177, page 102: The total weight decreased slowly, due to the evaporated water lost by sweating and, to a minor degree, due to the exhaled carbon bound in carbon dioxide.

Challenge 178, page 102: This is a challenge where the internet can help a lot. For a general introduction, see the book by LEE SIEGEL, *Net of Magic – Wonders and Deception in India*, University of Chicago Press, 1991.

Challenge 179, page 103: If the moving ball is not rotating, after the collision the two balls will depart with a *right* angle between them.

Challenge 180, page 103: As the block is heavy, the speed that it acquires from the hammer is small and easily stopped by the human body. This effect works also with an anvil instead of a concrete block. In another common variation the person does not lie on nails, but on air: he just keeps himself horizontal, with head and shoulders on one chair, and the feet on a second one.

Challenge 181, page 104: Yes, the definition of mass works also for magnetism, because the precise condition is not that the interaction is central, but that the interaction realizes a more general condition that includes accelerations such as those produced by magnetism. Can you deduce the condition from the definition of mass as that quantity that keeps momentum conserved?

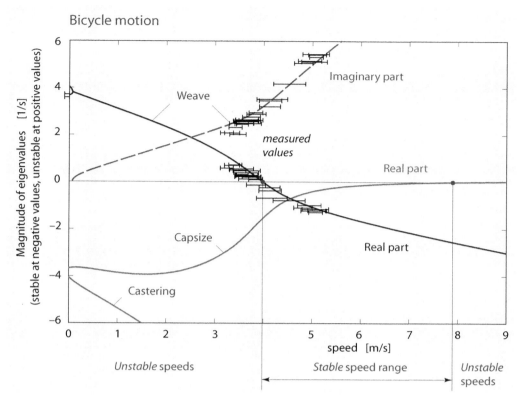

FIGURE 316 The measured (black bars) and calculated behaviour (coloured lines) – more precisely, the dynamical eigenvalues – of a bicycle as a function of its speed (© Arend Schwab).

Challenge 182, page 106: Rather than using the inertial effects of the Earth, it is easier to deduce its mass from its gravitational effects.

Challenge 187, page 107: At first sight, relativity implies that tachyons have imaginary mass; however, the imaginary factor can be extracted from the mass–energy and mass–momentum relation, so that one can define a real mass value for tachyons. As a result, faster tachyons have smaller energy and smaller momentum than slower ones. In fact, both tachyon momentum and tachyon energy can be a negative number of any size.

Challenge 188, page 109: The leftmost situation has a tiny effect, the second situation makes the car roll forward and backward, the right two pictures show ways to open wine bottles without bottle opener.

Challenge 189, page 109: Legs are never perfectly vertical; they would immediately glide away. Once the cat or the person is on the floor, it is almost impossible to stand up again.

Challenge 190, page 109: Momentum (or centre of mass) conservation would imply that the environment would be accelerated into the opposite direction. Energy conservation would imply that a huge amount of energy would be transferred between the two locations, melting everything in between. Teleportation would thus contradict energy and momentum conservation.

Challenge 191, page 110: The part of the tides due to the Sun, the solar wind, and the interactions between both magnetic fields are examples of friction mechanisms between the Earth and the Sun.

Challenge 192, page 110: With the factor 1/2, increase of (physical) kinetic energy is equal to the (physical) work performed on a system: total energy is thus conserved only if the factor 1/2 is added.

Challenge 194, page 112: It is a smart application of momentum conservation.

Challenge 195, page 112: Neither. With brakes on, the damage is higher, but still equal for both cars.

Challenge 196, page 113: Heating systems, transport engines, engines in factories, steel plants, electricity generators covering the losses in the power grid, etc. By the way, the richest countries in the world, such as Sweden or Switzerland, consume only half the energy per inhabitant as the USA. This waste is one of the reasons for the lower average standard of living in the USA.

Challenge 202, page 117: Just throw the brick into the air and compare the dexterity needed to make it turn around various axes.

Challenge 203, page 118: Use the definition of the moment of inertia and Pythagoras' theorem for every mass element of the body.

Challenge 204, page 119: Hang up the body, attaching the rope in two different points. The crossing point of the prolonged rope lines is the centre of mass.

Challenge 205, page 119: See Tables 19 and 20.

Challenge 206, page 119: Spheres have an orientation, because we can always add a tiny spot on their surface. This possibility is not given for microscopic objects, and we shall study this situation in the part on quantum theory.

Challenge 209, page 120: Yes, the ape can reach the banana. The ape just has to turn around its own axis. For every turn, the plate will rotate a bit towards the banana. Of course, other methods, like blowing at a right angle to the axis, peeing, etc., are also possible.

Challenge 210, page 121: Self-propelled linear motion contradicts the conservation of momentum; self-propelled change of orientation (as long as the motion stops again) does not contradict any conservation law. But the deep, underlying reason for the difference will be unveiled in the final part of our adventure.

Challenge 212, page 121: The points that move exactly along the radial direction of the wheel form a circle below the axis and above the rim. They are the points that are sharp in Figure 79 of page 121.

Challenge 213, page 122: Use the conservation of angular momentum around the point of contact. If all the wheel's mass is assumed in the rim, the final rotation speed is half the initial one; it is independent of the friction coefficient.

Challenge 215, page 123: Probably the 'rest of the universe' was meant by the writer. Indeed, a moving a part never shifts the centre of gravity of a closed system. But is the universe closed? Or a system? The last part of our adventure addresses these issues.

Challenge 219, page 124: Hint: energy and momentum conservation yield two equations; but in the case of three balls there are three variables. What else is needed? See F. HERRMANN & M. SEITZ, *How does the ball-chain work?*, American Journal of Physics 50, pp. 977–981, 1982. The bigger challenge is to build a high precision ball-chain, in which the balls behave as expected, minimizing spurious motion. Nobody seems to have built one yet, as the internet shows. Can you?

Challenge 220, page 124: The method allowed Phileas Fogg to win the central bet in the well-known adventure novel by JULES VERNE, *Around the World in Eighty Days*, translated from *Le tour du monde en quatre-vingts jours*, first published in 1872.

Challenge 221, page 125: The human body is more energy-efficient at low and medium power output. The topic is still subject of research, as detailed in the cited reference. The critical slope is estimated to be around 16° for uphill walkers, but should differ for downhill walkers.

Challenge 223, page 125: Hint: an energy per distance is a force.

Challenge 224, page 126: The conservation of angular momentum saves the glass. Try it.

Challenge 225, page 126: First of all, MacDougall's experimental data is flawed. In the six cases MacDougall examined, he did not know the exact timing of death. His claim of a mass decrease cannot be deduced from his own data. Modern measurements on dying sheep, about the same mass as humans, have shown no mass change, but clear weight pulses of a few dozen grams when the heart stopped. This temporary weight decrease could be due to the expelling of air or moisture, to the relaxing of muscles, or to the halting of blood circulation. The question is not settled.

Challenge 227, page 126: Assuming a square mountain, the height h above the surrounding crust and the depth d below are related by

$$\frac{h}{d} = \frac{\rho_m - \rho_c}{\rho_c} \qquad (134)$$

where ρ_c is the density of the crust and ρ_m is the density of the mantle. For the density values given, the ratio is 6.7, leading to an additional depth of 6.7 km below the mountain.

Challenge 229, page 127: The can filled with liquid. Videos on the internet show the experiment. Why is this the case?

Challenge 232, page 127: The matter in the universe could rotate – but not the universe itself. Measurements show that within measurement errors there is no mass rotation.

Challenge 233, page 128: The behaviour of the spheres can only be explained by noting that elastic waves propagate through the chain of balls. Only the propagation of these elastic waves, in particular their reflection at the end of the chain, explains that the same number of balls that hit on one side are lifted up on the other. For long times, friction makes all spheres oscillate in phase. Can you confirm this?

Challenge 234, page 128: When the short cylinder hits the long one, two compression waves start to run from the point of contact through the two cylinders. When each compression wave arrives at the end, it is reflected as an expansion wave. If the geometry is well chosen, the expansion wave coming back from the short cylinder can continue into the long one (which is still in his compression phase). For sufficiently long contact times, waves from the short cylinder can thus depose much of their energy into the long cylinder. Momentum is conserved, as is energy; the long cylinder is oscillating in length when it detaches, so that not all its energy is translational energy. This oscillation is then used to drive nails or drills into stone walls. In commercial hammer drills, length ratios of 1:10 are typically used.

Challenge 235, page 128: The momentum transfer to the wall is double when the ball rebounds perfectly.

Challenge 236, page 128: If the cork is in its intended position: take the plastic cover off the cork, put the cloth around the bottle or the bottle in the shoe (this is for protection reasons only) and repeatedly hit the bottle on the floor or a fall in an inclined way, as shown in Figure 72 on page 109. With each hit, the cork will come out a bit.

If the cork has fallen inside the bottle: put half the cloth inside the bottle; shake until the cork falls unto the cloth. Pull the cloth out: first slowly, until the cloth almost surround the cork, and then strongly.

Challenge 237, page 129: Indeed, the lower end of the ladder always touches the floor. Why?

Challenge 238, page 129: The atomic force microscope.

Challenge 240, page 129: Running man: $E \approx 0.5 \cdot 80\,\text{kg} \cdot (5\,\text{m/s})^2 = 1\,\text{kJ}$; rifle bullet: $E \approx 0.5 \cdot 0.04\,\text{kg} \cdot (500\,\text{m/s})^2 = 5\,\text{kJ}$.

Challenge 241, page 129: It almost doubles in size.

Challenge 242, page 130: At the highest point, the acceleration is $g \sin \alpha$, where α is the angle of the pendulum at the highest point. At the lowest point, the acceleration is v^2/l, where l is the length of the pendulum. Conservation of energy implies that $v^2 = 2gl(1 - \cos \alpha)$. Thus the problem requires that $\sin \alpha = 2(1 - \cos \alpha)$. This results in $\cos \alpha = 3/5$.

Challenge 243, page 130: One needs the mass change equation $dm/dt = \pi \rho_{\text{vapour}} r^2 |v|$ due to the mist and the drop speed evolution $m\,dv/dt = mg - v\,dm/dt$. These two equations yield

$$\frac{dv^2}{dr} = \frac{2g}{C} - 6\frac{v^2}{r} \qquad (135)$$

where $C = \rho_{\text{vapour}}/4\rho_{\text{water}}$. The trick is to show that this can be rewritten as

$$r\frac{d}{dr}\frac{v^2}{r} = \frac{2g}{C} - 7\frac{v^2}{r} \; . \qquad (136)$$

For large times, all physically sensible solutions approach $v^2/r = 2g/7C$; this implies that for large times,

$$\frac{dv}{dt}\frac{v^2}{r} = \frac{g}{7} \quad \text{and} \quad r = \frac{gC}{14}t^2 \; . \qquad (137)$$

About this famous problem, see for example, B. F. EDWARDS, J. W. WILDER & E. E. SCIME, *Dynamics of falling raindrops*, European Journal of Physics 22, pp. 113–118, 2001, or A. D. SOKAL, *The falling raindrop, revisited*, preprint at arxiv.org/abs/0908.0090.

Challenge 244, page 130: One is faster, because the moments of inertia differ. Which one?

Challenge 245, page 130: There is no simple answer, as aerodynamic drag plays an important role. There are almost no studies on the topic. By the way, competitive rope jumping is challenging; for example, a few people in the world are able to rotate the rope 5 times under their feet during a single jump. Can you do better?

Challenge 246, page 130: Weigh the bullet and shoot it against a mass hanging from the ceiling. From the mass and the angle it is deflected to, the momentum of the bullet can be determined.

Challenge 248, page 130: The curve described by the midpoint of a ladder sliding down a wall is a circle.

Challenge 249, page 130: The switches use the power that is received when the switch is pushed and feed it to a small transmitter that acts a high frequency remote control to switch on the light.

Challenge 250, page 131: A clever arrangement of bimetals is used. They move every time the temperature changes from day to night – and vice versa – and wind up a clock spring. The clock itself is a mechanical clock with low energy consumption.

Challenge 251, page 131: The weight of the lift does not change at all when a ship enters it. A twin lift, i.e., a system in which both lifts are mechanically or hydraulically connected, needs no engine at all: it is sufficient to fill the upper lift with a bit of additional water every time a ship enters it. Such ship lifts without engines at all used to exist in the past.

Challenge 254, page 133: This is not easy; a combination of friction and torques play a role. See for example the article J. SAUER, E. SCHÖRNER & C. LENNERZ, *Real-time rigid body simulation of some classical mechanical toys*, 10th European Simulation and Symposium and Exhibition (ESS '98) 1998, pp. 93–98, or www.lennerz.de/paper_ess98.pdf.

Challenge 257, page 135: See Figure 123 for an example. The pole is not at the zenith.

Challenge 258, page 135: Robert Peary had forgotten that on the date he claimed to be at the North Pole, 6th of April 1909, the Sun is very low on the horizon, casting very long shadows, about ten times the height of objects. But on his photograph the shadows are much shorter. (In fact, the picture is taken in such a way to hide all shadows as carefully as possible.) Interestingly, he had even convinced the US congress to officially declare him the first man on the North Pole in 1911. (A rival crook had claimed to have reached it before Peary, but his photograph has the same mistake.) Peary also cheated on the travelled distances of the last few days; he also failed to mention that the last days he was pulled by his partner, Matthew Henson, because he was not able to walk any more. In fact Matthew Henson deserves more credit for that adventure than Peary. Henson, however, did not know that Peary cheated on the position they had reached.

Challenge 260, page 136: Laplace and Gauss showed that the eastward deflection d of a falling object is given by

$$d = 2/3\,\Omega \cos\varphi \sqrt{2h^3/g} \ . \tag{138}$$

Here $\Omega = 72.92\,\mu$rad/s is the angular velocity of the Earth, φ is the latitude, g the gravitational acceleration and h is the height of the fall.

Challenge 261, page 139: The Coriolis effect can be seen as the sum two different effects of equal magnitude. The first effect is the following: on a rotating background, velocity changes over time. What an inertial (non-rotating) observer sees as a *constant* velocity will be seen a velocity that *changes* over time by the rotating observer. The acceleration seen by the rotating observer is negative, and is proportional to the angular velocity and to the velocity.

The second effect is change of velocity in space. In a rotating frame of reference, different points have different velocities. The effect is negative, and proportional to the angular velocity and to the velocity.

In total, the Coriolis acceleration (or Coriolis effect) is thus $\boldsymbol{a}_\text{C} = -2\boldsymbol{\omega} \times \boldsymbol{v}$.

Challenge 262, page 140: A *short* pendulum of length L that swings in two dimensions (with amplitude ρ and orientation φ) shows two additional terms in the Lagrangian $\mathcal{L}$:

$$\mathcal{L} = T - V = \tfrac{1}{2}m\dot\rho^2 \left(1 + \frac{\rho^2}{L^2}\right) + \frac{l_z^2}{2m\rho^2} - \tfrac{1}{2}m\omega_0^2 \rho^2 (1 + \frac{\rho^2}{4\,L^2}) \tag{139}$$

where as usual the basic frequency is $\omega_0^2 = g/L$ and the angular momentum is $l_z = m\rho^2\dot\varphi$. The two additional terms disappear when $L \to \infty$; in that case, if the system oscillates in an ellipse with semiaxes a and b, the ellipse is fixed in space, and the frequency is ω_0. For *finite* pendulum length L, the frequency changes to

$$\omega = \omega_0 \left(1 - \frac{a^2 + b^2}{16\,L^2}\right) \ . \tag{140}$$

The ellipse turns with a frequency

$$\Omega = \omega \frac{3}{8}\frac{ab}{L^2} \ . \tag{141}$$

These formulae can be derived using the least action principle, as shown by C. G. GRAY, G. KARL & V. A. NOVIKOV, *Progress in classical and quantum variational principles*, arxiv.org/abs/physics/0312071. In other words, a short pendulum in elliptical motion shows a precession even *without* the Coriolis effect. Since this precession frequency diminishes with $1/L^2$, the effect is small for long pendulums, where only the Coriolis effect is left over. To see the Coriolis effect in a short pendulum, one thus has to avoid that it starts swinging in an elliptical orbit by adding a mechanism that suppresses elliptical motion.

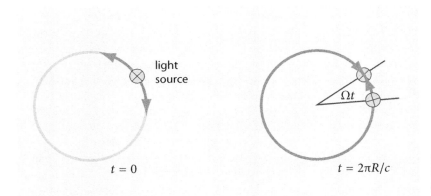

FIGURE 317 Deducing the expression for the Sagnac effect.

Challenge 263, page 141: The Coriolis acceleration is the reason for the deviation from the straight line. The Coriolis acceleration is due to the change of speed with distance from the rotation axis. Now think about a pendulum, located in Paris, swinging in the North-South direction with amplitude A. At the Southern end of the swing, the pendulum is further from the axis by $A \sin \varphi$, where φ is the latitude. At that end of the swing, the central support point overtakes the pendulum bob with a relative horizontal speed given by $v = 2\pi A \sin \varphi / 23\,\text{h}\,56\,\text{min}$. The period of precession is given by $T_F = v/2\pi A$, where $2\pi A$ is the circumference $2\pi A$ of the envelope of the pendulum's path (relative to the Earth). This yields $T_F = 23\,\text{h}\,56\,\text{min}/\sin \varphi$. Why is the value that appears in the formula not 24 h, but 23 h 56 min?

Challenge 264, page 141: Experiments show that the axis of the gyroscope stays fixed with respect to distant stars. No experiment shows that it stays fixed with respect to absolute space, because this kind of "absolute space" cannot be defined or observed at all. It is a useless concept.

Challenge 265, page 141: Rotation leads to a small frequency and thus colour changes of the circulating light.

Challenge 266, page 141: The weight changes when going east or when moving west due to the Coriolis acceleration. If the rotation speed is tuned to the oscillation frequency of the balance, the effect is increased by resonance. This trick was also used by Eötvös.

Challenge 267, page 141: The Coriolis acceleration makes the bar turn, as every moving body is deflected to the side, and the two deflections add up in this case. The direction of the deflection depends on whether the experiments is performed on the northern or the southern hemisphere.

Challenge 268, page 141: When rotated by π around an east–west axis, the Coriolis force produces a drift velocity of the liquid around the tube. It has the value

$$v = 2\omega r \sin \theta, \tag{142}$$

as long as friction is negligible. Here ω is the angular velocity of the Earth, θ the latitude and r the (larger) radius of the torus. For a tube with 1 m diameter in continental Europe, this gives a speed of about $6.3 \cdot 10^{-5}$ m/s.

The measurement can be made easier if the tube is restricted in diameter at one spot, so that the velocity is increased there. A restriction by an area factor of 100 increases the speed by the same factor. When the experiment is performed, one has to carefully avoid any other effects that lead to moving water, such as temperature gradients across the system.

Challenge 269, page 142: Imagine a circular light path (for example, inside a circular glass fibre) and two beams moving in opposite directions along it, as shown in Figure 317. If the fibre path rotates with rotation frequency Ω, we can deduce that, after one turn, the difference ΔL in path length is

$$\Delta L = 2R\Omega t = \frac{4\pi R^2 \Omega}{c} \,. \tag{143}$$

The phase difference is thus

$$\Delta \varphi = \frac{8\pi^2 \, R^2}{c\lambda}\Omega \tag{144}$$

if the refractive index is 1. This is the required formula for the main case of the Sagnac effect.

It is regularly suggested that the Sagnac effect can only be understood with help of general relativity; this is wrong. As just done, the effect is easily deduced from the invariance of the speed of light c. The effect is a consequence of special relativity.

Challenge 270, page 145: The metal rod is slightly longer on one side of the axis. When the wire keeping it up is burned with a candle, its moment of inertia decreases by a factor of 10^4; thus it starts to rotate with (ideally) 10^4 times the rotation rate of the Earth, a rate which is easily visible by shining a light beam on the mirror and observing how its reflection moves on the wall.

Challenge 272, page 152: The original result by Bessel was 0.3136 ″, or 657.7 thousand orbital radii, which he thought to be 10.3 light years or 97.5 Pm.

FIGURE 318 How the night sky, and our galaxy in particular, looks in the near infrared (NASA false colour image).

Challenge 274, page 155: The galaxy forms a stripe in the sky. The galaxy is thus a flattened structure. This is even clearer in the infrared, as shown more clearly in Figure 318. From the flattening (and its circular symmetry) we can deduce that the galaxy must be rotating. Thus other matter must exist in the universe.

Challenge 276, page 158: See page 189.

Challenge 278, page 158: The scale reacts to your heartbeat. The weight is almost constant over time, except when the heart beats: for a short duration of time, the weight is somewhat lowered at each beat. Apparently it is due to the blood hitting the aortic arch when the heart pumps it upwards. The speed of the blood is about 0.3 m/s at the maximum contraction of the left ventricle. The distance to the aortic arch is a few centimetres. The time between the contraction and the reversal of direction is about 15 ms. And the measured weight is not even constant for a dead person, as air currents disturb the measurement.

Challenge 279, page 158: Use Figure 97 on page 138 for the second half of the trajectory, and think carefully about the first half. The body falls down slightly to the west of the starting point.

Challenge 280, page 158: Hint: starting rockets at the Equator saves a lot of energy, thus of fuel and of weight.

Challenge 283, page 161: The flame leans towards the inside.

Challenge 284, page 161: Yes. There is no absolute position and no absolute direction. Equivalently, there is no preferred position, and no preferred direction. For time, only the big bang seems to provide an exception, at first; but when quantum effects are included, the lack of a preferred time scale is confirmed.

Challenge 285, page 161: For your exam it is better to say that centrifugal force does not exist. But since in each stationary system there is a force balance, the discussion is somewhat a red herring.

Challenge 288, page 162: Place the tea in cups on a board and attach the board to four long ropes that you keep in your hand.

Challenge 289, page 162: The ball leans in the direction it is accelerated to. As a result, one could imagine that the ball in a glass at rest pulls upwards because the floor is accelerated upwards. We will come back to this issue in the section of general relativity.

Challenge 290, page 162: The friction of the tides on Earth are the main cause.

Challenge 291, page 163: An earthquake with Richter magnitude of 12 is 1000 times the energy of the 1960 Chile quake with magnitude 10; the latter was due to a crack throughout the full 40 km of the Earth's crust along a length of 1000 km in which both sides slipped by 10 m with respect to each other. Only the impact of a large asteroid could lead to larger values than 12.

Challenge 293, page 163: Yes; it happens twice a year. To minimize the damage, dishes should be dark in colour.

Challenge 294, page 163: A rocket fired from the back would be a perfect defence against planes attacking from behind. However, when released, the rocket is effectively flying backwards with respect to the air, thus turns around and then becomes a danger to the plane that launched it. Engineers who did not think about this effect almost killed a pilot during the first such tests.

Challenge 295, page 163: Whatever the ape does, whether it climbs up or down or even lets himself fall, it remains at the same height as the mass. Now, what happens if there is friction at the wheel?

Challenge 296, page 163: Yes, if he moves at a large enough angle to the direction of the boat's motion.

Challenge 297, page 163: See the article by C. UCKE & H.-J. SCHLICHTING, *Faszinierendes Dynabee*, Physik in unserer Zeit 33, pp. 230–231, 2002.

Challenge 298, page 164: See the article by C. UCKE & H.-J. SCHLICHTING, *Die kreisende Büroklammer*, Physik in unserer Zeit 36, pp. 33–35, 2005.

Challenge 299, page 164: If a wedding ring rotates on an axis that is not a principal one, angular momentum and velocity are not parallel.

Challenge 300, page 164: The moment of inertia for a homogeneous sphere is $\Theta = \frac{2}{5}mr^2$.

Challenge 301, page 164: The three moments of inertia for the cube are equal, as in the case of the sphere, but the values are $\Theta = \frac{1}{6}ml^2$. The efforts required to put a sphere and a cube into rotationthe are thus different.

Challenge 304, page 165: Yes, the moon differs in this way. Can you imagine what happens for an observer on the Equator?

CHALLENGE HINTS AND SOLUTIONS

Challenge 305, page 167: A straight line at the zenith, and circles getting smaller at both sides. See an example on the website apod.nasa.gov/apod/ap021115.html.

Challenge 307, page 167: The plane is described in the websites cited; for a standing human the plane is the vertical plane containing the two eyes.

Challenge 308, page 168: If you managed, please send the author the video!

Challenge 309, page 169: As said before, legs are simpler than wheels to grow, to maintain and to repair; in addition, legs do not require flat surfaces (so-called 'streets') to work.

Challenge 310, page 170: The staircase formula is an empirical result found by experiment, used by engineers world-wide. Its origin and explanation seems to be lost in history.

Challenge 311, page 170: Classical or everyday nature is right-left symmetric and thus requires an even number of legs. Walking on two-dimensional surfaces naturally leads to a minimum of four legs. Starfish, snails, slugs, clams, eels and snakes are among the most important exceptions for which the arguments are not valid.

Challenge 313, page 173: The length of the day changes with latitude. So does the length of a shadow or the elevation of stars at night, facts that are easily checked by telephoning a friend. Ships appear at the horizon by showing their masts first. These arguments, together with the round shadow of the earth during a lunar eclipse and the observation that everything falls downwards everywhere, were all given already by Aristotle, in his text *On the Heavens*. It is now known that everybody in the last 2500 years knew that the Earth is a sphere. The myth that many people used to believe in a flat Earth was put into the world – as rhetorical polemic – by Copernicus. The story then continued to be exaggerated more and more during the following centuries, because a new device for spreading lies had just been invented: book printing. Fact is that for 2500 years the vast majority of people knew that the Earth is a sphere.

Challenge 314, page 177: The vector SF can be calculated by using $SC = -(GmM/E)\, SP/SP$ and then translating the construction given in the figure into formulae. This exercise yields

$$SF = \frac{K}{mE} \qquad (145)$$

where

$$K = p \times L - GMm^2 x/x \qquad (146)$$

is the so-called *Runge–Lenz vector*. The Runge-Lenz vector is directed along the line that connects the second focus to the first focus of the ellipse (the Sun). We have used $x = SP$ for the position of the orbiting body, p for its momentum and L for its angular momentum. The Runge–Lenz vector K is *constant* along the orbit of a body, thus has the *same* value for any position x on the orbit. (Prove it by starting from $xK = xK \cos \theta$.) The Runge–Lenz vector is thus a *conserved* quantity in universal gravity. As a result, the vector SF is also constant in time.

The Runge–Lenz vector is also often used in quantum mechanics, when calculating the energy levels of a hydrogen atom, as it appears in all problems with a $1/r$ potential. (In fact, the incorrect name 'Runge–Lenz vector' is due to Wolfgang Pauli; the discoverer of the vector was, in 1710, Jakob Hermann.)

Challenge 316, page 178: On orbits, see page 192.

Challenge 317, page 178: The low gravitational acceleration of the Moon, 1.6 m/s^2, implies that gas molecules at usual temperatures can escape its attraction.

Challenge 318, page 179: The tip of the velocity arrow, when drawn over time, produces a circle around the centre of motion.

Challenge 319, page 179: Draw a figure of the situation.

Challenge 320, page 179: Again, draw a figure of the situation.

Challenge 321, page 180: The value of the product GM for the Earth is $4.0 \cdot 10^{14}\,\mathrm{m^3/s^2}$.

Challenge 322, page 180: All points can be reached for general inclinations; but when shooting horizontally in one given direction, only points on the first half of the circumference can be reached.

Challenge 323, page 182: On the moon, the gravitational acceleration is $1.6\,\mathrm{m/s^2}$, about one sixth of the value on Earth. The surface values for the gravitational acceleration for the planets can be found on many internet sites.

Challenge 324, page 182: The Atwood machine is the answer: two almost equal masses m_1 and m_2 connected by a string hanging from a well-oiled wheel of negligible mass. The heavier one falls very slowly. Can show that the acceleration a of this 'unfree' fall is given by $a = g(m_1 - m_2)/(m_1 + m_2)$? In other words, the smaller the mass difference is, the slower the fall is.

Challenge 325, page 183: You should absolutely try to understand the origin of this expression. It allows understanding many important concepts of mechanics. The idea is that for small amplitudes, the acceleration of a pendulum of length l is due to gravity. Drawing a force diagram for a pendulum at a general angle α shows that

$$ma = -mg\sin\alpha$$
$$ml\frac{d^2\alpha}{dt^2} = -mg\sin\alpha$$
$$l\frac{d^2\alpha}{dt^2} = -g\sin\alpha\,. \tag{147}$$

For the mentioned small amplitudes (below 15°) we can approximate this to

$$l\frac{d^2\alpha}{dt^2} = -g\alpha\,. \tag{148}$$

This is the equation for a harmonic oscillation (i.e., a sinusoidal oscillation). The resulting motion is:

$$\alpha(t) = A\sin(\omega t + \varphi)\,. \tag{149}$$

The amplitude A and the phase φ depend on the initial conditions; however, the oscillation frequency is given by the length of the pendulum and the acceleration of gravity (check it!):

$$\omega = \sqrt{\frac{l}{g}}\,. \tag{150}$$

(For arbitrary amplitudes, the formula is much more complex; see the internet or special mechanics books for more details.)

Challenge 326, page 183: Walking speed is proportional to l/T, which makes it proportional to $l^{1/2}$. The relation is also true for animals in general. Indeed, measurements show that the maximum walking speed (thus not the running speed) across all animals is given by

$$v_{\mathrm{maxwalking}} = (2.2 \pm 0.2)\,\mathrm{m^{1/2}/s}\,\sqrt{l}\,. \tag{151}$$

Challenge 330, page 186: There is no obvious candidate formula. Can you find one?

Challenge 331, page 186: The acceleration due to gravity is $a = Gm/r^2 \approx 5\,\mathrm{nm/s^2}$ for a mass of 75 kg. For a fly with mass $m_{\mathrm{fly}} = 0.1\,\mathrm{g}$ landing on a person with a speed of $v_{\mathrm{fly}} = 1\,\mathrm{cm/s}$

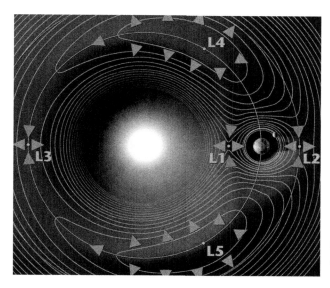

FIGURE 319 The Lagrangian points and the effective potential that produces them (NASA).

and deforming the skin (without energy loss) by $d = 0.3$ mm, a person would be accelerated by $a = (v^2/d)(m_{fly}/m) = 0.4\,\mu\text{m/s}^2$. The energy loss of the inelastic collision reduces this value at least by a factor of ten.

Challenge 332, page 187: The calculation shows that a surprisingly high energy value is stored in thermal motion.

Challenge 333, page 188: Yes, the effect has been measured for skyscrapers. Can you estimate the values?

Challenge 336, page 188: The easiest way to see this is to picture gravity as a flux emanating from a sphere. This gives a $1/r^{d-1}$ dependence for the force and thus a $1/r^{d-2}$ dependence of the potential.

Challenge 338, page 190: Since the paths of free fall are ellipses, which are curves lying in a plane, this is obvious.

Challenge 340, page 191: A flash of light is sent to the Moon, where several Cat's-eyes have been deposited by the Lunokhod and Apollo missions. The measurement precision of the time a flash take to go and come back is sufficient to measure the Moon's distance change. For more details, see challenge 8.

Challenge 342, page 193: A body having zero momentum at spatial infinity is on a parabolic path. A body with a lower momentum is on an elliptic path and one with a higher momentum is on a hyperbolic path.

Challenge 345, page 194: The Lagrangian points L4 and L5 are on the orbit, 60° before and behind the orbiting body. They are stable if the mass ratio of the central and the orbiting body is sufficiently large (above 24.9).

Challenge 346, page 194: The Lagrangian point L3 is located on the orbit, but precisely on the other side of the central body. The Lagrangian point L1 is located on the line connecting the planet with the central body, whereas L2 lies outside the orbit, on the same line. If R is the radius of the orbit, the distance between the orbiting body and the L1 and L2 point is $\sqrt[3]{m/3M}\,R$, giving around 4 times the distance of the Moon for the Sun-Earth system. L1, L2 and L3 are saddle points, but effectively stable orbits exist around them. Many satellites make use of these properties, including

FIGURE 320 The famous 'vomit comet', a KC-135, performing a parabolic flight (NASA).

the famous WMAP satellite that measured the ripples of the big bang, which is located at the 'quiet' point L2, where the Sun, the Earth and the Moon are easily shielded and satellite temperature remains constant.

Challenge 347, page 197: This is a resonance effect, in the same way that a small vibration of a string can lead to large oscillation of the air and sound box in a guitar.

Challenge 349, page 199: The expression for the strength of tides, namely $2GM/d^3$, can be rewritten as $(8/3)\pi G\rho(R/d)^3$. Now, R/d is roughly the same for Sun and Moon, as every eclipse shows. So the density ρ must be much larger for the Moon. In fact, the ratio of the strengths (height) of the tides of Moon and Sun is roughly 7 : 3. This is also the ratio between the mass densities of the two bodies.

Challenge 350, page 200: The total angular momentum of the Earth and the Moon must remain constant.

Challenge 352, page 201: Wait for a solar eclipse.

Challenge 354, page 203: Unfortunately, the myth of 'passive gravitational mass' is spread by many books. Careful investigation shows that it is measured in exactly the same way as inertial mass.

Both masses are measured with the same machines and set-ups. And all these experiments mix and require both inertial and passive gravitational mass effects. For example, a balance or bathroom scale has to dampen out any oscillation, which requires inertial mass. Generally speaking, it seems impossible to distinguish inertial mass from the passive gravitational mass due to all the masses in the rest of the universe. In short, the two concepts are in fact identical.

Challenge 356, page 203: These problems occur because gravitational mass determines potential energy and inertial mass determines kinetic energy.

Challenge 358, page 205: Either they fell on inclined snowy mountain sides, or they fell into high trees, or other soft structures. The record was over 7 km of survived free fall. A recent case made the news in 2007 and is told in www.bbc.co.uk/jersey/content/articles/2006/12/20/michael_holmes_fall_feature.shtml.

Challenge 360, page 207: For a few thousand Euros, you can experience zero-gravity in a parabolic flight, such as the one shown in Figure 320. (Many 'photographs' of parabolic flights found on the internet are in fact computer graphics. What about this one?)

How does zero-gravity *feel*? It feels similar to floating under water, but without the resistance of the water. It also feels like the time in the air when one is diving into water. However, for cosmonauts, there is an additional feeling; when they rotate their head rapidly, the sensors for orientation in our ear are not reset by gravity. Therefore, for the first day or two, most cosmonauts have feelings of vertigo and of nausea, the so-called *space sickness*. After that time, the body adapts and the cosmonaut can enjoy the situation thoroughly.

Challenge 361, page 207: The centre of mass of a broom falls with the usual acceleration; the end thus falls faster.

Challenge 362, page 207: Just use energy conservation for the two masses of the jumper and the string. For more details, including the comparison of experimental measurements and theory, see N. Dubelaar & R. Brantjes, *De valversnelling bij bungee-jumping*, Nederlands tijdschrift voor natuurkunde **69**, pp. 316–318, October 2003.

Challenge 363, page 207: About 1 ton.

Challenge 364, page 207: About 5 g.

Challenge 365, page 208: Your weight is roughly constant; thus the Earth must be round. On a flat Earth, the weight would change from place to place, depending on your distance from the border.

Challenge 366, page 208: Nobody ever claimed that the centre of mass is the same as the centre of gravity! The attraction of the Moon is negligible on the surface of the Earth.

Challenge 368, page 209: That is the mass of the Earth. Just turn the table on its head.

Challenge 370, page 209: The Moon will be about 1.25 times as far as it is now. The Sun then will slow down the Earth–Moon system rotation, this time due to the much smaller tidal friction from the Sun's deformation. As a result, the Moon will return to smaller and smaller distances to Earth. However, the Sun will have become a red giant by then, after having swallowed both the Earth and the Moon.

Challenge 372, page 210: As Galileo determined, for a swing (half a period) the ratio is $\sqrt{2}/\pi$. (See challenge 325). But not more than two, maybe three decimals of π can be determined in this way.

Challenge 373, page 210: Momentum conservation is not a hindrance, as any tennis racket has the same effect on the tennis ball.

Challenge 374, page 210: In fact, in velocity space, elliptic, parabolic and hyperbolic motions are all described by circles. In all cases, the hodograph is a circle.

Challenge 375, page 211: This question is old (it was already asked in Newton's times) and deep. One reason is that stars are kept apart by rotation around the galaxy. The other is that galaxies are kept apart by the momentum they got in the big bang. Without the big bang, all stars would have collapsed together. In this sense, the big bang can be deduced from the attraction of gravitation and the immobile sky at night. We shall find out later that the darkness of the night sky gives a second argument for the big bang.

Challenge 376, page 211: The choice is clear once you notice that there is no section of the orbit which is concave towards the Sun. Can you show this?

Challenge 378, page 212: The escape velocity, from Earth, to leave the Solar System – without help of the other planets – is 42 km/s. However, if help by the other planets is allowed, it can be less than half that value (why?).

If the escape velocity from a body were the speed of light, the body would be a black hole; not even light could escape. Black holes are discussed in detail in the volume on relativity.

Challenge 379, page 212: Using a maximal jumping height of $h = 0.5$ m on Earth and an estimated asteroid density of $\rho = 3$ Mg/m^3, we get a maximum radius of $R^2 = 3gh/4\pi G\rho$, or $R \approx 2.4$ km.

FIGURE 321 The analemma photographed, at local noon, from January to December 2002, at the Parthenon on Athen's Acropolis, and a precision sundial (© Anthony Ayiomamitis, Stefan Pietrzik).

Challenge 380, page 212: A handle of two bodies.

Challenge 382, page 212: In what does this argument differ from the more common argument that in the expression $ma = gMm/R^2$, the left m is inertial and the right m is gravitational?

Challenge 384, page 212: What counts is *local* verticality; with respect to it, the river always flows downhill.

Challenge 385, page 212: The shape of an analemma at local noon is shown in Figure 321. The shape is known since over 2000 years! The shape of the analemma also illustrates why the earliest sunrise is not at the longest day of the year.

The vertical extension of the analemma in the figure is due to the obliquity, i.e., the tilt of the Earth's axis (it is twice 23.45°). The horizontal extension is due to the combination of the obliquity and of the ellipticity of the orbit around the Sun. Both effects lead to roughly equal changes of the position of the Sun at local noon during the course of the year. The asymmetrical position of the central crossing point is purely due to the ellipticity of the orbit. The shape of the analemma, sometimes shown on globes, is built into the shadow pole or the reading curve of precision sundials. Examples are the one shown above and the one shown on page 45. For more details, see B. M. OLIVER, *The shape of the analemma*, Sky & Telescope 44, pp. 20–22, 1972, and the correction of the figures at 44, p. 303, 1972,

Challenge 386, page 215: Capture of a fluid body is possible if it is split by tidal forces.

Challenge 387, page 216: The tunnel would be an elongated ellipse in the plane of the Equator, reaching from one point of the Equator to the point at the antipodes. The time of revolution would not change, compared to a non-rotating Earth. See A. J. SIMONSON, *Falling down a hole through the Earth*, Mathematics Magazine 77, pp. 171–188, June 2004.

Challenge 389, page 216: The centre of mass of the Solar System can be as far as twice the radius from the centre of the Sun; it thus can be outside the Sun.

Challenge 390, page 216: First, during northern summer time the Earth moves faster around the Sun than during northern winter time. Second, shallow Sun's orbits on the sky give longer days because of light from when the Sun is below the horizon.

Challenge 391, page 216: Apart from the visibility of the Moon, no effect of the Moon on humans has ever been detected. Gravitational effects – including tidal effects – electrical effects, magnetic effects and changes in cosmic rays are all swamped by other effects. Indeed the gravity of passing trucks, factory electromagnetic fields, the weather and solar activity changes have larger influences on humans than the Moon. The locking of the menstrual cycle to the moon phase is a visual effect.

Challenge 392, page 216: Distances were difficult to measure. It is easy to observe a planet that is before the Sun, but it is hard to check whether a planet is behind the Sun. Phases of Venus are also predicted by the geocentric system; but the phases it predicts do not match the ones that are observed. Only the phases deduced from the heliocentric system match the observed ones. Venus orbits the Sun.

Challenge 393, page 216: See the mentioned reference.

Challenge 394, page 217: True.

Challenge 395, page 217: For each pair of opposite shell elements (drawn in yellow), the two attractions compensate.

Challenge 396, page 218: There is no practical way; if the masses on the shell could move, along the surface (in the same way that charges can move in a metal) this might be possible, provided that enough mass is available.

Challenge 400, page 219: Yes, one could, and this has been thought of many times, including by Jules Verne. The necessary speed depends on the direction of the shot with respect of the rotation of the Earth.

Challenge 401, page 219: Never. The Moon points always towards the Earth. The Earth changes position a bit, due to the ellipticity of the Moon's orbit. Obviously, the Earth shows phases.

Challenge 403, page 219: There are no such bodies, as the chapter of general relativity will show.

Challenge 405, page 221: The oscillation is a purely sinusoidal, or harmonic oscillation, as the restoring force increases linearly with distance from the centre of the Earth. The period T for a homogeneous Earth is $T = 2\pi\sqrt{R^3/GM} = 84$ min.

Challenge 406, page 221: The period is the same for all such tunnels and thus in particular it is the same as the 84 min period that is valid also for the pole to pole tunnel. See for example, R. H. ROMER, *The answer is forty-two – many mechanics problems, only one answer*, Physics Teacher **41**, pp. 286–290, May 2003.

Challenge 407, page 223: If the Earth were *not* rotating, the most general path of a falling stone would be an ellipse whose centre is the centre of the Earth. For a rotating Earth, the ellipse precesses. Simoson speculates that the spirographics swirls in the Spirograph Nebula, found at antwrp.gsfc.nasa.gov/apod/ap021214.html, might be due to such an effect. A special case is a path starting vertically at the equator; in this case, the path is similar to the path of the Foucault pendulum, a pointed star with about 16 points at which the stone resurfaces around the Equator.

Challenge 408, page 223: There is no simple answer: the speed depends on the latitude and on other parameters. The internet also provides videos of solar eclipses seen from space, showing how the shadow moves over the surface of the Earth.

Challenge 409, page 224: The centrifugal force must be equal to the gravitational force. Call the constant linear mass density d and the unknown length l. Then we have $GMd\int_R^{R+l} dr/r^2 = \omega^2 d\int_R^{R+l} r\,dr$. This gives $GMdl/(R^2 + Rl) = (2Rl + l^2)\omega^2 d/2$, yielding $l = 0.14$ Gm. For more on space elevators or lifts, see challenge 574.

Challenge 411, page 224: The inner rings must rotate faster than the outer rings. If the rings were solid, they would be torn apart. But this reasoning is true only if the rings are inside a certain limit, the so-called *Roche limit*. The Roche limit is that radius at which gravitational force F_g and tidal force F_t cancel on the surface of the satellite. For a satellite with mass m and radius r, orbiting a central mass M at distance d, we look at the forces on a small mass μ on its surface. We get the condition $Gm\mu/r^2 = 2GM\mu r/d^3$. A bit of algebra yields the approximate Roche limit value

$$d_{\text{Roche}} = R\left(2\frac{\rho_M}{\rho_m}\right)^{1/3}. \tag{152}$$

Below that distance from a central mass M, fluid satellites cannot exist. The calculation shown here is only an approximation; the actual Roche limit is about two times that value.

Challenge 414, page 228: The load is 5 times the load while standing. This explains why race horses regularly break their legs.

Challenge 415, page 228: At school, you are expected to answer that the weight is the same. This is a good approximation. But in fact the scale shows a slightly larger weight for the steadily running hourglass compared to the situation where the all the sand is at rest. Looking at the momentum flow explains the result in a simple way: the only issue that counts is the momentum of the sand in the upper chamber, all other effects being unimportant. That momentum slowly decreases during running. This requires a momentum flow from the scale: the effective weight increases. See also the experimental confirmation and its explanation by F. TUINSTRA & B. F. TUINSTRA, *The weight of an hourglass*, Europhysics News 41, pp. 25–28, March 2010, also available online.

If we imagine a photon bouncing up and down in a box made of perfect mirrors, the ideas from the hourglass puzzle imply that the scale shows an increased weight compared to the situation without a photon. The weight increase is Eg/c^2, where E is the energy of the photon, $g = 9.81$ m/s^2 and c is the speed of light. This story is told by E. HUGGINS, *Weighing photons using bathroom scales: a thought experiment*, The Physics Teacher 48, pp. 287–288, May 2010,

Challenge 416, page 229: The electricity consumption of a rising escalator indeed increases when the person on it walks upwards. By how much?

Challenge 417, page 229: Knowledge is power. Time is money. Now, power is defined as work per time. Inserting the previous equations and transforming them yields

$$\text{money} = \frac{\text{work}}{\text{knowledge}}, \qquad (153)$$

which shows that the less you know, the more money you make. That is why scientists have low salaries.

Challenge 418, page 229: In reality muscles keep an object above ground by continuously lifting and dropping it; that requires energy and work.

Challenge 421, page 234: Yes, because side wind increases the effective speed v in air due to vector addition, and because air resistance is (roughly) proportional to v^2.

Challenge 422, page 234: The lack of static friction would avoid that the fluid stays attached to the body; the so-called boundary layer would not exist. One then would have no wing effect.

Challenge 424, page 235: True?

Challenge 426, page 236: From $dv/dt = g - v^2(1/2c_w A\rho/m)$ and using the abbreviation $c = 1/2c_w A\rho$, we can solve for $v(t)$ by putting all terms containing the variable v on one side, all terms with t on the other, and integrating on both sides. We get $v(t) = \sqrt{gm/c} \tanh \sqrt{cg/m}\,t$.

Challenge 427, page 237: For extended deformable bodies, the intrinsic properties are given by the mass density – thus a function of space and time – and the state is described by the density of kinetic energy, local linear and angular momentum, as well as by its stress and strain distributions.

Challenge 428, page 237: Electric charge.

Challenge 429, page 238: The phase space has $3N$ position coordinates and $3N$ momentum coordinates.

Challenge 430, page 238: We recall that when a stone is thrown, the initial conditions summarize the effects of the thrower, his history, the way he got there etc.; in other words, initial conditions summarize the past of a system, i.e., the effects that the environment had during the history

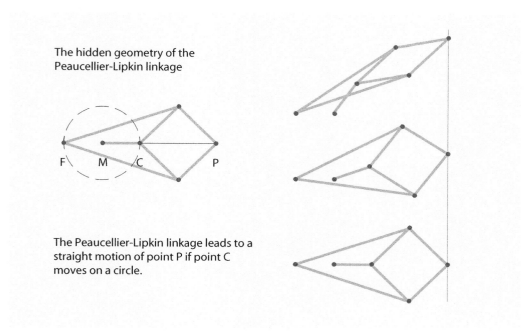

FIGURE 322 How to draw a straight line with a compass (drawn by Zach Joseph Espiritu).

of a system. Therefore, the universe has no initial conditions and no phase space. If you have found reasons to answer yes, you overlooked something. Just go into more details and check whether the concepts you used apply to the universe. Also define carefully what you mean by 'universe'.

Challenge 431, page 238: The light mill is an example.

Vol. III, page 122

Challenge 433, page 240: A system showing energy or matter motion faster than light would imply that for such systems there are observers for which the order between cause and effect are reversed. A space-time diagram (and a bit of exercise from the section on special relativity) shows this.

Challenge 434, page 240: If reproducibility would not exist, we would have difficulties in checking observations; also reading the clock is an observation. The connection between reproducibility and time shall become important in the final part of our adventure.

Challenge 435, page 241: Even if surprises were only rare, each surprise would make it impossible to define time just before and just after it.

Challenge 438, page 242: Of course; moral laws are summaries of what others think or will do about personal actions.

Challenge 439, page 243: The fastest glide path between two points, the *brachistochrone*, turns out to be the *cycloid*, the curve generated by a point on a wheel that is rolling along a horizontal plane.

The proof can be found in many ways. The simplest is by Johann Bernoulli and is given on en.wikipedia.org/wiki/Brachistochrone_problem.

Challenge 441, page 244: When F, C and P are aligned, this circle has a radius given by $R = \sqrt{FCFP}$; F is its centre. In other words, the Peaucellier-Lipkin linkage realizes an inversion at a circle.

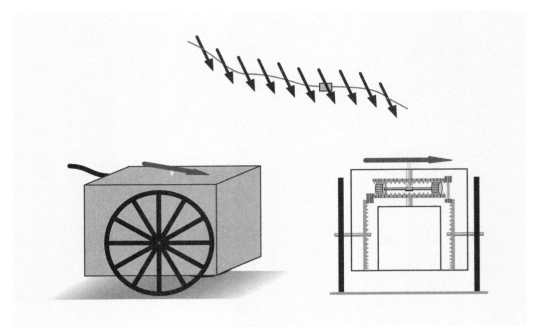

FIGURE 323 The mechanism inside the south-pointing carriage.

FIGURE 324 Falling brick chimneys – thus with limited stiffness – fall with a V shape (© John Glaser, Frank Siebner).

Challenge 442, page 244: When F, C and P are aligned, the circle to be followed has a radius given by half the distance FC; its centre lies midway between F and C. Figure 322 illustrates the situation.

Challenge 443, page 244: Figure 323 shows the most credible reconstruction of a south-pointing carriage.

Challenge 445, page 245: The water is drawn up along the sides of the spinning egg. The fastest way to empty a bottle of water is to spin the water while emptying it.

Challenge 446, page 246: The right way is the one where the chimney falls like a V, not like an inverted V. See challenge 361 on falling brooms for inspiration on how to deduce the answer.

Two examples are shown in Figure 324. It turns out that the chimney breaks (if it is not fastened to the base) at a height between half or two thirds of the total, depending at the angle at which this happens. For a complete solution of the problem, see the excellent paper G. VARESCHI & K. KAMIYA, *Toy models for the falling chimney*, American Journal of Physics 71, pp. 1025–1031, 2003.

Challenge 448, page 252: The definition of the integral given in the text is a simplified version of the so-called *Riemann integral*. It is sufficient for all uses in nature. Have a look at its exact definition in a mathematics text if you want more details.

Challenge 454, page 255: In one dimension, the expression $F = ma$ can be written as $-dV/dx = md^2x/dt^2$. This can be rewritten as $d(-V)/dx - d/dt[d/d\dot{x}(\frac{1}{2}m\dot{x}^2)] = 0$. This can be expanded to $\partial/\partial x(\frac{1}{2}m\dot{x}^2 - V(x)) - d/[\partial/\partial \dot{x}(\frac{1}{2}m\dot{x}^2 - V(x))] = 0$, which is Lagrange's equation for this case.

Challenge 456, page 256: Do not despair. Up to now, nobody has been able to imagine a universe (that is not necessarily the same as a 'world') different from the one we know. So far, such attempts have always led to logical inconsistencies.

Challenge 458, page 256: The two are equivalent since the equations of motion follow from the principle of minimum action and at the same time the principle of minimum action follows from the equations of motion.

Challenge 460, page 258: For gravity, all three systems exist: rotation in galaxies, pressure in planets and the Pauli pressure in stars that is due to Pauli's exclusion principle. Against the strong interaction, the exclusion principle acts in nuclei and neutron stars; in neutron stars maybe also rotation and pressure complement the Pauli pressure. But for the electromagnetic interaction there are no composites other than our everyday matter, which is organized by the Pauli's exclusion principle alone, acting among electrons.

Challenge 461, page 259: Aggregates often form by matter converging to a centre. If there is only a small asymmetry in this convergence – due to some external influence – the result is a final aggregate that rotates.

Challenge 462, page 262: Angular momentum is the change with respect to angle, whereas rotational energy is again the change with respect to time, as all energy is.

Challenge 463, page 262: Not in this way. A small change can have a large effect, as every switch shows. But a small change in the brain must be communicated outside, and that will happen roughly with a $1/r^2$ dependence. That makes the effects so small, that even with the most sensitive switches – which for thoughts do not exist anyway – no effects can be realized.

Challenge 465, page 262: This is a wrong question. $T - U$ is not minimal, only its average is.

Challenge 466, page 263: No. A system tends to a minimum potential only if it is dissipative. One could, however, deduce that conservative systems oscillate around potential minima.

Challenge 467, page 263: The relation is

$$\frac{c_1}{c_2} = \frac{\sin \alpha_1}{\sin \alpha_2} \,. \tag{154}$$

The particular speed ratio between air (or vacuum, which is almost the same) and a material gives the *index of refraction n*:

$$n = \frac{c_1}{c_0} = \frac{\sin \alpha_1}{\sin \alpha_0} \tag{155}$$

Challenge 468, page 263: The principle for the growth of trees is simply the minimum of potential energy, since the kinetic energy is negligible. The growth of vessels inside animal bodies is minimized for transport energy; that is again a minimum principle. The refraction of light is the path of shortest time; thus it minimizes change as well, if we imagine light as moving entities moving without any potential energy involved.

Challenge 469, page 264: Special relativity requires that an invariant measure of the action exist. It is presented later in the walk.

Challenge 470, page 264: The universe is not a physical system. This issue will be discussed in detail later on.

Challenge 471, page 264: Use either the substitution $u = \tan t/2$ or use the historical trick

$$\sec\varphi = \tfrac{1}{2}\left(\frac{\cos\varphi}{1+\sin\varphi} + \frac{\cos\varphi}{1-\sin\varphi}\right). \tag{156}$$

Challenge 472, page 264: A skateboarder in a cycloid has the same oscillation time independently of the oscillation amplitude. But a half-pipe needs to have vertical ends, in order to avoid jumping outside it. A cycloid never has a vertical end.

Challenge 475, page 267: We talk to a person because we know that somebody understands us. Thus we assume that she somehow sees the same things we do. That means that observation is partly viewpoint-independent. Thus nature is symmetric.

Challenge 476, page 270: Memory works because we recognize situations. This is possible because situations over time are similar. Memory would not have evolved without this reproducibility.

Challenge 477, page 271: Taste differences are not fundamental, but due to different viewpoints and – mainly – to different experiences of the observers. The same holds for feelings and judgements, as every psychologist will confirm.

Challenge 478, page 273: The integers under addition form a group. Does a painter's set of oil colours with the operation of mixing form a group?

Challenge 480, page 273: There is only one symmetry operation: a rotation about π around the central point. That is the reason that later on the group D_4 is only called the approximate symmetry group of Figure 202.

Challenge 486, page 278: Scalar is the magnitude of any vector; thus the speed, defined as $v = |\boldsymbol{v}|$, is a scalar, whereas the velocity $\boldsymbol{v}$ is not. Thus the length of any vector (or pseudo-vector), such as force, acceleration, magnetic field, or electric field, is a scalar, whereas the vector itself is not a scalar.

Challenge 489, page 278: The charge distribution of an extended body can be seen as a sum of a point charge, a charge dipole, a charge quadrupole, a charge octupole, etc. The quadrupole is described by a tensor.

Compare: The inertia against motion of an extended body can be seen as sum of a point mass, a mass dipole, a mass quadrupole, a mass octupole, etc. The mass quadrupole is described by the moment of inertia.

Challenge 493, page 281: The conserved charge for rotation invariance is angular momentum.

Challenge 497, page 285: The graph is a *logarithmic spiral* (can you show this?); it is illustrated in Figure 325. The travelled distance has a simple answer.

Challenge 498, page 285: An oscillation has a period in time, i.e., a discrete time translation symmetry. A wave has both discrete time and discrete space translation symmetry.

Challenge 499, page 285: Motion reversal is a symmetry for any closed system; despite the observations of daily life, the statements of thermodynamics and the opinion of several famous physicists (who form a minority though) all ideally closed systems are reversible.

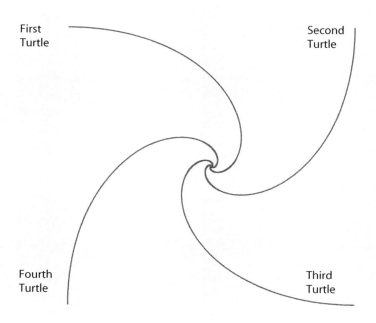

FIGURE 325 The motion of four turtles chasing each other (drawn by Zach Joseph Espiritu).

Challenge 500, page 285: The symmetry group is a Lie group and called U(1), for 'unitary group in 1 dimension'.

Challenge 501, page 285: See challenge 301

Challenge 502, page 285: There is no such thing as a 'perfect' symmetry.

Challenge 504, page 286: The rotating telephone dial had the digits 1 to 0 on the corners of a regular 14-gon. The even and the odd numbers were on the angles of regular heptagons.

Challenge 508, page 289: Just insert $x(t)$ into the Lagrangian $L = 0$, the minimum possible value for a system that transforms all kinetic energy into potential energy and vice versa.

Challenge 517, page 300: The potential energy is due to the 'bending' of the medium; a simple displacement produces no bending and thus contains no energy. Only the gradient captures the bending idea.

Challenge 519, page 301: The phase changes by π.

Challenge 520, page 301: A wave that carries angular momentum has to be transversal and has to propagate in three dimensions.

Challenge 521, page 301: Waves can be damped to extremely low intensities. If this is not possible, the observation is not a wave.

Challenge 522, page 303: The way to observe diffraction and interference with your naked fingers is told on page 101 in volume III.

Challenge 533, page 315: Interference can make radio signals unintelligible. Due to diffraction, radio signals are weakened behind a wall; this is valid especially for short wavelengths, such as those used in mobile phones. Refraction makes radio communication with submarines impossible for usual radio frequencies. Dispersion in glass fibres makes it necessary to add repeaters in sea-cables roughly every 100 km. Damping makes it impossible hear somebody speaking at larger distances. Radio signals can loose their polarisation and thus become hard to detect by usual Yagi antennas that have a fixed polarisation.

Challenge 535, page 320: Skiers scrape snow from the lower side of each bump towards the upper side of the next bump. This leads to an upward motion of ski bumps.

Challenge 536, page 320: If the distances to the loudspeaker is a few metres, and the distance to the orchestra is 20 m, as for people with enough money, the listener at home hears it first.

Challenge 537, page 320: As long as the amplitude is small compared to the length l, the period T is given by

$$T = 2\pi \sqrt{\frac{l}{g}} \,. \tag{157}$$

The formula does not contain the mass m at all. Independently of the mass m at its end, the pendulum has always the same period. In particular, for a length of 1 m, the period is about 2 s. Half a period, or one swing thus takes about 1 s. (This is the original reason for choosing the unit of metre.)

For an extremely long pendulum, the answer is a finite value though, and corresponds to the situation of challenge 26.

Challenge 538, page 320: In general, the body moves along an ellipse (as for planets around the Sun) but with the fixed point as centre. In contrast to planets, where the Sun is in a *focus* of the ellipse and there is a perihelion and an apohelion, such a body moves *symmetrically* around the *centre* of the ellipse. In special cases, the body moves back and forward along a straight segment.

Challenge 540, page 320: This follows from the formula that the frequency of a string is given by $f = \sqrt{T/\mu}/(2l)$, where T is the tension, μ is the linear mass density, and l is the length of a string. This is discussed in the beautiful paper by G. BARNES, *Physics and size in biological systems*, The Physics Teacher 27, pp. 234–253, 1989.

Challenge 542, page 321: The sound of thunder or of car traffic gets lower and lower in frequency with increasing distance.

Challenge 545, page 321: Neither; both possibilities are against the properties of water: in surface waves, the water molecules move in circles.

Challenge 546, page 322: Swimmers are able to cover 100 m in 48 s, or slightly better than 2 m/s. (Swimmer with fins achieve just over 3 m/s.) With a body length of about 1.9 m, the critical speed is 1.7 m/s. That is why short distance swimming depends on training; for longer distances the technique plays a larger role, as the critical speed has not been attained yet. The formula also predicts that on the 1500 m distance, a 2 m tall swimmer has a potential advantage of over 45 s on one with body height of 1.8 m. In addition, longer swimmers have an additional advantage: they swim shorter distances in pools (why?). It is thus predicted that successful long-distance swimmers will get taller and taller over time. This is a pity for a sport that so far could claim to have had champions of all sizes and body shapes, in contrast to many other sports.

Challenge 549, page 324: To reduce noise reflection and thus hall effects. They effectively diffuse the arriving wave fronts.

Challenge 551, page 324: Waves in a river are never elliptical; they remain circular.

Challenge 552, page 324: The lens is a cushion of material that is 'transparent' to sound. The speed of sound is faster in the cushion than in the air, in contrast to a glass lens, where the speed of light is slower in the glass. The shape is thus different: the cushion must look like a large biconcave optical lens.

Challenge 553, page 324: Experiments show that the sound does not depend on air flows (find out how), but does depend on external sound being present. The sound is due to the selective amplification by the resonances resulting from the geometry of the shell shape.

CHALLENGE HINTS AND SOLUTIONS

513

Challenge 554, page 324: The Sun is always at a different position than the one we observe it to be. What is the difference, measured in angular diameters of the Sun? Despite this position difference, the timing of the sunrise is determoned by the position of the horizon, not by the position of the Sun. (Imagine the it would not: in that case a room would not get dark when the window is closed, but eight minutes later ...) In short, there is no measurable effect of the speed of light on the sunrise.

Challenge 557, page 326: An overview of systems being tested at present can be found in K.-U. GRAW, *Energiereservoir Ozean*, Physik in unserer Zeit **33**, pp. 82–88, Februar 2002. See also *Oceans of electricity – new technologies convert the motion of waves into watts*, Science News **159**, pp. 234–236, April 2001.

Challenge 558, page 326: In everyday life, the assumption is usually justified, since each spot can be approximately represented by an atom, and atoms can be followed. The assumption is questionable in situations such as turbulence, where not all spots can be assigned to atoms, and most of all, in the case of motion of the vacuum itself. In other words, for gravity waves, and in particular for the quantum theory of gravity waves, the assumption is not justified.

Challenge 564, page 333: There are many. One would be that the transmission and thus reflection coefficient for waves would almost be independent of wavelength.

Challenge 565, page 334: A drop with a diameter of 3 mm would cover a surface of 7.1 m^2 with a 2 nm film.

Challenge 566, page 338: The wind will break tall trees that are too thin. For small and thus thin trees, the wind does not damage.

Challenge 567, page 338: The critical height for a column of material is given by $h_{crit}^4 = \frac{\beta}{4\pi g} m \frac{E}{\rho^2}$, where $\beta \approx 1.9$ is the constant determined by the calculation when a column buckles under its own weight.

Challenge 569, page 339: One possibility is to describe particles as clouds; another is given in the last part of the text.

Challenge 570, page 341: The results gives a range between 1 and $8 \cdot 10^{23}$.

Ref. 257

Challenge 572, page 345: Check your answers with the delightful text by P. GOLDRICH, S. MAHAJAN & S. PHINNEY, *Order-of-Magnitude Physics: Understanding the World with Dimensional Analysis, Educated Guesswork, and White Lies*, available on the internet.

Challenge 573, page 345: Glass shatters, glass is elastic, glass shows transverse sound waves, glass does not flow (in contrast to what many books state), not even on scale of centuries, glass molecules are fixed in space, glass is crystalline at small distances, a glass pane supported at the ends does not hang through.

Challenge 574, page 345: No metal wire allows building such a long wire or rope. Only the idea of carbon nanotubes has raised the hope again; some dream of wire material based on them, stronger than any material known so far. However, no such material is known yet. The system faces many dangers, such as fabrication defects, lightning, storms, meteoroids and space debris. All would lead to the breaking of the wires – if such wires will ever exist. But the biggest of all dangers is the lack of cash to build it. Nevertheless, numerous people are working towards the goal.

Challenge 575, page 346: The $3 \times 3 \times 3$ cube has a rigid system of three perpendicular axes, on which a square can rotate at each of the 6 ends. The other squares are attaches to pieces moving around theses axes. The $4 \times 4 \times 4$ cube is different though; just find out. From $7 \times 7 \times 7$ onwards, the parts do not all have the same size or shape. The present limit on the segment number in commercially available 'cubes' is $17 \times 17 \times 17$! It can be found at www.shapeways.com/shops/

oskarpuzzles. The website www.oinkleburger.com/Cube/applet allows playing with virtual cubes up to 100 × 100 × 100, and more.

Challenge 578, page 347: A medium-large earthquake would be generated.

Challenge 579, page 347: A stalactite contains a thin channel along its axis through which the water flows, whereas a stalagmite is massive throughout.

Challenge 580, page 347: About 1 part in a thousand.

Challenge 581, page 348: Even though the iron core of the Earth formed by collecting the iron from colliding asteroids which then sunk into the centre of the Earth, the scheme will not work today: in its youth, the Earth was much more liquid than today. The iron will most probably not sink. In addition, there is no known way to build a measurement probe that can send strong enough sound waves for this scheme. The temperature resistance is also an issue, but this may be solvable.

Challenge 583, page 350: Atoms are not infinitely hard, as quantum theory shows. Atoms are more similar to deformable clouds.

Challenge 586, page 360: If there is no friction, all three methods work equally fast – including the rightmost one.

Challenge 589, page 362: The constant k follows from the conservation of energy and that of mass:

$$k = \sqrt{\frac{2}{\rho(A_1^2/A_2^2 - 1)}} \ . \tag{158}$$

The cross sections are denoted by A and the subscript 1 refers to any point far from the constriction, and the subscript 2 to the constriction.

Challenge 592, page 369: The pressure destroys the lung. Snorkeling is only possible at the water surface, not below the water! This experiment is even dangerous when tried in your own bathtub! Breathing with a long tube is only possible if a pump at the surface pumps air down the tube at the correct pressure.

Challenge 594, page 369: Some people notice that in some cases friction is too high, and start sucking at one end of the tube to get the flow started; while doing so, they can inhale or swallow gasoline, which is poisonous.

Challenge 599, page 372: Calculation yields $N = J/j = (0.0001 \text{ m}^3/\text{s})/(7 \text{ μm}^2 0.0005 \text{ m/s})$, or about $6 \cdot 10^9$; in reality, the number is much larger, as most capillaries are closed at a given instant. The reddening of the face shows what happens when all small blood vessels are opened at the same time.

Challenge 600, page 373: Throwing the stone makes the level fall, throwing the water or the piece of wood leaves it unchanged.

Challenge 601, page 373: The ship rises higher into the sky. (Why?)

Challenge 603, page 373: The motion of a helium-filled balloon is opposite to that of an air-filled balloon or of people: the helium balloon moves towards the front when the car accelerates and to the back when the car decelerates. It also behaves differently in bends. Several films on the internet show the details.

Challenge 606, page 373: The pumps worked in suction; but air pressure only allows 10 m of height difference for such systems.

Challenge 607, page 373: This argument is comprehensible only when we remember that 'twice the amount' means 'twice as many molecules'.

Challenge 608, page 373: The alcohol is frozen and the chocolate is put around it.

TABLE 62 *Gaseous* composition of *dry* air, at *present* time[a] (sources: NASA, IPCC).

Gas	Symbol	Volume Part[b]
Nitrogen	N_2	78.084 %
Oxygen (pollution dependent)	O_2	20.946 %
Argon	Ar	0.934 %
Carbon dioxide (in large part due to human pollution)	CO_2	403 ppm
Neon	Ne	18.18 ppm
Helium	He	5.24 ppm
Methane (mostly due to human pollution)	CH_4	1.79 ppm
Krypton	Kr	1.14 ppm
Hydrogen	H_2	0.55 ppm
Nitrous oxide (mostly due to human pollution)	N_2O	0.3 ppm
Carbon monoxide (partly due to human pollution)	CO	0.1 ppm
Xenon	Xe	0.087 ppm
Ozone (strongly influenced by human pollution)	O_3	0 to 0.07 ppm
Nitrogen dioxide (mostly due to human pollution)	NO_2	0.02 ppm
Iodine	I_2	0.01 ppm
Ammonia (mostly due to human pollution)	NH_3	traces
Radon	Ra	traces
Halocarbons and other fluorine compounds (all being humans pollutants)	20 types	0.0012 ppm
Mercury, other metals, sulfur compounds, other organic compounds (all being human pollutants)	numerous	concentration varies

a. *Wet* air can contain up to 4 % water vapour, depending on the weather. *Apart from gases,* air can contain water droplets, ice, sand, dust, pollen, spores, volcanic ash, forest fire ash, fuel ash, smoke particles, pollutants of all kinds, meteoroids and cosmic ray particles. *During the history* of the Earth, the gaseous composition varied strongly. In particular, oxygen is part of the atmosphere only in the second half of the Earth's lifetime.
b. The abbreviation *ppm* means 'parts per million'.

Challenge 609, page 374: The author suggested in an old edition of this text that a machine should be based on the same machines that throw the clay pigeons used in the sports of trap shooting and skeet. In the meantime, Lydéric Bocquet and Christophe Clanet have built such a stone-skipping machine, but using a different design; a picture can be found on the website ilm-perso.univ-lyon1.fr/~lbocquet.

Challenge 610, page 374: The third component of *air* is the noble gas argon, making up about 1 %. A longer list of components is given in Table 62.

Challenge 611, page 374: The *pleural cavity* between the lungs and the thorax is permanently below atmospheric pressure, usually 5 mbar, but even 10 mbar at inspiration. A hole in it, formed for example by a bullet, a sword or an accident, leads to the collapse of the lung – the so-called *pneumothorax* – and often to death. Open chest operations on people have became possible only after the surgeon Ferdinand Sauerbruch learned in 1904 how to cope with the problem. Nowadays however, surgeons keep the lung under *higher* than atmospheric pressure until

FIGURE 326 A way to ride head-on against the wind using wind power (© Tobias Klaus).

everything is sealed again.

Challenge 612, page 374: The fountain shown in the figure is started by pouring water into the uppermost container. The fountain then uses the air pressure created by the water flowing downwards.

Challenge 613, page 374: Yes. The bulb will not resist two such cars though.

Challenge 614, page 375: Radon is about 8 times as heavy as air; it is he densest gas known. In comparison, Ni(CO) is 6 times, $SiCl_4$ 4 times heavier than air. Mercury vapour (obviously also a gas) is 7 times heavier than air. In comparison, bromine vapour is 5.5 times heavier than air.

Challenge 616, page 375: Yes, as the *ventomobil* shown in Figure 326 proves. It achieves the feat already for low wind speeds.

Challenge 617, page 376: None.

Challenge 619, page 376: He brought the ropes into the cabin by passing them through liquid mercury.

Challenge 621, page 376: There are no official solutions for these questions; just check your assumptions and calculations carefully. The internet is full of such calculations.

Challenge 622, page 377: The soap flows down the bulb, making it thicker at the bottom and thinner at the top, until it reaches the thickness of two molecular layers. Later, it bursts.

Challenge 623, page 377: The temperature leads to evaporation of the involved liquid, and the vapour prevents the direct contact between the two non-gaseous bodies.

Challenge 624, page 377: For this to happen, friction would have to exist on the microscopic scale and energy would have to disappear.

Challenge 625, page 378: The longer funnel is empty before the short one. (If you do not believe it, try it out.) In the case that the amount of water in the funnel outlet can be neglected, one can use energy conservation for the fluid motion. This yields the famous Bernoulli equation $p/\rho + gh + v^2/2 = \text{const}$, where p is pressure, ρ the density of water, and g is 9.81 m/s^2. Therefore, the speed v is higher for greater lengths h of the thin, straight part of the funnel: the longer funnel empties first.

But this is strange: the formula gives a simple free fall relation, as the *air* pressure is the same above and below and disappears from the calculation. The expression for the speed is thus independent of whether a tube is present or not. The real reason for the faster emptying of the tube is thus that a tube forces more water to flow out than the lack of a tube. Without tube, the diameter of the water flow *diminishes* during fall. With tube, it stays *constant*. This difference leads to the faster emptying for longer tubes.

Alternatively, you can look at the *water* pressure value *inside* the funnel. You will discover that the water pressure is lowest at the start of the exit tube. This internal water pressure is lower for longer tubes and sucks out the water faster in those cases.

Challenge 626, page 378: The eyes of fish are positioned in such a way that the pressure reduction by the flow is compensated by the pressure increase of the stall. By the way, their heart is positioned in such a way that it is helped by the underpressure.

Challenge 628, page 378: This feat has been achieved for lower mountains, such as the Monte Bianco in the Alps. At present however, there is no way to safely hover at the high altitudes of the Himalayas.

Challenge 630, page 378: Press the handkerchief in the glass, and lower the glass into the water with the opening first, while keeping the opening horizontal. This method is also used to lower people below the sea. The paper ball in the bottle will fly towards you. Blowing into a funnel will keep the ping-pong ball tightly into place, and the more so the stronger you blow. Blowing through a funnel towards a candle will make it lean towards you.

Challenge 637, page 388: In 5000 million years, the present method will stop, and the Sun will become a red giant. But it will burn for many more years after that.

Challenge 638, page 390: Bernoulli argued that the temperature describes the average kinetic energy of the constituents of the gas. From the kinetic energy he deduced the average momentum of the constituents. An average momentum leads to a pressure. Adding the details leads to the ideal gas relation.

Challenge 639, page 390: The answer depends on the size of the balloons, as the pressure is not a monotonous function of the size. If the smaller balloon is not too small, the smaller balloon wins.

Challenge 642, page 391: Measure the area of contact between tires and street (all four) and then multiply by 200 kPa, the usual tire pressure. You get the weight of the car.

Challenge 646, page 394: If the average square displacement is proportional to time, the liquid is made of smallest particles. This was confirmed by the experiments of Jean Perrin. The next step is to deduce the number of these particles from the proportionality constant. This constant, defined by $\langle d^2 \rangle = 4Dt$, is called the diffusion constant (the factor 4 is valid for random motion in two dimensions). The diffusion constant can be determined by watching the motion of a particle under the microscope.

We study a Brownian particle of radius a. In two dimensions, its square displacement is given by

$$\langle d^2 \rangle = \frac{4kT}{\mu} t \, , \tag{159}$$

where k is the Boltzmann constant and T the temperature. The relation is deduced by studying the motion of a particle with drag force $-\mu v$ that is subject to random hits. The linear drag coefficient μ of a sphere of radius a is given by

$$\mu = 6\pi\eta a \, , \tag{160}$$

where η is the kinematic viscosity. In other words, one has

$$k = \frac{6\pi\eta a}{4T} \frac{\langle d^2 \rangle}{t} \, . \tag{161}$$

All quantities on the right can be measured, thus allowing us to determine the Boltzmann constant k. Since the ideal gas relation shows that the ideal gas constant R is related to the Boltzmann constant by $R = N_A k$, the Avogadro constant N_A that gives the number of molecules in a mole is also found in this way.

Challenge 651, page 402: The possibility of motion inversion for all observed phenomena is indeed a fundamental property of nature. It has been confirmed for all interactions and all experiments every performed. Independent of this is the fact that realizing the inversion might be extremely hard, because inverting the motion of many atoms is usually not feasible.

Challenge 652, page 403: This is a trick question. To a good approximation, any tight box is an example. However, if we ask for complete precision, all systems radiate some energy, loose some atoms or particles and bend space; *ideal* closed systems do *not* exist.

Challenge 657, page 405: We will find out later that the universe is not a physical system; thus the concept of entropy does not apply to it. Thus the universe is neither isolated nor closed.

Challenge 659, page 406: Egg white starts to harden at lower temperature than yolk, but for complete hardening, the opposite is true. White hardens *completely* at 80°C, egg yolk hardens considerably at 66 to 68°C. Cook an egg at the latter temperature, and the feat is possible; the white remains runny, but does not remain transparent, though. Note again that the cooking time plays no role, only the precise temperature value.

Challenge 661, page 407: Yes, the effect is easily noticeable.

Challenge 664, page 407: Hot air is less dense and thus wants to rise.

Challenge 665, page 407: Keep the paper wet.

Challenge 666, page 407: Melting ice at 0°C to water at 0°C takes 334 kJ/kg. Cooling water by 1°C or 1 K yields 4.186 kJ/kgK. So the hot water needs to cool down to 20.2°C to melt the ice, so that the final mixing temperature will be 10.1°C.

Challenge 667, page 408: The air had to be dry.

Challenge 668, page 408: In general, it is impossible to draw a line through three points. Since absolute zero and the triple point of water are fixed in magnitude, it was practically a sure bet that the boiling point would not be at precisely 100°C.

FIGURE 327 A candle on Earth and in microgravity (NASA).

Challenge 669, page 408: No, as a water molecule is heavier than that. However, if the water is allowed to be dirty, it is possible. What happens if the quantum of action is taken into account?

Challenge 670, page 408: The danger is not due to the amount of energy, but due to the time in which it is available.

Challenge 671, page 409: The internet is full of solutions.

Challenge 672, page 409: There are 2^n possible sequences of n coin throws. Of those, $n!/(\frac{n}{2}!)^2$ contain $n/2$ heads and $n/2$ tails. For a fair coin, the probability p of getting $n/2$ heads in n throws is thus

$$p = \frac{n!}{2^n \left(\frac{n}{2}!\right)^2} \; . \qquad (162)$$

We approximate this result with the help of Gosper's formula $n! \approx \sqrt{(2n+\frac{1}{3})\pi}\, (\frac{n}{e})^n$ and get

$$p \approx \frac{\sqrt{(2n+\frac{1}{3})\pi}\, (\frac{n}{e})^n}{2^n \left(\sqrt{(n+\frac{1}{3})\pi}\, (\frac{n}{2e})^{\frac{n}{2}} \right)^2} = \frac{\sqrt{2n+\frac{1}{3}}}{(n+\frac{1}{3})\sqrt{\pi}} \; . \qquad (163)$$

For $n = 1\,000\,000$, we get a probability $p \approx 0.0007979$, thus a rather small value between $\frac{1}{1254}$ and $\frac{1}{1253}$.

Challenge 673, page 409: The entropy can be defined for the universe as a whole only if the universe is a closed system. But is the universe closed? Is it a system? This issue is discussed in the final part of our adventure.

Challenge 676, page 410: For such small animals the body temperature would fall too low. They could not eat fast enough to get the energy needed to keep themselves warm.

Challenge 679, page 410: The answer depends on the volume, of course. But several families have died overnight because they had modified their mobile homes to be airtight.

Challenge 680, page 410: The metal salts in the ash act as catalysts, and the sugar burns instead of just melting. Watch the video of the experiment at www.youtube.com/watch?v=BfBgAaeaVgk.

Challenge 685, page 411: It is about 10^{-9} that of the Earth.

Challenge 687, page 411: The thickness of the folds in the brain, the bubbles in the lung, the density of blood vessels and the size of biological cells.

Challenge 688, page 411: The mercury vapour above the liquid gets saturated.

Challenge 689, page 411: A dedicated NASA project studied this question. Figure 327 gives an example comparison. You can find more details on their website.

Challenge 690, page 411: The risks due to storms and the financial risks are too high.

Challenge 691, page 412: The vortex inside the tube is cold near its axis and hot in the regions away from the axis. Through the membrane in the middle of the tube (shown in Figure 285 on page 412) the air from the axis region is sent to one end and the air from the outside region to the other end. The heating of the outside region is due to the work that the air rotating inside has to do on the air outside to get a rotation that consumes angular momentum. For a detailed explanation, see the beautiful text by MARK P. SILVERMAN, *And Yet it Moves: Strange Systems and Subtle Questions in Physics*, Cambridge University Press, 1993, p. 221.

Challenge 692, page 412: No.

Challenge 693, page 412: At the highest possible mass concentration, entropy is naturally the highest possible.

Challenge 694, page 412: The units do not match.

Challenge 695, page 412: In the case of water, a few turns mixes the ink, and turning backwards increases the mixing. In the case of glycerine, a few turns *seems* to mix the ink, and turning backwards undoes the mixing.

Challenge 696, page 413: Put them in clothes.

Challenge 700, page 413: Negative temperatures are a conceptual crutch definable only for systems with a few discrete states; they are not real temperatures, because they do not describe equilibrium states, and indeed never apply to systems with a continuum of states.

Challenge 701, page 415: This is also true for the shape of human bodies, the brain control of human motion, the growth of flowers, the waves of the sea, the formation of clouds, the processes leading to volcano eruptions, etc.

Challenge 704, page 422: See the puzzle about the motion of ski moguls.

Challenge 709, page 425: First, there are many more butterflies than tornadoes. Second, tornadoes do not rely on small initial disturbances for their appearance. Third, the belief in the butterfly 'effect' completely neglects an aspect of nature that is essential for self-organization: friction and dissipation. The butterfly 'effect', assumed that it existed, would require that dissipation in the air should have completely unrealistic properties. This is not the case in the atmosphere. But most important of all, there is no experimental basis for the 'effect': it has never been observed. Thus it does not exist.

Challenge 719, page 436: No. Nature does not allow more than about 20 digits of precision, as we will discover later in our walk. That is not sufficient for a standard book. The question whether such a number can be part of its own book thus disappears.

Challenge 720, page 436: All three statements are hogwash. A drag coefficient implies that the cross area of the car is known to the same precision. This is actually extremely difficult to measure and to keep constant. In fact, the value 0.375 for the Ford Escort was a cheat, as many other measurements showed. The fuel consumption is even more ridiculous, as it implies that fuel volumes and distances can be measured to that same precision. Opinion polls are taken by phoning at most 2000 people; due to the difficulties in selecting the right representative sample, that gives a precision of at most 3 % for typical countries.

Challenge 722, page 437: Space-time is defined using matter; matter is defined using space-time.

Challenge 723, page 437: Fact is that physics has been based on a circular definition for hundreds of years. Thus it is possible to build even an exact science on sand. Nevertheless, the elimination of the circularity is an important aim.

Challenge 724, page 438: Every measurement is a comparison with a standard; every comparison requires light or some other electromagnetic field. This is also the case for time measurements.

Challenge 725, page 438: Every mass measurement is a comparison with a standard; every comparison requires light or some other electromagnetic field.

Challenge 726, page 438: Angle measurements have the same properties as length or time measurements.

Challenge 728, page 454: Mass is a measure of the amount of energy. The 'square of mass' makes no sense.

Challenge 731, page 456: About 10 μg.

Challenge 732, page 457: Probably the quantity with the biggest variation is mass, where a prefix for $1\,\text{eV}/c^2$ would be useful, as would be one for the total mass in the universe, which is about 10^{90} times larger.

Challenge 733, page 458: The formula with $n-1$ is a better fit. Why?

Challenge 736, page 459: No! They are much too precise to make sense. They are only given as an illustration for the behaviour of the Gaussian distribution. Real measurement distributions are not Gaussian to the precision implied in these numbers.

Challenge 737, page 459: About 0.3 m/s. It is *not* 0.33 m/s, it is *not* 0.333 m/s and it is *not* any longer strings of threes!

Challenge 739, page 465: The slowdown goes *quadratically* with time, because every new slowdown adds to the old one!

Challenge 740, page 466: No, only properties of parts of the universe are listed. The universe itself has no properties, as shown in the last volume.

Challenge 741, page 526: For example, speed inside materials is slowed, but between atoms, light still travels with vacuum speed.

BIBLIOGRAPHY

> " Aiunt enim multum legendum esse, non multa. "
> Plinius, *Epistulae*.*

1   For a history of science in antiquity, see Lucio Russo, *La rivoluzione dimenticata*, Feltrinelli, 1996, also available in several other languages. Cited on page 15.

2   If you want to catch up secondary school physics, the clearest and shortest introduction world-wide is a free school text, available in English and several other languages, written by a researcher who has dedicated all his life to the teaching of physics in secondary school, together with his university team: Friedrich Herrmann, *The Karlsruhe Physics Course*, free to download in English, Spanish, Russian, Italian and Chinese at www.physikdidaktik. uni-karlsruhe.de/index_en.html. It is one of the few secondary school texts that captivates and surprises even professional physicists. (The 2013 paper on this book by C. Strunk & K. Rincke, *Zum Gutachten der Deutschen Physikalischen Gesellschaft über den Karlsruher Physikkurs*, available on the internet, makes many interesting points and is enlightening for every physicist.) This can be said even more of the wonderfully daring companion text Friedrich Herrmann & Georg Job, *Historical Burdens on Physics*, whose content is also freely available on the Karlsruhe site, in English and in several other languages.

   A beautiful book explaining physics and its many applications in nature and technology vividly and thoroughly is Paul G. Hewitt, John Suchocki & Leslie A. Hewitt, *Conceptual Physical Science*, Bejamin/Cummings, 1999.

   A great introduction is Klaus Dransfeld, Paul Kienle & Georg Kalvius, *Physik 1: Mechanik und Wärme*, Oldenburg, 2005.

   A book series famous for its passion for curiosity is Richard P. Feynman, Robert B. Leighton & Matthew Sands, *The Feynman Lectures on Physics*, Addison Wesley, 1977. The volumes can now be read online for free at www.feynmanlectures. info.

   A lot can be learned about motion from quiz books. One of the best is the well-structured collection of beautiful problems that require no mathematics, written by Jean-Marc Lévy-Leblond, *La physique en questions – mécanique*, Vuibert, 1998.

   Another excellent quiz collection is Yakov Perelman, *Oh, la physique*, Dunod, 2000, a translation from the Russian original.

   A good problem book is W. G. Rees, *Physics by Example: 200 Problems and Solutions*, Cambridge University Press, 1994.

   A good history of physical ideas is given in the excellent text by David Park, *The How*

---

* 'Read much, but not anything.' Ep. 7, 9, 15. Gaius Plinius Secundus (b. 23/4 Novum Comum, d. 79 Vesuvius eruption), Roman writer, especially famous for his large, mainly scientific work *Historia naturalis*, which has been translated and read for almost 2000 years.

and the Why, Princeton University Press, 1988.

An excellent introduction into physics is ROBERT POHL, *Pohl's Einführung in die Physik*, Klaus Lüders & Robert O. Pohl editors, Springer, 2004, in two volumes with CDs. It is a new edition of a book that is over 70 years old; but the didactic quality, in particular of the experimental side of physics, is unsurpassed.

Another excellent Russian physics problem book, the so-called *Saraeva*, seems to exist only as Spanish translation: B.B. BÚJOVTSEV, V.D. KRÍVCHENKOV, G.YA. MIÁKISHEV & I.M. SARÁEVA *Problemas seleccionados de física elemental*, Mir, 1979.

Another good physics problem book is GIOVANNI TONZIG, *Cento errori di fisica pronti per l'uso*, Sansoni, third edition, 2006. See also his www.giovannitonzig.it website. Cited on pages 15, 120, 219, 324, and 531.

3   An overview of motion illusions can be found on the excellent website www.michaelbach.de/ot. The complex motion illusion figure is found on www.michaelbach.de/ot/mot_rotsnake/index.html; it is a slight variation of the original by Kitaoka Akiyoshi at www.ritsumei.ac.jp/~akitaoka/rotsnake.gif, published as A. KITAOKA & H. ASHIDA, *Phenomenal characteristics of the peripheral drift illusion*, Vision 15, pp. 261–262, 2003. A common scam is to claim that the illusion is due to or depends on stress. Cited on page 16.

4   These and other fantastic illusions are also found in AKIYOSHI KITAOKA, *Trick Eyes*, Barnes & Noble, 2005. Cited on page 16.

5   A well-known principle in the social sciences states that, given a question, for every possible answer, however weird it may seem, there is somebody – and often a whole group – who holds it as his opinion. One just has to go through literature (or the internet) to confirm this.

About group behaviour in general, see R. AXELROD, *The Evolution of Cooperation*, Harper Collins, 1984. The propagation and acceptance of ideas, such as those of physics, are also an example of human cooperation, with all its potential dangers and weaknesses. Cited on page 16.

6   All the known texts by Parmenides and Heraclitus can be found in JEAN-PAUL DUMONT, *Les écoles présocratiques*, Folio-Gallimard, 1988. Views about the non-existence of motion have also been put forward by much more modern and much more contemptible authors, such as in 1710 by Berkeley. Cited on page 17.

7   An example of people worried by Zeno is given by WILLIAM MCLAUGHLIN, *Resolving Zeno's paradoxes*, Scientific American pp. 66–71, November 1994. The actual argument was not about a hand slapping a face, but about an arrow hitting the target. See also Ref. 65. Cited on page 17.

8   The full text of *La Beauté* and the other poems from *Les fleurs du mal*, one of the finest books of poetry ever written, can be found at the hypermedia.univ-paris8.fr/bibliotheque/Baudelaire/Spleen.html website. Cited on page 18.

9   A famous collection of interesting examples of motion in everyday life is the excellent book by JEARL WALKER, *The Flying Circus of Physics*, Wiley, 1975. Its website is at www.flyingcircusofphysics.com. Another beautiful book is CHRISTIAN UCKE & H. JOACHIM SCHLICHTING, *Spiel, Physik und Spaß – Physik zum Mitdenken und Mitmachen*, Wiley-VCH, 2011. For more interesting physical effects in everyday life, see ERWEIN FLACHSEL, *Hundertfünfzig Physikrätsel*, Ernst Klett Verlag, 1985. The book also covers several clock puzzles, in puzzle numbers 126 to 128. Cited on page 19.

10 A concise and informative introduction into the history of classical physics is given in the first chapter of the book by FLOYD KARKER RICHTMYER, EARLE HESSE KENNARD

& John N. Cooper, *Introduction to Modern Physics*, McGraw–Hill, 1969. Cited on page 19.

11 An introduction into perception research is E. Bruce Goldstein, *Perception*, Books/Cole, 5th edition, 1998. Cited on pages 21 and 26.

12 A good overview over the arguments used to prove the existence of god from motion is given by Michael Buckley, *Motion and Motion's God*, Princeton University Press, 1971. The intensity of the battles waged around these failed attempts is one of the tragicomic chapters of history. Cited on page 21.

13 Thomas Aquinas, *Summa Theologiae* or *Summa Theologica*, 1265–1273, online in Latin at www.newadvent.org/summa, in English on several other servers. Cited on page 21.

14 For an exploration of 'inner' motions, see the beautiful text by Richard Schwartz, *Internal Family Systems Therapy*, The Guilford Press, 1995. Cited on page 21.

15 For an authoritative description of proper motion development in babies and about how it leads to a healthy character see Emmi Pikler, *Laßt mir Zeit - Die selbstständige Bewegungsentwicklung des Kindes bis zum freien Gehen*, Pflaum Verlag, 2001, and her other books. See also the website www.pikler.org. Cited on page 21.

16 See e.g. the fascinating text by David G. Chandler, *The Campaigns of Napoleon - The Mind and Method of History's Greatest Soldier*, Macmillan, 1966. Cited on page 21.

17 Richard Marcus, *American Roulette*, St Martin's Press, 2003, a thriller and a true story. Cited on page 21.

18 A good and funny book on behaviour change is the well-known text Richard Bandler, *Using Your Brain for a Change*, Real People Press, 1985. See also Richard Bandler & John Grinder, *Frogs into princes - Neuro Linguistic Programming*, Eden Grove Editions, 1990. Cited on pages 21 and 32.

19 A beautiful book about the mechanisms of human growth from the original cell to full size is Lewis Wolpert, *The Triumph of the Embryo*, Oxford University Press, 1991. Cited on page 21.

20 On the topic of grace and poise, see e.g. the numerous books on the Alexander technique, such as M. Gelb, *Body Learning - An Introduction to the Alexander Technique*, Aurum Press, 1981, and Richard Brennan, *Introduction to the Alexander Technique*, Little Brown and Company, 1996. Among others, the idea of the Alexander technique is to return to the situation that the muscle groups for sustainment and those for motion are used only for their respective function, and not vice versa. Any unnecessary muscle tension, such as neck stiffness, is a waste of energy due to the use of sustainment muscles for movement and of motion muscles for sustainment. The technique teaches the way to return to the natural use of muscles.

Motion of animals was discussed extensively already in the seventeenth century by G. Borelli, *De motu animalium*, 1680. An example of a more modern approach is J. J. Collins & I. Stewart, *Hexapodal gaits and coupled nonlinear oscillator models*, Biological Cybernetics 68, pp. 287–298, 1993. See also I. Stewart & M. Golubitsky, *Fearful Symmetry*, Blackwell, 1992. Cited on pages 23 and 122.

21 The results on the development of children mentioned here and in the following have been drawn mainly from the studies initiated by Jean Piaget; for more details on child development, see later on. At www.piaget.org you can find the website maintained by the Jean Piaget Society. Cited on pages 24, 40, and 42.

22 The reptilian brain (eat? flee? ignore?), also called the R-complex, includes the brain stem, the cerebellum, the basal ganglia and the thalamus; the old mammalian (emotions) brain,

also called the limbic system, contains the amygdala, the hypothalamus and the hippocampus; the human (and primate) (rational) brain, called the neocortex, consists of the famous grey matter. For images of the brain, see the atlas by JOHN NOLTE, *The Human Brain: An Introduction to its Functional Anatomy*, Mosby, fourth edition, 1999. Cited on page 25.

23 The lower left corner film can be reproduced on a computer after typing the following lines in the Mathematica software package: Cited on page 26.

```
« Graphics'Animation'
Nxpixels=72; Nypixels=54; Nframes=Nxpixels 4/3;
Nxwind=Round[Nxpixels/4]; Nywind=Round[Nypixels/3];
front=Table[Round[Random[]],{y,1,Nypixels},{x,1,Nxpixels}];
back =Table[Round[Random[]],{y,1,Nypixels},{x,1,Nxpixels}];
frame=Table[front,{nf,1,Nframes}];
Do[ If[ x>n-Nxwind && x<n && y>Nywind && y<2Nywind,
    frame[[n,y,x]]=back[[y,x-n]] ],
       {x,1,Nxpixels}, {y,1,Nypixels}, {n,1,Nframes}];
film=Table[ListDensityPlot[frame[[nf]], Mesh-> False,
    Frame-> False, AspectRatio-> N[Nypixels/Nxpixels],
    DisplayFunction-> Identity],   {nf,1,Nframes}]
ShowAnimation[film]
```

But our motion detection system is much more powerful than the example shown in the lower left corners. The following, different film makes the point.

```
« Graphics'Animation'
Nxpixels=72; Nypixels=54; Nframes=Nxpixels 4/3;
Nxwind=Round[Nxpixels/4]; Nywind=Round[Nypixels/3];
front=Table[Round[Random[]],{y,1,Nypixels},{x,1,Nxpixels}];
back =Table[Round[Random[]],{y,1,Nypixels},{x,1,Nxpixels}];
frame=Table[front,{nf,1,Nframes}];
Do[ If[ x>n-Nxwind && x<n && y>Nywind && y<2Nywind,
    frame[[n,y,x]]=back[[y,x]] ],
       {x,1,Nxpixels}, {y,1,Nypixels}, {n,1,Nframes}];
film=Table[ListDensityPlot[frame[[nf]], Mesh-> False,
    Frame-> False, AspectRatio-> N[Nypixels/Nxpixels],
    DisplayFunction-> Identity],   {nf,1,Nframes}]
ShowAnimation[film]
```

Similar experiments, e.g. using randomly changing random patterns, show that the eye perceives motion even in cases where all Fourier components of the image are practically zero; such image motion is called *drift-balanced* or *non-Fourier* motion. Several examples are presented in J. ZANKER, *Modelling human motion perception I: Classical stimuli*, Naturwissenschaften **81**, pp. 156–163, 1994, and J. ZANKER, *Modelling human motion perception II: Beyond Fourier motion stimuli*, Naturwissenschaften **81**, pp. 200–209, 1994. Modern research has helped to find the corresponding neuronal structures, as shown in S. A. BACCUS, B. P. OLVECZKY, M. MANU & M. MEISTER, *A retinal circuit that computes object motion*, Journal of Neuroscience **28**, pp. 6807–6817, 2008.

24 All fragments from Heraclitus are from JOHN MANSLEY ROBINSON, *An Introduction to Early Greek Philosophy*, Houghton Muffin 1968, chapter 5. Cited on page 27.

25 On the block and tackle, see the explanations by Donald Simanek at http://www.lhup.edu/~dsimanek/TTT-fool/fool.htm. Cited on page 31.

26 An overview over these pretty puzzles is found in E. D. Demaine, M. L. Demaine, Y. N. Minski, J. S. B. Mitchell, R. L. Rivest & M. Patrascu, *Picture-hanging puzzles*, preprint at arxiv.org/abs/1203.3602. Cited on page 33.

27 An introduction to Newton the alchemist are the books by Betty Jo Teeter Dobbs, *The Foundations of Newton's Alchemy*, Cambridge University Press, 1983, and *The Janus Face of Genius*, Cambridge University Press, 1992. Newton is found to be a sort of highly intellectual magician, desperately looking for examples of processes where gods interact with the material world. An intense but tragic tale. A good overview is provided by R. G. Keesing, *Essay Review: Newton's Alchemy*, Contemporary Physics **36**, pp. 117–119, 1995.

Newton's infantile theology, typical for god seekers who grew up without a father, can be found in the many books summarizing the letter exchanges between Clarke, his secretary, and Leibniz, Newton's rival for fame. Cited on page 34.

28 An introduction to the story of classical mechanics, which also destroys a few of the myths surrounding it – such as the idea that Newton could solve differential equations or that he introduced the expression $F = ma$ – is given by Clifford A. Truesdell, *Essays in the History of Mechanics*, Springer, 1968. Cited on pages 34, 178, and 228.

29 The slowness of the effective speed of light inside the Sun is due to the frequent scattering of photons by solar matter. The best estimate of its value is by R. Mitalas & K. R. Sills, *On the photon diffusion time scale for the Sun*, The Astrophysical Journal **401**, pp. 759–760, 1992. They give an average speed of 0.97 cm/s over the whole Sun and a value about 10 times smaller at its centre. Cited on page 36.

30 C. Liu, Z. Dutton, C. H. Behroozi & L. Vestergaard Hau, *Observation of coherent optical information storage in an atomic medium using halted light pulses*, Nature **409**, pp. 490–493, 2001. There is also a comment on the paper by E. A. Cornell, *Stopping light in its track*, **409**, pp. 461–462, 2001. However, despite the claim, the light pulses of course have *not* been halted. Can you give at least two reasons without even reading the paper, and maybe a third after reading it?

The work was an improvement on the previous experiment where a group velocity of light of 17 m/s had been achieved, in an ultracold gas of sodium atoms, at nanokelvin temperatures. This was reported by L. Vestergaard Hau, S. E. Harris, Z. Dutton & C. H. Behroozi, *Light speed reduction to 17 meters per second in an ultracold atomic gas*, Nature **397**, pp. 594–598, 1999. Cited on page 36.

31 Rainer Flindt, *Biologie in Zahlen – Eine Datensammlung in Tabellen mit über 10.000 Einzelwerten*, Spektrum Akademischer Verlag, 2000. Cited on page 36.

32 Two jets with that speed have been observed by I. F. Mirabel & L. F. Rodríguez, *A superluminal source in the Galaxy*, Nature **371**, pp. 46–48, 1994, as well as the comments on p. 18. Cited on page 36.

33 A beautiful introduction to the slowest motions in nature, the changes in landscapes, is Detlev Busche, Jürgen Kempf & Ingrid Stengel, *Landschaftsformen der Erde – Bildatlas der Geomorphologie*, Primus Verlag, 2005. Cited on page 37.

34 To build your own sundial, see the pretty and short Arnold Zenkert, *Faszination Sonnenuhr*, VEB Verlag Technik, 1984. See also the excellent and complete introduction into this somewhat strange world at the www.sundials.co.uk website. Cited on page 42.

35 An introduction to the sense of time as a result of clocks in the brain is found in R. B. Ivry & R. Spencer, *The neural representation of time*, Current Opinion in Neurobiology **14**, pp. 225–232, 2004. The chemical clocks in our body are described in John D. Palmer,

The Living Clock, Oxford University Press, 2002, or in A. AHLGREN & F. HALBERG, *Cycles of Nature: An Introduction to Biological Rhythms*, National Science Teachers Association, 1990. See also the www.msi.umn.edu/~halberg/introd website. Cited on page 43.

36 This has been shown among others by the work of Anna Wierzbicka that is discussed in more detail in one of the subsequent volumes. The passionate best seller by the Chomskian author STEVEN PINKER, *The Language Instinct – How the Mind Creates Language*, Harper Perennial, 1994, also discusses issues related to this matter, refuting amongst others on page 63 the often repeated false statement that the *Hopi* language is an exception. Cited on page 43.

37 For more information, see the excellent and freely downloadable books on biological clocks by Wolfgang Engelmann on the website www.uni-tuebingen.de/plantphys/bioclox. Cited on page 44.

38 B. GÜNTHER & E. MORGADO, *Allometric scaling of biological rhythms in mammals*, Biological Research 38, pp. 207–212, 2005. Cited on page 44.

39 Aristotle rejects the idea of the flow of time in chapter IV of his *Physics*. See the full text on the classics.mit.edu/Aristotle/physics.4.iv.html website. Cited on page 48.

40 Perhaps the most informative of the books about the 'arrow of time' is HANS DIETER ZEH, *The Physical Basis of the Direction of Time*, Springer Verlag, 4th edition, 2001. It is still the best book on the topic. Most other texts exist – have a look on the internet – but lack clarity of ideas.

 A typical conference proceeding is J. J. HALLIWELL, J. PÉREZ-MERCADER & WOJCIECH H. ZUREK, *Physical Origins of Time Asymmetry*, Cambridge University Press, 1994. Cited on page 49.

41 On the issue of absolute and relative motion there are many books about few issues. Examples are JULIAN BARBOUR, *Absolute or Relative Motion? Vol. 1: A Study from the Machian Point of View of the Discovery and the Structure of Spacetime Theories*, Cambridge University Press, 1989, JULIAN BARBOUR, *Absolute or Relative Motion? Vol. 2: The Deep Structure of General Relativity*, Oxford University Press, 2005, or JOHN EARMAN, *World Enough and Spacetime: Absolute vs Relational Theories of Spacetime*, MIT Press, 1989. A speculative solution on the alternative between absolute and relative motion is presented in volume VI. Cited on page 52.

42 Coastlines and other fractals are beautifully presented in HEINZ-OTTO PEITGEN, HARTMUT JÜRGENS & DIETMAR SAUPE, *Fractals for the Classroom*, Springer Verlag, 1992, pp. 232–245. It is also available in several other languages. Cited on page 54.

43 R. DOUGHERTY & M. FOREMAN, *Banach–Tarski decompositions using sets with the property of Baire*, Journal of the American Mathematical Society 7, pp. 75–124, 1994. See also ALAN L.T. PATERSON, *Amenability*, American Mathematical Society, 1998, and ROBERT M. FRENCH, *The Banach–Tarski theorem*, The Mathematical Intelligencer 10, pp. 21–28, 1998. Finally, there are the books by BERNARD R. GELBAUM & JOHN M. H. OLMSTED, *counter-examples in Analysis*, Holden–Day, 1964, and their *Theorems and counter-examples in Mathematics*, Springer, 1993. Cited on page 57.

44 The beautiful but not easy text is STAN WAGON, *The Banach Tarski Paradox*, Cambridge University Press, 1993. Cited on pages 58 and 479.

45 About the shapes of salt water bacteria, see the corresponding section in the interesting book by BERNARD DIXON, *Power Unseen – How Microbes Rule the World*, W.H. Freeman, 1994. The book has about 80 sections, in which as many microorganisms are vividly presented. Cited on page 59.

46 OLAF MEDENBACH & HARRY WILK, *Zauberwelt der Mineralien*, Sigloch Edition, 1977. It combines beautiful photographs with an introduction into the science of crystals, minerals and stones. About the largest crystals, see P. C. RICKWOOD, *The largest crystals*, 66, pp. 885–908, 1981, also available on www.minsocam.org/MSA/collectors_corner/arc/large_crystals.htm. For an impressive example, the Naica cave in Mexico, see www.naica.com.mx/ingles/index.htm Cited on page 59.

47 See the websites www.weltbildfrage.de/3frame.htm and www.lhup.edu/~dsimanek/hollow/morrow.htm. Cited on page 60.

48 The smallest distances are probed in particle accelerators; the distance can be determined from the energy of the particle beam. In 1996, the value of 10^{-19} m (for the upper limit of the size of quarks) was taken from the experiments described in F. ABE & al., *Measurement of dijet angular distributions by the collider detector at Fermilab*, Physical Review Letters 77, pp. 5336–5341, 1996. Cited on page 66.

49 More on the Moon illusion can be found at the website science.nasa.gov/science-news/science-at-nasa/2008/16jun_moonillusion/. All the works of Ptolemy are found online at www.ptolemaeus.badw.de. Cited on page 69.

50 These puzzles are taken from the puzzle collection at www.mathematische-basteleien.de. Cited on page 70.

51 ALEXANDER K. DEWDNEY, *The Planiverse – Computer Contact with a Two-dimensional World*, Poseidon Books/Simon & Schuster, 1984. See also EDWIN A. ABBOTT, *Flatland: A romance of many dimensions*, 1884. Several other fiction authors had explored the option of a two-dimensional universe before, always answering, incorrectly, in the affirmative. Cited on page 71.

52 J. BOHR & K. OLSEN, *The ancient art of laying rope*, preprint at arxiv.org/abs/1004.0814 Cited on page 71.

53 For an overview and references see www.pbrc.hawaii.edu/~petra/animal_olympians.html. Cited on page 72.

54 P. PIERANSKI, S. PRZYBYL & A. STASIAK, *Tight open knots*, European Physical Journal E 6, pp. 123–128, 2001, preprint at arxiv.org/abs/physics/0103016. Cited on page 73.

55 On the world of fireworks, see the frequently asked questions list of the usenet group rec.pyrotechnics, or search the web. A simple introduction is the article by J. A. CONKLING, *Pyrotechnics*, Scientific American pp. 66–73, July 1990. Cited on page 75.

56 There is a whole story behind the variations of g. It can be discovered in CHUJI TSUBOI, *Gravity*, Allen & Unwin, 1979, or in WOLFGANG TORGE, *Gravimetry*, de Gruyter, 1989, or in MILAN BURŠA & KAREL PĚČ, *The Gravity Field and the Dynamics of the Earth*, Springer, 1993. The variation of the height of the soil by up to 0.3 m due to the Moon is one of the interesting effects found by these investigations. Cited on pages 76 and 197.

57 STILLMAN DRAKE, *Galileo: A Very Short Introduction*, Oxford University Press, 2001. Cited on page 76.

58 ANDREA FROVA, *La fisica sotto il naso – 44 pezzi facili*, Biblioteca Universale Rizzoli, Milano, 2001. Cited on page 77.

59 On the other hands, other sciences enjoy studying usual paths in all detail. See, for example, HEINI HEDIGER, editor, *Die Straßen der Tiere*, Vieweg & Sohn, 1967. Cited on page 77.

60 H. K. ERIKSEN, J. R. KRISTIANSEN, Ø. LANGANGEN & I. K. WEHUS, *How fast could Usain Bolt have run? A dynamical study*, American Journal of Physics 77, pp. 224–228, 2009. See also the references at en.wikipedia.org/wiki/Footspeed. Cited on page 78.

61 This was discussed in the *Frankfurter Allgemeine Zeitung*, 2nd of August, 1997, at the time of the world athletics championship. The values are for the fastest part of the race of a 100 m sprinter; the exact values cited were called the running speed world records in 1997, and were given as 12.048 m/s = 43.372 km/h by Ben Johnson for men, and 10.99 m/s = 39.56 km/h for women. Cited on page 78.

62 Long jump data and literature can be found in three articles all entitled *Is a good long jumper a good high jumper?*, in the American Journal of Physics **69**, pp. 104–105, 2001. In particular, world class long jumpers run at 9.35 ± 0.15 m/s, with vertical take-off speeds of 3.35 ± 0.15 m/s, giving take-off angles of about (only) 20°. A new technique for achieving higher take-off angles would allow the world long jump record to increase dramatically. Cited on page 78.

63 The study of shooting faeces (i.e., shit) and its mechanisms is a part of modern biology. The reason that caterpillars do this was determined by M. WEISS, *Good housekeeping: why do shelter-dwelling caterpillars fling their frass?*, Ecology Letters **6**, pp. 361–370, 2003, who also gives the present record of 1.5 m for the 24 mg pellets of *Epargyreus clarus*. The picture of the flying frass is from S. CAVENEY, H. MCLEAN & D. SURRY, *Faecal firing in a skipper caterpillar is pressure-driven*, The Journal of Experimental Biology **201**, pp. 121–133, 1998. Cited on page 79.

64 H. C. BENNET-CLARK, *Scale effects in jumping animals*, pp. 185–201, in T. J. PEDLEY, editor, *Scale Effects in Animal Locomotion*, Academic Press, 1977. Cited on page 80.

65 The arguments of Zeno can be found in ARISTOTLE, *Physics*, VI, 9. It can be found translated in almost any language. The classics.mit.edu/Aristotle/physics.6.vi.html website provides an online version in English. Cited on pages 83 and 523.

66 See, for exaple, K. V. KUMAR & W. T. NORFLEET, *Issues of human acceleration tolerance after long-duration space flights*, NASA Technical Memorandum 104753, pp. 1–55, 1992, available at ntrs.nasa.gov. Cited on page 85.

67 Etymology can be a fascinating topic, e.g. when research discovers the origin of the German word 'Weib' ('woman', related to English 'wife'). It was discovered, via a few texts in Tocharian – an extinct Indo-European language from a region inside modern China – to mean originally 'shame'. It was used for the female genital region in an expression meaning 'place of shame'. With time, this expression became to mean 'woman' in general, while being shortened to the second term only. This connection was discovered by the linguist Klaus T. Schmidt; it explains in particular why the word is not feminine but neutral, i.e., why it uses the article 'das' instead of 'die'. Julia Simon, private communication.

Etymology can also be simple and plain fun, for example when one discovers in the *Oxford English Dictionary* that 'testimony' and 'testicle' have the same origin; indeed in Latin the same word 'testis' was used for both concepts. Cited on pages 86 and 101.

68 An overview of the latest developments is given by J. T. ARMSTRONG, D. J. HUNTER, K. J. JOHNSTON & D. MOZURKEWICH, *Stellar optical interferometry in the 1990s*, Physics Today pp. 42–49, May 1995. More than 100 stellar diameters were known already in 1995. Several dedicated powerful instruments are being planned. Cited on page 87.

69 A good biology textbook on growth is ARTHUR F. HOPPER & NATHAN H. HART, *Foundations of Animal Deveopment*, Oxford University Press, 2006. Cited on page 89.

70 This is discussed for example in C. L. STONG, *The amateur scientist – how to supply electric power to something which is turning*, Scientific American pp. 120–125, December 1975. It also discusses how to make a still picture of something rotating simply by using a few prisms, the so-called *Dove prisms*. Other examples of attaching something to a rotating body are given by E. RIEFLIN, *Some mechanisms related to Dirac's strings*, American Journal of Physics 47, pp. 379–381, 1979. Cited on page 89.

71 JAMES A. YOUNG, *Tumbleweed*, Scientific American 264, pp. 82–87, March 1991. The tumbleweed is in fact quite rare, except in Hollywood westerns, where all directors feel obliged to give it a special appearance. Cited on page 90.

72 The classic book on the topic is JAMES GRAY, *Animal Locomotion*, Weidenfeld & Nicolson, 1968. Cited on page 90.

73 About *N. decemspinosa*, see R. L. CALDWELL, *A unique form of locomotion in a stomatopod – backward somersaulting*, Nature 282, pp. 71–73, 1979, and R. FULL, K. EARLS, M. WONG & R. CALDWELL, *Locomotion like a wheel?*, Nature 365, p. 495, 1993. About rolling caterpillars, see J. BRACKENBURY, *Caterpillar kinematics*, Nature 330, p. 453, 1997, and J. BRACKENBURY, *Fast locomotion in caterpillars*, Journal of Insect Physiology 45, pp. 525–533, 1999. More images around legs can be found on rjf9.biol.berkeley.edu/twiki/bin/view/PolyPEDAL/LabPhotographs. Cited on page 90.

74 The locomotion of the spiders of the species *Cebrennus villosus* has been described by Ingo Rechenberg from Berlin. See the video at www.youtube.com/watch?v=Aayb_h31RyQ. Cited on page 90.

75 The first experiments to prove the rotation of the flagella were by M. SILVERMAN & M. I. SIMON, *Flagellar rotation and the mechanism of bacterial motility*, Nature 249, pp. 73–74, 1974. For some pretty pictures of the molecules involved, see K. NAMBA, *A biological molecular machine: bacterial flagellar motor and filament*, Wear 168, pp. 189–193, 1993, or the website www.nanonet.go.jp/english/mailmag/2004/011a.html. The present record speed of rotation, 1700 rotations per second, is reported by Y. MAGARIYAMA, S. SUGIYAMA, K. MURAMOTO, Y. MAEKAWA, I. KAWAGISHI, Y. IMAE & S. KUDO, *Very fast flagellar rotation*, Nature 371, p. 752, 1994.

More on bacteria can be learned from DAVID DUSENBERY, *Life at a Small Scale*, Scientific American Library, 1996. Cited on page 92.

76 S. CHEN & al., *Structural diversity of bacterial flagellar motors*, EMBO Journal 30, pp. 2972–2981, 2011, also online at emboj.embopress.org/content/30/14/2972. Cited on page 92.

77 M. P. BRENNER, S. HILGENFELDT & D. LOHSE, *Single bubble sonoluminescence*, Reviews of Modern Physics 74, pp. 425–484, 2002. Cited on page 96.

78 K. R. WENINGER, B. P. BARBER & S. J. PUTTERMAN, *Pulsed Mie scattering measurements of the collapse of a sonoluminescing bubble*, Physical Review Letters 78, pp. 1799–1802, 1997. Cited on page 96.

79 On shadows, see the agreeable popular text by ROBERTO CASATI, *Alla scoperta dell'ombra – Da Platone a Galileo la storia di un enigma che ha affascinato le grandi menti dell'umanità*, Oscar Mondadori, 2000, and his website located at www.shadowes.org. Cited on page 98.

80 There is also the beautiful book by PENELOPE FARRANT, *Colour in Nature*, Blandford, 1997. Cited on page 98.

81 The 'laws' of cartoon physics can easily be found using any search engine on the internet. Cited on page 98.

82 For the curious, an overview of the illusions used in the cinema and in television, which lead to some of the strange behaviour of images mentioned above, is given in BERNARD WILKIE, *The Technique of Special Effects in Television*, Focal Press, 1993, and his other books, or in the *Cinefex* magazine. On digital cinema techniques, see PETER C. SLANSKY, editor, *Digitaler film – digitales Kino*, UVK Verlag, 2004. Cited on page 99.

83 AETIUS, *Opinions*, I, XXIII, 3. See JEAN-PAUL DUMONT, *Les écoles présocratiques*, Folio Essais, Gallimard, p. 426, 1991. Cited on page 99.

84 GIUSEPPE FUMAGALLI, *Chi l'ha detto?*, Hoepli, 1983. It is from Pappus of Alexandria's opus *Synagoge*, book VIII, 19. Cited on pages 100 and 237.

85 See www.straightdope.com/classics/a5_262.html and the more dubious en.wikipedia.org/wiki/Guillotine. Cited on page 102.

86 See the path-breaking paper by A. DISESSA, *Momentum flow as an alternative perspective in elementary mechanics*, 48, p. 365, 1980, and A. DISESSA, *Erratum: "Momentum flow as an alternative perspective in elementary mechanics" [Am. J. Phys. 48, 365 (1980)]*, 48, p. 784, 1980. Also the wonderful free textbook by FRIEDRICH HERRMANN, *The Karlsruhe Physics Course*, makes this point extensively; see Ref. 2. Cited on pages 109, 228, 231, and 506.

87 For the role and chemistry of adenosine triphosphate (ATP) in cells and in living beings, see any chemistry book, or search the internet. The uncovering of the mechanisms around ATP has led to Nobel Prizes in Chemistry in 1978 and in 1997. Cited on page 109.

88 A picture of this unique clock can be found in the article by A. GARRETT, *Perpetual motion – a delicious delirium*, Physics World pp. 23–26, December 1990. Cited on page 110.

89 ESGER BRUNNER, *Het ongelijk van Newton – het kleibakexperiment van 's Gravesande nagespeld*, Nederland tijdschrift voor natuurkunde pp. 95–96, Maart 2012. The paper contains photographs of the mud imprints. Cited on page 111.

90 A Shell study estimated the world's total energy consumption in 2000 to be 500 EJ. The US Department of Energy estimated it to be around 416 EJ. We took the lower value here. A discussion and a breakdown into electricity usage (14 EJ) and other energy forms, with variations per country, can be found in S. BENKA, *The energy challenge*, Physics Today 55, pp. 38–39, April 2002, and in E. J. MONITZ & M. A. KENDERDINE, *Meeting energy challenges: technology and policy*, Physics Today 55, pp. 40–46, April 2002. Cited on pages 113 and 114.

91 L. M. MILLER, F. GANS & A. KLEIDON, *Estimating maximum global land surface wind power extractability and associated climatic consequences*, Earth System Dynamics 2, pp. 1–12, 2011. Cited on page 114.

92 For an overview, see the paper by J. F. MULLIGAN & H. G. HERTZ, *An unpublished lecture by Heinrich Hertz: 'On the energy balance of the Earth'*, American Journal of Physics 65, pp. 36–45, 1997. Cited on page 114.

93 For a beautiful photograph of this feline feat, see the cover of the journal and the article of J. DARIUS, *A tale of a falling cat*, Nature 308, p. 109, 1984. Cited on page 121.

94 NATTHI L. SHARMA, *A new observation about rolling motion*, European Journal of Physics 17, pp. 353–356, 1996. Cited on page 121.

95 C. SINGH, *When physical intuition fails*, American Journal of Physics 70, pp. 1103–1109, 2002. Cited on page 121.

96 There is a vast literature on walking. Among the books on the topic, two well-known introductions are ROBERT MCNEILL ALEXANDER, *Exploring Biomechanics: Animals in Motion*, Scientific American Library, 1992, and STEVEN VOGEL, *Comparative Biomechanics - Life's Physical World*, Princeton University Press, 2003. Cited on page 122.

97 SERGE GRACOVETSKY, *The Spinal Engine*, Springer Verlag, 1990. It is now also known that human gait is chaotic. This is explained by M. PERC, *The dynamics of human gait*, European Journal of Physics 26, pp. 525–534, 2005. On the physics of walking and running, see also the respective chapters in the delightful book by WERNER GRUBER, *Unglaublich einfach, einfach unglaublich: Physik für jeden Tag*, Heyne, 2006. Cited on page 123.

98 M. LLOBERA & T.J. SLUCKIN, *Zigzagging: theoretical insights on climbing strategies*, Journal of Theoretical Biology 249, pp. 206–217, 2007. Cited on page 125.

99 This description of life and death is called the concept of *maximal metabolic scope*. Look up details in your favourite library. A different phrasing is the one by M. YA. AZBEL, *Universal biological scaling and mortality*, Proceedings of the National Academy of Sciences of the USA 91, pp. 12453–12457, 1994. He explains that every atom in an organism consumes, on average, 20 oxygen molecules per life-span. Cited on page 125.

100 DUNCAN MACDOUGALL, *Hypothesis concerning soul substance together with experimental evidence of the existence of such substance*, American Medicine 2, pp. 240–243, April 1907, and DUNCAN MACDOUGALL, *Hypothesis concerning soul substance*, American Medicine 2, pp. 395–397, July 1907. Reading the papers shows that the author has little practice in performing reliable weight and time measurements. Cited on page 126.

101 A good roulette prediction story from the 1970s is told by THOMAS A. BASS, *The Eudaemonic Pie* also published under the title *The Newtonian Casino*, Backinprint, 2000. An overview up to 1998 is given in the paper EDWARD O. THORP, *The invention of the first wearable computer*, Proceedings of the Second International Symposium on Wearable Computers (ISWC 1998), 19-20 October 1998, Pittsburgh, Pennsylvania, USA (IEEE Computer Society), pp. 4–8, 1998, downloadable at csdl.computer.org/comp/proceedings/iswc/1998/9074/00/9074toc.htm. Cited on page 127.

102 This and many other physics surprises are described in the beautiful lecture script by JOSEF ZWECK, *Physik im Alltag*, the notes of his lectures held in 1999/2000 at the Universität Regensburg. Cited on pages 128 and 133.

103 The equilibrium of ships, so important in car ferries, is an interesting part of shipbuilding; an introduction was already given by LEONHARD EULER, *Scientia navalis*, 1749. Cited on page 129.

104 THOMAS HEATH, *Aristarchus of Samos – the Ancient Copernicus*, Dover, 1981, reprinted from the original 1913 edition. Aristarchus' treaty is given in Greek and English. Aristarchus was the first proposer of the heliocentric system. Aristarchus had measured the length of the day (in fact, by determining the number of days per year) to the astonishing precision of less than one second. This excellent book also gives an overview of Greek astronomy before Aristarchus, explained in detail for each Greek thinker. Aristarchus' text is also reprinted in ARISTARCHUS, *On the sizes and the distances of the Sun and the Moon, c.* 280 BCE in MICHAEL J. CROWE, *Theories of the World From Antiquity to the Copernican Revolution*, Dover, 1990, especially on pp. 27–29. No citations.

105 T. GERKEMA & L. GOSTIAUX, *A brief history of the Coriolis force*, Europhysics News 43, pp. 14–17, 2012. Cited on page 138.

106 See for example the videos on the Coriolis effect at techtv.mit.edu/videos/3722 and techtv.mit.edu/videos/3714, or search for videos on youtube.com. Cited on page 139.

107 The influence of the Coriolis effect on icebergs was studied most thoroughly by the physicist turned oceanographer Walfrid Ekman (b. 1874 Stockholm, d. 1954 Gostad); the topic was suggested by the great explorer Fridtjof Nansen, who also made the first observations. In his honour, one speaks of the Ekman layer, Ekman transport and Ekman spirals. Any text on oceanography or physical geography will give more details about them. Cited on page 139.

108 An overview of the effects of the Coriolis acceleration $a = -2\omega \times v$ in the rotating frame is given by EDWARD A. DESLOGE, *Classical Mechanics*, Volume 1, John Wiley & Sons, 1982. Even the so-called *Gulf Stream*, the current of warm water flowing from the Caribbean to the North Sea, is influenced by it. Cited on page 139.

109 The original publication is by A. H. SHAPIRO, *Bath-tub vortex*, Nature **196**, pp. 1080–1081, 1962. He also produced two films of the experiment. The experiment has been repeated many times in the northern and in the southern hemisphere, where the water drains clockwise; the first southern hemisphere test was L. M. TREFETHEN & al., *The bath-tub vortex in the southern hemisphere*, Nature **201**, pp. 1084–1085, 1965. A complete literature list is found in the letters to the editor of the American Journal of Physics **62**, p. 1063, 1994. Cited on page 140.

110 The tricks are explained by H. RICHARD CRANE, *Short Foucault pendulum: a way to eliminate the precession due to ellipticity*, American Journal of Physics **49**, pp. 1004–1006, 1981, and particularly in H. RICHARD CRANE, *Foucault pendulum wall clock*, American Journal of Physics **63**, pp. 33–39, 1993. The Foucault pendulum was also the topic of the thesis of HEIKE KAMERLING ONNES, *Nieuwe bewijzen der aswenteling der aarde*, Universiteit Groningen, 1879. Cited on page 140.

111 The reference is J. G. HAGEN, *La rotation de la terre : ses preuves mécaniques anciennes et nouvelles*, Sp. Astr. Vaticana Second. App. Rome, 1910. His other experiment is published as J. G. HAGEN, *How Atwood's machine shows the rotation of the Earth even quantitatively*, International Congress of Mathematics, Aug. 1912. Cited on page 141.

112 The original papers are A. H. COMPTON, *A laboratory method of demonstrating the Earth's rotation*, Science **37**, pp. 803–806, 1913, A. H. COMPTON, *Watching the Earth revolve*, Scientific American Supplement no. 2047, pp. 196–197, 1915, and A. H. COMPTON, *A determination of latitude, azimuth and the length of the day independent of astronomical observations*, Physical Review (second series) **5**, pp. 109–117, 1915. Cited on page 141.

113 The G-ring in Wettzell is so precise, with a resolution of less than 10^{-8}, that it has detected the motion of the poles. For details, see K. U. SCHREIBER, A. VELIKOSELTSEV, M. ROTHACHER, T. KLUEGEL, G. E. STEDMAN & D. L. WILTSHIRE, *Direct measurement of diurnal polar motion by ring laser gyroscopes*, Journal of Geophysical Research **109** B, p. 06405, 2004, an a review article at T. KLÜGEL, W. SCHLÜTER, U. SCHREIBER & M. SCHNEIDER, *Großringlaser zur kontinuierlichen Beobachtung der Erdrotation*, Zeitschrift für Vermessungswesen **130**, pp. 99–108, February 2005. Cited on page 142.

114 R. ANDERSON, H. R. BILGER & G. E. STEDMAN, *The Sagnac-effect: a century of Earth-rotated interferometers*, American Journal of Physics **62**, pp. 975–985, 1994.

See also the clear and extensive paper by G. E. STEDMAN, *Ring laser tests of fundamental physics and geophysics*, Reports on Progress in Physics **60**, pp. 615–688, 1997. Cited on page 144.

115 About the length of the day, see the maia.usno.navy.mil website, or the books by K. LAMBECK, *The Earth's Variable Rotation: Geophysical Causes and Consequences*, Cambridge University Press, 1980, and by W. H. MUNK & G. J. F. MACDONALD, *The Rotation of the Earth*, Cambridge University Press, 1960. For a modern ring laser set-up, see www.wettzell.ifag.de. Cited on pages 145 and 200.

116 H. BUCKA, *Zwei einfache Vorlesungsversuche zum Nachweis der Erddrehung*, Zeitschrift für Physik **126**, pp. 98–105, 1949, and H. BUCKA, *Zwei einfache Vorlesungsversuche zum Nachweis der Erddrehung. II. Teil*, Zeitschrift für Physik **128**, pp. 104–107, 1950. Cited on page 145.

117 One example of data is by C. P. SONETT, E. P. KVALE, A. ZAKHARIAN, M. A. CHAN & T. M. DEMKO, *Late proterozoic and paleozoic tides, retreat of the moon, and rotation of the Earth*, Science **273**, pp. 100–104, 5 July 1996. They deduce from tidal sediment analysis that days were only 18 to 19 hours long in the Proterozoic, i.e., 900 million years ago; they assume that the year was 31 million seconds long from then to today. See also C. P. SONETT

& M. A. CHAN, *Neoproterozoic Earth-Moon dynamics – rework of the 900 MA Big Cottonwood canyon tidal laminae*, Geophysical Research Letters **25**, pp. 539–542, 1998. Another determination was by G. E. WILLIAMS, *Precambrian tidal and glacial clastic deposits: implications for precambrian Earth-Moon dynamics and palaeoclimate*, Sedimentary Geology **120**, pp. 55–74, 1998. Using a geological formation called *tidal rhythmites*, he deduced that about 600 million years ago there were 13 months per year and a day had 22 hours. Cited on page 145.

118 For the story of this combination of history and astronomy see RICHARD STEPHENSON, *Historical Eclispes and Earth's Rotation*, Cambridge University Press, 1996. Cited on page 146.

119 B. F. CHAO, *Earth Rotational Variations excited by geophysical fluids*, IVS 2004 General Meeting proceedings/ pages 38-46. Cited on page 146.

120 On the rotation and history of the Solar System, see S. BRUSH, *Theories of the origin of the solar system 1956-1985*, Reviews of Modern Physics **62**, pp. 43–112, 1990. Cited on page 146.

121 The website hpiers.obspm.fr/eop-pc shows the motion of the Earth's axis over the last ten years. The International Latitude Service founded by Küstner is now part of the International Earth Rotation Service; more information can be found on the www.iers.org website. The latest idea is that two-thirds of the circular component of the polar motion, which in the USA is called 'Chandler wobble' after the person who attributed to himself the discovery by Küstner, is due to fluctuations of the ocean pressure at the bottom of the oceans and one-third is due to pressure changes in the atmosphere of the Earth. This is explained by R. S. GROSS, *The excitation of the Chandler wobble*, Geophysical Physics Letters **27**, pp. 2329–2332, 2000. Cited on page 147.

122 S. B. LAMBERT, C. BIZOUARD & V. DEHANT, *Rapid variations in polar motion during the 2005-2006 winter season*, Geophysical Research Letters **33**, p. L13303, 2006. Cited on page 147.

123 For more information about Alfred Wegener, see the (simple) text by KLAUS ROHRBACH, *Alfred Wegener – Erforscher der wandernden Kontinente*, Verlag Freies Geisteslebeen, 1993; about plate tectonics, see the www.scotese.com website. About earthquakes, see the www.geo.ed.ac.uk/quakexe/quakes and the www.iris.edu/seismon website. See the vulcan.wr.usgs.gov and the www.dartmouth.edu/~volcano websites for information about volcanoes. Cited on page 150.

124 J. JOUZEL & al., *Orbital and millennial Antarctic climate variability over the past 800,000 years*, Science **317**, pp. 793–796, 2007, takes the data from isotope concentrations in ice cores. In contrast, J. D. HAYS, J. IMBRIE & N. J. SHACKLETON, *Variations in the Earth's orbit: pacemaker of the ice ages*, Science **194**, pp. 1121–1132, 1976, confirmed the connection with orbital parameters by literally digging in the mud that covers the ocean floor in certain places. Note that the web is full of information on the ice ages. Just look up 'Milankovitch' in a search engine. Cited on pages 153 and 154.

125 R. HUMPHREYS & J. LARSEN, *The sun's distance above the galactic plane*, Astronomical Journal **110**, pp. 2183–2188, November 1995. Cited on page 153.

126 C. L. BENNET, M. S. TURNER & M. WHITE, *The cosmic rosetta stone*, Physics Today **50**, pp. 32–38, November 1997. Cited on page 155.

127 The website www.geoffreylandis.com/vacuum.html gives a description of what happened. See also the www.geoffreylandis.com/ebullism.html and imagine.gsfc.nasa.gov/docs/ask_astro/answers/970603.html websites. They all give details on the effects of vacuum on humans. Cited on page 161.

128 R. McN. Alexander, *Leg design and jumping technique for humans, other vertebrates and insects*, Philosophical Transactions of the Royal Society in London B **347**, pp. 235–249, 1995. Cited on page 169.

129 J. W. Glasheen & T. A. McMahon, *A hydrodynamic model of locomotion in the basilisk lizard*, Nature **380**, pp. 340–342, For pictures, see also New Scientist, p. 18, 30 March 1996, or Scientific American, pp. 48–49, September 1997, or the website by the author at rjf2.biol.berkeley.edu/Full_Lab/FL_Personnel/J_Glasheen/J_Glasheen.html.

Several shore birds also have the ability to run over water, using the same mechanism. Cited on page 169.

130 A. Fernandez-Nieves & F. J. de las Nieves, *About the propulsion system of a kayak and of Basiliscus basiliscus*, European Journal of Physics **19**, pp. 425–429, 1998. Cited on page 170.

131 Y. S. Song, S. H. Suhr & M. Sitti, *Modeling of the supporting legs for designing biomimetic water strider robot*, Proceedings of the IEEE International Conference on Robotics and Automation, Orlando, USA, 2006. S. H. Suhr, Y. S. Song, S. J. Lee & M. Sitti, *Biologically inspired miniature water strider robot*, Proceedings of the Robotics: Science and Systems I, Boston, USA, 2005. See also the website www.me.cmu.edu/faculty1/sitti/nano/projects/waterstrider. Cited on page 170.

132 J. Iriarte-Díaz, *Differential scaling of locomotor performance in small and large terrestrial mammals*, The Journal of Experimental Biology **205**, pp. 2897–2908, 2002. Cited on pages 171 and 559.

133 M. Wittlinger, R. Wehner & H. Wolf, *The ant odometer: stepping on stilts and stumps*, Science **312**, pp. 1965–1967, 2006. Cited on page 171.

134 P. G. Weyand, D. B. Sternlight, M. J. Bellizzi & S. Wright, *Faster top running speeds are achieved with greater ground forces not more rapid leg movements*, Journal of Applied Physiology **89**, pp. 1991–1999, 2000. Cited on page 171.

135 The material on the shadow discussion is from the book by Robert M. Pryce, *Cook and Peary*, Stackpole Books, 1997. See also the details of Peary's forgeries in Wally Herbert, *The Noose of Laurels*, Doubleday 1989. The sad story of Robert Peary is also told in the centenary number of *National Geographic*, September 1988. Since the National Geographic Society had financed Peary in his attempt and had supported him until the US Congress had declared him the first man at the Pole, the (partial) retraction is noteworthy. (The magazine then changed its mind again later on, to sell more copies, and now again claims that Peary reached the North Pole.) By the way, the photographs of Cook, who claimed to have been at the North Pole even before Peary, have the same problem with the shadow length. Both men have a history of cheating about their 'exploits'. As a result, the first man at the North Pole was probably Roald Amundsen, who arrived there a few years later, and who was also the first man at the South Pole. Cited on page 135.

136 The story is told in M. Nauenberg, *Hooke, orbital motion, and Newton's Principia*, American Journal of Physics **62**, 1994, pp. 331–350. Cited on page 178.

137 More details are given by D. Rawlins, in *Doubling your sunsets or how anyone can measure the Earth's size with wristwatch and meter stick*, American Journal of Physics **47**, 1979, pp. 126–128. Another simple measurement of the Earth radius, using only a sextant, is given by R. O'Keefe & B. Ghavimi-Alagha, in *The World Trade Center and the distance to the world's center*, American Journal of Physics **60**, pp. 183–185, 1992. Cited on page 179.

138 More details on astronomical distance measurements can be found in the beautiful little book by A. van Helden, *Measuring the Universe*, University of Chicago Press, 1985, and in NIGEL HENBEST & HEATHER COOPER, *The Guide to the Galaxy*, Cambridge University Press, 1994. Cited on page 179.

139 A lot of details can be found in M. JAMMER, *Concepts of Mass in Classical and Modern Physics*, reprinted by Dover, 1997, and in *Concepts of Force, a Study in the Foundations of Mechanics*, Harvard University Press, 1957. These eclectic and thoroughly researched texts provide numerous details and explain various philosophical viewpoints, but lack clear statements and conclusions on the accurate description of nature; thus are not of help on fundamental issues.

Jean Buridan (*c.* 1295 to *c.* 1366) criticizes the distinction of sublunar and translunar motion in his book *De Caelo*, one of his numerous works. Cited on page 180.

140 D. TOPPER & D. E. VINCENT, *An analysis of Newton's projectile diagram*, European Journal of Physics 20, pp. 59–66, 1999. Cited on page 180.

141 The absurd story of the metre is told in the historical novel by KEN ALDER, *The Measure of All Things : The Seven-Year Odyssey and Hidden Error that Transformed the World*, The Free Press, 2003. Cited on page 183.

142 H. CAVENDISH, *Experiments to determine the density of the Earth*, Philosophical Transactions of the Royal Society 88, pp. 469–526, 1798. In fact, the first value of the gravitational constant G found in the literature is only from 1873, by Marie-Alfred Cornu and Jean-Baptistin Baille, who used an improved version of Cavendish's method. Cited on page 185.

143 About the measurement of spatial dimensions via gravity – and the failure to find any hint for a number different from three – see the review by E. G. ADELBERGER, B. R. HECKEL & A. E. NELSON, *Tests of the gravitational inverse-square law*, Annual Review of Nuclear and Particle Science 53, pp. 77–121, 2003, also arxiv.org/abs/hep-ph/0307284, or the review by J. A. HEWETT & M. SPIROPULU, *Particle physics probes of extra spacetime dimensions*, Annual Review of Nuclear and Particle Science 52, pp. 397–424, 2002, arxiv.org/abs/hep-ph/0205106. Cited on page 188.

144 There are many books explaining the origin of the precise shape of the Earth, such as the pocket book S. ANDERS, *Weil die Erde rotiert*, Verlag Harri Deutsch, 1985. Cited on page 188.

145 The shape of the Earth is described most precisely with the World Geodetic System. For a presentation, see the en.wikipedia.org/wiki/World_Geodetic_System and www.dqts.net/wgs84.htm websites. See also the website of the *International Earth Rotation Service* at hpiers.obspm.fr. Cited on page 188.

146 G. HECKMAN & M. VAN HAANDEL, *De vele beweijzen van Kepler's wet over ellipsenbanen: een nieuwe voor 'het Boek'?*, Nederlands tijdschrift voor natuurkunde 73, pp. 366–368, November 2007. Cited on page 177.

147 W. K. HARTMAN, R. J. PHILLIPS & G. J. TAYLOR, editors, *Origin of the Moon*, Lunar and Planetary Institute, 1986. Cited on page 191.

148 If you want to read about the motion of the Moon in all its fascinating details, have a look at MARTIN C. GUTZWILLER, *Moon–Earth–Sun: the oldest three body problem*, Reviews of Modern Physics 70, pp. 589–639, 1998. Cited on page 191.

149 DIETRICH NEUMANN, *Physiologische Uhren von Insekten – Zur Ökophysiologie lunarperiodisch kontrollierter Fortpflanzungszeiten*, Naturwissenschaften 82, pp. 310–320, 1995. Cited on page 192.

150 The origin of the duration of the menstrual cycle is not yet settled; however, there are explanations on how it becomes synchronized with other cycles. For a general explanation see

ARKADY PIKOVSKY, MICHAEL ROSENBLUM & JÜRGEN KURTHS, *Synchronization: A Universal Concept in Nonlinear Science*, Cambridge University Press, 2002. Cited on page 192.

151 J. LASKAR, F. JOUTEL & P. ROBUTEL, *Stability of the Earth's obliquity by the moon*, Nature **361**, pp. 615–617, 1993. However, the question is not completely settled, and other opinions exist. Cited on page 192.

152 NEIL F. COMINS, *What if the Moon Did not Exist? – Voyages to Earths that Might Have Been*, Harper Collins, 1993. Cited on page 192.

153 A recent proposal is M. ĆUK, D. P. HAMILTON, S. J. LOCK & S. T. STEWART, *Tidal evolution of the Moon from a high-obliquity, high-angular-momentum Earth*, Nature **539**, pp. 402–406, 2016. Cited on page 192.

154 M. CONNORS, C. VEILLET, R. BRASSER, P. A. WIEGERT, P. W. CHODAS, S. MIKKOLA & K. A. INNANEN, *Discovery of Earth's quasi-satellite*, Meteoritics & Planetary Science **39**, pp. 1251–1255, 2004, and R. BRASSER, K. A. INNANEN, M. CONNORS, C. VEILLET, P. A. WIEGERT, S. MIKKOLA & P. W. CHODAS, *Transient co-orbital asteroids*, Icarus **171**, pp. 102–109, 2004. See also the orbits frawn in M. CONNORS, C. VEILLET, R. BRASSER, P. A. WIEGERT, P. W. CHODAS, S. MIKKOLA & K. A. INNANEN, *Horseshoe asteroids and quasi-satellites in Earth-like orbits*, Lunar and Planetary Science **35**, p. 1562, 2004,, preprint at www.lpi.usra.edu/meetings/lpsc2004/pdf/1565.pdf. Cited on page 195.

155 P. A. WIEGERT, K. A. INNANEN & S. MIKKOLA, *An asteroidal companion to the Earth*, Nature **387**, pp. 685–686, 12 June 1997, together with the comment on pp. 651–652. Details on the orbit and on the fact that Lagrangian points do not always form equilateral triangles can be found in F. NAMOUNI, A. A. CHRISTOU & C. D. MURRAY, *Coorbital dynamics at large eccentricity and inclination*, Physical Review Letters **83**, pp. 2506–2509, 1999. Cited on page 194.

156 SIMON NEWCOMB, Astronomical Papers of the American Ephemeris **1**, p. 472, 1882. Cited on page 197.

157 For an animation of the tides, have a look at www.jason.oceanobs.com/html/applications/marees/m2_atlantique_fr.html. Cited on page 197.

158 A beautiful introduction is the classic G. FALK & W. RUPPEL, *Mechanik, Relativität, Gravitation – ein Lehrbuch*, Springer Verlag, Dritte Auflage, 1983. Cited on page 197.

159 J. SOLDNER, *Berliner Astronomisches Jahrbuch auf das Jahr 1804*, 1801, p. 161. Cited on page 201.

160 The equality was first tested with precision by R. VON EÖTVÖS, Annalen der Physik & Chemie **59**, p. 354, 1896, and by R. VON EÖTVÖS, V. PEKÁR, E. FEKETE, *Beiträge zum Gesetz der Proportionalität von Trägheit und Gravität*, Annalen der Physik 4, Leipzig **68**, pp. 11–66, 1922. He found agreement to 5 parts in 10^9. More experiments were performed by P. G. ROLL, R. KROTKOW & R. H. DICKE, *The equivalence of inertial and passive gravitational mass*, Annals of Physics (NY) **26**, pp. 442–517, 1964, one of the most interesting and entertaining research articles in experimental physics, and by V. B. BRAGINSKY & V. I. PANOV, Soviet Physics – JETP **34**, pp. 463–466, 1971. Modern results, with errors less than one part in 10^{12}, are by Y. SU & al., *New tests of the universality of free fall*, Physical Review **D50**, pp. 3614–3636, 1994. Several experiments have been proposed to test the equality in space to less than one part in 10^{16}. Cited on page 202.

161 H. EDELMANN, R. NAPIWOTZKI, U. HEBER, N. CHRISTLIEB & D. REIMERS, *HE 0437-5439: an unbound hyper-velocity B-type star*, The Astrophysical Journal **634**, pp. L181–

L184, 2005. Cited on page 210.

162 This is explained for example by D.K. FIRPIĆ & I.V. ANIÇIN, *The planets, after all, may run only in perfect circles – but in the velocity space!*, European Journal of Physics **14**, pp. 255–258, 1993. Cited on pages 210 and 503.

163 See L. HODGES, *Gravitational field strength inside the Earth*, American Journal of Physics **59**, pp. 954–956, 1991. Cited on page 211.

164 The controversial argument is proposed in A. E. CHUBYKALO & S. J. VLAEV, *Theorem on the proportionality of inertial and gravitational masses in classical mechanics*, European Journal of Physics **19**, pp. 1–6, 1998, preprint at arXiv.org/abs/physics/9703031. This paper might be wrong; see the tough comment by B. JANCOVICI, European Journal of Physics **19**, p. 399, 1998, and the reply in arxiv.org/abs/physics/9805003. Cited on page 212.

165 P. MOHAZZABI & M. C. JAMES, *Plumb line and the shape of the Earth*, American Journal of Physics **68**, pp. 1038–1041, 2000. Cited on page 212.

166 From NEIL DE GASSE TYSON, *The Universe Down to Earth*, Columbia University Press, 1994. Cited on page 213.

167 G. D. QUINLAN, *Planet X: a myth exposed*, Nature **363**, pp. 18–19, 1993. Cited on page 213.

168 See en.wikipedia.org/wiki/90377_Sedna. Cited on page 214.

169 See R. MATTHEWS, *Not a snowball's chance ...*, New Scientist 12 July 1997, pp. 24–27. The original claim is by LOUIS A. FRANK, J. B. SIGWARTH & J. D. CRAVEN, *On the influx of small comets into the Earth's upper atmosphere*, parts I and II, Geophysical Research Letters **13**, pp. 303–306, pp. 307–310, 1986. The latest observations have disproved the claim. Cited on page 215.

170 The ray form is beautifully explained by J. EVANS, *The ray form of Newton's law of motion*, American Journal of Physics **61**, pp. 347–350, 1993. Cited on page 216.

171 This is a small example from the beautiful text by MARK P. SILVERMAN, *And Yet It Moves: Strange Systems and Subtle Questions in Physics*, Cambridge University Press, 1993. It is a treasure chest for anybody interested in the details of physics. Cited on page 217.

172 G.-L. LESAGE, *Lucrèce Newtonien*, Nouveaux mémoires de l'Académie Royale des Sciences et Belles Lettres pp. 404–431, 1747, or www3.bbaw.de/bibliothek/digital/struktur/03-nouv/1782/jpg-0600/00000495.htm. See also en.wikipedia.org/wiki/Le_Sage's_theory_of_gravitation. In fact, the first to propose the idea of gravitation as a result of small particles pushing masses around was Nicolas Fatio de Duillier in 1688. Cited on page 218.

173 J. LASKAR, *A numerical experiment on the chaotic behaviour of the solar system*, Nature **338**, pp. 237–238, 1989, and J. LASKAR, *The chaotic motion of the solar system - A numerical estimate of the size of the chaotic zones*, Icarus **88**, pp. 266–291, 1990. The work by Laskar was later expanded by Jack Wisdom, using specially built computers, following only the planets, without taking into account the smaller objects. For more details, see G. J. SUSSMAN & J. WISDOM, *Chaotic Evolution of the Solar System*, Science **257**, pp. 56–62, 1992. Today, such calculations can be performed on your home PC with computer code freely available on the internet. Cited on page 219.

174 B. DUBRULLE & F. GRANER, *Titius-Bode laws in the solar system. 1: Scale invariance explains everything*, Astronomy and Astrophysics **282**, pp. 262–268, 1994, and *Titius-Bode laws in the solar system. 2: Build your own law from disk models*, Astronomy and Astrophysics **282**, pp. 269–-276, 1994. Cited on page 221.

175 M. LECAR, *Bode's Law*, Nature 242, pp. 318–319, 1973, and M. HENON, *A comment on "The resonant structure of the solar system" by A.M. Molchanov*, Icarus 11, pp. 93–94, 1969. Cited on page 221.

176 CASSIUS DIO, *Historia Romana*, c. 220, book 37, 18. For an English translation, see the site penelope.uchicago.edu/Thayer/E/Roman/Texts/Cassius_Dio/37*.html. Cited on page 221.

177 See the beautiful paper A. J. SIMOSON, *Falling down a hole through the Earth*, Mathematics Magazine 77, pp. 171–188, 2004. See also A. J. SIMOSON, *The gravity of Hades*, 75, pp. 335–350, 2002. Cited on pages 223 and 505.

178 M. BEVIS, D. ALSDORF, E. KENDRICK, L. P. FORTES, B. FORSBERG, R. MALLEY & J. BECKER, *Seasonal fluctuations in the mass of the Amazon River system and Earth's elastic response*, Geophysical Research Letters 32, p. L16308, 2005. Cited on page 223.

179 D. HESTENES, M. WELLS & G. SWACKHAMER, *Force concept inventory*, Physics Teacher 30, pp. 141–158, 1982. The authors developed tests to check the understanding of the concept of physical force in students; the work has attracted a lot of attention in the field of physics teaching. Cited on page 229.

180 For a general overview on friction, from physics to economics, architecture and organizational theory, see N. ÅKERMAN, editor, *The Necessity of Friction – Nineteen Essays on a Vital Force*, Springer Verlag, 1993. Cited on page 234.

181 See M. HIRANO, K. SHINJO, R. KANECKO & Y. MURATA, *Observation of superlubricity by scanning tunneling microscopy*, Physical Review Letters 78, pp. 1448–1451, 1997. See also the discussion of their results by SERGE FAYEULLE, *Superlubricity: when friction stops*, Physics World pp. 29–30, May 1997. Cited on page 234.

182 C. DONALD AHRENS, *Meteorology Today: An Introduction to the Weather, Climate, and the Environment*, West Publishing Company, 1991. Cited on page 235.

183 This topic is discussed with lucidity by J. R. MUREIKA, *What really are the best 100 m performances?*, Athletics: Canada's National Track and Field Running Magazine, July 1997. It can also be found as arxiv.org/abs/physics/9705004, together with other papers on similar topics by the same author. Cited on page 235.

184 F. P. BOWDEN & D. TABOR, *The Friction and Lubrication of Solids*, Oxford University Press, Part I, revised edition, 1954, and part II, 1964. Cited on page 236.

185 A powerful book on human violence is JAMES GILLIGAN, *Violence – Our Deadly Epidemic and its Causes*, Grosset/Putnam, 1992. Cited on page 236.

186 The main tests of randomness of number series – among them the gorilla test – can be found in the authoritative paper by G. MARSAGLIA & W. W. TSANG, *Some difficult-to-pass tests of randomness*, Journal of Statistical Software 7, p. 8, 2002. It can also be downloaded from www.jstatsoft.org/v07/i03. Cited on page 239.

187 See the interesting book on the topic by JAROSLAW STRZALKO, JULIUSZ GRABSKI, PRZEMYSLAW PERLIKOWSKI, ANDRZEIJ STEFANSKI & TOMASZ KAPITANIAK, *Dynamics of Gambling: Origin of Randomness in Mechanical Systems*, Springer, 2009, as well as the more recent publications by Kapitaniak. Cited on page 239.

188 For one aspect on free will, see the captivating book by BERT HELLINGER, *Zweierlei Glück*, Carl Auer Systeme Verlag, 1997. The author explains how to live serenely and with the highest possible responsibility for one's actions, by reducing entanglements with the destiny of others. He describes a powerful technique to realise this goal.

A completely different viewpoint is given by AUNG SAN SUU KYI, *Freedom from Fear*, Penguin, 1991. One of the bravest women on Earth, she won the Nobel Peace Price in 1991.

An effective personal technique is presented by Phil Stutz & Barry Michels, *The Tools*, Spiegel & Grau, 2012. Cited on page 242.

189 Henrik Walter, *Neurophilosophie der Willensfreiheit*, Mentis Verlag, Paderborn 1999. Also available in English translation. Cited on page 242.

190 Giuseppe Fumagalli, *Chi l'ha detto?*, Hoepli, 1983. Cited on page 242.

191 See the tutorial on the Peaucellier-Lipkin linkage by D.W. Henderson and D. Taimina found on kmoddl.library.cornell.edu/tutorials/11/index.php. The internet contains many other pages on the topic. Cited on page 243.

192 The beautiful story of the south-pointing carriage is told in Appendix B of James Foster & J. D. Nightingale, *A Short Course in General Relativity*, Springer Verlag, 2nd edition, 1998. Such carriages have existed in China, as told by the great sinologist Joseph Needham, but their construction is unknown. The carriage described by Foster and Nightingale is the one reconstructed in 1947 by George Lancaster, a British engineer. Cited on page 244.

193 T. Van de Kamp, P. Vagovic, T. Baumbach & A. Riedel, *A biological screw in a beetle's leg*, Science 333, p. 52, 2011. Cited on page 244.

194 M. Burrows & G. P. Sutton, *Interacting gears synchronise propulsive leg movements in a jumping insect*, Science 341, pp. 1254–1256, 2013. Cited on page 244.

195 See for example Z. Ghahramani, *Building blocks of movement*, Nature 407, pp. 682–683, 2000. Researchers in robot control are also interested in such topics. Cited on page 244.

196 G. Gutierrez, C. Fehr, A. Calzadilla & D. Figueroa, *Fluid flow up the wall of a spinning egg*, American Journal of Physics 66, pp. 442–445, 1998. Cited on page 244.

197 A historical account is given in Wolfgang Yourgray & Stanley Mandelstam, *Variational Principles in Dynamics and Quantum Theory*, Dover, 1968. Cited on pages 248 and 256.

198 C. G. Gray & E. F. Taylor, *When action is not least*, American Journal of Physics 75, pp. 434–458, 2007. Cited on page 253.

199 Max Päsler, *Prinzipe der Mechanik*, Walter de Gruyter & Co., 1968. Cited on page 254.

200 The relations between possible Lagrangians are explained by Herbert Goldstein, *Classical Mechanics*, 2nd edition, Addison-Wesley, 1980. Cited on page 255.

201 The Hemingway statement is quoted by Marlene Dietrich in Aaron E. Hotchner, *Papa Hemingway*, Random House, 1966, in part 1, chapter 1. Cited on page 256.

202 C. G. Gray, G. Karl & V. A. Novikov, *From Maupertius to Schrödinger. Quantization of classical variational principles*, American Journal of Physics 67, pp. 959–961, 1999. Cited on page 257.

203 J. A. Moore, *An innovation in physics instruction for nonscience majors*, American Journal of Physics 46, pp. 607–612, 1978. Cited on page 257.

204 See e.g. Alan P. Boss, *Extrasolar planets*, Physics Today 49, pp. 32–38. September 1996. The most recent information can be found at the 'Extrasolar Planet Encyclopaedia' maintained at www.obspm.fr/planets by Jean Schneider at the Observatoire de Paris. Cited on page 261.

205 A good review article is by David W. Hughes, *Comets and Asteroids*, Contemporary Physics 35, pp. 75–93, 1994. Cited on page 261.

206 G. B. West, J. H. Brown & B. J. Enquist, *A general model for the origin of allometric scaling laws in biology*, Science 276, pp. 122–126, 4 April 1997, with a comment on page 34 of the same issue. The rules governing branching properties of blood vessels, of lymph

systems and of vessel systems in plants are explained. For more about plants, see also the paper G. B. WEST, J. H. BROWN & B. J. ENQUIST, *A general model for the structure and allometry of plant vascular systems*, Nature **400**, pp. 664–667, 1999. Cited on page 263.

207 J. R. BANAVAR, A. MARTIN & A. RINALDO, *Size and form in efficient transportation networks*, Nature **399**, pp. 130–132, 1999. Cited on page 263.

208 N. MOREIRA, *New striders - new humanoids with efficient gaits change the robotics landscape*, Science News Online 6th of August, 2005. Cited on page 264.

209 WERNER HEISENBERG, *Der Teil und das Ganze*, Piper, 1969. Cited on page 266.

210 See the clear presenttion by E. H. LOCKWOOD & R. H. MACMILLAN, *Geometric Symmetry*, Cambridge University Press, 1978. Cited on page 266.

211 JOHN MANSLEY ROBINSON, *An Introduction to Early Greek Philosophy*, Houghton Muffin 1968, chapter 5. Cited on page 270.

212 See e.g. B. BOWER, *A child's theory of mind*, Science News **144**, pp. 40–41. Cited on page 271.

213 The most beautiful book on this topic is the text by BRANKO GRÜNBAUM & G. C. SHEPHARD, *Tilings and Patterns*, W.H. Freeman and Company, New York, 1987. It has been translated into several languages and republished several times. Cited on page 273.

214 About tensors and ellipsoids in three-dimensional space, see mysite.du.edu/~jcalvert/phys/ellipso.htm. In four-dimensional space-time, tensors are more abstract to comprehend. With emphasis on their applications in relativity, such tensors are explained in R. FROSCH, *Four-tensors, the mother tongue of classical physics*, vdf Hochschulverlag, 2006, partly available on books.google.com. Cited on page 278.

215 U. NIEDERER, *The maximal kinematical invariance group of the free Schrödinger equation*, Helvetica Physica Acta **45**, pp. 802–810, 1972. See also the introduction by O. JAHN & V. V. SREEDHAR, *The maximal invariance group of Newton's equations for a free point particle*, arxiv.org/abs/math-ph/0102011. Cited on page 279.

216 The story is told in the interesting biography of Einstein by A. PAIS, *'Subtle is the Lord...' – The Science and the Life of Albert Einstein*, Oxford University Press, 1982. Cited on page 280.

217 W. ZÜRN & R. WIDMER-SCHNIDRIG, *Globale Eigenschwingungen der Erde*, Physik Journal **1**, pp. 49–55, 2002. Cited on page 290.

218 N. GAUTHIER, *What happens to energy and momentum when two oppositely-moving wave pulses overlap?*, American Journal of Physics **71**, pp. 787–790, 2003. Cited on page 300.

219 An informative and modern summary of present research about the ear and the details of its function is www.physicsweb.org/article/world/15/5/8. Cited on page 303.

220 A renowned expert of the physics of singing is Ingo Titze. Among his many books and papers is the popular introduction I. R. TITZE, *The human instrument*, Scientific American pp. 94–101, January 2008. Several of his books, papers and presentation are free to download on the website www.ncvs.org of the National Center of Voice & Speech. They are valuable to everybody who has a passion for singing and the human voice. See also the article and sound clips at www.scientificamerican.com/article.cfm?id=sound-clips-human-instrument. An interesting paper is also M. KOB & al., *Analysing and understanding the singing voice: recent progress and open questions*, Current Bioinformatics **6**, pp. 362–374, 2011. Cited on page 308.

221 S. ADACHI, *Principles of sound production in wind instruments*, Acoustical Science and Technology **25**, pp. 400–404, 2004. Cited on page 309.

222 The literature on tones and their effects is vast. For example, people have explored the differences and effects of various intonations in great detail. Several websites, such as bellsouthpwp.net/j/d/jdelaub/jstudio.htm, allow listening to music played with different intonations. People have even researched whether animals use just or chromatic intonation. (See, for example, K. LEUTWYLER, *Exploring the musical brain*, Scientific American January 2001.) There are also studies of the effects of low frequencies, of beat notes, and of many other effects on humans. However, many studies mix serious and non-serious arguments. It is easy to get lost in them. Cited on page 310.

223 M. FATEMI, P. L. OGBURN & J. F. GREENLEAF, *Fetal stimulation by pulsed diagnostic ultrasound*, Journal of Ultrasound in Medicine 20, pp. 883–889, 2001. See also M. FATEMI, A. ALIZAD & J. F. GREENLEAF, *Characteristics of the audio sound generated by ultrasound imaging systems*, Journal of the Acoustical Society of America 117, pp. 1448–1455, 2005. Cited on page 313.

224 I know a female gynecologist who, during her own pregnancy, imaged her child *every day* with her ultrasound machine. The baby was born with strong hearing problems that did not go away. Cited on page 313.

225 R. MOLE, *Possible hazards of imaging and Doppler ultrasound in obstetrics*, Birth Issues in Perinatal Care 13, pp. 29–37, 2007. Cited on page 314.

226 A. L. HODGKIN & A. F. HUXLEY, *A quantitative description of membrane current and its application to conduction and excitation in nerve*, Journal of Physiology 117, pp. 500–544, 1952. This famous paper of theoretical biology earned the authors the Nobel Prize in Medicine in 1963. Cited on page 315.

227 T. FILIPPOV, *The Versatile Soliton*, Springer Verlag, 2000. See also J. S. RUSSEL, *Report of the Fourteenth Meeting of the British Association for the Advancement of Science*, Murray, London, 1844, pp. 311–390. Cited on pages 316 and 318.

228 R. S. WARD, *Solitons and other extended field configurations*, preprint at arxiv.org/abs/hep-th/0505135. Cited on page 318.

229 D. B. BAHR, W. T. PFEFFER & R. C. BROWNING, *The surprising motion of ski moguls*, Physics Today 62, pp. 68–69, November 2009. Cited on page 319.

230 N. J. ZABUSKY & M. D. KRUSKAL, *Interaction of solitons in a collisionless plasma and the recurrence of initial states*, Physical Review Letters 15, pp. 240–243, 1965. Cited on page 317.

231 O. MUSKENS, *De kortste knal ter wereld*, Nederlands tijdschrift voor natuurkunde pp. 70–73, 2005. Cited on page 318.

232 E. HELLER, *Freak waves: just bad luck, or avoidable?*, Europhysics News pp. 159–161, September/October 2005, downloadable at www.europhysicsnews.org. Cited on page 320.

233 See the beautiful article by D. AARTS, M. SCHMIDT & H. LEKKERKERKER, *Directe visuele waarneming van thermische capillaire golven*, Nederlands tijdschrift voor natuurkunde 70, pp. 216–218, 2004. Cited on page 321.

234 For more about the ocean sound channel, see the novel by TOM CLANCY, *The Hunt for Red October*. See also the physics script by R. A. MULLER, *Government secrets of the oceans, atmosphere, and UFOs*, web.archive.org/web/*/muller.lbl.gov/teaching/Physics10/chapters/9-SecretsofUFOs.html 2001. Cited on page 322.

235 B. WILSON, R. S. BATTY & L. M. DILL, *Pacific and Atlantic herring produce burst pulse sounds*, Biology Letters 271, number S3, 7 February 2004. Cited on page 323.

236 A. CHABCHOUB & M. FINK, *Time-reversal generation of rogue Wwaves*, Physical Review Letters 112, p. 124101, 2014, preprint at arxiv.org/abs/1311.2990. See also the cited references. Cited on page 324.

237 See for example the article by G. Fritsch, *Infraschall*, Physik in unserer Zeit **13**, pp. 104–110, 1982. Cited on page 325.

238 Wavelet transformations were developed by the French mathematicians Alex Grossmann, Jean Morlet and Thierry Paul. The basic paper is A. Grossmann, J. Morlet & T. Paul, *Integral transforms associated to square integrable representations*, Journal of Mathematical Physics **26**, pp. 2473–2479, 1985. For a modern introduction, see Stéphane Mallat, *A Wavelet Tour of Signal Processing*, Academic Press, 1999. Cited on page 326.

239 P. Manogg, *Knall und Superknall beim Überschallflug*, Der mathematische und naturwissenschaftliche Unterricht **35**, pp. 26–33, 1982, my physics teacher in secondary school. Cited on page 326.

240 See the excellent introduction by L. Ellerbreok & L. van den Hoorn, *In het kielzog van Kelvin*, Nederlands tijdschrift voor natuurkunde **73**, pp. 310–313, 2007. About exceptions to the Kelvin angle, see www.graingerdesigns.net/oshunpro/design-technology/wave-cancellation. Cited on page 327.

241 Jay Ingram, *The Velocity of Honey - And More Science of Everyday Life*, Viking, 2003. See also W. W. L. Au & J. A. Simmons, *Echolocation in dolphins and bats*, Physics Today **60**, pp. 40–45, 2007. Cited on page 328.

242 M. Boiti, J.-P. Leon, L. Martina & F. Pempinelli, *Scattering of localized solitons in the plane*, Physics Letters A **132**, pp. 432–439, 1988, A. S. Fokas & P. M. Santini, *Coherent structures in multidimensions*, Physics Review Letters **63**, pp. 1329–1333, 1989, J. Hietarinta & R. Hirota, *Multidromion solutions to the Davey–Stewartson equation*, Physics Letters A **145**, pp. 237–244, 1990. Cited on page 328.

243 For some of this fascinating research, see J. L. Hammack, D. M. Henderson & H. Segur, *Progressive waves with persistent two-dimensional surface patterns in deep water*, Journal of Fluid Mechanics **532**, pp. 1–52, 2005. For a beautiful photograph of crossing cnoidal waves, see A. R. Osborne, *Hyperfast Modeling of Shallow-Water Waves: The KdV and KP Equations*, International Geophysics **97**, pp. 821–856, 2010. See also en.wikipedia.org/wiki/Waves_and_shallow_water, en.wikipedia.org/wiki/Cnoidal_wave and en.wikipedia.org/wiki/Tidal_bore for a first impression. Cited on page 329.

244 The sound frequency change with bottle volume is explained on hyperphysics.phy-astr.gsu.edu/Hbase/Waves/cavity.html. Cited on page 329.

245 A passionate introduction is Neville H. Fletcher & Thomas D. Rossing, *The Physics of Musical Instruments*, second edition, Springer 2000. Cited on page 330.

246 M. Ausloos & D. H. Berman, *Multivariate Weierstrass–Mandelbrot function*, Proceedings of the Royal Society in London A **400**, pp. 331–350, 1985. Cited on page 333.

247 Catechism of the Catholic Church, Part Two, Section Two, Chapter One, Article 3, statements 1376, 1377 and 1413, found at www.vatican.va/archive/ENG0015/__P41.HTM or www.vatican.va/archive/ITA0014/__P40.HTM with their explanations on www.vatican.va/archive/compendium_ccc/documents/archive_2005_compendium-ccc_en.html and www.vatican.va/archive/compendium_ccc/documents/archive_2005_compendium-ccc_it.html. Cited on page 336.

248 The original text of the 1633 conviction of Galileo can be found on it.wikisource.org/wiki/Sentenza_di_condanna_di_Galileo_Galilei. Cited on page 336.

249 The retraction that Galileo was forced to sign in 1633 can be found on it.wikisource.org/wiki/Abiura_di_Galileo_Galilei. Cited on page 336.

250 M. Artigas, *Un nuovo documento sul caso Galileo: EE 291*, Acta Philosophica **10**, pp. 199–214, 2001. Cited on page 337.

251 Most of these points are made, directly or indirectly, in the book by ANNIBALE FANTOLI, *Galileo: For Copernicanism and for the Church*, Vatican Observatory Publications, second edition, 1996, and by George Coyne, director of the Vatican observatory, in his speeches and publications, for example in G. COYNE, *Galileo: for Copernicanism and for the church*, Zwoje 3/36, 2003, found at www.zwoje-scrolls.com/zwoje36/text05p.htm. Cited on page 337.

252 THOMAS A. MCMAHON & JOHN T. BONNER, *On Size and Life*, Scientific American/Freeman, 1983. Another book by John Bonner, who won the Nobel Prize in Biology, is JOHN T. BONNER, *Why Size Matters: From Bacteria to Blue Whales*, Princeton University Press, 2011. Cited on page 337.

253 G. W. KOCH, S. C. SILLETT, G. M. JENNINGS & S. D. DAVIS, *The limits to tree height*, Nature 428, pp. 851–854, 2004. Cited on page 338.

254 A simple article explaining the tallness of trees is A. MINEYEV, *Trees worthy of Paul Bunyan*, Quantum pp. 4–10, January–February 1994. (Paul Bunyan is a mythical giant lumberjack who is the hero of the early frontier pioneers in the United States.) Note that the transport of liquids in trees sets no limits on their height, since water is pumped up along tree stems (except in spring, when it is pumped up from the roots) by evaporation from the leaves. This works almost without limits because water columns, when nucleation is carefully avoided, can be put under tensile stresses of over 100 bar, corresponding to 1000 m. See also P. NOBEL, *Plant Physiology*, Academic Press, 2nd Edition, 1999. An artificial tree – though extremely small – using the same mechanism was built and studied by T. D. WHEELER & A. D. STROOCK, *The transpiration of water at negative pressures in a synthetic tree*, Nature 455, pp. 208–212, 2008. See also N. M. HOLBROOK & M. A. ZWIENIECKI, *Transporting water to the top of trees*, Physics Today pp. 76–77, 2008. Cited on pages 338 and 359.

255 Such information can be taken from the excellent overview article by M. F. ASHBY, *On the engineering properties of materials*, Acta Metallurgica 37, pp. 1273–1293, 1989. The article explains the various general criteria which determine the selection of materials, and gives numerous tables to guide the selection. Cited on page 338.

256 See the beautiful paper by S. E. VIRGO, *Loschmidt's number*, Science Progress 27, pp. 634–649, 1933. It is also freely available in HTML format on the internet. Cited on pages 340 and 342.

257 See the delightful paper by PETER PESIC, *Estimating Avogadro's number from skylight and airlight*, European Journal of Physics 26, pp. 183–187, 2005. The mistaken statement that the blue colour is due to density fluctuations is dispelled in C. F. BOHREN & A. B. FRASER, *Color of the Sky*, Physics Teacher 238, pp. 267–272, 1985. It also explains that the variations of the sky colour, like the colour of milk, are due to multiple scattering. Cited on pages 341 and 513.

258 For a photograph of a single barium *atom* – named Astrid – see HANS DEHMELT, *Experiments with an isolated subatomic particle at rest*, Reviews of Modern Physics 62, pp. 525–530, 1990. For another photograph of a barium *ion*, see W. NEUHAUSER, M. HOHENSTATT, P. E. TOSCHEK & H. DEHMELT, *Localized visible Ba^+ mono-ion oscillator*, Physical Review A 22, pp. 1137–1140, 1980. See also the photograph on page 344. Cited on page 344.

259 Holograms of atoms were first produced by HANS-WERNER FINK & al., *Atomic resolution in lens-less low-energy electron holography*, Physical Review Letters 67, pp. 1543–1546, 1991. Cited on page 344.

260 A single-atom laser was built in 1994 by K. AN, J. J. CHILDS, R. R. DASARI & M. S. FELD, *Microlaser: a laser with one atom in an optical resonator*, Physical Review Letters **73**, p. 3375, 1994. Cited on page 344.

261 The photograph on the left of Figure 240 on page 344 is the first image that showed subatomic structures (visible as shadows on the atoms). It was published by F. J. GIESSIBL, S. HEMBACHER, H. BIELEFELDT & J. MANNHART, *Subatomic features on the silicon (111)-(7x7) surface observed by atomic force microscopy*, Science **289**, pp. 422 – 425, 2000. Cited on page 344.

262 See for example C. SCHILLER, A. A. KOOMANS, T. L. VAN ROOY, C. SCHÖNENBERGER & H. B. ELSWIJK, *Decapitation of tungsten field emitter tips during sputter sharpening*, Surface Science Letters **339**, pp. L925–L930, 1996. Cited on page 344.

263 U. WEIERSTALL & J. C. H. SPENCE, *An STM with time-of-flight analyzer for atomic species identification*, MSA 2000, Philadelphia, Microscopy and Microanalysis **6**, Supplement 2, p. 718, 2000. Cited on page 345.

264 P. KREHL, S. ENGEMANN & D. SCHWENKEL, *The puzzle of whip cracking – uncovered by a correlation of whip-tip kinematics with shock wave emission*, Shock Waves **8**, pp. 1–9, 1998. The authors used high-speed cameras to study the motion of the whip. A new aspect has been added by A. GORIELY & T. MCMILLEN, *Shape of a cracking whip*, Physical Review Letters **88**, p. 244301, 2002. This article focuses on the tapered shape of the whip. However, the neglection of the tuft – a piece at the end of the whip which is required to make it crack – in the latter paper shows that there is more to be discovered still. Cited on page 349.

265 Z. SHENG & K. YAMAFUJI, *Realization of a Human Riding a Unicycle by a Robot*, Proceedings of the 1995 IEEE International Conference on Robotics and Automation, Vol. 2, pp. 1319–1326, 1995. Cited on page 349.

266 On human unicycling, see JACK WILEY, *The Complete Book of Unicycling*, Lodi, 1984, and SEBASTIAN HOEHER, *Einradfahren und die Physik*, Reinbeck, 1991. Cited on page 349.

267 W. THOMSON, *Lecture to the Royal Society of Edinburgh*, 18 February 1867, Proceedings of the Royal Society in Edinborough **6**, p. 94, 1869. Cited on page 350.

268 S. T. THORODDSEN & A. Q. SHEN, *Granular jets*, Physics of Fluids **13**, pp. 4–6, 2001, and A. Q. SHEN & S. T. THORODDSEN, *Granular jetting*, Physics of Fluids **14**, p. S3, 2002, Cited on page 350.

269 M. J. HANCOCK & J. W. M. BUSH, *Fluid pipes*, Journal of Fluid Mechanics **466**, pp. 285–304, 2002. A. E. HOSOI & J. W. M. BUSH, *Evaporative instabilities in climbing films*, Journal of Fluid Mechanics **442**, pp. 217–239, 2001. J. W. M. BUSH & A. E. HASHA, *On the collision of laminar jets: fluid chains and fishbones*, Journal of Fluid Mechanics **511**, pp. 285–310, 2004. Cited on page 354.

270 The present record for negative pressure in water was achieved by Q. ZHENG, D. J. DURBEN, G. H. WOLF & C. A. ANGELL, *Liquids at large negative pressures: water at the homogeneous nucleation limit*, Science **254**, pp. 829–832, 1991. Cited on page 359.

271 H. MARIS & S. BALIBAR, *Negative pressures and cavitation in liquid helium*, Physics Today **53**, pp. 29–34, 2000. Sebastien Balibar has also written several popular books that are presented at his website www.lps.ens.fr/~balibar. Cited on page 359.

272 The present state of our understanding of turbulence is described by G. FALKOVICH & K. P. SREENIVASAN, *Lessons from hydrodynamic turbulence*, Physics Today **59**, pp. 43–49, 2006. Cited on page 363.

273 K. WELTNER, *A comparison of explanations of aerodynamical lifting force*, American Journal of Physics **55**, pp. 50–54, 1987, K. WELTNER, *Aerodynamic lifting force*, The Physics

Teacher 28, pp. 78–82, 1990. See also the user.uni-frankfurt.de/~weltner/Flight/PHYSIC4.htm and the www.av8n.com/how/htm/airfoils.html websites. Cited on page 363.

274 L. LANOTTE, J. MAUER, S. MENDEZ, D. A. FEDOSOV, J.-M. FROMENTAL, V. CLAVERIA, F. NICOUD, G. GOMPPER & M. ABKARIAN, *Red cells' dynamic morphologies govern blood shear thinning under microcirculatory flow conditions*, Proceedings of the National Academy of Sciences of the United States of America 2016, preprint at arxiv.org/abs/1608.03730. Cited on page 368.

275 S. GEKLE, I. R. PETERS, J. M. GORDILLO, D. VAN DER MEER & D. LOHSE, *Supersonic air flow due to solid-liquid impact*, Physical Review Letters 104, p. 024501, 2010. Films of the effect can be found at physics.aps.org/articles/v3/4. Cited on page 370.

276 See the beautiful book by RAINER F. FOELIX, *Biologie der Spinnen*, Thieme Verlag, 1996, also available in an even newer edition in English as RAINER F. FOELIX, *Biology of Spiders*, Oxford University Press, third edition, 2011. Special fora dedicated only to spiders can be found on the internet. Cited on page 370.

277 See the website www.esa.int/esaCP/SEMER89U7TG_index_0.html. Cited on page 370.

278 For a fascinating account of the passion and the techniques of apnoea diving, see UMBERTO PELIZZARI, *L'Homme et la mer*, Flammarion, 1994. Palizzari cites and explains the saying by Enzo Maiorca: 'The first breath you take when you come back to the surface is like the first breath with which you enter life.' Cited on page 371.

279 LYDÉRIC BOCQUET, *The physics of stone skipping*, American Journal of Physics 17, pp. 150–155, 2003. The present recod holder is Kurt Steiner, with 40 skips. See pastoneskipping.com/steiner.htm and www.stoneskipping.com. The site www.yeeha.net/nassa/guin/g2.html is by the a previous world record holder, Jerdome Coleman-McGhee. Cited on page 374.

280 S. F. KISTLER & L. E. SCRIVEN, *The teapot effect: sheetforming flows with deflection, wetting, and hysteresis*, Journal of Fluid Mechanics 263, pp. 19–62, 1994. Cited on page 376.

281 J. WALKER, *Boiling and the Leidenfrost effect*, a chapter from DAVID HALLIDAY, ROBERT RESNICK & JEARL WALKER, *Fundamentals of Physics*, Wiley, 2007. The chapter can also be found on the internet as pdf file. Cited on page 377.

282 E. HOLLANDER, *Over trechters en zo ...*, Nederlands tijdschrift voor natuurkunde 68, p. 303, 2002. Cited on page 378.

283 S. DORBOLO, H. CAPS & N. VANDEWALLE, *Fluid instabilities in the birth and death of antibubbles*, New Journal of Physics 5, p. 161, 2003. Cited on page 379.

284 T. T. LIM, *A note on the leapfrogging between two coaxial vortex rings at low Reynolds numbers*, Physics of Fluids 9, pp. 239–241, 1997. Cited on page 380.

285 P. AUSSILLOUS & D. QUÉRÉ, *Properties of liquid marbles*, Proc. Roy. Soc. London 462, pp. 973–999, 2006, and references therein. Cited on page 380.

286 Thermostatics and thermodynamics are difficult to learn also because the fields were not discovered in a systematic way. See C. TRUESDELL, *The Tragicomical History of Thermodynamics 1822–1854*, Springer Verlag, 1980. An excellent advanced textbook on thermostatics and thermodynamics is LINDA REICHL, *A Modern Course in Statistical Physics*, Wiley, 2nd edition, 1998. Cited on page 383.

287 Gas expansion was the main method used for the definition of the official temperature scale. Only in 1990 were other methods introduced officially, such as total radiation thermometry (in the range 140 K to 373 K), noise thermometry (2 K to 4 K and 900 K to 1235 K), acoustical thermometry (around 303 K), magnetic thermometry (0.5 K to 2.6 K) and optical

radiation thermometry (above 730 K). Radiation thermometry is still the central method in the range from about 3 K to about 1000 K. This is explained in detail in R. L. RUSBY, R. P. HUDSON, M. DURIEUX, J. F. SCHOOLEY, P. P. M. STEUR & C. A. SWENSON, *The basis of the ITS-90*, Metrologia **28**, pp. 9–18, 1991. On the water boiling point see also Ref. 315. Cited on pages 385, 549, and 553.

288 Other methods to rig lottery draws made use of balls of different mass or of balls that are more polished. One example of such a scam was uncovered in 1999. Cited on page 384.

289 See for example the captivating text by GINO SEGRÈ, *A Matter of Degrees: What Temperature Reveals About the Past and Future of Our Species, Planet and Universe*, Viking, New York, 2002. Cited on page 385.

290 D. KARSTÄDT, F. PINNO, K.-P. MÖLLMANN & M. VOLLMER, *Anschauliche Wärmelehre im Unterricht: ein Beitrag zur Visualisierung thermischer Vorgänge*, Praxis der Naturwissenschaften Physik 5-48, pp. 24–31, 1999, K.-P. MÖLLMANN & M. VOLLMER, *Eine etwas andere, physikalische Sehweise - Visualisierung von Energieumwandlungen und Strahlungsphysik für die (Hochschul-)lehre*, Physikalische Blätter **56**, pp. 65–69, 2000, D. KARSTÄDT, K.-P. MÖLLMANN, F. PINNO & M. VOLLMER, *There is more to see than eyes can detect: visualization of energy transfer processes and the laws of radiation for physics education*, The Physics Teacher **39**, pp. 371–376, 2001, K.-P. MÖLLMANN & M. VOLLMER, *Infrared thermal imaging as a tool in university physics education*, European Journal of Physics **28**, pp. S37–S50, 2007. Cited on page 387.

291 See for example the article by H. PRESTON-THOMAS, *The international temperature scale of 1990 (ITS-90)*, Metrologia **27**, pp. 3–10, 1990, and the errata H. PRESTON-THOMAS, *The international temperature scale of 1990 (ITS-90)*, Metrologia **27**, p. 107, 1990, Cited on page 391.

292 For an overview, see CHRISTIAN ENSS & SIEGFRIED HUNKLINGER, *Low-Temperature Physics*, Springer, 2005. Cited on page 391.

293 The famous paper on Brownian motion which contributed so much to Einstein's fame is A. EINSTEIN, *Über die von der molekularkinetischen Theorie der Wärme geforderte Bewegung von in ruhenden Flüssigkeiten suspendierten Teilchen*, Annalen der Physik **17**, pp. 549–560, 1905. In the following years, Einstein wrote a series of further papers elaborating on this topic. For example, he published his 1905 Ph.D. thesis as A. EINSTEIN, *Eine neue Bestimmung der Moleküldimensionen*, Annalen der Physik **19**, pp. 289–306, 1906, and he corrected a small mistake in A. EINSTEIN, *Berichtigung zu meiner Arbeit: 'Eine neue Bestimmung der Moleküldimensionen'*, Annalen der Physik **34**, pp. 591–592, 1911, where, using new data, he found the value $6.6 \cdot 10^{23}$ for Avogadro's number. However, five years before Smoluchowski and Einstein, a much more practically-minded man had made the *same* calculations, but in a different domain: the mathematician Louis Bachelier did so in his PhD about stock options; this young financial analyst was thus smarter than Einstein. Cited on page 392.

294 The first experimental confirmation of the prediction was performed by J. PERRIN, Comptes Rendus de l'Académie des Sciences **147**, pp. 475–476, and pp. 530–532, 1908. He masterfully sums up the whole discussion in JEAN PERRIN, *Les atomes*, Librarie Félix Alcan, Paris, 1913. Cited on page 394.

295 PIERRE GASPARD & al., *Experimental evidence for microscopic chaos*, Nature **394**, p. 865, 27 August 1998. Cited on page 394.

296 An excellent introduction into the physics of heat is the book by LINDA REICHL, *A Modern Course in Statistical Physics*, Wiley, 2nd edition, 1998. Cited on page 395.

297 On the bicycle speed record, see the website fredrompelberg.com. It also shows details of the bicycle he used. On the drag effect of a motorbike behind a bicycle, see B. BLOCKEN,

Y. Toparlar & T. Andrianne, *Aerodynamic benefit for a cyclist by a following motorcycle*, Journal of Wind Engineering & Industrial Aerodynamics 2016, available for free download at www.urbanphysics.net. Cited on page 381.

298 F. Herrmann, *Mengenartige Größen im Physikunterricht*, Physikalische Blätter **54**, pp. 830–832, September 1998. See also his lecture notes on general introductory physics on the website www.physikdidaktik.uni-karlsruhe.de/skripten. Cited on pages 232, 354, and 395.

299 These points are made clearly and forcibly, as is his style, by N.G. van Kampen, *Entropie*, Nederlands tijdschrift voor natuurkunde **62**, pp. 395–396, 3 December 1996. Cited on page 398.

300 This is a disappointing result of all efforts so far, as Grégoire Nicolis always stresses in his university courses. Seth Lloyd has compiled a list of 31 proposed definitions of complexity, containing among others, fractal dimension, grammatical complexity, computational complexity, thermodynamic depth. See, for example, a short summary in Scientific American p. 77, June 1995. Cited on page 398.

301 Minimal entropy is discussed by L. Szilard, *Über die Entropieverminderung in einem thermodynamischen System bei Eingriffen intelligenter Wesen*, Zeitschrift für Physik **53**, pp. 840–856, 1929. This classic paper can also be found in English translation in his collected works. Cited on page 399.

302 G. Cohen-Tannoudji, *Les constantes universelles*, Pluriel, Hachette, 1998. See also L. Brillouin, *Science and Information Theory*, Academic Press, 1962. Cited on page 399.

303 H.W. Zimmermann, *Particle entropies and entropy quanta IV: the ideal gas, the second law of thermodynamics, and the P-t uncertainty relation*, Zeitschrift für physikalische Chemie **217**, pp. 55–78, 2003, and H.W. Zimmermann, *Particle entropies and entropy quanta V: the P-t uncertainty relation*, Zeitschrift für physikalische Chemie **217**, pp. 1097–1108, 2003. Cited on pages 399 and 400.

304 See for example A.E. Shalyt-Margolin & A.Ya. Tregubovich, *Generalized uncertainty relation in thermodynamics*, arxiv.org/abs/gr-qc/0307018, or J. Uffink & J. van Lith-van Dis, *Thermodynamic uncertainty relations*, Foundations of Physics **29**, p. 655, 1999. Cited on page 399.

305 B. Lavenda, *Statistical Physics: A Probabilistic Approach*, Wiley-Interscience, 1991. Cited on pages 399 and 400.

306 The quote given is found in the introduction by George Wald to the text by Lawrence J. Henderson, *The Fitness of the Environment*, Macmillan, New York, 1913, reprinted 1958. Cited on page 400.

307 A fascinating introduction to chemistry is the text by John Emsley, *Molecules at an Exhibition*, Oxford University Press, 1998. Cited on page 401.

308 B. Polster, *What is the best way to lace your shoes?*, Nature **420**, p. 476, 5 December 2002. Cited on page 402.

309 L. Boltzmann, *Über die mechanische Bedeutung des zweiten Hauptsatzes der Wärmetheorie*, Sitzungsberichte der königlichen Akademie der Wissenschaften in Wien **53**, pp. 155–220, 1866. Cited on page 403.

310 See for example, the web page www.snopes.com/science/cricket.asp. Cited on page 406.

311 H. de Lang, *Moleculaire gastronomie*, Nederlands tijdschrift voor natuurkunde **74**, pp. 431–433, 2008. Cited on page 406.

312 Emile Borel, *Introduction géométrique à la physique*, Gauthier-Villars, 1912. Cited on page 406.

313 See V. L. TELEGDI, *Enrico Fermi in America*, Physics Today **55**, pp. 38–43, June 2002. Cited on page 407.

314 K. SCHMIDT-NIELSEN, *Desert Animals: Physiological Problems of Heat and Water*, Oxford University Press, 1964. Cited on page 408.

315 Following a private communication by Richard Rusby, this is the value of 1997, whereas it was estimated as 99.975°C in 1989, as reported by GARETH JONES & RICHARD RUSBY, *Official: water boils at 99.975°C*, Physics World **2**, pp. 23–24, September 1989, and R. L. RUSBY, *Ironing out the standard scale*, Nature **338**, p. 1169, March 1989. For more details on temperature measurements, see Ref. 287. Cited on pages 408 and 547.

316 Why entropy is created when information is erased, but not when it is acquired, is explained in C. H. BENNETT & R. LANDAUER, *Fundamental Limits of Computation*, Scientific American **253**:1, pp. 48–56, 1985. The conclusion: we should pay to throw the newspaper away, not to buy it. Cited on page 409.

317 See, for example, G. SWIFT, *Thermoacoustic engines and refrigerators*, Physics Today **48**, pp. 22–28, July 1995. Cited on page 412.

318 Quoted in D. CAMPBELL, J. CRUTCHFIELD, J. FARMER & E. JEN, *Experimental mathematics: the role of computation in nonlinear science*, Communications of the Association of Computing Machinery **28**, pp. 374–384, 1985. Cited on page 415.

319 For more about the shapes of snowflakes, see the famous book by W. A. BENTLEY & W. J. HUMPHREYS, *Snow Crystals*, Dover Publications, New York, 1962. This second printing of the original from 1931 shows a large part of the result of Bentley's lifelong passion, namely several thousand photographs of snowflakes. Cited on page 415.

320 K. SCHWENK, *Why snakes have forked tongues*, Science **263**, pp. 1573–1577, 1994. Cited on page 418.

321 Human hands do not have five fingers in around 1 case out of 1000. How does nature ensure this constancy? The detailed mechanisms are not completely known yet. It is known, though, that a combination of spatial and temporal self-organization during cell proliferation and differentiation in the embryo is the key factor. In this self-regulating system, the GLI3 transcription factor plays an essential role. Cited on page 419.

322 E. MARTÍNEZ, C. PÉREZ-PENICHET, O. SOTOLONGO-COSTA, O. RAMOS, K. J. MÅLØY, S. DOUADY, E. ALTSHULER, *Uphill solitary waves in granular flows*, Physical Review **75**, p. 031303, 2007, and E. ALTSHULER, O. RAMOS, E. MARTÍNEZ, A. J. BATISTA-LEYVA, A. RIVERA & K. E. BASSLER, *Sandpile formation by revolving rivers*, Physical Review Letters **91**, p. 014501, 2003. Cited on page 420.

323 P. B. UMBANHOWAR, F. MELO & H. L. SWINNEY, *Localized excitations in a vertically vibrated granular layer*, Nature **382**, pp. 793–796, 29 August 1996. Cited on page 421.

324 D. K. CAMPBELL, S. FLACH & Y. S. KIVSHAR, *Localizing energy through nonlinearity and discreteness*, Physics Today **57**, pp. 43–49, January 2004. Cited on page 421.

325 B. ANDREOTTI, *The song of dunes as a wave-particle mode locking*, Physical Review Letters **92**, p. 238001, 2004. Cited on page 421.

326 D. C. MAYS & B. A. FAYBISHENKO, *Washboards in unpaved highways as a complex dynamic system*, Complexity **5**, pp. 51–60, 2000. See also N. TABERLET, S. W. MORRIS & J. N. MCELWAINE, *Washboard road: the dynamics of granular ripples formed by rolling wheels*, Physical Review Letters **99**, p. 068003, 2007. Cited on pages 422 and 561.

327 K. KÖTTER, E. GOLES & M. MARKUS, *Shell structures with 'magic numbers' of spheres in a swirled disk*, Physical Review E **60**, pp. 7182–7185, 1999. Cited on page 422.

328 A good introduction is the text by DANIEL WALGRAEF, *Spatiotemporal Pattern Formation, With Examples in Physics, Chemistry and Materials Science*, Springer 1996. Cited on page 422.

329 For an overview, see the Ph.D. thesis by JOCELINE LEGA, *Défauts topologiques associés à la brisure de l'invariance de translation dans le temps*, Université de Nice, 1989. Cited on page 424.

330 An idea of the fascinating mechanisms at the basis of the heart beat is given by A. BABLOYANTZ & A. DESTEXHE, *Is the normal heart a periodic oscillator?*, Biological Cybernetics **58**, pp. 203–211, 1989. Cited on page 425.

331 For a short, modern overview of turbulence, see L. P. KADANOFF, *A model of turbulence*, Physics Today **48**, pp. 11–13, September 1995. Cited on page 426.

332 For a clear introduction, see T. SCHMIDT & M. MAHRL, *A simple mathematical model of a dripping tap*, European Journal of Physics **18**, pp. 377–383, 1997. Cited on page 426.

333 The mathematics of fur patterns has been studied in great detail. By varying parameters in reaction–diffusion equations, it is possible to explain the patterns on zebras, leopards, giraffes and many other animals. The equations can be checked by noting, for example, how the calculated patterns continue on the tail, which usually looks quite different. In fact, most patterns look differently if the fur is not flat but curved. This is a general phenomenon, valid also for the spot patterns of ladybugs, as shown by S. S. LIAW, C. C. YANG, R. T. LIU & J. T. HONG, *Turing model for the patterns of lady beetles*, Physical Review E **64**, p. 041909, 2001. Cited on page 426.

334 An overview of science humour can be found in the famous anthology compiled by R. L. WEBER, edited by E. MENDOZA, *A Random Walk in Science*, Institute of Physics, 1973. It is also available in several expanded translations. Cited on page 426.

335 W. DREYBRODT, *Physik von Stalagmiten*, Physik in unserer Zeit pp. 25–30, Physik in unserer Zeit February 2009. Cited on page 427.

336 K. MERTENS, V. PUTKARADZE & P. VOROBIEFF, *Braiding patterns on an inclined plane*, Nature **430**, p. 165, 2004. Cited on page 428.

337 These beautifully simple experiments were published in G. MÜLLER, *Starch columns: analog model for basalt columns*, Journal of Geophysical Research **103**, pp. 15239–15253, 1998, in G. MÜLLER, *Experimental simulation of basalt columns*, Journal of Volcanology and Geothermal Research **86**, pp. 93–96, 1998, and in G. MÜLLER, *Trocknungsrisse in Stärke*, Physikalische Blätter **55**, pp. 35–37, 1999. Cited on page 429.

338 To get a feeling for viscosity, see the fascinating text by STEVEN VOGEL, *Life in Moving Fluids: the Physical Biology of Flow*, Princeton University Press, 1994. Cited on page 430.

339 B. HOF, C. W. H. VAN DOORNE, J. WESTERWEEL, F. T. M. NIEUWSTADT, H. WEDIN, R. KERSWELL, F. WALEFFE, H. FAISST & B. ECKHARDT, *Experimental observation of nonlinear traveling waves in turbulent pipe flow*, Science **305**, pp. 1594–1598, 2004. See also B. HOF & al., *Finite lifetime of turbulence in shear flows*, Nature **443**, p. 59, 2006. Cited on page 430.

340 A fascinating book on the topic is KENNETH LAWS & MARTHA SWOPE, *Physics and the Art of Dance: Understanding Movement*, Oxford University Press 2002. See also KENNETH LAWS & M. LOTT, *Resource Letter PoD-1: The Physics of Dance*, American Journal of Physics **81**, pp. 7–13, 2013. Cited on page 430.

341 The fascinating variation of snow crystals is presented in C. MAGONO & C. W. LEE, *Meteorological classification of natural snow crystals*, Journal of the Faculty of Science, Hokkaido

University Ser. VII, II, pp. 321–325, 1966, also online at the eprints.lib.hokudai.ac.jp website. Cited on page 431.

342 Josef H. Reichholf, *Eine kurze Naturgeschichte des letzten Jahrtausends*, Fischer Verlag, 2007. Cited on page 431.

343 See for example, E. F. Bunn, *Evolution and the second law of thermodynamics*, American Journal of Physics **77**, pp. 922–925, 2009. Cited on page 432.

344 See the paper J. Maldacena, S. H. Shenker & D. Stanford, *A bound on chaos*, free preprint at www.arxiv.org/abs/1503.01409. The bound has not be put into question yet. Cited on page 432.

345 A good introduction of the physics of bird swarms is T. Feder, *Statistical physics is for the birds*, Physics Today **60**, pp. 28–30, October 2007. Cited on page 433.

346 The Nagel-Schreckenberg model for vehicle traffic, for example, explains how simple fluctuations in traffic can lead to congestions. Cited on page 433.

347 J. J. Lissauer, *Chaotic motion in the solar system*, Reviews of Modern Physics **71**, pp. 835–845, 1999. Cited on page 434.

348 See Jean-Paul Dumont, *Les écoles présocratiques*, Folio Essais, Gallimard, 1991, p. 426. Cited on page 435.

349 For information about the number π, and about some other mathematical constants, the website oldweb.cecm.sfu.ca/pi/pi.html provides the most extensive information and references. It also has a link to the many other sites on the topic, including the overview at mathworld.wolfram.com/Pi.html. Simple formulae for π are

$$\pi + 3 = \sum_{n=1}^{\infty} \frac{n 2^n}{\binom{2n}{n}} \qquad (164)$$

or the beautiful formula discovered in 1996 by Bailey, Borwein and Plouffe

$$\pi = \sum_{n=0}^{\infty} \frac{1}{16^n} \left(\frac{4}{8n+1} - \frac{2}{8n+4} - \frac{1}{8n+5} - \frac{1}{8n+6} \right). \qquad (165)$$

The mentioned site also explains the newly discovered methods for calculating specific binary digits of π without having to calculate all the preceding ones. The known digits of π pass all tests of randomness, as the mathworld.wolfram.com/PiDigits.html website explains. However, this property, called *normality*, has never been proven; it is the biggest open question about π. It is possible that the theory of chaotic dynamics will lead to a solution of this puzzle in the coming years.

Another method to calculate π and other constants was discovered and published by D. V. Chudnovsky & G. V. Chudnovsky, *The computation of classical constants*, Proceedings of the National Academy of Sciences (USA) **86**, pp. 8178–8182, 1989. The Chudnowsky brothers have built a supercomputer in Gregory's apartment for about 70 000 euros, and for many years held the record for calculating the largest number of digits of π. They have battled for decades with Kanada Yasumasa, who held the record in 2000, calculated on an industrial supercomputer. From 2009 on, the record number of (consecutive) digits of π has been calculated on a desktop PC. The first was Fabrice Bellard, who needed 123 days and used a Chudnovsky formula. Bellard calculated over 2.7 million million digits, as told on bellard.org. Only in 2019 did people start to use cloud computers. The present record is 31.415 million million digits. For the most recent records, see en.wikipedia.org/wiki/Chronology_of_computation_of_%CF%80. New formulae to calculate π are still occasionally discovered.

For the calculation of Euler's constant γ see also D. W. DeTemple, *A quicker convergence to Euler's constant*, The Mathematical Intelligencer, pp. 468–470, May 1993. Cited on pages 436 and 467.

350 The Johnson quote is found in William Seward, *Biographiana*, 1799. For details, see the story in quoteinvestigator.com/2014/11/08/without-effort/. Cited on page 440.

351 The first written record of the letter U seems to be Leon Battista Alberti, *Grammatica della lingua toscana*, 1442, the first grammar of a modern (non-latin) language, written by a genius that was intellectual, architect and the father of cryptology. The first written record of the letter J seems to be Antonio de Nebrija, *Gramática castellana*, 1492. Before writing it, Nebrija lived for ten years in Italy, so that it is possible that the I/J distinction is of Italian origin as well. Nebrija was one of the most important Spanish scholars. Cited on page 442.

352 For more information about the letters thorn and eth, have a look at the extensive report to be found on the website www.everytype.com/standards/wynnyogh/thorn.html. Cited on page 442.

353 For a modern history of the English language, see David Crystal, *The Stories of English*, Allen Lane, 2004. Cited on page 442.

354 Hans Jensen, *Die Schrift*, Berlin, 1969, translated into English as *Sign, Symbol and Script: an Account of Man's Efforts to Write*, Putnam's Sons, 1970. Cited on page 442.

355 David R. Lide, editor, *CRC Handbook of Chemistry and Physics*, 78th edition, CRC Press, 1997. This classic reference work appears in a new edition every year. The full Hebrew alphabet is given on page 2-90. The list of abbreviations of physical quantities for use in formulae approved by ISO, IUPAP and IUPAC can also be found there.

However, the ISO 31 standard, which defines these abbreviations, costs around a thousand euro, is not available on the internet, and therefore can safely be ignored, like any standard that is supposed to be used in teaching but is kept inaccessible to teachers. Cited on pages 444 and 446.

356 See the mighty text by Peter T. Daniels & William Bright, *The World's Writing Systems*, Oxford University Press, 1996. Cited on page 445.

357 The story of the development of the numbers is told most interestingly by Georges Ifrah, *Histoire universelle des chiffres*, Seghers, 1981, which has been translated into several languages. He sums up the genealogy of the number signs in ten beautiful tables, one for each digit, at the end of the book. However, the book itself contains factual errors on every page, as explained for example in the review found at www.ams.org/notices/200201/rev-dauben.pdf and www.ams.org/notices/200202/rev-dauben.pdf. Cited on page 445.

358 See the for example the fascinating book by Steven B. Smith, *The Great Mental Calculators – The Psychology, Methods and Lives of the Calculating Prodigies*, Columbia University Press, 1983. The book also presents the techniques that they use, and that anybody else can use to emulate them. Cited on page 446.

359 See for example the article 'Mathematical notation' in the *Encyclopedia of Mathematics*, 10 volumes, Kluwer Academic Publishers, 1988–1993. But first all, have a look at the informative and beautiful jeff560.tripod.com/mathsym.html website. The main source for all these results is the classic and extensive research by Florian Cajori, *A History of Mathematical Notations*, 2 volumes, The Open Court Publishing Co., 1928–1929. The square root sign is used in Christoff Rudolff, *Die Coss*, Vuolfius Cephaleus Joanni Jung: Argentorati, 1525. (The full title was *Behend vnnd Hubsch Rechnung durch die kunstreichen regeln Algebre* *so gemeinlicklich die Coss genent werden. Darinnen alles so treülich an tag gegeben, das auch allein auss vleissigem lesen on allen mündtlichē vnterricht mag begriffen werden, etc.*) Cited on page 446.

360 J. TSCHICHOLD, *Formenwamdlungen der et-Zeichen*, Stempel AG, 1953. Cited on page 448.

361 MALCOLM B. PARKES, *Pause and Effect: An Introduction to the History of Punctuation in the West*, University of California Press, 1993. Cited on page 448.

362 This is explained by BERTHOLD LOUIS ULLMAN, *Ancient Writing and its Influence*, 1932. Cited on page 448.

363 PAUL LEHMANN, *Erforschung des Mittelalters – Ausgewählte Abhandlungen und Aufsätze*, Anton Hiersemann, 1961, pp. 4–21. Cited on page 448.

364 BERNARD BISCHOFF, *Paläographie des römischen Altertums und des abendländischen Mittelalters*, Erich Schmidt Verlag, 1979, pp. 215–219. Cited on page 448.

365 HUTTON WEBSTER, *Rest Days: A Study in Early Law and Morality*, MacMillan, 1916. The discovery of the unlucky day in Babylonia was made in 1869 by George Smith, who also rediscovered the famous *Epic of Gilgamesh*. Cited on page 449.

366 The connections between Greek roots and many French words – and thus many English ones – can be used to rapidly build up a vocabulary of ancient Greek without much study, as shown by the practical collection by J. CHAINEUX, *Quelques racines grecques*, Wetteren – De Meester, 1929. See also DONALD M. AYERS, *English Words from Latin and Greek Elements*, University of Arizona Press, 1986. Cited on page 450.

367 In order to write well, read WILLIAM STRUNK & E. B. WHITE, *The Elements of Style*, Macmillan, 1935, 1979, or WOLF SCHNEIDER, *Deutsch für Kenner – Die neue Stilkunde*, Gruner und Jahr, 1987. Cited on page 451.

368 *Le Système International d'Unités*, Bureau International des Poids et Mesures, Pavillon de Breteuil, Parc de Saint Cloud, 92310 Sèvres, France. All new developments concerning SI units are published in the journal *Metrologia*, edited by the same body. Showing the slow pace of an old institution, the BIPM launched a website only in 1998; it is now reachable at www.bipm.fr. See also the www.utc.fr/~tthomass/Themes/Unites/index.html website; this includes the biographies of people who gave their names to various units. The site of its British equivalent, www.npl.co.uk/npl/reference, is much better; it provides many details as well as the English-language version of the SI unit definitions. Cited on page 452.

369 The bible in the field of time measurement is the two-volume work by J. VANIER & C. AUDOIN, *The Quantum Physics of Atomic Frequency Standards*, Adam Hilge, 1989. A popular account is TONY JONES, *Splitting the Second*, Institute of Physics Publishing, 2000.

The site opdaf1.obspm.fr/www/lexique.html gives a glossary of terms used in the field. For precision *length* measurements, the tools of choice are special lasers, such as mode-locked lasers and frequency combs. There is a huge literature on these topics. Equally large is the literature on precision *electric current* measurements; there is a race going on for the best way to do this: counting charges or measuring magnetic forces. The issue is still open. On *mass* and atomic mass measurements, see the volume on relativity. On high-precision *temperature* measurements, see Ref. 287. Cited on page 453.

370 The unofficial SI prefixes were first proposed in the 1990s by Jeff K. Aronson of the University of Oxford, and might come into general usage in the future. See New Scientist **144**, p. 81, 3 December 1994. Other, less serious proposals also exist. Cited on page 454.

371 The most precise clock built in 2004, a caesium fountain clock, had a precision of one part in 10^{15}. Higher precision has been predicted to be possible soon, among others by M. Takamoto, F.-L. Hong, R. Higashi & H. Katori, *An optical lattice clock*, Nature 435, pp. 321–324, 2005. Cited on page 456.

372 J. Bergquist, ed., *Proceedings of the Fifth Symposium on Frequency Standards and Metrology*, World Scientific, 1997. Cited on page 456.

373 J. Short, *Newton's apples fall from grace*, New Scientist 2098, p. 5, 6 September 1997. More details can be found in R. G. Keesing, *The history of Newton's apple tree*, Contemporary Physics 39, pp. 377–391, 1998. Cited on page 457.

374 The various concepts are even the topic of a separate international standard, ISO 5725, with the title *Accuracy and precision of measurement methods and results*. A good introduction is John R. Taylor, *An Introduction to Error Analysis: the Study of Uncertainties in Physical Measurements*, 2nd edition, University Science Books, Sausalito, 1997. Cited on page 458.

375 The most recent (2010) recommended values of the fundamental physical constants are found only on the website physics.nist.gov/cuu/Constants/index.html. This set of constants results from an international adjustment and is recommended for international use by the Committee on Data for Science and Technology (CODATA), a body in the International Council of Scientific Unions, which brings together the International Union of Pure and Applied Physics (IUPAP), the International Union of Pure and Applied Chemistry (IUPAC) and other organizations. The website of IUPAC is www.iupac.org. Cited on page 460.

376 Some of the stories can be found in the text by N. W. Wise, *The Values of Precision*, Princeton University Press, 1994. The field of high-precision measurements, from which the results on these pages stem, is a world on its own. A beautiful introduction to it is J. D. Fairbanks, B. S. Deaver, C. W. Everitt & P. F. Michaelson, eds., *Near Zero: Frontiers of Physics*, Freeman, 1988. Cited on page 460.

377 For details see the well-known astronomical reference, P. Kenneth Seidelmann, *Explanatory Supplement to the Astronomical Almanac*, 1992. Cited on page 465.

378 F.F. Stanaway & al., *How fast does the Grim Reaper walk? Receiver operating characteristic curve analysis in healthy men aged 70 and over*, British Medical Journal 343, p. 7679, 2011. This paper by an Australian research team, was based on a study of 1800 older men that were followed over several years; the paper was part of the 2011 Christmas issue and is freely downloadable at www.bmj.com. Additional research shows that walking and training to walk rapidly can indeed push death further away, as summarized by K. Jahn & T. Brandt, *Wie Alter und Krankheit den Gang verändern*, Akademie Aktuell 03, pp. 22–25, 2012, The paper also shows that humans walk upright since at least 3.6 million years and that walking speed decreases about 1 % per year after the age of 60. Cited on page 479.

CREDITS

ACKNOWLEDGEMENTS

Many people who have kept their gift of curiosity alive have helped to make this project come true. Most of all, Peter Rudolph and Saverio Pascazio have been – present or not – a constant reference for this project. Fernand Mayné, Anna Koolen, Ata Masafumi, Roberto Crespi, Serge Pahaut, Luca Bombelli, Herman Elswijk, Marcel Krijn, Marc de Jong, Martin van der Mark, Kim Jalink, my parents Peter and Isabella Schiller, Mike van Wijk, Renate Georgi, Paul Tegelaar, Barbara and Edgar Augel, M. Jamil, Ron Murdock, Carol Pritchard, Richard Hoffman, Stephan Schiller, Franz Aichinger and, most of all, my wife Britta have all provided valuable advice and encouragement.

Many people have helped with the project and the collection of material. Most useful was the help of Mikael Johansson, Bruno Barberi Gnecco, Lothar Beyer, the numerous improvements by Bert Sierra, the detailed suggestions by Claudio Farinati, the many improvements by Eric Sheldon, the detailed suggestions by Andrew Young, the continuous help and advice of Jonatan Kelu, the corrections of Elmar Bartel, and in particular the extensive, passionate and conscientious help of Adrian Kubala.

Important material was provided by Bert Peeters, Anna Wierzbicka, William Beaty, Jim Carr, John Merrit, John Baez, Frank DiFilippo, Jonathan Scott, Jon Thaler, Luca Bombelli, Douglas Singleton, George McQuarry, Tilman Hausherr, Brian Oberquell, Peer Zalm, Martin van der Mark, Vladimir Surdin, Julia Simon, Antonio Fermani, Don Page, Stephen Haley, Peter Mayr, Allan Hayes, Norbert Dragon, Igor Ivanov, Doug Renselle, Wim de Muynck, Steve Carlip, Tom Bruce, Ryan Budney, Gary Ruben, Chris Hillman, Olivier Glassey, Jochen Greiner, squark, Martin Hardcastle, Mark Biggar, Pavel Kuzin, Douglas Brebner, Luciano Lombardi, Franco Bagnoli, Lukas Fabian Moser, Dejan Corovic, Paul Vannoni, John Haber, Saverio Pascazio, Klaus Finkenzeller, Leo Volin, Jeff Aronson, Roggie Boone, Lawrence Tuppen, Quentin David Jones, Arnaldo Uguzzoni, Frans van Nieuwpoort, Alan Mahoney, Britta Schiller, Petr Danecek, Ingo Thies, Vitaliy Solomatin, Carl Offner, Nuno Proença, Elena Colazingari, Paula Henderson, Daniel Darre, Wolfgang Rankl, John Heumann, Joseph Kiss, Martha Weiss, Antonio González, Antonio Martos, André Slabber, Ferdinand Bautista, Zoltán Gácsi, Pat Furrie, Michael Reppisch, Enrico Pasi, Thomas Köppe, Martin Rivas, Herman Beeksma, Tom Helmond, John Brandes, Vlad Tarko, Nadia Murillo, Ciprian Dobra, Romano Perini, Harald van Lintel, Andrea Conti, François Belfort, Dirk Van de Moortel, Heinrich Neumaier, Jarosław Królikowski, John Dahlman, Fathi Namouni, Paul Townsend, Sergei Emelin, Freeman Dyson, S.R. Madhu Rao, David Parks, Jürgen Janek, Daniel Huber, Alfons Buchmann, William Purves, Pietro Redondi, Damoon Saghian, Frank Sweetser, Markus Zecherle, Zach Joseph Espiritu, Marian Denes, Miles Mutka, plus a number of people who wanted to remain unnamed.

The software tools were refined with extensive help on fonts and typesetting by Michael Zedler and Achim Blumensath and with the repeated and valuable support of Donald Arseneau; help came also from Ulrike Fischer, Piet van Oostrum, Gerben Wierda, Klaus Böhncke, Craig Upright, Herbert Voss, Andrew Trevorrow, Danie Els, Heiko Oberdiek, Sebastian Rahtz, Don Story, Vincent Darley, Johan Linde, Joseph Hertzlinger, Rick Zaccone, John Warkentin, Ulrich Diez, Uwe Siart, Will Robertson, Joseph Wright, Enrico Gregorio, Rolf Niepraschk, Alexander Grahn, Werner Fabian and Karl Köller.

The typesetting and book design is due to the professional consulting of Ulrich Dirr. The typography was much improved with the help of Johannes Küster and his Minion Math font. The design of the book and its website also owe much to the suggestions and support of my wife Britta.

I also thank the lawmakers and the taxpayers in Germany, who, in contrast to most other countries in the world, allow residents to use the local university libraries.

From 2007 to 2011, the electronic edition and distribution of the Motion Mountain text was generously supported by the Klaus Tschira Foundation.

Film credits

The beautiful animations of the rotating attached dodecaeder on page 90 and of the embedded ball on page 168 are copyright and courtesy of Jason Hise; he made them for this text and for the Wikimedia Commons website. Several of his animations are found on his website www.entropygames.net. The clear animation of a suspended spinning top, shown on page 148, was made for this text by Lucas Barbosa. The impressive animation of the Solar System on page 155 was made for this text by Rhys Taylor and is now found at his website www.rhysy.net. The beautiful animation of the lunation on page 190 was calculated from actual astronomical data and is copyright and courtesy by Martin Elsässer. It can be found on his website www.mondatlas.de/lunation.html. The beautiful film of geostationary satellites on page 195 is copyright and courtesy by Michael Kunze and can be found on his beautiful site www.sky-in-motion.de/en. The beautiful animation of the planets and planetoids on page 220 is copyright and courtesy by Hans-Christian Greier. It can be found on his wonderful website www.parallax.at. The film of an oscillating quartz on page 291 is copyright and courtesy of Micro Crystal, part of the Swatch Group, found at www.microcrystal.com. The animation illustrating group and wave velocity on page 299 and the animation illustrating the molecular motion in a sound wave on page 311 are courtesy and copyright of the ISVR at the University of Southampton. The film of the rogue wave on page 323 is courtesy and copyright of Amin Chabchoub; details can be found at journals.aps.org/prx/abstract/10.1103/PhysRevX.2.011015. The films of solitons on page 317 and of dromions on page 329 are copyright and courtesy by Jarmo Hietarinta. They can be found on his website users.utu.fi/hietarin. The film of leapfrogging vortex rings on page 380 is copyright and courtesy by Lim Tee Tai. It can be found via his fluid dynamics website serve.me.nus.edu.sg. The film of the growing snowflake on page 422 is copyright and courtesy by Kenneth Libbrecht. It can be found on his website www.its.caltech.edu/~atomic/snowcrystals.

Image credits

The photograph of the east side of the Langtang Lirung peak in the Nepalese Himalayas, shown on the front cover, is courtesy and copyright by Kevin Hite and found on his blog thegettingthere.com. The lightning photograph on page 14 is courtesy and copyright by Harald Edens and found on the www.lightningsafety.noaa.gov/photos.htm and www.weather-photography.com websites. The motion illusion on page 18 is courtesy and copyright by Michael Bach and found on his website www.michaelbach.de/ot/mot_rotsnake/index.html. It is a variation of the illusion by Kitaoka Akiyoshi found on www.ritsumei.ac.jp/~akitaoka and used here with his permission. The figures on pages 20, 57 and 207 were made especially for this text and are copyright by Luca Gastaldi. The high speed photograph of a bouncing tennis ball on page 20 is courtesy and copyright by the

International Tennis Federation, and were provided by Janet Page. The figure of Etna on pages 22 and 150 is copyright and courtesy of Marco Fulle and found on the wonderful website www.stromboli.net. The famous photograph of the Les Poulains and its lighthouse by Philip Plisson on page 23 is courtesy and copyright by Pechêurs d'Images; see the websites www.plisson.com and www.pecheurs-d-images.com. It is also found in Plisson's magnus opus *La Mer*, a stunning book of photographs of the sea. The picture on page 23 of Alexander Tsukanov jumping from one ultimate wheel to another is copyright and courtesy of the Moscow State Circus. The photograph of a deer on page 25 is copyright and courtesy of Tony Rodgers and taken from his website www.flickr.com/photos/moonm. The photographs of speed measurement devices on page 35 are courtesy and copyright of the Fachhochschule Koblenz, of Silva, of Tracer and of Wikimedia. The graph on page 38 is redrawn and translated from the wonderful book by HENK TENNEKES, *De wetten van de vliegkunst - Over stijgen, dalen, vliegen en zweven*, Aramith Uitgevers, 1993. The photographs of the ping-pong ball on page 40 and of the dripping water tap on page 355 are copyright and courtesy of Andrew Davidhazy and found on his website www.rit.edu/~andpph. The photograph of the bouncing water droplet on page 40 are copyright and courtesy of Max Groenendijk and found on the website www.lightmotif.nl. The photograph of the precision sundial on page 45 is copyright and courtesy of Stefan Pietrzik and found at commons.wikimedia.org/wiki/Image:Präzissions-Sonnenuhr_mit_Sommerwalze.jpg The other clock photographs in the figure are from public domain sources as indicated. The graph on the scaling of biological rhythms on page 47 is drawn by the author using data from the European Molecular Biology Organization found at www.nature.com/embor/journal/v6/n1s/fig_tab/7400425_f3.html and Enrique Morgado. The drawing of the human ear on page page 50 and on page 325 are copyright of Northwestern University and courtesy of Tim Hain; it is found on his website www.dizziness-and-balance.com/disorders/bppv/otoliths.html. The illustrations of the vernier caliper and the micrometer screw on page 54 and 65 are copyright of Medien Werkstatt, courtesy of Stephan Bogusch, and taken from their instruction course found on their website www.medien-werkstatt.de. The photo of the tiger on page 54 is copyright of Naples zoo (in Florida, not in Italy), and courtesy of Tim Tetzlaff; see their website at www.napleszoo.com. The other length measurement devices on page 54 are courtesy and copyright of Keyence and Leica Geosystems, found at www.leica-geosystems.com. The curvimeter photograph on page 55 is copyright and courtesy of Frank Müller and found on the www.wikimedia.org website. The crystal photograph on the left of page 58 is copyright and courtesy of Stephan Wolfsried and found on the www.mindat.org website. The crystal photograph on the right of page 58 is courtesy of Tullio Bernabei, copyright of Arch. Speleoresearch & Films/La Venta and found on the www.laventa.it and www.naica.com.mx websites. The hollow Earth figure on pages 60 is courtesy of Helmut Diel and was drawn by Isolde Diel. The wonderful photographs on page 69, page 150, page 179, page 217, page 213 and page 504 are courtesy and copyright by Anthony Ayiomamitis; the story of the photographs is told on his beautiful website at www.perseus.gr. The anticrepuscular photograph on page 71 is courtesy and copyright by Peggy Peterson. The rope images on page 72 are copyright and courtesy of Jakob Bohr. The image of the tight knot on page 73 is courtesy and copyright by Piotr Pieranski. The firing caterpillar figure of page 79 is courtesy and copyright of Stanley Caveney. The photograph of an airbag sensor on page 86 is courtesy and copyright of Bosch; the accelerometer picture is courtesy and copyright of Rieker Electronics; the three drawings of the human ear are copyright of Northwestern University and courtesy of Tim Hain and found on his website www.dizziness-and-balance.com/disorders/bppv/otoliths.html. The photograph of Orion on page 87 is courtesy and copyright by Matthew Spinelli; it was also featured on antwrp.gsfc.nasa.gov/apod/ap030207.html. On page 88, the drawing of star sizes is courtesy and copyright Dave Jarvis. The photograph of Regulus and Mars on page 88 is courtesy and copyright of Jürgen Michelberger and found on www.jmichelberger.de. On page 91, the millipede photograph is courtesy and copyright of David Parks and found on his website www.mobot.org/mobot/madagascar/image.asp?relation=A71. The photograph of the gecko climbing the bus window on page 91 is courtesy and copyright of Marcel Berendsen, and found on his website www.flickr.com/photos/berendm. The photograph of the amoeba is courtesy and copyright of Antonio Guillén Oterino and is taken from his wonderful website *Proyecto Agua* at www.flickr.com/photos/microagua. The photograph of N. decemspinosa on page 91 is courtesy and copyright of Robert Full, and found on his website rjf9.biol.berkeley.edu/twiki/bin/view/PolyPEDAL/LabPhotographs. The photograph of P. ruralis on page 91 is courtesy and copyright of John Brackenbury, and part of his wonderful collection on the website www.sciencephoto.co.uk. The photograph of the rolling spider on page 91 is courtesy and copyright of Ingo Rechenberg and can be found at www.bionik.tu-berlin.de, while the photo of the child somersaulting is courtesy and copyright of Karva Javi, and can be found at www.flickr.com/photos/karvajavi. The photographs of flagellar motors on page 93 are copyright and courtesy by Wiley & Sons and found at emboj.embopress.org/content/30/14/2972. The two wonderful films about bacterial flagella on page 94 and on page 94 are copyright and courtesy of the Graduate School of Frontier Biosciences at Osaka University. The beautiful photograph of comet McNaught on page 95 is courtesy and copyright by its discoveror, Robert McNaught; it is taken from his website at www.mso.anu.edu.au/~rmn and is found also on antwrp.gsfc.nasa.gov/apod/ap070122.html. The sonoluminsecence picture on page 96 is courtesy and copyright of Detlef Lohse. The photograph of the standard kilogram on page 100 is courtesy and copyright by the Bureau International des Poids et Mesures (BIPM). On page 108, the photograph of Mendeleyev's balance is copyright of Thinktank Trust and courtesy of Jack Kirby; it can be found on the www.birminghamstories.co.uk website. The photograph of the laboratory balance is copyright and courtesy of Mettler-Toledo. The photograph of the cosmonaut mass measurement device is courtesy of NASA. On page 115, the photographs of the power meters are courtesy and copyright of SRAM, Laser Components and Wikimedia. The measured graph of the walking human on page 122 is courtesy and copyright of Ray McCoy. On page 130, the photograph of the stacked gyros is cortesy of Wikimedia. The photograph of the clock that needs no winding up is copyright Jaeger-LeCoultre and courtesy of Ralph Stieber. Its history and working are described in detail in a brochure available from the company. The company's website is www.Jaeger-LeCoultre.com. The photograph of the ship lift at Strépy-Thieux on page 132 is courtesy and copyright of Jean-Marie Hoornaert and found on Wikimedia Commons. The photograph of the Celtic wobble stone on page 133 is courtesy and copyright of Ed Keath and found on Wikimedia Commons. The photograph of the star trails on page 137 is courtesy and copyright of Robert Schwartz; it was featured on apod.nasa.gov/apod/ap120802.html. The photograph of Foucault's gyroscope on page 142 is courtesy and copyright of the museum of the CNAM, the Conservatoire National des Arts et Métiers in Paris, whose website is at www.arts-et-metiers.net. The photograph of the laser gyroscope on page 142 is courtesy and copyright of JAXA, the Japan Aerospace Exploration Agency, and found on their website at jda.jaxa.jp. On page 143, the three-dimensional model of the gyroscope is copyright and courtesy of Zach Joseph Espiritu. The drawing of the precision laser gyroscope on page 144 is courtesy of Thomas Klügel and copyright of the Bundesamt für Kartographie und Geodäsie. The photograph of the instrument is courtesy and copyright of Carl Zeiss. The machine is located at the Fundamentalstation Wettzell, and its website is found at www.wettzell.ifag.de. The illustration of plate tectonics on page 149 is from a film produced by NASA's HoloGlobe project and can be found on svs.gsfc.nasa.gov/cgi-bin/details.cgi?aid=1288 The graph of the temperature record on page 154 is copyright and courtesy Jean Jouzel and Science/AAAS. The photograph of a car driving through the snow on page 156 is copyright and courtesy by Neil Provo at neilprovo.com. The photographs of a crane fly and of a hevring fly with their halteres on page 160 are by Pinzo, found on Wikimedia Commons, and by Sean McCann from his website ibycter.com. The MEMS photograph and graph is copyright and courtesy of ST Microelectronics. On page 166, Figure 123 is courtesy and copyright of the international Gemini project (Gemini Observatory/Association of Universities for Research in Astronomy) at www.ausgo.unsw.edu.au and www.gemini.edu; the photograph with the geostationary satellites is copyright and courtesy of Michael Kunze and can be found just before his equally beautiful film at www.sky-in-motion.de/de/zeitraffer_einzel.php?NR=12. The photograph of the Earth's shadow on page 167 is courtesy and copyright by Ian R. Rees and found on his website at weaknuclearforce.wordpress.com/2014/03/30/earths-shadow. The basilisk running over water, page 169 and on the back cover, is courtesy and copyright by the Belgian group TERRA vzw and found on their website www.terravzw.org. The water strider photograph on page 170 is courtesy and copyright by Charles Lewallen. The photograph of the water robot on page 170 is courtesy and copyright by the American Institute of Physics. The allometry graph about running speed in mammals is courtesy and copyright of José Iriarte-Díaz and of The Journal of Experimental Biology; it is reproduced and adapted with their permission from the original article, Ref. 132, found at jeb.biologists.org/content/205/18/2897. The illustration of the motion of Mars on page 175 is courtesy and copyright of Tunc Tezel. The photograph of the precision pendulum clock on page 181 is copyright of Erwin Sattler OHG,Sattler OHG, Erwin and courtesy of Ms. Stephanie Sattler-Rick; it can be found at the www.erwinsattler.de website. The figure on the triangulation of the meridian of Paris on page 184 is copyright and courtesy of Ken Alder and found on his website www.kenalder.com. The photographs of the home version of the Cavendish experiment on page 185 are courtesy and copyright by John Walker and found on his website www.fourmilab.ch/gravitation/foobar. The photographs of the precision Cavendish experiment on page 186 are courtesy and copyright of the Eöt-Wash Group at the University of Washington and found at www.npl.washington.edu/eotwash. The geoid of page 189 is courtesy and copyright by the GeoForschungsZentrum Potsdam, found at www.gfz-potsdam.de. The moon maps on page 191 are courtesy of the USGS Astrogeology Research Program, astrogeology.usgs.gov, in particular Mark Rosek and Trent Hare. The graph of orbits on page 193 is courtesy and copyright of Geoffrey Marcy. On page 196, the asteroid orbit is courtesy and copyrigt of Seppo Mikkola. The photograph of the tides on page 197 is copyright and courtesy of Gilles Régnier and found on his website www.gillesregnier.com; it also shows an animation of that tide over the whole day. The wonderful meteor shower photograph on page 206 is courtesy and copyright Brad Goldpaint and found on its website goldpaintphotography.com; it was also featured on apod.nasa.gov/apod/ap160808.html. The meteor photograph on page 206 is courtesy and copyright of Robert Mikaelyan and found on his website www.fotoarena.nl/tag/robert-mikaelyan/. The pictures of fast descents on snow on page 209 are copyright and courtesy of Simone Origone, www.simoneorigone.it, and of Éric Barone, www.ericbarone.com. The photograph of the Galilean satellites on page 210 is courtesy and copyright by Robin Scagell and taken from his website www.galaxypix.com. On page 217, the photographs of Venus are copyrigt of Wah! and courtesy of Wikimedia Commons; see also apod.nasa.gov/apod/ap060110.html. On page 218, the old drawing of Le Sage is courtesy of Wikimedia. The picture of the celestial bodies on page 222 is copyright and courtesy of Alex Cherney and was featured on apod.nasa.gov/apod/ap160816.html.

The pictures of solar eclipses on page 223 are courtesy and copyright by the Centre National d'Etudes Spatiales, at www.cnes.fr, and of Laurent Laveder, from his beautiful site at www.PixHeaven.net. The photograph of water parabolae on page 227 is copyright and courtesy of Oase GmbH and found on their site www.oase-livingwater.com. The photograph of insect gear on page 245 is copyright and courtesy of Malcolm Burrows; it is found on his website www.zoo.cam.ac.uk/departments/insect-neuro. The pictures of daisies on page 246 are copyright and courtesy of Giorgio Di Iorio, found on his website www.flickr.com/photos/gioischia, and of Thomas Lüthi, found on his website www.tiptom.ch/album/blumen/. The photograph of fireworks in Chantilly on page 249 is courtesy and copyright of Christophe Blanc and taken from his beautiful website at christopheblanc.free.fr. On page 258, the beautiful photograph of M74 is copyright and courtesy of Mike Hankey and found on his beautiful website cdn.mikesastrophotos.com. The figure of myosotis on page 267 is courtesy and copyright by Markku Savela. The image of the wallpaper groups on page page 268 is copyright and courtesy of Dror Bar-Natan, and is taken from his fascinating website at www.math.toronto.edu/~drorbn/Gallery. The images of solid symmetries on page page 269 is copyright and courtesy of Jonathan Goss, and is taken from his website at www.phys.ncl.ac.uk/staff.njpg/symmetry. Also David Mermin and Neil Ashcroft have given their blessing to the use. On page 292, the Fourier decomposition graph is courtesy Wikimedia. The drawings of a ringing bell on page 292 are courtesy and copyright of H. Spiess. The image of a vinyl record on page 293scale=1 is copyright of Chris Supranowitz and courtesy of the University of Rochester; it can be found on his expert website at www.optics.rochester.edu/workgroups/cml/opt307/spr05/chris. On page 296, the water wave photographs are courtesy and copyright of Eric Willis, Wikimedia and allyhook. The interference figures on page 303 are copyright and courtesy of Rüdiger Paschotta and found on his free laser encyclopedia at www.rp-photonics.com. On page 308, the drawings of the larynx are courtesy Wikimedia. The images of the microanemometer on page page 312 are copyright of Microflown and courtesy of Marcin Korbasiewicz. More images can be found on their website at www.microflown.com. The image of the portable ultrasound machine on page 313 is courtesy and copyright General Electric. The ultrasound image on page 313 courtesy and copyright Wikimedia. The figure of the soliton in the water canal on page 316 is copyright and courtesy of Dugald Duncan and taken from his website on www.ma.hw.ac.uk/solitons/soliton1.html. The photograph on page 319 is courtesy and copyright Andreas Hallerbach and found on his website www.donvanone.de. The image of Rubik's cube on page 346 is courtesy of Wikimedia. On page 327, the photographs of shock waves are copyright and courtesy of Andrew Davidhazy, Gary Settles and NASA. The photographs of wakes on page 328 are courtesy Wikimedia and courtesy and copyright of Christopher Thorn. On page 330, the photographs un unusual water waves are copyright and courtesy of Diane Henderson, Anonymous and Wikimedia. The fractal mountain on page 334 is courtesy and copyright by Paul Martz, who explains on his website www.gameprogrammer.com/fractal.html how to program such images. The photograph of the oil droplet on a water surface on page 335 is courtesy and copyright of Wolfgang Rueckner and found on sciencedemonstrations.fas.harvard.edu/icb. The soap bubble photograph on page page 342 is copyright and courtesy of LordV and found on his website www.flickr.com/photos/lordv. The photographs of silicon carbide on page 343 are copyright and courtesy of Dietmar Siche. The photograph of a single barium ion on page 344 is copyright and courtesy of Werner Neuhauser at the Universität Hamburg. The AFM image of silicon on page 344 is copyright of the Universität Augsburg and is used by kind permission of German Hammerl. The figure of helium atoms on metal on page 344 is copyright and courtesy of IBM. The photograph of an AFM on page 345 is copyright of Nanosurf (see www.nanosurf.ch) and used with kind permission of Robert Sum. The photograph of the tensegrity tower on page 349 is copyright and courtesy of Kenneth Snelson. The photograph of the Atomium on page 351 is courtesy and copyright by the Asbl Atomium Vzw and used with their permission, in cooperation with SABAM in Belgium. Both the picture and the Atomium itself are under copyright. The photographs of the granular jet on page 352 in sand are copyright and courtesy of Amy Shen, who discovered the phenomenon together with Sigurdur Thoroddsen. The photographs of the machines on page 352 are courtesy and copyright ASML and Voith. The photograph of the bucket-wheel excavator on page 353 is copyright and courtesy of RWE and can be found on their website www.rwe.com. The photographs of fluid motion on page 355 are copyright and courtesy of John Bush, Massachusetts Institute of Technology, and taken from his website www-math.mit.edu/~bush. On page 360, the images of the fluid paradoxa are courtesy and copyright of IFE. The im-

CREDITS 561

ages of the historic Magdeburg experiments by Guericke on page 361 are copyright of Deutsche Post, Otto-von-Guericke-Gesellschaft at www.ovgg.ovgu.de, and the Deutsche Fotothek at www.deutschefotothek.de; they are used with their respective permissions. On page page 362, the laminar flow photograph is copyright and courtesy of Martin Thum and found on his website at www.flickr.com/photos/39904644@N05; the melt water photograph is courtesy and copyright of Steve Butler and found on his website at www.flickr.com/photos/11665506@N00. The sailing boat on page 363 is courtesy and copyright of Bladerider International. The illustration of the atmosphere on page 365 is copyright of Sebman81 and courtesy of Wikimedia. The impressive computer image about the amount of water on Earth on page 371 is copyright and courtesy of Jack Cook, Adam Nieman, Woods Hole Oceanographic Institution, Howard Perlman and USGS; it was featured on apod.nasa.gov/apod/ap160911.html. The figures of wind speed measurement systems on page 375 are courtesy and copyright of AQSystems, at www.aqs.se, and Leosphere at www.leosphere.fr On page 377, the Leidenfrost photographs are courtesy and copyright Kenji Lopez-Alt and found on www.seriouseats.com/2010/08/how-to-boil-water-faster-simmer-temperatures.html. The photograph of the smoke ring at Etna on page 379 is courtesy and copyright by Daniela Szczepanski and found at her extensive websites www.vulkanarchiv.de and www.vulkane.net. On page 381, the photographs of rolling droplets are copyright and courtesy of David Quéré and taken from iusti.polytech.univ-mrs.fr/~aussillous/marbles.htm. The thermographic images of a braking bicycle on page 384 are copyright Klaus-Peter Möllmann and Michael Vollmer, Fachhochschule Brandenburg/Germany, and courtesy of Michael Vollmer and Frank Pinno. The image of page 384 is courtesy and copyright of ISTA. The images of thermometers on page 387 are courtesy and copyright Wikimedia, Ron Marcus, Braun GmbH, Universum, Wikimedia and Thermodevices. The balloon photograph on page 390 is copyright Johan de Jong and courtesy of the Dutch Balloon Register found at www.dutchballoonregister.nl.ballonregister, nederlands The pollen image on page 392 is from the Dartmouth College Electron Microscope Facility and courtesy Wikimedia. The scanning tunnelling microscope picture of gold on page 401 is courtesy of Sylvie Rousset and copyright by CNRS in France. The photograph of the Ranque–Hilsch vortex tube is courtesy and copyright Coolquip. The photographs and figure on page 419 are copyright and courtesy of Ernesto Altshuler, Claro Noda and coworkers, and found on their website www.complexperiments.net. The road corrugation photo is courtesy of David Mays and taken from his paper Ref. 326. The oscillon picture on page 421 is courtesy and copyright by Paul Umbanhowar. The drawing of swirled spheres on page 421 is courtesy and copyright by Karsten Kötter. The pendulum fractal on page 425 is courtesy and copyright by Paul Nylander and found on his website bugman123.com. The fluid flowing over an inclined plate on page 427 is courtesy and copyright by Vakhtang Putkaradze. The photograph of the Belousov-Zhabotinski reaction on page 428 is courtesy and copyright of Yamaguchi University and found on their picture gallery at www.sci.yamaguchi-u.ac.jp/sw/sw2006/. The photographs of starch columns on page 429 are copyright of Gerhard Müller (1940–2002), and are courtesy of Ingrid Hörnchen. The other photographs on the same page are courtesy and copyright of Raphael Kessler, from his websitewww.raphaelk.co.uk, of Bob Pohlad, from his websitewww.ferrum.edu/bpohlad, and of Cédric Hüsler. On page 431, the diagram about snow crystals is copyright and courtesy by Kenneth Libbrecht; see his website www.its.caltech.edu/~atomic/snowcrystals. The photograph of a swarm of starlings on page 432 is copyright and courtesy of Andrea Cavagna and Physics Today. The table of the alphabet on page 441 is copyright and courtesy of Matt Baker and found on his website at usefulcharts.com. The photograph of the bursting soap bubble on page 475 is copyright and courtesy by Peter Wienerroither and found on his website homepage.univie.ac.at/Peter.Wienerroither. The photograph of sunbeams on page 478 is copyright and courtesy by Fritz Bieri and Heinz Rieder and found on their website www.beatenbergbilder.ch. The drawing on page 483 is courtesy and copyright of Daniel Hawkins. The photograph of a slide rule on page 487 is courtesy and copyright of Jörn Lütjens, and found on his website www.joernluetjens.de. On page 491, the bicycle diagram is courtesy and copyright of Arend Schwab. On page 504, the sundial photograph is courtesy and copyright of Stefan Pietrzik. On page 508 the chimney photographs are copyright and courtesy of John Glaser and Frank Siebner. The photograph of the ventomobil on page 516 is courtesy and copyright Tobias Klaus. All drawings are copyright by Christoph Schiller. If you suspect that your copyright is not correctly given or obtained, this has not been done on purpose; please contact me in this case.

SUBJECT INDEX

Symbols
(, open bracket 447
), closed bracket 447
+, plus 446
−, minus 446
·, multiplied by 447
×, times 447
:, divided by 447
<, is smaller than 447
=, is equal to 446
>, is larger than 447
!, factorial 447
[ ], measurement unit of 448
∅, empty set 448
≠, different from 447
√⎯, square root of 446
@ , at sign 448
Δ, Laplace operator 447
∩, set intersection 447
∪, set union 447
∈, element of 447
∞, infinity 447
∇, nabla/gradient of 447
⊗, dyadic product 447
⊂, subset of/contained in 447
⊃, superset of/contains 447
⟨ |, bra state vector 447
| ⟩, ket state vector 447
|x|, absolute value 447
∼, similar to 447

A
a^n 447
a_n 447
a (year) 41
abacus 445
aberration 135, 152
abjad 444
abugida 444
Academia del Cimento 77
acausality 239, 240
acceleration 85
    angular 118
    animal record 169
    centrifugal 161
    centripetal 161, 179
    dangers of 96
    due to gravity, table of values 182
    effects of 85
    highest 96
    of continents 489
    sensors, table 86
    table of values 84
    tidal 199
accelerometers 476
    photographs of 86
accents
    in Greek language 444
accumulation 356
accuracy 436, 458
    limits to 459
    why limited? 436
Acetabularia 46
Acinonyx jubatus 36
Ackermann steering 483
acoustical thermometry 546
action 248–265, 276
    as integral over time 251
    definition 251
    is effect 252
    is not always action 248
    measured values 250
    measurement unit 252
    physical 248
    principle 242
    principle of least 190
    quantum of 399
action, quantum of, $\hbar$
    physics and 8
actuator 233
    table 233
addition 273
additivity 37, 43, 51, 106
    of area and volume 56
adenosine triphosphate 531
aeroplane
    flight puzzle 96
    speed unit 103
    speed, graph of 38
    toilet 380
aerostat 376
AFM *see* microscope, atomic force
aggregate
    of matter 257
    overview 259
    table 259
air 515
    composition table 515
    jet, supersonic 370
    pressure 110
    resistance 80
Airbus 478
airflow instruments 309
alchemy 34
Aldebaran 87, 261
aleph 444
algebraic surfaces 472
Alice 404
Alpha Centauri 261

A

ALPHABET

alphabet
    Cyrillic 444
    first 440
    Greek 442, 443
    Hebrew 444
    Latin 440
    modern Latin 442
    phonemic 444
    Runic 444
    story of 441
    syllabic 444
alphasyllabaries 444
Alps 189
    weight of the 219
Altair 261
alveoli 347, 369
Amazon 148
Amazon River 321, 379
Amoeba proteus 91
ampere
    definition 452
amplitude 289
anagyre 133
analemma 212, 504
angels 21, 99
angle 118
    and knuckles 67
    in the night sky 486
    plane 67
    solid 67
angular momentum 117
    aspect of state 118
    conservation 190
    pseudo-vector 162
    values, table 118
anharmonicity 315
anholonomic constraints 254
Antares 87, 261
anti-bubbles 379
antigravity 194, 206
    device 207
antimatter 106
Antiqua 444
antisymmetry 275
ape
    puzzle 163
apex angle 486
aphelion 465
Aphistogoniulus
    erythrocephalus 91
Apis mellifera 105
apnoea 370–371
apogee 464
Apollo 501
apple
    and fall 100
    standard 457
    trees 457
approximation
    continuum 89
Aquarius 213
aqueducts 369
Arabic digit *see* number
    system, Indian
Arabic number *see* number
    system, Indian
Arabidopsis 46
Arcturus 261
area
    additivity 56
    existence of 56, 58
argon 515
Aries 213
Aristarchus of Samos 532
Aristotle 499
arithmetic sequence 476
arm
    swinging 122
Armillaria ostoyae 53, 105
arrow of time 48
artefact 92
ash 442
associativity 272
Asterix 125
asteroid 261
    difficulty of noticing 205
    falling on Earth 205
    puzzle 212
    Trojan 194
Astrid, an atom 544
astrology 178, 213
astronaut *see* cosmonaut
astronomer
    smallest known 192
astronomical unit 220, 465
astronomy
    picture of the day 472
at-sign 448
athletics 171
    and drag 235
atmosphere 159, 364
    angular momentum 118
    composition 515
    composition table 515
    layer table 366, 367
    of the Moon 411
atmospheric pressure 464
atom
    manipulating single 344
Atomic Age 385
atomic clock 45
atomic force microscope 494
atomic mass unit 461
Atomium 351
atoms
    and breaking 338
    and Galileo 336
    are not indivisible 406
    arranging helium 344
    explain dislocations 343
    explain round crystal
        reflection 343
    explain steps 343
    Galileo and 335–338
    Greeks thinkers and 339
    image of silicon 344
    in ferritic steel 351
    Lego and 339
    photo of levitated 344
ATP 109, 531
atto 454
attraction
    types of physical 258
Atwood machine 500
auricola 324
austenitic steels 348
Avogadro's number 341, 376, 462
    from the colour of the sky 341
axioms 35
axis
    Earth's, motion of 146
    Earth's, precession 153
axle
    impossibility in living beings 489

B

Babylonians 221, 449
background 26, 27, 49
bacterium 92, 114
badminton smash
    record 36, 481
Balaena mysticetus 371
Balaenoptera musculus 105, 320
ball
    rotating in mattress 167
balloon
    inverted, puzzle 390
    puzzle 389
    rope puzzle 376
Banach measure 56
Banach–Tarski paradox or
    theorem 57, 334
banana
    catching puzzle 120
barycentre 216
baryon number density 466
basal metabolic rate 114
base units 452
Basiliscus basiliscus 169
basilisk 169
bath-tub vortex 140, 533
bathroom scales 158, 347
BCE 449
bear
    puzzle 62
beauty 18, 415, 423, 424
    origin of 415
becquerel 454
beer 379
beer mat 63
beetle
    click 169
before the Common Era 449
behaviour 20
belief
    collection 22
belief systems 99
bell
    resonance 291
    vibration patterns 292
Belousov-Zhabotinski
    reaction 428
Bernoulli equation 361, 362, 378
Bessel functions 150
Betelgeuse 87, 261
beth 444
bets
    how to win 141
bicycle
    research 490
    riding 100
    stability, graph of 491
    weight 391
bifurcation 423
billiards 103
bimorphs 233
biographies of
    mathematicians 472
biological evolution 418
biology 233
BIPM 452
bird
    fastest 36
    singing 328
    speed, graph of 38
bismuth 41
bits to entropy conversion 463
black holes 239
block and tackle 30, 31
blood
    circulation, physics of 367
    dynamic viscosity 367
    supply 89
board divers 121
boat 109
    sailing 363
Bode's rule 219
body
    connected 89
    definition 26
    extended, existence of 333
    extended, non-rigid 244
    fluids 367
    rigid 243
Bohr magneton 462
Bohr radius 462
Boltzmann constant 390, 460
Boltzmann constant k 399
    and heat capacity 394
    definition 392, 399
    minimum entropy 408
physics and 8
bone
    and jumping 338
    human 348
book
    and physics paradox 437
    definition 440
    information and entropy 398
boom
    sonic, due to supersonic 327
boost 282
bore in river 329
bottle 109, 128
    empty rapidly 245
bottom quark mass 461
boundaries 98
boundary layer 506
    planetary 367
bow, record distance 78
brachistochrone 243, 507
braid pattern 428
brain 409
    and physics paradox 437
    stem 25, 524
brass instruments 309
bread 336, 479
breath
    physics of 367
breathings 444
Bronshtein cube 8
Bronze Age 385
brooms 207, 508
brown dwarfs 261
Brownian motion 391, 394
    typical path 393
browser 469
bubble
    soap 376, 475
    soap and molecular size 342
bucket
    experiment, Newton's 158
    puzzle 19
bullet
    speed 36
    speed measurement 61, 480

B

BUNSEN

Bunsen burner 386
Bureau International des Poids et Mesures 452
bushbabies 169
butterfly effect 425
button
    future 404

C

c. 449
caesium 41
calculating prodigies 446
calculus
    of variation 255
calendar 448, 449
    Gregorian 449
    Julian 449
    modern 449
caliper 54, 65
Callisto 210
calorie 456
Calpodes ethlius 79
camera 66
Cancer 213
candela
    definition 453
candle 411
    in space 519
    motion 32
canoeing 170
canon
    puzzle 158
Canopus 261
cans of peas 127
cans of ravioli 127
capacity 357
Capella 261
capillary
    human 372
Capricornus 213
capture
    in universal gravity 215
car
    engine 388
    on lightbulb 374
    parking 65
    weight 391
    wheel angular momentum 118

carbon dioxide 391
Carlson, Matt 471
Carparachne 92
carriage
    south-pointing 244, 508, 540
cars 436
Cartesian 51
cartoon physics, 'laws' of 98
cat 109
    falling 120
Cataglyphis fortis 171
catenary 486
caterpillars 79
catholicism 336
causality
    of motion 239
cavitation 96, 314
    and knuckle cracking 331
cavity resonance 329
CD
    angular momentum 118
Cebrennus villosus 90, 91, 530
celestrocentric system 479
Celsius temperature scale 385
Celtic wobble stone 133
cement kiln 490
cementite 350
centi 454
centre of gravity 212
centre of mass 212
centrifugal acceleration 161, 180
centripetal acceleration 161
cerebrospinal fluid 367
Ceres 214
cerussite 58
Cetus 213
CGPM 453
chain
    fountain 133
    hanging shape 69
challenge
    classification 9
chandelier 183
change
    and transport 24
    measure of 251
    measuring 248–265

quantum of, precise value 460
    types of 20
chaos 424
    and initial conditions 425
    in magnetic pendulum 425
chapter sign 448
charge 281
    elementary e, physics and 8
    positron or electron, value of 460
charm quark mass 461
chaturanga 32
cheetah 36, 84
chemistry 435
chess 32
chest operations 515
child's mass 126
childhood 36, 49
Chimborazo, Mount 189
chocolate 373
    does not last forever 333
chocolate bars 333
Chomolungma, Mount 189, 378
circalunar 192
circle packing 486
circular definition
    in physics 437
Clay Mathematics Institute 363
click beetles 169
clock 44–48, 110, 183
    air pressure powered 110
    definition 42
    exchange of hands 64
    precision 44
    puzzle 523
    puzzles 64
    summary 437
    types, table 46
clock puzzles 63
clockwise
    rotation 48
closed system 403, 405
cloud
    mass of 418
Clunio 192

coastline
  infinite 53
  length 53
CODATA 554
coffee machines 402
coin
  puzzle 62, 205, 346
collision
  and acceleration 233
  and momentum 110
  and motion 99
  open issues in 435
comet 192
  as water source 370
  Halley's 192
  mini 215
  origin of 214
comic books 347
Commission Internationale
  des Poids et Mesures 452
compass 244
completeness 43, 51, 106
complexity 398
Compton tube 141
Compton wavelength 462
Compton wheel 141
computer 409
concatenation 272
concave Earth theory 479
concepts 436
conditions
  initial, definition of 237
conductance quantum 462
Conférence Générale des
  Poids et Mesures 452, 457
configuration space 77
Conférence Générale des
  Poids et Mesures 453
conic sections 192
conservation 17, 106, 286
  of momentum 108
  of work 31
  principles 133
constant
  gravitational 177
  ideal gas 385
constants
  table of astronomical 463
  table of basic physical 460

table of cosmological 466
table of derived physical
  462
constellations 87, 213
constraints 254
contact 435
  and motion 99
container 49
continent
  acceleration of 489
  motion of 147
continuity 43, 50, 51, 106
  equation 231
  in symmetry
    transformations 276
  limits of 74
continuum 37
  as approximation 89
  mechanics 245
  physics 354
convection 405
Convention du Mètre 452
conventions
  used in text 440
cooking 435
cooperative structures 422
coordinates 51, 75
  generalized 254, 256
Copernicus 532
Corallus caninus 387
cords, vocal 308
Coriolis acceleration 138, 139,
  495, 532
Coriolis effect 138, 495, 532
  and navigation 159–161
  and rivers 159
Coriolis force 295
cork 109, 125, 128
corn starch 429, 430
corner film
  lower left 26, 27
corpuscle 87
corrugation
  road 420
corrugations
  road 420
corrugations, road 422
cortex 25
cosmological constant 466

cosmonaut 107, 108, 369, 501
  space sickness 503
cosmonauts
  health issues 204
cosmos 27
coulomb 454
countertenors 331
crackle 489
crackpots 473
cradle
  Mariotte 125
  Newton 125
Crassula ovata 273
creation 17
  of motion 108
crest
  of a wave 301
cricket bowl 481
cricket chirping
  and temperature 406
crooks 432
cross product 115
crossbow 78
crust
  growth of deep sea
    manganese 36
crying
  sounds of 331
crystal
  and straight line 59
  class 269
  symmetry table 269
cube
  Bronshtein 8
  physics 8
  Rubik's 346
cumulonimbus 322
curiosity 18, 241
curvature 64
curve
  of constant width 33
curvimeter 55
cycle 315
  limit 423
  menstrual 192, 504
cycloid 264, 507
cyclotron frequency 462
Cyrillic alphabet 444

D

∂, partial differential 447
dx 447
daleth 444
damping 234, 289, 301
dancer
    angular momentum 118
    rotations 79
Davey–Stewartson equation 328
day
    length in the past 41
    length of 145, 200
    mean solar 151
    sidereal 151, 463
    time unit 454
death 280, 437
    and energy consumption 125
    conservation and 110
    energy and 125
    mass change with 126
    origin 15
    rotation and 116
death sentence 336
deca 454
decay 409
deci 454
deer 25
definition
    circular, in physics 437
degree
    angle unit 454
    Celsius 454
delta function 305
Denier 457
denseness 43, 51
derivative 82
    definition 82
description 75
    accuracy of 436
Desmodium gyrans 46
details
    and pleasure 19
determinism 239, 240, 426
deviation
    standard, illustration 458
devil 24
Devonian 200

diagram
    state space 78
diamond
    breaking 347
die throw 239
Diet Coca Cola 379
differential 82
diffraction 301
diffusion 409, 411
digamma 443
digits
    Arabic *see* number system, Indian
    history of 445
    Indian 445
dihedral angles 57
dilations 279
dimensionality 37, 51
dimensionless 462
dimensions 50
    number of 304
dinosaurs 205
diptera 159
direction 37
disappearance
    of motion 108
dispersion 301
    and dimensionality 304
dissection
    of volumes 57
dissipative systems 423
distance 52
    measurement devices, table 55
    values, table 53
distinguish 25
distinguishability 37, 43, 51, 106
distribution 305
    Gaussian 458
    Gaussian normal 394
    normal 458
divergence 188
diving 370
DNA 53, 457
DNA (human) 261
DNA, ripping apart 227
Dolittle
    nature as Dr. 255

donate
    to this book 10
doublets 276
Dove prisms 529
down quark mass 461
drag 234
    coefficient 234, 436
    viscous 235
drift
    effect 456
dromion 329
    film of motion 329
drop
    of rain 235
Drosophila melanogaster 46
duck
    swimming 322
    wake behind 326
duration
    definition 41
Dutch Balloon Register 561
duvet 405
dwarf
    stars 261
dwarf planets 214
dx 447
dyadic product 447
dynabee 163
dynamics 189, 226

E

e, natural exponential 447, 467
e.g. 450
ear 50, 53, 303, 340, 541
    as atomic force microscope 345
    human 324
    illustration 325
    problems 324
    wave emission 303
Earth 405
    age 41, 464
    angular momentum 118
    average density 464
    axis tilt 153
    crust 126
    density 185
    dissection 58

SUBJECT INDEX

equatorial radius 464
flat 489
flattened 136, 159
flattening 464
from space 472
gravitational length 464
hollow 60
humming 331
mass 464
mass measurement 184
mass, time variation 205
motion around Sun 152
normal gravity 464
radius 464
rise 219
rotation 135–150, 455
change of 145
rotation data, table 151
rotation speed 145
shadow of 165
shape 188–189, 212
speed 36
speed through the
universe 155
stops rotating 189
earthquake 162, 189
and Moon 192
energy 162
information 534
starting 347
triggered by humans 347
eccentricity 194
of Earth's axis 153
echo 303
ecliptic 175
illustration of 222
effort 229
everyday 229
egg
cooking 406
eigenvalue 275
Ekman layer 532
elasticity 233
Elea 339
Electric Age 385
electric effects 233
electrodynamics 354
electromagnetism 186
electron *see also* positron

g-factor 463
magnetic moment 463
mass 461
speed 36
electron charge *see also*
positron charge
electron radius
classical 462
electron volt
value 463
element of set 37, 43, 51, 106
elephant
and physics 415
sound use 322
elevator
space 224, 345
ellipse 192
as orbit 182
email 469
emergence 426
emission
otoacoustic 303
EMS98 162
Encyclopedia of Earth 473
energy 110, 111, 231, 278
as change per time 111
conservation 187, 190, 281
conservation and time 113
consumption in First
World 113
definition 111
flows 108
from action 262
kinetic 110
observer independence
278
of a wave 300
rotational, definition 119
scalar generalization of 278
thermal 388
values, table 112
engine
car 388
enlightenment 201
entropy 357
definition 395, 398
flow 404
measuring 396
quantum of 399

random motion and 396
smallest in nature 399
specific, value table 397
state of highest 412
values table 397
environment 26
Epargyreus clarus 79, 529
ephemeris time 42
eponym 451
equation
differential 480
Equator
wire 224
equilibrium 402
thermal 402
Eris 214
eros 186
error
example values 435
in measurements 458
random 458
relative 458
systematic 458
total 458
ESA 472
escape velocity 212
Escherichia coli 92
et al. 450
eth 442, 552
ethel 442
ethics 242
Eucalyptus regnans 338
Eucharist 336
Euclidean space 51
Euclidean vector space 35, 37
Euler's wobble 147
Europa 210
evaporation 409
evening
lack of quietness of 322
event
definition 41
Everest, Mount 189, 378
everything flows 17
evolution 27
and running 123
and thermodynamics 432
biological 21
cosmic 21

E

EVOLUTION

569

E

Exa

equation, definition 239
Exa 454
exclamation mark 448
exclusion principle 258
exobase 366
exoplanet
    discoveries 216
exosphere
    details 366
expansion
    of gases 384
    of the universe 156
expansions 279
Experience Island 16, 22
exponential notation 66
exponents
    notation, table 66
extension 351
    and waves 293
extrasolar planets 261
eye 26
    blinking after guillotine 102
    of fish 378
eye motion 50
eyelid 64

F

φx 447
$f(x)$ 447
$f'(x)$ 447
F. spectabilis 46
F. suspensa 46
F. viridissima 46
faeces 79
Falco peregrinus 36
fall 180
    and flight are independent 76
    is not vertical 136
    is parabolic 77
    of light 201
    of Moon 180
familiarity 25
family names 450
fantasy
    and physics 438
farad 454
Faraday's constant 462
farting
    for communication 323
fear of formulae 32
Fedorov groups 267
femto 454
Fermi coupling constant 460
Fermi problem 407
Fermilab 66
ferritic steels 348
Fiat Cinquecento 374
fifth
    augmented 309
    perfect 309
figures
    in corners 26
filament 386
film
    Hollyoood 333
    Hollywood, and action 248
    lower left corner 27
fine-structure constant 460, 461
finite 395
fire 386
    pump 395
firework 67
fish
    eyes of 378
    farting 323
flagella 92
flagellar motor 114
flame 407
flame puzzle 161
flatness 59
flattening
    of the Earth 136, 159
fleas 169
flies 26
flip film 26
    explanation of 75
floor
    are not gavitational 226
flow 354, 356
    minimum, value table 401
    of eveything 17
    of time 48
fluid
    and self-organization 418
    ink motion puzzle 413
    mechanics 244
    motion examples 355
    motion know-how 370
fluids
    body 367
flute 288
fly
    common 186
focal point 488
focus 488
foetus 313
folds, vocal 308
fool's tackle 31
footbow 78
force 118
    acting through surfaces 231
    central 104
    centrifugal 189
    definition 104
    definition of 228
    is momentum flow 228
    measurement 228
    normal 236
    physical 228
    use of 227–236
    values, table 227
Ford and precision 436
forest 25
forget-me-not 267
formulae
    ISO 552
    liking them 32
    mathematical 472
Forsythia europaea 46
Foucault's pendulum
    web cam 141
fountain
    chain 133
    Heron's 374
    water 95
Fourier analysis 292
Fourier decomposition 292
Fourier transformation 326
fourth
    augmented 309
    perfect 309
fractal
    landscapes 334
fractals 53, 83, 333

frame
    of refeence 266
Franz Aichinger 555
frass 79
freak wave 323
free fall
    speed of 36
frequency 293
    values for sound, table 290
friction 109, 110, 234, 402, 404
    between planets and the
        Sun 110
    dynamic 234
    importance of 402
    not due to gravity 226
    picture of heat 384
    produced by tides 199
    static 234, 236
    sticking 234
froghopper 84
Froude number 322
fuel consumption 436
full width at half maximum
    458
funnel 378
    puzzle 378
Futhark 442
Futhorc 442
futhorc 444
future
    button 404
    fixed 239
    remembering 404
fx 447

G

γ, Euler's constant 467
Gaia 424
gait
    animal 500
    human 123
galaxy
    and Sun 153
    centre 153
    collision 104
    rotation 155
    size 53
galaxy cluster 260
galaxy group 260

galaxy supercluster 260
Galilean physics 29, 34, 226,
    434
    highlight 247
    in six statements 29
    research in 434
Galilean satellites 210
Galilean space-time
    summary 74
Galilean time
    definition 43
    limitations 44
Galilean transformations 278
Galilean velocity 35
galvanometer 233
Ganymede 210
gas 389
    as particle collection 393
    constant, universal 462
    ideal 384
gas constant
    ideal 390
gasoline
    dangers of 369
gauge
    change 272
    symmetry 280, 283
    theory 49
Gaussian distribution 394, 458
gearbox
    differential 244
gears
    in nature 244
Gemini 213
geocentric system 216
geocorona 366
geodesics 236
geoid 188
geological maps 472
geology of rocks 472
geometric sequence 476
geometry
    plane 73
geostationary satellites 194
Gerridae 170
geyser 417
ghosts 31, 99
Giant's Causeway 59, 429
giants 338

Gibbs' paradox 283
Giga 454
gimel 444
giraffe 368
glass
    and π 61
    is not a liquid 345
Global Infrasound Network
    325
global warming 323, 405
gluon 461
gnomonics 42
gods
    and conservation 133
    and energy 111
    and freedom 256
    and Laplace 196
    and motion 524
    and Newton 34, 526
    and walking on water 170
gold 479
    surface atoms 401
golden rule of mechanics 31
golf balls 481
Gondwanaland 147
gorilla test for random
    numbers 239
Gosper's formula 409
Gothic alphabet 444
Gothic letters 444
GPS 42
grace 23, 122, 430, 524
gradient 187
granular jet 350
gravitation *see also* universal
    gravitation, 173, 233
    and measurement 438
    and planets 174–178
    as momentum pump 228
    Earth acceleration value
        182
    essence of 218
    properties of 182–186
    sideways action of 183
    summary of universal 224
    universal *see* universal
        gravitation, 177
    value of constant 184
gravitational acceleration,

G

GRAVITATIONAL

standard 84
gravitational constant 177, 460
   geocentric 464
   heliocentric 464
gravitational constant G
   physics and 8
gravitational coupling constant 460
gravitational field 188
gravitons 219
gravity *see* gravitation
   as limit to motion 173
   centre of 212
   inside matter shells 217
   is gravitation 173
   lack of 204
   wave 295
gray (unit) 454
Greek alphabet 442, 443
greenhouse effect 405
Greenland 41
Gregorian calendar 65, 449
grim reaper 61
group
   Abelian, or commutative 273
   crystal 269
   crystallographic point 269
   mathematical 272
growth 21, 246
   as self-organization 418
   human 36
Gulf Stream 532
guns and the Coriolis effect 139
gymnasts 121
gynaecologist 313
gyroscope 120, 141
   laser 144
   vibrating Coriolis 159

H

hafnium carbide 386
Hagedorn temperature 386
hair
   clip 234
   diameter 53
   growth 36
halfpipe 264
haltere
   and insect navigation 159
Hamilton's principle 248
hammer drills 128
hands of clock 63, 64
hank 457
hard discs
   friction in 236
heartbeat 183
heat 233
   'creation' 387
   engine 388
   in everyday life 395
   in physics 395
Hebrew abjad
   table 444
Hebrew alphabet 444
hecto 454
helicity
   of stairs 48
helicopter 378, 473
heliocentric system 216
helium 260, 386, 409
   danger of inhaling 318
   superfluid 376
helix
   in Solar System 154
hemispheres
   Magdeburg 361
henry 454
heptagon
   in everyday life 511
heresy 336
Hermitean 275
Heron's fountain 374
herpolhode 164, 165
herring
   farting 323
hertz 454
heterosphere
   details 366
hiccup 431
Higgs mass 461
Himalaya age 41
hips 122
hoax
   overview 473
hodograph 77, 179, 210
hole
   system or not? 26
   through the Earth 221
Hollywood
   films 333
   films and action 248
   thriller 255
   westerns 530
holonomic 255
holonomic systems 254
holonomic–rheonomic 254
holonomic–scleronomic 254
Homo sapiens 46
homogeneity 43, 51
homomorphism 275
homopause 366
homosphere
   details 366
honey bees 105
Hopi 527
horse 114, 228
   power 114
   speed of 78
horse power 114
hour 454
hourglass puzzle 61
Hubble parameter 466
Hubble space telescope 472
Hudson Bay 41
humour 426
   physics 473
Huygens' principle 304
   illustration of 304
   illustration of consequence 305
hydrofoils 363
Hydroptère 36
hyperbola 192

I

i, imaginary unit 447
i.e. 450
ibid. 450
IBM 560
ice ages 153, 534
iceberg 105
icicles 347
icosane 397
ideal gas 390

ideal gas constant 518
ideal gas relation 390
idem 450
Illacme plenipes 96
illness 367
illusions
    of motion 16
    optical 472
image 98
    definition 26
    difference from object 98
imagination
    and symmetry 271
impenetrability 106
    of matter 34
inclination
    of Earth's axis 153
indeterminacy relation 314, 315
    of thermodynamics 399
index of refraction 509
Indian numbers 445
individuality 28, 237
induction 281–283
inertia 99
    moment of 164, 285
inf. 450
infinite
    coastline 53
infinity 37, 43, 51, 106
    and motion 17
    in physics 83
information 398
    erasing of 409
    quantum of 399
infrasound 325
infusion 368
initial condition
    unfortunate term 240
initial conditions 237
injectivity 275
inner world theory 479
Inquisition 336
insect
    breathing 411
    navigation 160
    speed, graph of 38
    water walking 300
instant 34

definition 41
    single 17
instant, human 41
insulation power 405
integral
    definition 252
integration 56, 251
integration, symbolic 472
interaction 27, 227
    symmetry 285
interface
    waves on an 294
interference 301
    illustration 303
interferometer 141
International Astronomical
    Union 465
International Earth Rotation
    Service 455, 534, 536
International Geodesic Union 465
International Latitude Service 146, 534
International Phonetic
    Alphabet 440
internet 469
    list of interesting websites 470
interval
    musical 309
intonation
    equal 310
    just 310
    well-tempered 310
invariance 106, 272
    mirror 284
    parity 284
inverse element 272
inversion
    at circle 244, 507
    spatial 284
invisibility
    of loudspeaker 409
    of objects 89
Io 200, 210
ionosphere
    details 366
    shadow of 165
IPA 440

Iron Age 385
irreducible 276
irreversibility 383
    of motion 239
irreversible 402
Island, Experience 16
ISO 80000 446
isolated system 405
isomorphism 275
isotomeograph 141
Issus coleoptratus 244
Istiophorus platypterus 36, 372
Isua Belt 41
IUPAC 554
IUPAP 554

J
Jarlskog invariant 460
jerk 85, 237, 489
Jesus 170
joints
    cracking 331
Josephson frequency ratio 462
joule 454
juggling 202
    record 488
    robot 488
jump 169
    height of animals 80
    long 529
Jupiter 217, 261
    angular momentum 118
    moons of 220
    properties 464

K
k-calculus 477
Kadomtsev–Petviashvili
    equation 329
Kapitza pendulum 319
kefir 260
kelvin
    definition 452
Kepler's laws 176
ketchup
    motion 36
kilo 454
kilogram
    definition 452

K — KILOTONNE

kilotonne 112
kinematics 75, 226
    summary 97
kinetic energy 110
Klitzing, von – constant 462
knot 73
knowledge
    humorous definition 506
knuckle
    angles 67
    cracking 331
koppa 443
Korteweg–de Vries equation 307, 317, 329
Kuiper belt 213–215, 260, 370
Kurdish alphabet 442
Kármán line 364

L

ladder 129
    puzzles 70
    sliding 129
Lagrange
    equations of motion 255
Lagrangian
    average 255
    examples, table 257
    is not unique 255
    points in astronomy 501
Lagrangian (function) 251–253
Lake Nyos 391
laminarity 361
languages on Earth 445
Laplace operator 188
laser gyroscopes 144
laser loudspeaker 409
Latin 450
Latin alphabet 440
laughter 426
lawyers 100
laziness
    cosmic, principle of 243, 253
    of nature 255
lea 457
lead 375
leaf
    falling 379

leap day 448
learning 25
    best method for 9
    mechanics 229
    without markers 9
    without screens 9
Lebesgue measure 58
lecture scripts 471
leg 183
    number record 96
    performance 169
Lego 339, 344
legs 31
    advantages 168
    and plants 31
    efficiency of 170
    in nature 89
    on water 169
    vs. wheels 168
Leidenfrost effect 377
length 50, 52, 81, 438
    assumptions 56
    issues 56
    measurement devices, table 55
    puzzle 55
    scale 411
    values, table 53
Leo 213
Leptonychotes weddellii 371
letters
    origin of 441
Leucanthemum vulgare 246
levitation 194
lex parsimoniae 262
Libra 213
libration 191
    Lagrangian points 194
lichen growth 36
lidar 376
life
    everlasting 110
    shortest 41
lifespan
    animal 125
lift
    for ships 131
    into space 345
light 233

deflection near masses 201
    fall of 201
    mill 233
    slow group velocity 526
    year 464, 465
light speed
    measurement 61
lightbulb
    below car 374
    temperature 386
lighthouse 36, 71
lightning speed 36
Lilliputians 320
limbic system 525
limit cycle 423
limits
    to precision 459
line
    straight, in nature 59
Listing's 'law' 167
litre 454
living thing, heaviest 105
living thing, largest 53
lizard 169
ln 2 467
ln 10 467
local time 42
locusts 169
logarithms 176
long jump 78, 529
    record 112
Loschmidt's number 341, 376, 462
lottery
    and temperature 384
    rigging 383
loudspeaker
    invisible 409
    laser-based 409
loudspeaker, invisible 409
love
    physics of 186
Love number 151
low-temperature physics 391
luggage 224
lumen 454
lunar calendar 42
Lunokhod 501
lux 454

SUBJECT INDEX

Lyapounov exponent 424
Lyapunov exponent
    largest 432
lymph 367

M
M82, galaxy 84
mach 103
Mach number 159
Mach's principle 103
Machaeropterus deliciosus 290
machine 92
    example of high-tech 352
    moving giant 353
    servants 113
Magdeburg hemispheres 361
magic 473
Magna Graecia 339
magnetic effects 233
magnetic flux quantum 462
magnetic thermometry 546
magnetism 104
magnetization
    of rocks 149
magneton, nuclear 463
magnetosphere 366
    details 366
magnitude 81, 277
man 255
    wise old 255
manakin
    club winged 290
many-body problem 194–197, 434
marble
    oil covered 109
marker
    bad for learning 9
Mars 455
martensitic steels 348
mass 101, 106, 231, 438
    centre of 119, 212
    concept of 101
    definition 102
    definition implies momentum conservation 104
    deflects light 201
    gravitational, definition 202
    gravitational, puzzle 212
    identity of gravitational and inertial 202–204
    inertial 99
    inertial, definition 202
    inertial, puzzle 212
    is conserved 102
    measures motion difficulty 101
    measures quantity of matter 102
    negative 106, 107
    no passive gravitational 203
    of children 126
    of Earth, time variation 205
    point 87
    properties, table 106
    sensors, table 107
    values, table 105
mass ratio
    muon–electron 463
    neutron–electron 463
    neutron–proton 463
    proton–electron 463
match
    lighting 276
math forum 472
math problems 472
math videos 472
mathematicians 472
mathematics 35
    problem of the week 472
matrix
    adjoint 275
    anti-Hermitean 275
    anti-unitary 275
    antisymmetric 275
    complex conjugate 275
    Hermitean 275
    orthogonal 275
    self-adjoint 275
    singular 275
    skew-symmetric 275
    symmetric 275
    transposed 275
matter 86, 99
    impenetrability of 34
    quantity 106
    quantity of 102
    shell, gravity inside 217
mattress
    with rotating steel ball 167
meaning
    of curiosity 19
measurability 37, 43, 51, 106
measure 438
measurement
    and gravitation 438
    comparison 455
    definition 37, 452, 455
    error definition 458
    errors, example values 435
    irreversibility 455
    meaning 455
    process 455
mechanics 243
    classical 226
    continuum 245
    definition 226
    fluid 244
    learning 229
    quantum 226
Medicean satellites 210
medicines 31
Mega 454
megatonne 112
memory 25, 31, 402
menstrual cycle 536
Mentos 379
Mercalli scale 162
Mercury 196
mercury 342
meridian
    and metre definition 183
mesopause 366
mesopeak 366
Mesopotamia 440
mesosphere
    details 366
metabolic rate 125
metabolic scope, maximal 532
metallurgy
    oldest science 435
meteorites 205
meteoroid 261

M

METEORS

meteors 325
metre
   definition 452
   stick 49
metre sticks 437
metricity 37, 43, 51, 106
micro 454
microanemometer 310
microphone 310
microscope 321
   atomic force 236, 344, 345, 494
   atomic force, in the ear 345
   lateral force 236
microwave background temperature 466
mile 455
military
   anemometer 311
   sonic boom 326
   underwater motion 372
   underwater sound 322, 324
   useful website 456
milk carton
   puzzle 369
Milky Way
   age 465
   angular momentum 118
   mass 465
   size 465
milli 454
mind
   change 21
minerals 472
minimum
   of curve, definition 253
Minion Math font 556
minute 454
   definition 465
Mir 223
mirror
   cosmic 472
   invariance 30, 284
   invariance of everyday motion 284
   symmetry 285
mixing matrix
   CKM quark 460
   PMNS neutrino 461

mogul, ski 319
mol 341
   definition 341
   from the colour of the sky 341
molar volume 462
molecules 340
moment 42
   of inertia 116, 118, 278, 285
   of inertia, extrinsic 118
   of inertia, intrinsic 118
momentum 102, 108, 231
   angular, extrinsic 119, 120
   angular, intrinsic 119
   as a substance 228
   as change per distance 111
   as fluid 109
   change 228
   conservation 108
   conservation follows from mass definition 104
   flow is force 228–236
   flows 108
   from action 262
   of a wave 300
   total 102
   values, table 104
momentum conservation
   and force 231
   and surface flow 231
momentum flow 231
momentum, angular 119
money
   humorous definition 506
monster wave 323
month 448
Moon 190–192
   angular momentum 118
   atmosphere 411
   calculation 180
   dangers of 200
   density 464
   density and tides 199
   fall of 180
   hidden part 191
   illusion 69
   orbit 192
   path around Sun 211
   phase 216

   properties 464
   size illusion 69
   size, angular 69
   size, apparent 179
   weight of 219
moons
   limit for 505
   observed 260
moped 133
morals 242
morning
   quietness of 322
moth 363
motion 27, 437
   and change 20
   and collision 99
   and contact 99
   and dimensions 189
   and infinity 17
   and measurement units 453
   Aristotelian view 229
   as an illusion 17
   as change of state of permanent objects 28
   as illusion 17
   as opposite of rest 17
   based on friction 234
   books on, table 21
   conditions for its existence 17
   creation 108
   detector 32
   disappearance 108
   does not exist 16
   drift-balanced 525
   everyday, is mirror-invariant 30
   existence of 16
   faster than light 404
   global descriptions 242–247
   harmonic 288
   has six properties 29–30, 33
   illusions, figures showing 16
   in configuration space 424
   infinite 438

is and must be predictable 242
is change of position with time 75
is conserved 29
is continuous 29
is due to particles 24
is everywhere 15
is fundamental 15, 33, 453
is important 15
is lazy 30
is mysterious 15
is never inertial 174
is never uniform 173
is part of being human 15
is relative 26, 29
is reversible 30
is simple 264
is transport 24
limits of 438
linear–rotational correspondence table 118
manifestations 16
minimizes action 30
mirror invariance 284
non-Fourier 525
of continents 36, 147
parity invariance 284
passive 22, 234
polar 146
possibly infinite? 31
predictability 434
predictability of 226–247
relative, through snow 156
reversal invariance 30
reversibility 284
simplest 24
stationary fluid 361
turbulent fluid 361
types of 20
unlimited 438
volitional 22
voluntary 234, 238
Motion Mountain 16
  aims of book series 7
  helping the project 10
  supporting the project 10
motor
  definition 232
  electrostatic 233
  linear 233
  table 233
  type table 233
motor bike 100
mountain
  surface 333
moustache 54
movement 22
multiplet 273, 274
multiplication 272
muon
  g-factor 463
muon magnetic moment 463
muon mass 461
Musca domestica 46, 84
music 307–310
  notes and frequencies 310
  search online 473
Myosotis 267
myosotis 266
mystery
  of motion 15

N

N. decemspinosa 530
nabla 447
nail puzzle 33
names
  of people 450
Nannosquilla decemspinosa 91
nano 454
Nanoarchaeum equitans 260
NASA 455
NASA 472
natural unit 462
nature 27
  is lazy 30, 255, 264
Navier–Stokes equations 363
needle on water 379
negative vector 81
neocortex 525
Neptune 213, 217
nerve
  signal propagation 315
  signal speed 36
Neurospora crassa 46
neutral element 272
neutrino 351
  masses 461
  PMNS mixing matrix 461
neutron
  Compton wavelength 463
  magnetic moment 463
  mass 461
  star 260
newspaper 549
Newton
  his energy mistake 111
newton 454
Newtonian physics 34, 226
NGC 2240 386
Niagara 426
Niagara Falls 426
nitrogen 374
Noether charge 281
Noether's theorem 280
noise 341, 392, 456
  physical 315
  shot 340
  thermometry 546
nonholonomic constraints 254
nonius 65
norm 81
normal distribution 394
normality 551
North Pole 48, 147, 189
notation 440
  scientific 440
nuclear explosions 322
nuclear magneton 463
nuclei 233
nucleon 351
null vector 81
number
  Arabic *see* number system, Indian
  Indian 445
  real 51
Number English 457
number system
  Greek 445
  Indian 445
  Roman 445
numbers
  and time 43
  necessity of 277

nutation 152
Nyos, Lake 391

O

object 26, 27, 98
   definition 40
   defintion 26
   difference from image 98
   immovable 100
   invisibility 89
   movable 100
   summary 437
obliquity 153, 504
oboe 313
observable
   definition 28
   physical 276
observables
   discrete 277
observation
   physical 271
   sequence and time 40
ocean
   angular momentum 118
   origin of 370
oceanography 326
octave 309
   definition 318
octet 276
odometer 53
ohm 454
oil 109, 377
oil film experiment 334
Olympic year counting 449
Oort cloud 214, 215, 260
op. cit. 450
openness 106
operator 188
Ophiuchus 213
optical radiation
   thermometry 546
optics picture of the day 471
optimist 253
orbit
   elliptical 177
order 43, 51, 106, 415
   appearance examples 418
   appearance, mathematics of 423
   of tensor 279
   table of observed phenomena 415
order appearance
   examples 415
order parameter 422
orgasm
   against hiccup 431
orientation
   change needs no background 120
origin
   human 15
Orion 87
ornament
   Hispano-Arabic 274
orthogonality 81
oscillation 288
   damped 289
   definition 288
   harmonic and anharmonic 289
   harmonic or linear 288
oscillons 421
osmosis 233, 409
oxygen 374
   bottle 371
ozone layer 366

P

φx 447
π and glasses 61
π and gravity 182
π, circle number 447, 467, 551
paerlite 350
painting puzzle 33
paper
   cup puzzle 407
   aeroplanes 473
   boat contest 375
parabola 69, 77, 182, 192
   of safety 95
parachte
   and drag 236
parachute
   needs friction 234
paradox
   about physics books 437
   hydrodynamic 360
   hydrostatic 360
parallax 135, 150
parallelepiped 116
parenthesis 448
parity
   invariance 30, 284
   inversion 284
parking
   car statistics 428
   challenge 65
   mathematics 482
   parallel 65, 482
parsec 464
part
   everything is made of 401
   of systems 27
particle 87
   elementary 351
particle data group 470
parts 85
pascal 454
passim 450
passive motion 238
past
   of a system 237
path 75
pattern
   braid 428
   in corners 26
   random 26
Paul trap 344
pea
   dissection 58
   in can 127
pearls 346
Peaucellier-Lipkin linkage 244, 507
pee
   research 428
Peirce's puzzle 63
pencil 96
   invention of 34
   puzzle 62
pendulum 183
   and walking 122
   inverse 319
   short 495
penguins 122
people names 450

peplosphere 367
perigee 464
perihelion 197, 465
    shift 153, 194
period 289
periodic table
    with videos 473
permanence 25
    of nature 17
permeability
    vacuum 460
permittivity
    vacuum 460
permutation
    symmetry 272
perpetuum mobile 107, 110
    first and second kind 110
    of the first kind 388
    of the second kind 413
Peta 454
phase 289, 293
    velocity 293
phase space 27, 238
    diagram 77
PhD
    enjoying it 430
Philaenus spumarius 84
photoacoustic effect 409, 410
photon
    mass 461
    number density 466
Physeter macrocephalus 371
physicists 471
physics 15
    circular definition in 437
    continuum 354
    everyday 29
    Galilean 29
    Galilean, highlight 247
    map of 8
    outdated definition 22
    problems 470
    school 522
    scourge of 450
physics cube 8
pico 454
piezoelectricity 233
pigeons 122
ping-pong ball 162

pinna 324
Pioneer satellites 84
Pisces 213
Pitot–Prandtl tube 35
Planck constant
    value of 460
planet 110
    and universal gravitation 174
    distance values, table 220
    dwarf 214
    gas 261
    minor 261
    orbit periods, Babylonian table 221
planet–Sun friction 110
planetoid 261
planetoids 200, 260
plants
    and sound 323
plate tectonics 149
plats
    and legs 31
play 22
Pleiades star cluster 175
pleural cavity 359, 515
Pleurotya ruralis 91
plumb-line 59
Pluto 213
pneumothorax 515
point
    in space 34, 89
    mass 87
    mathematical 51
    particle 87
point-likeness 89
poisons 30
Poisson equation 188
Polaris 261
polarization 301
polhode 148, 164, 165
pollen 392
pollutants 515
pollution 515
pool
    filled with corn starch and water 430
pop 489
Pororoca 321

Porpoise Cove 41
positions 81
positivity 106
positron charge
    specific 462
    value of 460
possibility of knots 51
postcard 62
    stepping through 480
potassium 315, 411
potential
    gravitational 186
    sources of 188
potential energy 187
power
    humorous definition 506
    in flows 357
    physical 113, 229
    sensors, table 115
    values, table 114
ppm 515
pralines 373
praying effects 262
precession 153, 165
    Earth's axis 153
    equinoctial 146
    of a pendulum 141
precision 18, 35, 75, 436, 458
    limits to 459
    measuring it 31
    why limited? 436
predictability
    of motion 226–247, 434
prefixes 454, 553
    SI, table 454
prefixes, SI 454
preprints 469, 470
pressure 231, 340, 389
    air, strength of 361
    definition 357
    measured values 359
    puzzle 360
    second definition 362
principle
    conservation 133
    extremal 248
    of cosmic laziness 253
    of gauge invariance 272
    of laziness 243

P

PRINCIPLE

of least action 136, 190, 243, 253, 254
history of 261
of relativity 272, 477
of the straightest path 236
variational 242, 248, 254
principle of thermodynamics
    second 403
prism 529
prize
    one million dollar 363
problem
    collection 470
    many-body 194–197, 434
process
    change and action in a 252
    in thermodynamics 402
    sudden 241
Procyon 261
prodigy
    calculating 446
product
    dyadic 447
    outer 447
    vector 115
pronunciation
    Erasmian 443
proof 35
propagation
    velocity 293
propeller 489
    in living beings 89
property
    emergent 426
    intrinsic 28
    invariant 266
    permanent 28
protestantism 336
proton
    age 41
    Compton wavelength 463
    g factor 463
    gyromagnetic ratio 463
    magnetic moment 463
    mass 461
    specific charge 463
pseudo-scalar 284
pseudo-tensor 284
pseudo-vector 162, 284

PSR 1257+12 261
PSR 1913+16 41
psychokinesis 22
pulley 233
pulsar period 41
pulse 304
    circular 305
    spherical 305
puzzle
    about clocks 64
    ape 163
    bear 63
    bucket 19
    coin 62, 205
    five litre 62
    flame 161
    hourglass 61
    knot 73
    length 55
    liquid container 369
    milk carton 369
    nail 33
    painting 33
    reel 19
    snail and horse 63
    sphere 206
    sunrise 324
    train 94
    turtle 285
pyramid 116
Pythagoras' theorem 66

Q

Q-factor 289
qoppa 443
quanta
    smallest 339
quanti, piccolissimi 335–337
quantity
    conserved 17
    extensive 108, 354
    extensive, table 356
    necessity in nature 277
quantum
    of entropy 399
    of information 399
Quantum Age 385
quantum mechanics 226
quantum of action

    precise value 460
quantum of circulation 462
quantum theory 26, 337
quark
    mixing matrix 460
quartets 276
quartz 291
    oscillator film 291
quasar 260
quasi-satellite 204
quietness
    of mornings 322
quint
    false 309
Quito 140
quotations
    mathematical 473

R

R-complex 524
radian 67, 453
radiation 27, 99, 405
    as physical system 26
    thermometry 546
radio speed 36
rain drops 79, 235
rain speed 36
rainbow 278
randomness 239
rank
    of tensor 279
Ranque–Hilsch vortex tube 412
ratio between the electron magnetic moment and the Bohr magneton 435
rattleback 133
ravioli 127
ray form 216
reactance
    inertive 309
reaper, grim 61
recognition 25
record
    human running speed 78
    vinyl 293
reducible 276
reed instruments 309
reel

puzzle 19
reflection
    of waves 301
refraction 263, 301
    and minimum time 263
    and sound 322
    illustration 321
refractive index 263
Regulus 261
relation
    among nature's parts 27
    indeterminacy 314
relativity 243
    and ships 156
    challenges 158
    Galilean 157
    Galilean, summary 172
    Galileo's principle of 157
    of rotation 158
    principle 477
reluctance
    to rotation 116
representation 275
    definition 275
    faithful 275
    of observable 279
    unitary 275
reproducibility 270
reproduction 15
repulsion
    types of physical 258
research
    in classical mechanics 434
resolution
    of measurements is finite 29
resonance 291–293
    definition 291
    direct 318
    parametric 319
rest
    as opposite of motion 17
    Galilean 82
    is relative 157
Reuleaux curves 477
reversibility 383
    of everyday motion 284
    of motion 239
reversible 402

Reynolds number
    definition 430
Rhine 379
rhythmites, tidal 534
Richter magnitude 162
riddle 16
Riemann integral 509
Rigel 261
right-hand rule 115
ring
    planetary 505
ring laser 533
rings
    astronomical, and tides 198
ripple
    water 299
road
    corrguated or washboard 422
    corrugations 420
Roadrunner 347
robot 540
    juggling 488
    walking 264
    walking on water 170
    walking or running 170
Roche limit 505
rock
    magnetization 149
    self-organization 429
rocket
    launch site puzzle 158
rocket motor 233
rogue wave 323
roll
    speed 32
roller coasters 224
rolling 121
    motion 91
    puzzle 33
    wheels 121
rope
    around Earth 481
    geometry of 71
    motion of hanging 430
rosette groups 267
Rossby radius 159
Roswell incident 322
rotation

    absolute or relative 158
    and arms 158
    as vector 120
    cat 120
    change of Earth 145
    clockwise 48
    frequency values, table 117
    in dance 79
    of the Earth 135–150, 455
    rate 116
    reluctance 116
    sense in athletic stadia 48
    snake 121
    speed 116
    tethered 89, 90, 168
roulette and Galilean mechanics 127
Rubik's cube 346
rugby 111
rum 410
Runge–Lenz vector 499
Runic script 442
running
    backwards 405
    human 123
    on water 535
    reduces weight 209
Rutherford, Ernest 24
Rydberg constant 435, 462

S

Sagarmatha, Mount 189, 378
Sagittarius 153, 213
Sagnac effect 142, 497
sailfish 36, 372
sailing 109
    and flow 363
Salmonella 92
sampi 443
san 443
sand 420
    Cuban 421
    granular jets in 352
    self-organization 419
    singing 420
    table of patterns in 420
Saraeva 523
Sarcoman robot 488
satellite

artificial 88
definition 88
geostationary 163
limit 505
of the Earth 204
satellites 215
observable artificial 472
Saturn 198, 215
saucer
flying 322
scalar 81, 270, 277
true 284
scalar product 51, 81
scale invariance 338
scales
bathroom 158, 347
scaling
'law' 44
allometric 37
school physics 522
science
main result 400
time line 19
scientist
time line 19
Scorpius 213
scourge
of physics 450
screw
and friction 234
in nature 244
scripts
complex 444
scuba diving 369
sea 297
seal
Weddell 371
season 175
second 454
definition 452, 465
Sedna 214
seiche 300
selenite 58
self-adjoint 275
self-organization 246, 415–433
self-similarity 53
sensor
acoustic vector 310
sequence 43, 51, 106

Sequoiadendron giganteum
105
serpent bearer 213
Serpentarius 213
servant
machines 113
set
connected 83
sexism in physics 339
sextant 535
SF_6 318
sha 443
shadow
and attraction of bodies
218
book about 530
Earth, during lunar eclipse
179
motion 98
of ionosphere 165
of sundials 48
of the Earth 165
shadows 98
shampoo, jumping 380
shape 51, 87
and atoms 351
deformation and motion
90
of the Earth 212
sharp s 442
Shaw, George Bernard
alphabet of 445
sheep
Greek 443
shell
gravity inside matter 217
Shell study 531
shift
perihelion 153
ship 105
and relativity 157
critical speed 322
leaving river 373
lift 131
mass of 131
pulling riddle 376
speed limit 322
wake behind 326
shit 529

shock wave
in supersonic motion 326,
327
shoe size 456
shoelaces 50, 402
shore birds 535
shortest measured time 41
shot
noise 340
small lead 375
shoulders 122
shutter 66
shutter time 66
SI
prefixes 457
table of 454
units 452, 460
SI prefixes
infinite number 457
SI units
definition 452
prefixes 454
supplementary 453
siemens 454
sievert 454
signal
decomposition in
harmonic components 292
types, table 306
Simon, Julia 529
sine curve 288
singing 307
singlets 276
singular 275
sink vortex 533
siren 320
Sirius 87, 261
size 49
skateboarding 264
ski moguls 319, 420
skin 89
skipper 79
sky
colour and density
fluctuations 544
colour and molecule
counting 341
moving 146
nature of 125

slide rule 486, 487
slingshot effect 210
Sloan Digital Sky Survey 472
smallest experimentally
    probed distance 66
smartphone
    bad for learning 9
smiley 448
smoke 393
    ring picture 379
snake
    rotation 121
snap 489
snorkeling
    dangers of 514
snowflake 549
    self-organization film 422
    speed 36
snowflakes 549
soap
    bubble bursting 31
    bubbles 376
    bubbles and atoms 342
soccer ball 92
sodar 376
sodium 315, 385, 411, 526
solar data 472
Solar System 260
    angular momentum 118
    as helix 154
    formation 146
    future of 219
    motion 153
    small bodies in 261
Solar System simulator 472
solitary wave
    definition 316
    in sand 420
soliton 300, 317
    film 317
solitons 316
sonar 328
sonic boom 326
sonoluminescence 96
soprano
    male 331
soul 126
sound 294
    channel 322

frequency table 290
in Earth 331
in plants 323
intensity table 311
measurement of 310–313
none in the high
    atmosphere 364
speed 36
threshold 311
source term 188
South Pole 188
south-pointing carriage 244
space 437
    absolute 141
    absoluteness of 51
    as container 52
    definition 49
    elevator 224, 345
    empty 188, 301
    Euclidean 51
    Galilean 51
    is necessary 51
    lift 345
    physical 49
    point 89
    properties, table 51
    relative or absolute 52
    sickness 503
    white 448
Space Station
    International 216
Space Station, International
    204
space travel 161
space-time 27
    Galilean, summary 74
    relative or absolute 52
space-time diagram 77
Spanish burton 31
spatial inversion 279
special relativity before the
    age of four 271
specific mechanical energy
    488
spectrum
    photoacoustic 410
speed 82
    critical ship 322
    infinite 438

limit 439
lowest 31
of light c
physics and 8
of light at Sun centre 36
values, table 36
speed of sound 103, 145
speed record
    human running 78
    under water 372
speed, highest 36
sperm
    motion 36
sphere
    puzzle 206
    swirled, self-organized 421
Spica 261
spider
    jumping 370
    leg motion 370
    rolling 90
    water walking 170
spin
    group 266
spinal engine 123
spine
    human 348
spinning top
    angular momentum 118
spinors 277
spiral
    logarithmic 510
spirituality 241
Spirograph Nebula 505
split personality 240
sponsor
    this book 10
spoon 133
sport
    and drag 235
spring 288
    constant 288
spring tides 199
sprint
    training 171
squark 555
St. Louis arch 486
stadia 48
stainless steel 350

staircase
    formula 170
stairs 48
stalactites 347, 427
stalagmites 36
standard apple 457
standard clock 210
standard deviation 436, 458
    illustration 458
standard kilogram 101, 131
standard pitch 290
standing wave 303
star
    age 41
    dark 261
    twinkle 88
    twinkling 88
star classes 261
starch
    self-organization 429
stargazers 472
stars 87, 261
state 27, 28
    allows to distinguish 28
    definition 28
    in flows 356
    initial 237
    of a mass point, complete 237
    of a system 237
    of motion 27
    physical 27
state space 78
states 437
stationary 253
statistical mechanics 246
statistical physics
    definition 383
steel 348
    stainless 350
    types table 350
Stefan–Boltzmann black body
    radiation constant 463
Steiner's parallel axis theorem 118
steradian 67, 453
stick
    metre 49
stigma 443

Stirling's formula 409
stone
    falling into water 299
    skipping 373
stones 24, 40, 76, 80, 87, 109, 133, 136, 176, 178, 179, 182, 207, 210, 221, 304, 324, 402
straight
    lines in nature 59
straight line
    drawing with a compass 244
straightness 50, 59, 64
strange quark mass 461
stratopause 366
stratosphere
    details 366
stream
    dorsal 29
    ventral 29
streets 499
strong coupling constant 460
structure
    highest human-built 53
stuff
    in flows 354
subgroup 273
submarines 322
subscripts 444
Sun 105, 110, 217, 261
    angular momentum 118
    density and tides 199
    stopping 347
    will rise tomorrow 189
Sun size, angular 69
Sun's age 465
Sun's lower photospheric
    pressure 465
Sun's luminosity 464
Sun's mass 464
Sun's surface gravity 465
Sun–planet friction 110
sunbeams 478
sundial 45, 504, 526
    story of 48
sundials 42
sunrise
    puzzle 324
superboom 326

supercavitation 372
superfluidity 376
supergiants 261
superlubrication 234
supermarket 48
supernovae 96
superposition 301
support
    this book 10
surface
    area 50
surface tension 235
    dangers of 372
    wave 295
surfing 321
surjectivity 275
surprises 281
    in nature 241
swan
    wake behind 326
swarms 433
    self-organization in starling 432
swell 297
swimming
    olympic 322
swimming speed
    underwater 372
swing 318
Swiss cheese 333
switching
    off the lights 305
syllabary 444
symbols
    mathematical 446
    other 446
symmetry 272
    and talking 270
    classification, table 267
    colour 266
    comparison table 277
    crystal, full list 269
    discrete 272, 284
    external 272
    geometric 266
    internal 272
    low 266
    mirror 285
    of interactions 285

of Lagrangian 279
of wave 285
parity 284
permutation 272
scale 266
summary on 286
types in nature 281–283
wallpaper, full list 268
symmetry operations 272
Système International
　d'Unités (SI) 452
system
　conservative 187, 234, 254
　definition 28
　dissipative 234, 236, 254
　extended 293
　geocentric 216
　heliocentric 216
　motion in simple 426
　physical 26
　positional number 445
　writing 444
system, isolated 405

T

Talitrus saltator 46
talking
　and symmetry 270
tantalum 41
tau mass 461
Taurus 213
tax collection 452
teaching
　best method for 9
teapot 376
technology 445
tectonics 149
　plate 534
tectonism 192
teeth growth 418
telekinesis 262
telephone speed 36
teleportation
　impossibility of 109
telescope 321
television 25
temperature 357, 384, 385
　absolute 385
　and cricket chirping 406

granular 427
introduction 383
lowest 385
lowest in universe 385
measured values table 385
negative 385, 413
scale 546
tensegrity
　example of 349
　structures 348
tensor 277, 278
　and rotation 282
　definition 278
　moment of inertia 285
　order of 279
　product 447
　rank 279
　rank of 278, 279
　web sites 541
Tera 454
tesla 454
testicle 529
testimony 529
tether 89
tetrahedron 65, 116
Theophrastus 40, 44
theorem
　Noether's 280
thermoacoustic engines 412
thermodynamic degree of
　freedom 394
thermodynamics 354, 383, 389
　first law 113
　first principle 388
　indeterminacy relation of
　　399
　second law 113
　second principle 403
　second principle of 409
　third principle 385
　third principle of 400
thermometer 387
thermometry 546, 547
thermopause 366
thermosphere
　details 366
thermostatics 389
third, major 309
thorn 442, 552

thought
　processes 433
thriller 255
throw
　and motion 24
　record 53
throwing
　importance of 24
　speed record 36, 64
thumb 69
tide
　and Coriolis effect 139
　and friction 199
　and hatching insects 192
　as wave 296
　importance of 197
　once or twice per day 224
　slowing Moon 191, 533
　spring 192
time 437, 438
　coordinate, universal 42
　absolute 43
　absoluteness of 43
　arrow of 48
　deduced from clocks 42
　deduction 42
　definition 41
　definition of 281
　flow of 48–49
　interval 43
　is absolute 43
　is necessary 43
　is unique 43
　measured with real
　　numbers 43
　measurement, ideal 44
　proper 42
　properties, table 43
　relative or absolute 52
　reversal 48, 284
　translation 279
　travel, difficulty of 156
　values, table 41
time-bandwidth product 314
Titius's rule 219
TNT energy content 463
Tocharian 529
tog 405
toilet

T

TOKAMAK

aeroplane 380
  research 428
tokamak 386
tonne, or ton 454
top quark mass 461
topology 51
torque 118
touch 98
tower
  leaning, in Pisa 76
toys
  physical 472
train 105
  motion puzzle 158
  puzzle 94
trajectory 75
transformation
  of matter and bodies 20
  of motion in engines 109
  symmetry 272
translation invariance 43, 51
transport 20
  and change 24
  is motion 24
transubstantiation 336
travel
  space, and health 161
tree 457
  family, of physical
    concepts 27
  growth 36
  height, limits to 338
  leaves and Earth rotation
    145
  pumping of water in 544
  strength 338
triangle
  geometry 73
triboscopes 236
Trigonopterus oblongus 244
tripod 233
tritone 309
Trojan asteroids 194, 319
tropical year 463
tropopause 367
troposphere
  details 367
truth 336
tsunami 299
tuft 545
tumbleweed 90
tunnel
  through the Earth 221
turbopause 366
turbulence 362, 426
  in pipes 430
  is not yet understood 435
Turkish mathematics 442
turtle
  puzzle 285
twinkle
  of stars 88
typeface
  italic 440
Tyrannosaurus rex 169

U

U(1) 511
udeko 454
Udekta 454
UFOs 322
ultracentrifuges 209
ultrasound
  imaging 290, 311, 313
  motor 233
umbrella 79
unboundedness 37, 43, 51, 106, 276
uncertainty
  relation in
    thermodynamics 399
  relative 458
  total 458
underwater
  sound 322, 324
  speed records 372
  swimming 372
  wings 363
Unicode 445
unicycle 99, 349
uniqueness 43, 51
unit 405
  astronomical 464
unitary 275
units 51, 452
  non-SI 455
  provincial 455
  SI, definition 452

universal gravitation 177
  origin of name 180
universal time coordinate 456
universality of gravity 195
universe 27, 238
  accumulated change 264
  angular momentum 118
  description by universal
    gravitation 211
  two-dimensional 71
unpredictability
  practical 239
up quark mass 461
urination
  time for 380
URL 469
UTC 42

V

vacuum
  human exposure to 161
  permeability 460
  permittivity 460
  wave resistance 462
variability 25
variance 458
variation 27
Varuna 260
vases, communicating 359
vector 278
  axial 284
  definition 80
  polar 284
  product 115
  space 81
  space, Euclidean 81
Vega 261
  at the North pole 146
velars 443
velocimeters 476
velocity 35, 82, 438
  angular 116, 118
  as derivative 83
  escape 212
  first cosmic 212
  Galilean 35
  in space 37
  is not Galilean 37
  measurement devices,

table 39
of birds 37
phase 293
propagation 293
proper 36
properties, table 37
second cosmic 212
third cosmic 212
values, table 36
wave 293
vena cava 367
vendeko 454
Vendekta 454
ventomobil 516
Venturi gauge 362
Venus 217
vernier 65
vessels, communicating 359
video
 bad for learning 9
 recorder, dream 404
Vietnam 224
viewpoint-independence 270
viewpoints 266
vinyl record 293
Virgo 213
vis viva 110
viscosity
 dynamic 341, 430
 kinematic 430, 518
voice
 human 307
void 339
volcanoes 534
volt 454
volume 57
 additivity 56
 assumptions 58
vomit comet 502
vortex
 at wing end 363
 in bath tub or sink 533
 in fluids 382
 in the sea 382
 ring film 380
 ring leapfrogging 380
 rings 380
 tube 412
Vulcan 197

Vulcanoids 261
W
W boson mass 461
wafer
 consecrated 344
 silicon 344
wake 297
 angle 326, 327
 behind ship or swan 326, 328
walking
 and friction 402
 and pendulum 122
 angular momentum 118
 human 122
 maximum animal speed 500
 on water 169, 300
 robots 264
 speed 183
wallpaper groups 268
washboard roads 422
water 544
 dead 294
 origin of oceans 370
 patterns in 427
 ripple 299
 speed records under 372
 strider 170
 tap 426
 triple point 385
 walking on 169, 300
 walking robot 170
water wave 295, 329
 formation of deep 295
 group velocity of 322
 photographs of types 296
 solitary 316
 spectrum, table 297
 thermal capillary 321
 types and properties 298
watt 454
wave 291, 293
 capillary 299
 circular 305, 307
 cnoidal 329
 crest 323
 deep water 295

 dispersion, films 299
 emission 303
 energy 300
 freak 323
 gravity water 295–300
 group 304
 harmonic 293
 highest sea 320
 in deep water 327
 infra-gravity 296
 internal 300
 linear 293, 307
 long water 295
 long-period 296
 longitudinal 294
 momentum 300
 monster 323
 motion 300
 plane 307
 pulse 304
 reflection 301
 rogue 323
 sea, energy 326
 secondary 304
 shallow 299
 shallow water 295
 shock 326
 short water 295
 sine 293
 solitary 316
 solitary, in sand 420
 speed
  measured values, table 294
 spherical 305, 307
 standing 303, 307
 summary 331
 surface tension 295–300
 symmetry 285
 thermal capillary 321
 trans-tidal 296
 transverse 294
 ultra-gravity 296
 velocity 293
 water *see* water wave, 295, 329
wave equation
 definition 306
wave motion
 six main properties and

WAVELENGTH

effects 302
wavelength 293, 301
wavelet transformation 326
waw 443
weak mixing angle 460
weather 159, 431
  unpredictability of 239
web
  largest spider 53
  list of interesting sites 470
  world-wide 469
web cam
  Foucault's pendulum 141
weber 454
week 449
  days, order of 221
Weib 529
weight 202, 228
  of the Moon 219
weightlessness
  effects on cosmonauts 204
weightlessness, feeling of 503
weko 454
Wekta 454
whale
  blue 320
  constellation 213
  ears 324
  sounds 322
  sperm 371
wheel
  axle, effect of 489
  biggest ever 92
  in bacteria 92
  in living beings 89–92
  rolling 121
  ultimate 23, 557
  vs. leg 168

whelps 329
whip
  cracking 349
  speed of 36
whirl 154
white colour
  does not last 403
Wien's displacement constant 463
wife 529
Wikimedia 561
will, free 241–242
wind generator
  angular momentum 118
wind resistance 234
wine 336
  arcs 428
  bottle 109, 128
  puzzle 62
wing
  and flow 363
  underwater 363
Wirbelrohr 412
wire
  at Equator 224
WMAP 502
wobble
  Euler's 147
  falsely claimed by Chandler 147, 534
women 192, 529
  dangers of looking after 167
wonder
  impossibility of 388
work 31, 110, 254
  conservation 31
  definition in physics 111

physical 229
world 15, 27
  question center 473
World Geodetic System 465, 536
writing 440
  system 444
wrong 436
wyn 442

X

xenno 454
Xenta 454
xylem 359

Y

yard 457
year
  number of days in the past 200
yo-yo 182
yocto 454
yogh 442
yot 443
Yotta 454

Z

Z boson mass 461
zenith 163
  angle 179
zepto 454
zero
  Indian digit 445
zero gravity 502
Zetta 454
zippo 425
zodiac 175
  illustration of 222

Printed in Great Britain
by Amazon

78163063R00337